Rabelais

SPRINGER
LAB MANUALS

Springer

Berlin
Heidelberg
New York
Barcelona
Hong Kong
London
Milan
Paris
Singapore
Tokyo

Rochelle A. Diamond   Susan DeMaggio (Eds.)

In Living Color

Protocols in Flow Cytometry and Cell Sorting

With 199 Figures and 26 Tables

Springer

ROCHELLE A. DIAMOND
Member of the Professional Staff
California Institute
of Technology
Division of Biology
1200 E. California Blvd.
Pasadena, CA 91125, USA
Tel: 626 395 – 4947
Fax: 626 449-0756
E-mail: diamond@its.caltech.edu

SUSAN DEMAGGIO
MS BSMT (ASCP)CQ

Optical Biology Core Facility
Developmental Biology Center
University of California
Irvine, CA 92687-2275, USA
Tel: (949) 824-4110
Fax: (949) 824-3571
E-mail: suedemag@uci.edu

ISBN 3-540-65149-7   Springer-Verlag Berlin Heidelberg New York

Library of Congress Cataloging-in-Publication Data
In living color: protocols in flow cytometry and cell sorting; with 26 tables /
Rochelle A. Diamond ; Susan DeMaggio (ed.). – Berlin ; Heidelberg ; New York ; Barcelona ;
Hong Kong ; London ; Milan ; Paris ; Singapore ; Tokyo : Springer, 2000
   (Springer lab manual)
   ISBN 3-540-65149-7

Production: PRO EDIT GmbH, 61126 Heidelberg, Germany
Cover design: design & production GmbH, 69121 Heidelberg, Germany
Typesetting: Mitterweger & Partner GmbH, 68723 Plankstadt, Germany
Printed on acid free paper   SPIN: 10680098   27/3136/So 5 4 3 2 1 0

Acknowledgements

The Editors wish to thank the generous encouragement and support of many people during the process of preparing this manual. Dr. Peter J. Bryant, Director of the Developmental Biology Center, and Dr. Michael Cahalan, Director of the Optical Biology Core Facility at the University of California, Irvine, who provided the allowance of time necessary to complete this project. This is especially true of Dr. Ellen Rothenberg at Caltech, who deserves a special thank you for her enthusiasm, advice, and personal kidness towards this, project. Without these peoples generosity this manual would not exist.

We would also like to thank all the authors who so generously offered their protocols, and spent many hours preparing their individual manuscripts for inclusion in the manual, and a special thanks to our preface author, Dr. Harry Crissman. Without the assistance of Paul Robinson and Steve Kelley and the Purdue Cytometry Listserv, we would not have been able to recruit authors and research new protocols for inclusion in the text. The Table of Contents illustrates the number and caliber of each of these authors. We would like to also thank those who prepared appendix materials and vendors who supported the printing of color plates for the illustration of the protocols.

Technical assistance for typing, computer problems, and secretarial tasks was invaluable. Much appreciation is expressed to Bob Turing of the Caltech Graphic Arts Facility and Stephanie Canada for their help in downloading unusual files and scanning figures. Our thanks especially go to Rochelle's personal assistant, Catherine Springer, and to Amanda and Ronald Diamond for loan of a laptop computer to make life easier. We want to thank Barbara Belmont for her help proofreading and in preparation of the glossary as well as personal assistance with software support and computer problems, her cheerful encouragement, and dinner a few nights, too!!

I, Sue, would like to thank my family left at home, Mark and Laura DeMaggio for their understanding of the long hours at work and the late and missing meals. And Rochelle, for being a wonderful compliment to my skills. She is a delight to work with and we seemed to be a great team. She also rewarded me with wonderful meals on long evenings in Pasadena working on the manuscript.

I, Rochelle, could never have taken on such an endeavor without Sue's organizational and FTP skills. I never realized how much work goes into one of these projects, even when you think it is done. For my part, I dedicate this work to my family and friends whose encouragement keeps me flowing.

Pasadena/Irvine   THE EDITORS
September 1999

Guest Preface

Advances in the field of cell biology have always been closely related to the development of quantitative analytical methods that can be applied to individual cells or cell organelles. Almost from the early stages following the invention of the microscope, the investigator has been keenly interested in obtaining information on the functionality of single cells and how cells perform under different sets of experimental conditions. Although cells could be viewed in the microscope for a few hundred years, only since the relatively recent application of autoradiography did we come to realize that, although cells may visually appear very much alike, they are quite different in their functional capacity. The quest to understand these differences in a cell population lead to a new series of techniques for labeling and quantitating DNA content and similar approaches have driven the development of methods for analyzing various other cellular properties.

The development of new analytical techniques follows the age old pattern of applying successes of the past with current innovation, logic and new biological information. Results from autoradiography expanded the concept of the cell cycle from interphase and mitosis to the more definitive G0/G1, S and G2/M phases. This new knowledge lead to the development of technology to measure and analyze various parameters related to the cell cycle. The results of such studies indicated that the dynamic nature of cycle progression, cell maturation and other physiological processes, involves changes in the size, the shape and function of the various cell organelles that are rapid and under well coordinated, temporal regulation. From numerous observations it has become clear that a better understanding of cell physiology can only be obtained by the measurement of as many functional properties as possible in single cells. The concept is now well-founded and has been more accurately described as the multiparameter analysis approach. From such an analysis it is possible to obtain a more accurate understanding about the normal operations of the cell machinery and thereby be better positioned to assess changes that occur under experimental or pathological conditions.

When attempting to develop new analytical techniques, one quickly realizes the limitations of our physical attributes to make rapid quantitative measurements. Under a microscope one cannot distinguish a G1 from a G2 phase cell stained with a fluorescent dye, although there is a relatively two fold difference in intensity. For obvious reasons then, investigators have pursued the parallel development of cell probes and instrumentation that can measure, quantitatively, particular markers of interest. Although significant strides have been achieved, it is clear that much of the methodology is still quite complex. Most instrumentation is still complicated and not always user friendly nor universal in their application to all types of measurements. Some methods are routine and well-tested, while others require well defined conditions and some careful consideration in adapting them to different biological material. It is important to consider all the pertinent aspects when undertaking the applications of new analytical procedures and locating the relevant information in a detailed, but easy to grasp context, can be of considerable benefit.

The editors developed this manual with consideration of the needs of the beginner and the more experienced cytometrist attempting to expand both their range of applications and/or their area of investigations. The applications, both immediate and potential, have been selected and presented by authors who have in-depth knowledge of the well controlled use and limitations of the various techniques. This compilation of chapters is intended to save the investigator from the tedious task of searching the literature for those details on the particular analytical technique that are, in fact, often lacking in the journal publications.

The manual is divided into nine sections with a detailed appendix. The early chapters provide the fundamentals on the operation of a flow cytometer, data display and analysis. These details will guide the reader into limitations and realistic expectations of the instrument and interpretation of results from complicated measurements. Sample preparation for single and multiple parameter analysis, an important, critical consideration in any experiment, is addressed in considerable detail, along with labeling and multi-color fluorescence analysis of cell surface markers. Analysis of reporter genes, a relatively new application, particularly adapted to cytometry while cell tracking methods have become more popular for determining the accountability of particular subpopulations of interest. Many studies continue to include cell cycle phase-specific as well as RNA content determinations, and a section is provided on appropriate nucleic acid labeling and analysis for both plant and animal cells and the use of selected fluorochromes for identifying cells in apoptosis. Techniques for analysis of cell physiology involving probes for intracellular pH, Ca determinations and cytoenzyme systems provide the reader with approaches for examining the functionality of single cells. Recovery of viable cells by physical sorting remains of interest to many investigators, but actually employing the technique is not an easy task. However, the manual provides a straight forward approach for performing sorting of viable cells under sterile conditions. The successful and safe operation of the core facility is presented in some detail by individuals who have really researched the subject and obtained the input of a number of facility operators. The appendix contains a wealth of valuable information for quick reference to the availability and the commercial sources that can provide lasers, fluorescent compounds, computer software and specific antibodies. In general, the editors have made every attempt to compile a very comprehensive manual that should be extremely useful to the reader. They have selected the various sections based on their long standing experience in the field of cytometry and on the questions and analytical problems that have been brought to them by investigators wishing to analyze a wide range of biological material and often times using probes that require much adaptation of the methodology.

HARRY CRISSMAN

Contents

Section 6

Section 9

Appendices

List of Contributors

AULT, KENNETH A.
Maine Medical Center Research Institute, MMCRI,
125 John Roberts Rd., Suite 8, South Portland, ME 04106, USA

BACHANT, JEFF
Dept. of Biochemistry, Howard Hughes Medical Institute,
Baylor College of Medicine, Houston, TX 77030, USA

BAGWELL, BRUCE
Verity Software House, Inc., PO Box 247, 45A Augusta Rd.,
Topsham, ME 04086, USA

BEAVIS, ANDY
Flow Cytometry Core Facility, Princeton University,
Department of Molecular Biology, Lewis Thomas Laboratory,
Washington Road, Princeton, NJ 08544–1014, USA

BEHRENS-JUNG, UTE
Miltenyi Biotec GmbH, Friedrich-Ebert-Strasse 68,
51429 Bergisch Gladbach, Germany

BIGOS, MARTIN
Senior Scientific Programmer, Department of Genetics,
Stanford University School of Medicine, Beckman B-007,
Stanford, CA 94305, USA

CHEN, FEI
California Institute of Technology, Division of Biology,
1200 E. California Blvd., Pasadena, CA 91125, USA

CHEW, KAREN
University of California at San Francisco, Cancer Center,
Room S435, 2340 Sutter Street, San Francisco, CA 94115, USA

CODER, DAVID M.
University of Washington, School of Medicine,
Dept. of Immunology, Box 357650, Seattle, WA 98195-7650, USA

DAYN, ANDREW
Director of Development, Ingenex, Inc., 400 Oyster Point Blvd.,
Suite 505, South San Francisco, CA 94080-1921, USA

DAVIES, DEREK
FACS Laboratory (Room 1B1), Imperial Cancer Reserach Fund,
44 Lincoln's Inn Fields, London WC2A 3PX, UK

DEMAGGIO, SUSAN
Optical Biology Core Facility, Developmental Biology Center,
University of California, Irvine, CA 92687-2275, USA

DIAMOND, ROCHELLE A.
California Institute of Technology, Division of Biology,
1200 E. California B., Pasadena, CA 91125, USA

DURACK, GARY
Biotechnology Center Flow Cytometry Facility,
University of Illinois at Urbana-Champaign, 231 ERML,
1201 W. Gregory Dr., Urbana, IL 61801, USA

DYNLACHT, JOSEPH R.
Indiana University School of Medicine, Dept. of Radiation
Oncology, Indiana Cancer Pavilion, RT 041, 535 Barnhill Dr.,
Indianapolis, IN 46202, USA

FIERING, STEVEN
Microbiology Department, Dartmouth Medical School,
6 West Boswell, Dartmouth-Hitchcock Medical Center,
Lebanon, NH 03756, USA

FOX, MICHAEL H.
Professor, Radiological Health Sciences, Chairman,
Graduate Program in Cell and Molecular Biology,
Colorado State Univ., Fort Collins, CO 80523 – 1673, USA

FREY, TOM
Becton Dickenson, Becton Dickinson Immunocytometry
Systems, 2350 Qume Road, San Jose, CA 95131, USA

GINOUVES, PAUL
Coherent, Laser Group, 5100 Patrick Henry Dr.,
Santa Clara, CA 95054, USA

GIVANS, ALICE L.
Englert Cell Analysis Laboratory of the Norris Cotton Cancer
Center, Dartmouth Medical School, Dept. of Physiology, Borwell
Building, Lebanon, NH 03756-0001, USA

GOTTLIEB, ROBERTA
The Scripps Research Institute, 10550 N. Torrey Pines Rd. NX7,
La Jolla, CA 92037, USA

GUILLOU, LAURE
CNRS-UPR 9042, Station Biologique de Roscoff,
Place Georges Teissier, 29680 Roscoff, France

HARVEY, JEFF
Bio-Rad Laboratories Inc., Flow Cytometry,
4000 Alfred Nobel Dr., Hercules, CA 94547, USA

HUGHES, CLARE
FACS Laboratory (Room 1B1), Imperial Cancer Research Fund,
44 Lincoln's Inn Fields, London WC2A 3PX, UK

KAIN, PH.D., STEVEN R.
Clontech Laboratories Inc, Cell Biology Group,
1020 E. Meadow Circle, Palo Alto, CA 94303-4230, USA

KNOWLES, CATHY
Maine Medical Center Research Institute, MMCRI 125 John
Roberts Rd., Suite 8, South Portland, ME 04106, USA

KRUEGER, TOM
Coulter, Mail Code: 42 – B01, P.O. Box 169015,
Miami, FL 33116-9015, USA

LEWIS, DOROTHY
Baylor College of Medicine, Department of Microbiolgoy and
Immunology M910, One Baylor Plaza, Houston, TX 77030, USA

LEIF, ROBERT C.
Vice President, Newport Instruments, 5648 Toyon Rd,
San Diego, CA 92115-1022, USA

LOPEZ, PETER A.
Cytomation, Inc., 4850 Innovation Dr.,
Fort Collins, CO 80525, USA

MARIE, DOMINIQUE
CNRS-UPR 9042, Station Biologique de Roscoff,
Place Georges Teissier, 29680 Roscoff, France

MCFARLAND, DAVID C.
Howard Hughes Medical Institute Flow Cytometry Facility,
807 Light Hall, Vanderbilt University Medical Center,
Nashville, TN 37232-0295, USA

MEISENHOLDER, GRANT W.
Scripps Research Institute, 10550 N. Torrey Pines Rd.,
La Jolla, CA 92037, USA

MERLIN, STEVEN
Flow Cytometry Facility, Hospital for Special Surgery, Research
Bldg, Room 312, 535 E. 70th St., New York, NY 10021, USA

MORRISON, SEAN J.
Division of Molecular Medicine and Genetics, Department of
Internal Medicine, University of Michigan, 3215 Comprehensive
Cancer Center, 1500 E. Medical Center Drive,
Ann Arbor, MI, 48109-0934 USA

MOTTLEY, JOHN
Cellular and Molecular Biology Research Unit, Department
of Life Sciences, University of East London, Romford Road,
London E15 4LZ, UK

MUNN, MALCOLM
Chromaprobe Inc., 897 Independance Ave 4C,
Mountain View, CA 94043, USA

MURPHY, ROBERT F.
Department of Biological Sciences, Carnegie Mellon University,
440 Fifth Avenue, Pittsburgh, PA 15213, USA

NORBERG, JUDITH
Research Flow Cytometry Lab, VA San Diego Medical
Care Health System, 3350 La Jolla Village Dr. V151,
La Jolla, CA 92161, USA

OSBORNE, GEOFFREY W.
Flow Cytometry Laboratory, EM/HISTO/FACS Unit,
John Curtin School of Medicine, Australian National University,
PO Box 334, National Australian Capital, Canberra, 2601,
Australia

PAN, ALISON CHINHUEI
PPD Discovery, Inc. 1505 O'Brien Drive, Suite 13,
Menlo Park, CA 94025-1435, USA

PARK, SUK
University of California, Davis School of Medicine,
Davis, CA 95616, USA

PARKS, DAVID R.
Department of Genetics, Stanford University School
of Medicine, Beckman B-007, Stanford, CA 94305 – 5125, USA

PARTENSKY, FREDERIC
CNRS-UPR 9042, Station Biologique de Roscoff,
Place Georges Teissier, 29680 Roscoff, France

POON, REBECCA
Sigma Chemicals, R&D Department 47, 3600 S. 2nd St.,
St. Louis, MO 63118, USA

POULOS, NICHOLAS
Southside Neurosurgical Associates, 183 S. Main St.,
Danville, VA 24541, USA

RAGHU, GANAPATHIRAMA
Discovery Research, Protein Science Dept., INCYTE Pharmaceuticals, Inc., 3174 Porter Drive, Palo Alto, CA 94304, USA

ROBERTS, ANDREW V.
Cellular and Molecular Biology Research Unit, Department
of Life Sciences, University of East London, Romford Road,
London E15 4LZ, UK

RYBACK, SHEREE L.
Department of Biological Sciences, Carnegie Mellon University,
4400 Fifth Avenue, Pittsburgh, PA 15213, USA

SCHMID, INGRID
Flow Cytometry Specialist, UCLA School of Medicine, Division
of Hematology-Oncology, 10833 Le Conte Ave.,
Los Angeles, CA 90095, USA

SCHOBER-DITMORE, WENDY
MLT (ASCP), QCYM, FLow Cytometry Specialist, Baylor College
of Medicine, One Baylor Plaza, Houston, TX 77030

SEAMER, LARRY
University of New Mexico, Albuquerque, NM 87131, USA

SHAPIRO, HOWARD
283 Highland Ave., West Newton, MA 02165 – 6044, USA

SIMON, NATALIE
CNRS-UPR 9042, Alfred Wegener-Institut für Polar- und
Meeresforschung, Am Handelshafen 12, 27570 Bremerhaven,
Germany

SKLAR, LARRY
Professor of Pathology, UNM Health Sciences Center, Cancer
Research Facility Bldg. 229, UNM Health Science Center,
Albuquerque, NM 87131, USA

STANBRIDGE, ERIC
Department of Microbiology and Molecular Genetics,
Mail Code 4025 B235, B210 MedSci I, University of California,
Irvine, CA 92687-2275, USA

STOVEL, RICHARD T.
Department of Genetics, Stanford University School
of Medicine, Beckman B-007, Stanford, CA 94305-5125, USA

VAULOT, DANIEL
CNRS-UPR 9042, Station Biologique de Roscoff, Place Georges
Teissier, 29680 Roscoff, France

WANG, HUA
Stowers Medical Research Institute, Caltech Division of Biology,
1200 E. California Blvd., Pasadena, CA 91125, USA

WARING, PAUL
Apoptosis Laboratory, Division of Immunology
and Cell Biology, John Curtin School of Medicine,
Australian National University, PO Box 334,
National Australian Capitol Territory, 2601 Australia

WILKE-DOUGLAS, MINDY
Miltenyi Biotec, 251 Auburn Ravine Road, Suite 208,
Auburn, CA 95603, USA

WOO, GARY
Amgen Inc., 1840 DeHavilland Dr.,
Thousand Oaks, CA 91320, USA

YOKOYA, KAZUTOMO
Cellular and Molecular Biology Research Unit, Department
of Life Sciences, University of East London, Romford Road,
London E15 4LZ, UK

YUI, MARY
California Institue of Technology, Division of Biology 156 – 29,
1200 E. California Blvd., Pasadena, CA 91125, USA

Section 1

A Palette of Living Colors

ROCHELLE DIAMOND

Introduction

Flow Cytometry and cell sorting are becoming increasingly important in many facets of biological research. In particular, developmental, cellular, and molecular biologists are fast becoming integrated users of flow cytometry core facilities which used to be the sole bastion of immunologists and clinical researchers. The vision for this book stems from the editors' experiences managing flow cytometry/cell sorting core facilities for these emerging researchers. Many first time facility users are as unacquainted with the instrumentation as they are innocent in the concepts of what can be done with the technology. We have attempted to set a framework for understanding the instrumentation residing in these core facilities, whether they are located at universities or at sea. In addition, we provide a step by step approach to the methodology for measuring various cellular attributes demonstrated in the particular cells of interest. We also present a myriad of resources to fuel the curiosity and answer questions for both the new and adept user.

Investigators utilizing flow cytometry are no longer just defining cells by their phenotypic surface markers. Physiological responses[1], functional expression of intracellular products[2], and a cell's apoptotic status[3] are just three attributes that can be measured simultaneously with each other and/or with cell surface markers. Powerful new genetic engineering techniques have come into use with flow cytometry. "Enhancer-trap" marker and reporter gene constructs[4] are now commercially available which can be used to define endogenous patterns of gene expression, help determine lineage commitment and differentiation, and look for regulatory mechanisms involved in generating gene expression patterns[5]. Teamed with electronic cell sorting these reagents can help to answer many unanswered biological questions by allowing the investigator to, for example, purify stem cells for use in *in vivo* reinsertion, *in vitro* cultivation experiments, or clonal expansion assays[6]. Now investigators are no longer tied to the flow cytometer for hours sorting enough cells to make cDNA libraries from rare populations. The advent of post-sorting reverse transcription-dependent polymerase chain reaction (RT-PCR) has reduced the number of cells required to test for expression of mRNAs by orders of magnitude[7]. Advances in high speed sorting have allowed for all kinds of rare cell sorting, for example, purification on the basis of reporter gene expression in transfected or transgenic animal cells[8]. There are new ways of tracking cells, both *in vivo* and *in vitro*, by measuring their proliferative expansion[9] as well as their cell death. These and many other new and exciting ways to analyze and sort out cells of interest are fast becoming commercial products and off-the-shelf reagents.

As this technology has become more accessible and applicable to many biologists, it has also become more powerful. Our ability to measure multiple parameters simultaneously on single cells in a heterogeneous population has dramatically improved with the research on new fluorochromes and tandem-conjugated dyes. Where we once measured just the scatter profiles and one or two colors for a population of cells, most facilities now routinely perform three and four color data acquisition with a single laser and more colors with the addition of other lasers and electronics.

The goal of this book is to make this increasingly valuable technology accessible to researchers, including the students, post-doctoral scholars, and technicians who labor in the fields harvesting the grains of knowledge inherent in this integration between analysis and physical isolation/purification methodologies. We want to make the rich resources now available through institutional, commercial, and on-line sources known.

On the institutional level, many universities and some companies now have or are forming "core" facilities that help distribute the expense and management of these sophisticated instruments. In many cases, knowledgeable operators and managers are willing to help researchers interface their laboratory's specific needs with the versatile setup and operation of the instruments. Their knowledge and dedication to the instruments entrusted to their care are extremely valuable assets and should not be overlooked. As more and more researchers get together to form user groups for core facilities they will need to know everything that they never wanted to know about how a flow facility is set up, operated, and maintained, both physically and financially. We have attempted to provide this in our section on core facilities. This kind of information is important for new users as well as for forming consortiums. It will give insight into how and why a facility is managed, in order to provide the information necessary for a successful interaction between the user and the facility. A new user should be able to ask the kinds of questions needed to plan for his/her experiments and budgets, and, in turn be ready to answer the core facility's questions about what will be required for each experiment.

A few words of caution: the protocols that we have provided are an array of common and/or interesting procedures that many core facilities perform or have the capability of setting up for researchers. Many protocols may need to be adapted for particular cell types or instrument requirements. Cell preparation is individual to each cell type and you should refer to relevant literature for your particular cell or tissue type as discussed in the section on cell preparation. Pilot experiments are highly recommended to determine feasibility for first time procedures and to make sure that reagents are working properly and that the method is fully understood. There are multiple ways of performing some measurements, and for good reason. Some instruments and systems are better suited for one protocol than another, due to available reagents, filter set-ups, or laser configurations. We have attempted to provide some examples of overlapping protocols for just such purposes but they are not an exhaustive comparison. New methods and techniques appear daily. Use the references as a starting point to move out and up to new heights.

We want to bring all types of resources to the attention of researchers. Time is a very valuable commodity in these days of competitive research arenas. There are many commercial enterprises and entrepreneurs who are attempting to provide not only instrumentation but also "off-the-shelf" reagents and protocols for many of the commonly used flow cytometric methodologies. These include cell cycle and apoptosis kits, intracellular cytokine and enzyme expression packages, reporter gene vector and construction kits, and contract service providers for custom antibody preparations and conjugations. These products and services are readily available and in many cases provide easy and reproducible assays which researchers can build upon. For this reason you will find that commercial research and development scientists contribute to some of our sections and we have attempted to provide sources in our appendices for researching these products as well as their on-line services. In addition we have listed academic flow cytometry on-line resources for core facilities, discussion lists, users' groups, societies, academic courses, meetings, and national resource facilities.

We wish to thank all of our contributing authors for sharing their expertise and valuable insight on these protocols. It is just as important to understand the underlying principles of these methods as it is to generate the data, for the interpretation of the data relies on that fundamental knowledge. We hope that when you use some of the protocols and approaches we have provided, by themselves and in combination with each other, you will be struck by the insight that if "you can see it – you can sort it" in living color. We want researchers to be inspired to seek out additional information and protocols through the resources that we have provided. It is our hope that this will encourage each of us to be inventive and to strike out on our own to initiate new advances and protocols, because it is within our own creativity and imagination to fill our palette with living colors.

References

1. Griffioen AW, Rijkers GT, Cambier JC. Flow cytometric analysis of intracellular calcium: the polyclonal and antigen-specific response in human B lymphocytes. Methods: A Companion to Methods in Enzymology 1991; 2(3): 219-226.
2. Bani L, David D, Moreau JL et al. Expression of the Il-2 receptor gamma subunit in resting human CD4 T-lymphocytes-messenger RNA is constitutively transcribed and the protein stored as an intracellular component. Int Immunol. 1997; 9(4): 573-580.
3. Darzynkiewicz Z, Juan G, Li X et al. Cytometry in cell necrobiology-analysis of apoptosis and accidental cell death (necrosis). 1997; 27(1): 1-20.
4. Roederer M, Fiering S, Herzenberg LA. FACS-Gal: flow cytometric analysis and sorting of cells expressing reporter gene constructs. Methods: A Companion to Methods in Enzymology 1991; 2(3): 248-260.
5. Reddy S, Rayburn H, Vonmelchner H. Fluorescence-activated sorting of totipotent embryonic stem-cells expressing developmentally regulated lac z fusion genes. P-NAS-US 1992 89:6721-6725.

6. Leary JF. Strategies in rare cell detection and isolation. Methods in Cell Biology 1994; 42:331-358.

7. Hu M, Krause D, Greaves M et al. Multilineage gene-expression precedes commitment in the hematopoeitic system. Genes and Development 1997; 11(6): 774-785.

8. Lorincz M, Roederer M, Diwu Z et al. Enzyme-generated intracellular fluorescence for single-cell reporter gene analysis utilizing *Escherichia coli* β-glucuronidase. Cytometry 1996; 24:321-329.

9. Young JC, Varma A, Diguisto D. Retention of quiescent hematopoeitic-cells with high proliferative potential. Blood 1996; 87(2): 545-556.

The Colorful History of Flow Cytometry

SUSAN DEMAGGIO

To write an exhaustive history of Flow Cytometry in this manual is not really necessary. There are many very good texts and articles on the history of flow cytometry, many of them listed in the references of this manual. You can read them as you wish, at your leisure. However, it is a good idea to look at the varied and colorful past of this technology and how it has developed into the tool we use today. A quick look at the historical influences will give you a better idea of what a remarkable rainbow of applications and uses this instrument has had over the almost 3 decades since it was introduced to the world commercially.

The first "flowing" system for measurement of cells was described by Andrew Moldavan in Montreal in 1934. It was designed to count red blood cells and neutral-red stained yeast cells as they passed through a capillary tube on a microscope stage, passing a photodetector attached to the eyepiece. It isn't clear that it was ever really produced or functional; however, the idea was born. A stream of colorful inventors and researchers followed. Louis Kamentsky, Myron Melamed, Dittrich and Gohde, the Hertzenbergs, Mack Fulwyler, Marvin Van Dilla, and Howard Shapiro, just to name a few, all influenced the development of the technology and instruments over the next 40 years. Even inventors of peripheral technologies such a Kohler and Milsteins monoclonal antibody technology, Wallace Coulter's understanding of cell measurement and the use of Coulter volume in early research, Crosland-Taylor's hydrodynamic focusing principles, and the invention of the ink-jet printer by RG Sweet at Stanford contributed to the development of instruments as we know them. The insight and creativity of each of these has shaped the instruments we now use. The complexity of today's instruments and our ability to analyze such a variety of parameters on each cell that passes through these instruments, is directly attributable to these pioneers of flow technology.

Louis Kamentsky in the mid 1960s developed a system based on the microscope spectrophotometer pattern of Moldavan to be used in cervical cytology screening using UV absorption and the scatter of blue light. In 1967, he and Myron Melamed elaborated on the original design using a syringe system to retrieve suspicious cells from the flow based on the ratio of absorption to scatter. About the same time, Lenora and Leonard Hertzenberg at Stanford were modifiying a system for particle sorting used at Los Alamos National Laboratory, into a fluorescence sorting system. Dittrich and Gohde introduced the generation of histograms describing the data of ethidium bromide stained cells the first DNA histograms. Fulwyler elaborated on the inkjet design of Sweet to develop a charged stream deflection system of sorting in the mid 1960s. Los Alamos Laboratory was busy developing a system which utilized the first laminar flow (hydrodynamic focusing) and which incidently used an argon ion laser as the light source; the development of a system which could perform multiparameter fluorescence, scatter and volume measurements simultaneously as well as sort cells was accomplished. Commercial cytometers appeared on the market shortly thereafter, in the early 70s. Elaborations on the theme have continued and we find today a variety of instruments and a multitude of applications - with new ones being added daily. Commercial development has brought this technology into the reach of most labs and has provided us with the simpler operation of push-button bench top analyzers and highly complex multi-stream high speed sorters at the same time. The differences in the instruments have contributed to the variety of applications and techniques we utilize daily in a flow facility. Both the simplicity of the bench-top analyzers and the complexity of the multi-laser, multi-stream high speed sorters have a place in our labs. The development of new techniques is now being passed on to those of us who are using these instruments today. With open minds, and a little creativity, the horizon is bright with possibility - there are a myriad of colors to explore and a rainbow of possibilities yet to develop. The future of flow is going to be as colorful as the past has been.

Section 2

Getting Started

ROCHELLE DIAMOND

Jumping in and getting started with any technology requires a certain familiarity with the instrumentation. In this section we get started with Tom Frey who presents an in depth description of two types of commercial flow cytometer/cell sorters and their differences. We get to know the various subsystems and how they interact with each other to provide us with interpretable end products – data and sorted cells. Howard Shapiro, one of the gurus of flow instrumentation, follows up this discussion with some interesting caveats, take home lessons, and pointers for getting the best out of our instruments. Data output is the most crucial aspect of flow cytometry for users and proves to be the most frequently asked question category for new users. "What are we looking at?" Larry Seamer and Susan DeMaggio respectively discuss the standardized data file and the varied data displays, which are common to commercial cytometers. Wendy Schober and Dorothy Lewis try to keep us out of trouble by giving us some guidelines to common problems that occur in displaying, interpreting, and publishing data. Dave Coder then provides answers to the question of what to do with all the data we generate, suggesting ways to network and archive information. Lastly, Rochelle Diamond presents a comprehensive view of the spectra and compensation issues, as well as quality control problems we deal with in Flow Cytometry including guidelines on choosing filter setups, setting compensation and assuring the quality of our analyses.

We would like to refer you to our appendix for further information about commercial instruments currently available. We are not endorsing particular suppliers or manufacturers but rather setting your pointer for your information searches.

How the FACSCalibur and FACS Vantage Work and Why It Matters

TOM FREY

Introduction

Cytometry is the measurement of properties of cells. In a flow cytometer this measurement is done as the cells are delivered to the measurement area in a flowing stream. The current generation of flow cytometers generally are able to measure light scatter from the cells and fluorescence from the cells themselves or from fluorescent labeled molecules attached to the cells. To accomplish this analysis, three main subsystems of the cytometer are required. The fluidics subsystem delivers the cells to the measurement area. The optics subsystem guides light from the excitation source or sources into the measurement area and collects the light scatter and emitted fluorescence, splitting the collected light into spectral bands (colors) for multiparameter analysis. The electronics subsystem converts the detected light into digital values for computer analysis of the data. In addition, flow cytometers can use the information about the properties of the cell to sort cells of interest. The following sections describe the fluidics, optics, electronics, and sorting mechanisms of the Becton Dickinson FACSCalibur (a benchtop flow cytometer for research and clinical users) and FACS Vantage (a research cell sorter). Similar types of systems are available from other commercial vendors.

Fluidics

Flow cytometers usually use a stream-within-a-stream method for delivering cells for analysis. A large flow of carrier fluid is used to move a small inner stream that contains the sample. The major flow of fluid surrounds the sample stream on all sides and is therefore called the sheath stream. This sheath flow carries the cells through the measurement area. On a FACSCalibur the sheath fluid comes from a 4-liter pressurized tank at 18 ml per minute. This gives a stream velocity through the measurement area of 6 meters/second. The sample is also pressurized, and is injected into the sheath flow. On a FACSCalibur the injection rate can be 60, 35, or 12 μL/min. Note that even at the highest fixed rate the sample flow is less than 1% of the sheath flow!

As the sample is injected, it is caught up in sheath flow. The high velocity of sheath flow through a small region reduces mixing across the sheath stream, so the sample continues as a well defined internal stream surrounded by the sheath. (Fluid flow in which layers remain well defined without mixing is known as laminar flow.) The internal stream is called the sample stream or core stream. A well controlled sheath stream with minimal mixing of the sample stream allows the sample to be delivered to a well defined and predictable location for measurement (Fig. 1). Since the sample flow is such a small fraction of the sheath

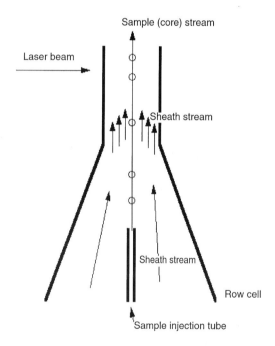

Fig. 1. Fluid flow in the FACSCalibur flow cell. An outer (sheath) stream carries the sample stream past the laser intercept. The sample remains in a narrow stream in a well defined location and with a well defined velocity. The mixing between the sample (core) stream and the sheath is minimal.

flow, increasing the injection rate has a negligible effect on the velocity of the particles in the sample as they pass the measurement area. Increasing the sample injection rate, however, does cause the sample stream to take up a larger fraction of the total flow. This widens the sample stream as the sample injection rate is increased. This can have an impact on the precision of fluorescence measurements from the sample, as discussed below.

The FACS Vantage uses a similar fluidics system. Sheath flow is user-adjustable and depends on the pressure on the sheath tank and on the size of the nozzle through which the flow must pass. Unlike on the FACSCalibur, pressure on the sample is continuously adjustable. The velocity of the sample as it passes the interrogation point on a FACS Vantage is usually around 10 meters/sec, but depends to some extent on sheath pressure.

Optics

The optical system must guide and shape the light from the excitation source, collect the light of interest from the sample, and direct the collected light to the detectors. The optics on the FACS-Calibur are easier to describe because they are not user adjustable, so they will be the basis for the first part of this section.

Excitation

The primary excitation source in a FACSCalibur is an air-cooled Argon-ion laser that emits 15 mW of 488 nm (blue) light. The laser beam first passes through a set of prisms, which shape the originally circular beam so that its width will be three times its height when focused onto the sample (Fig. 2). The reshaped beam is then focused by a lens. There is a small optical element just after the lens that allows fine adjustment of the beam path. The beam enters the flow cell and reaches focus at the core stream (Fig. 3A) . At focus, the beam is 20 μm high (along the direction of flow) and 60 μm wide (front to back in the flow cell). The laser is well-focused for at least 100 μm along its direction of travel (left to right through the flow cell).

The secondary source on the FACSCalibur is a diode laser that emits red light around 635 nm (diode laser emission can vary from laser to laser and with temperature). This laser is part of

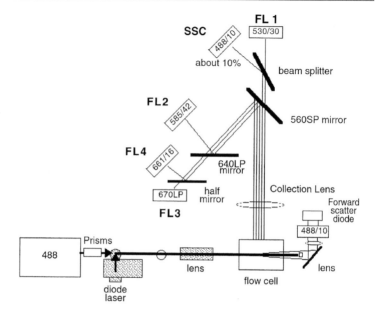

Fig. 2. Optical components for excitation and emission. The layout for the 4-color FACSCalibur is shown.

the optional 4-color upgrade. The laser is mounted at right angles to the 488 nm beam. The red beam hits a mirror which reflects this beam along the path of the blue beam (Fig. 2). This occurs after the beam shaping prisms because the diode laser beam is already shaped to give an appropriate height to width ratio. The two beams are focused by the lens to different spots on the flow stream. Cells pass through the red beam first and then the blue beam.

The beam dimensions (Fig. 3) have some impact on experiment design and interpretation. First, the 20 μm height is intended to be large enough to include the entire target particle within the laser spot. When this is true, the whole particle is excited and fluoresces. The fluorescence pulse height is then proportional to the total fluorescence and the pulse height is used to measure total fluorescence. In the case of very large particles this assumption does not hold. Second, the 60 μm width allows fairly uniform excitation across the center of the beam. Laser beams shaped by lenses have an intensity profile that falls off gradually (Gaussian) as you move out from the center of the beam. The

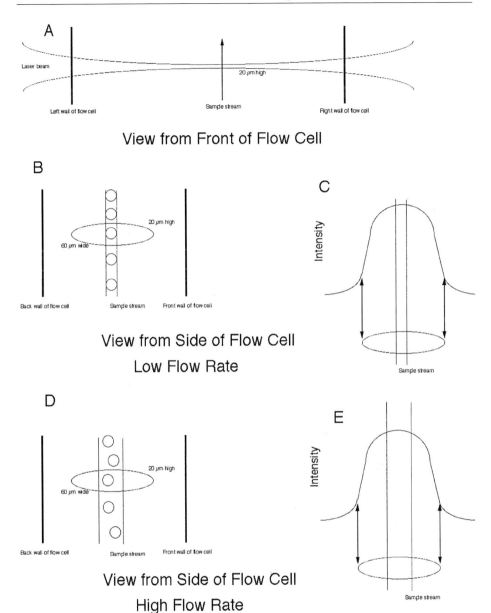

Fig. 3. Laser profile on the FACSCalibur. (**A**) Laser is focused along its path. (**B, C**) The dimensions at focus are 20 × 60 µm. At low flow, the sample stream is narrow, keeping the cells in the most intense and uniform part of the Gaussian beam. The beam width (60 µm) is measured at the point at which the intensity has dropped to about 13% of the peak intensity. (**D, E**) Similar to B and C, but showing the increase in sample stream width at high flow rate.

widths quoted are for the point at which the beam intensity is down to 13% of the peak intensity. The fact that the beam is not exactly the same intensity across the entire width means that if particles are allowed to appear at different locations within the stream the intensity of excitation (and therefore emission) will vary. The low flow rate (12 µL/min) gives the narrowest core stream width, and this forces all of the events to pass through the center of the laser spot where the intensity is most uniform. Very uniform excitation is important for particles that have a very narrow distribution of fluorescence to begin with, for example when measuring the DNA content of cells. When samples are run on high the core stream is wider and particles experience more of the beam profile (Fig. 3). This can result in higher variability in the fluorescence measurements. The biological variability in the number of cell surface antigens present on cells is usually much higher than any variability introduced by the wider core stream, and so these types of measurements are almost always done using the high flow rate.

There is a second area in which the stream width can have an impact on experimental results. In some experimental situations it is desirable to avoid washing the sample before running it. This can be either to avoid re-equilibration of a dye used to stain the sample, to avoid selective loss of some particle types during the wash, or merely for convenience. In unwashed samples there will be free fluorochrome (dye or antibody) in the sample fluid, not just the bound fluorochrome on the cell or particle. As the sample stream widens more of this unbound material is measured at the time that the cell is being measured. This leads to an increase in fluorescence, which is most noticeable for particles that would normally give a low signal because they have no bound fluorochrome. Thus, the discrimination between negative and dimly positive events can be compromised as the sample stream widens and measurements that depend on the exact magnitude of the separation between stained and unstained events are also compromised.

The excitation optics on a FACS Vantage are considerably more complex. The main beam and any secondary beams must be shaped and steered to the flow stream. As on the FACS-Calibur, the main beam and secondary beams intercept the sample stream at different locations. An adjustable lens is also available to change the beam shape at the intercept. Beam spacing is adjustable to allow pulses from secondary beams to occur during the appropriate time interval for measurement. However, many of the same factors must be considered. For example decreased sample injection rates will lead to lower variability in fluorescence intensity and for lowest variability the excitation beam is defocused to allow for a flatter beam profile and more likelihood of the particles experiencing identical illumination intensity.

Collection

The collection optics on the FACSCalibur were designed for high light collection efficiency, which allows the use of low-power excitation sources. The quartz flow cell is optically coupled to a high numerical aperture lens to minimize light loss through the system. The collection lens system is designed to focus the collected light to create an image of the measurement area and the particle being measured. Because of the focal length of the lens system, this image is formed just in front of the detectors. On the FACS Vantage the steam-in-air system allows high sorting rates, but the air-stream interface and the fact that the stream can't be optically coupled to the collection lens make light collection less efficient. The FACS Vantage, therefore, uses higher power excitation sources to generate more signal in order to obtain sensitivity comparable to that of the FACSCalibur.

Emission

Flow cytometers use dichroic mirrors (which reflect some wavelengths and pass others) and filters (which pass only certain wavelengths) to separate the collected light into different wavelength bands. These bands are directed to specific detectors. The choice of the upper and lower wavelengths of these bands is usually dictated by the fluorescence spectrum of the molecules that are being detected, and by the need to minimize the detection of fluorescence from other molecules present in the sample. The emission optics are also designed to reject light from locations that should not contain the cells of interest. This is done by physically blocking light from inappropriate regions, see the "Image forming optics and spatial filtering" section below.

Forward Scatter

Light hitting cells is defracted and scattered in all directions. Light scatter only slightly off of the axis of the exciting beam is called low angle light scatter, forward angle light scatter (FALS), or forward scatter (FSC). This scattering occurs when a particle is in the laser, and is often used as an indicator of the presence of a particle. Larger particles generate more intense forward scatter, although differences in the index of refraction of different particle types also changes the intensity of the forward scatter. Microspheres of known diameter (usually made of some type of plastic) are not good calibrators of cell size because their index of refraction is quite different from that of cells.

In order to detect the forward scattered light near the laser beam it is necessary to block the laser light that is not scattered. A blocker, called an obscuration bar, is usually placed in the path of the beam to prevent it from reaching the forward scatter detector. This obscuration bar is fixed on the FACSCalibur and adjustable on the FACS Vantage. Light that gets around the bar then goes through a hole (aperture) that limits the angle of scatter that is detected. This is fixed to less than about $10°$ off-axis on the FACSCalibur but is adjustable with an iris on the FACS Vantage. This adjustment is important because particle size is one factor that can determine the intensity and angular distribution of light scatter, and detection of forward light scatter from small particles can often be optimized by adjusting the obscuration bar and iris on the FACS Vantage.

Side scatter and fluorescence

Light scattered at right angle to the laser beam is called orthogonal light scatter (OLS) or side scatter (SSC). Side scatter is often most intense from cell types with complex internal structure (although refractive index can influence the signals from different particle types). When used with forward scatter, the side scatter signal can be very useful in distinguishing the populations present in blood samples. Side scattered light is collected by the same lens that collects fluorescence emission and is collected through a large angle section centered around 90° from the beam path. Since fluorescence is emitted at all angles, a larger angle of collection allows more of the fluorescence to be brought to the detectors. At the point of collection there is an expanding cone of light from all of the wavelengths. The lens system converts this to a decreasing cone of light that reaches focus near the detectors. The dichroic mirrors sort out the wavelengths into bands. The wavelength bands are often referred to as channels, and are labeled by the detector that is used for that wavelength band. The different fluorescence channels on the FACSCalibur are diagrammed in Fig. 2, and discussed in some detail below. The information is also summarized in Table 1.

Table 1. Emission filters in the FACSCalibur.

Channel	color	label	3-color	4-color
Forward Scatter	blue (488 nm)	FSC	no filter	488/10
Side Scatter	blue (488 nm)	SSC	no filter	488/10
low wavelength	green	FL1	530/30	530/30
medium wavelength	yellow/orange	FL2	585/42	585/42
long wavelength	blue excited red	FL3	650LP	670LP
long wavelength	red excited red	FL4		661/16

In a FACSCalibur, the first mirror that the collected light encounters is a 560SP (short pass) mirror. This mirror passes light shorter than 560 nm and reflects light longer than that to other optical components. The light that passes this mirror then reaches a beam splitter that reflects about 10% of the light to one detector and lets the other 90% pass. Since light scatter is usually more intense than fluorescence, the detector that gets 10% of the light generates the side scatter signal (light scatter is the same wavelength as the 488 nm excitation and therefore passes the 560SP mirror). In a single laser instrument there is no optical filter in this path since the light scatter intensity is usually much greater than any fluorescence. In a dual laser system an optical filter that passes only 488 nm ($\pm$ 10 nm) light is placed in front of the detector to eliminate light scatter from the secondary beam.

Side scatter and short wavelength (green) fluorescence

The remaining 90% of the light reaches the detector intended for short wavelength fluorescence (FL1). A 530/30 band pass filter (530 nm center wavelength, 30 nm width) is standard on the FL1 detector. This is optimized for fluorescein, but other dyes also emit in this wavelength range. These include, for example, rhodamine 123, $DiOC_6(3)$, BODIPY, and thiazole orange. The filter is mounted on removable carrier and it is possible to replace it with an alternative filter.

Medium wavelength emission (FL2)

Light reflected by the 560SP mirror next reaches a 640LP (long pass) dichroic. Wavelengths shorter than 640 nm are reflected to the FL2 detector and longer wavelengths are passed. The FL2 detector has a 585/42 band pass filter which is optimized for detection of phycoerythrin (PE) but can also be used for a wide variety of other dyes that includes many rhodamines, Cy3, $DiIC_{18}(3)$, and propidium iodide.

Long wavelength blue excited (FL3) and red excited (FL4) emission

On a 3-color FACSCalibur, light passing the 640LP dichroic reaches the FL3 detector through a 650LP filter. Peridinin chlorophyll protein (PerCP), PE/Cy5, PE/Texas Red, LDS-751, propidium iodide, and 7-aminoactinomycin D are among the fluorochromes that emit in this range.

On a FACSCalibur with the 4-color option, another detector is added along with a removable optical block that contains a mirror and filters. The mirror reflects the FL4 emission and passes FL3. This mirror relies on the physical separation of the two signals, which is discussed below. The block also contains removable filters for each detector. A 670LP filter is in the path to the FL3 detector. This filter should work well for almost all of the dyes discussed above for FL3 in the three color system, but will reduce the signal from PE/Texas Red. For the FL4 detector, a pair of filters create a 661/16 band pass with strong rejection of light scatter from the diode laser. This FL4 filter was chosen to optimize the detection of allophycocyanin (APC) while rejecting the emission from PerCP that is excited by the red laser. However, many red excited dyes are also usable with this filter. These include the nucleic acid dye TO-PRO-3, the carbocyanine Cy5, and many oxa-, indo-, and thia-carbocyanine dyes such as $DiIC_{18}(5)$.

Image forming optics and spatial filtering

The laser excitation passes through the walls of the flow cell and a large amount of sheath fluid before and after it hits the sample stream (Fig. 4A). Particles and surfaces in the sheath stream and flow cell can generate light scatter and sometimes even fluorescence (Fig. 4B). These spurious signals can occur throughout the flow cell. The use of a lens in the optics system allows an image of the measurement area to form at the focus of the lens (Fig. 4C). This image is magnified (and a little blurry). Since there is an image, light from different sources (Fig. 4B) in the flow cells end up at different places (Fig. 4D) in the image plane. This fact is used for two different purposes. First, light that originates from parts of the flow cell far away from the focus of the laser can be rejected. This is done by putting a blocking plate in the image plane that only allows light from the central area of the image to pass through a slit in the plate (Fig. 4D). Thus only light from the sample stream reaches the PMT (Fig. 4E). Second, light from the different excitation sources (which hit the sample stream in different spots) is imaged to different locations. This means that a normal mirror that reflects all light can be placed in the path of the emission from only one spot, and that emission from the other spot can pass physically over or under the mirror. (These mirrors that block only part of the image are often referred to as half-mirrors). This is how the FL3 and FL4 emission are separated in the FACSCalibur. The slit in the image plane can also be designed to physically reject light from areas of the sample stream that are not of interest, usually those being excited by a different laser. This strategy is also used in the optical block in the FACSCalibur.

A similar approach using half mirrors and apertures is used in the FACS Vantage. The mirrors and filters are more flexible, they can more easily be removed and changed by the user.

Electronics

The electronic subsystem has several main functions. The detectors convert the light collected to an electrical signal. The threshold circuits let the system know when a relevant particle is in the measurement area. The electrical signal from the PMT must be further amplified and processed to obtain data that can be dis-

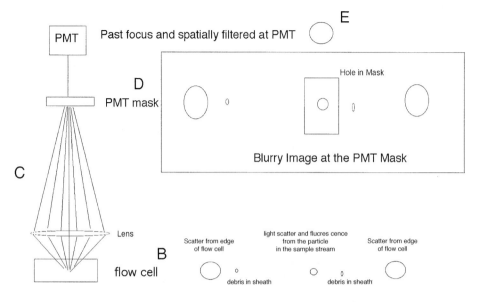

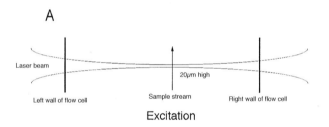

Fig. 4. Spatial filtering of the collected light. (*A, B*) Fluorescence and light scatter occur at various locations as the laser beam crosses the flow cell. (*C*) The light is collected by a lens and focused to form an image. (*D*) The image is formed on a metal plate (mask) with a hole (slit). Only light originating near the sample stream passes through the slit to reach the PMT.

played for the user. Much of the electronics subsystem is devoted to amplifying and producing a computer-usable digital value from the analog electrical signal produced by the detectors. Modern flow cytometers also have compensation circuits which help make data more interpretable when the fluorochromes used emit in more than one of the fluorescence channels of the instrument (see Compensation). Special circuitry is also included to produce an area and width measurements of the pulses generated from the cells.

Detectors and amplifiers

The side scatter and fluorescence detectors on the FACSCalibur and FACS Vantage are photomultiplier tubes (PMTs). These provide the sensitivity needed for the low light level signals that must be detected. The more intense forward scatter signal is detected using a photodiode.

The amount of electrical signal that is produced for each photon striking the PMT is controlled by the PMT voltage. This is the first stage of amplification. When all the amplification is finished, the signal ranges from 1 mV to 10 V. There are also two types of electronic amplifiers present in the flow cytometer. The linear amplifier merely provides a fine adjustment on the signal strength. On the other hand, it is possible to route the signal through a logarithmic amplifier (log amp). This amplifies low amplitude signals more than it amplifies high amplitude ones. The four decade log amp maps each 10-fold range of intensity (1-10 mV, 10-100 mV, 100 mV to 1 V, 1-10 V) to its own quarter of the scale (see Fig. 5). The significance of this is discussed below under "Digitization".

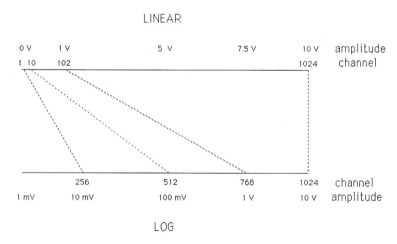

Fig. 5. How signal amplitude maps to channel. The amplitude (in millivolts or Volts) is shown. The channel values are indicated along the axes, and the dotted lines show where a signal of given amplitude would appear on both the linear (top) and log scale (bottom).

Threshold

The amplified analog signal from one of the channels is continuously monitored by the threshold circuitry. Which channel is monitored can be selected by the user. When the signal level goes above a user selected level, the system knows that a relevant particle has entered the laser excitation spot. The digitizing electronics are then committed to measuring the signals from all of the detectors and converting them to digital information. This takes some time, the system is not available for another threshold event until this process has been completed. The time difference from the occurrence of a threshold to the time that the next threshold can be accepted is usually known as the system dead time.

Both the FACSCalibur and FACS Vantage have abort circuitry, which checks for events that have suspicious pulse signatures. Large dips or deformations in the pulse, or the appearance of a second threshold during the dead time, are indications that an incorrect value may be generated for the particle. These events are tagged as aborts and are usually not processed. (In some situations it is possible to ignore the abort signal for sorting purposes.)

The 4-color upgrade to the FACSCalibur provides a secondary threshold. This can be used to limit acquisition to events that meet both threshold criteria. This is potentially useful in some rare event applications in which the majority of event must be ignored to allow adequate event throughput.

Digitization

The digitization system first uses analog circuits to track the incoming pulse from the amplifiers and produce a signal that is proportional to the highest value seen. These peak detection circuits are often called peak-and-holds. This analog value is then processed by an analog-to-digital converter (ADC). The ADC produces a digital value in the range of 0-1023 (10 bits) from the analog signal. The digital values for each channel are collected and packaged as an event. This group of digital values is sent to the host computer. Sort decisions based on the digital data (see Sorting) are made before the data is passed to the host computer to avoid any delay or throughput limit caused by the data transfer.

The fact that the ADC is a 10-bit converter is of some importance. The fact that the measurement can only be assigned to one of about 1000 bins means that cells 10% as bright as the brightest cells must fall in the bins between 0 and 100 and that cells 1% as bright can only be assigned to the first 10 bins. This means that resolution at these low levels is very limited and conversion from intensity to channel must be very accurate. For immunofluorescence, differences in intensity can be even greater than the 100-fold! This is why log amps are used. Now, cells from 1% to 10% as bright as the brightest cells can fall between 512 and 768 on the 1024 scale, those between 0.1% and 1% can fall between 256 and 512, and the first 256 channels can be used for cells less than 0.1% as bright as those in the uppermost channels (see Fig. 5).

The experiment being done determines the need for log or linear amplification. If the experiment requires optimum resolution of two events with similar intensity and narrow intensity distributions (like DNA content measurements), then linear amplification should be chosen. For example, events that differ by two-fold in intensity will be about 80 channels apart with a 4-decade log amp, but can be several hundred channels apart when using linear amplification. (The actual separation will depend on the amplification of the signals). For samples with a wide dynamic range of signals (like immunofluorescence) it is necessary to use log amplification to keep signals on scale with adequate resolution across the whole range of signals.

Compensation

Most fluorochromes have relatively broad emission spectra that are detected in the wavelength bands assigned to different detectors. Many new users have the impression that detection of a fluorochrome in more than one channel is an artifact of the instrumentation, but it should be stressed that these fluorochromes truly emit in these wavelength regions (see the Appendix for the spectra of commonly used fluorochromes). However, for ease of interpretation it is very often useful to assign one fluorochrome to one channel of the cytometer, for example thinking of FL1 as the fluorescein channel. Since fluorescein truly emits in the FL2 channel, some subtraction of the signals is needed to eliminate this if the FL2 channel is to be considered, for example, the PE signal. This is done by the compensation circuits (Fig. 6).

Since each fluorescein molecule has a fixed probability of emitting FL1 and FL2 photons (good old quantum mechanics) it should be possible to figure out how much FL2 should appear every time there is fluorescein emission in FL1. This could then be subtracted from the FL2 signal that is actually observed. Thus some percentage of the FL1 signal (%FL1) is subtracted from the FL2 signal and the compensation is referred to as FL2-%FL1. Now if quantum mechanics is so reliable, why does the FL2-%FL1 value need to be adjusted for fluorescein? The main reason is that the relative number of photons in each channel can be predicted, but the electrical signal produced for each photon can be altered independently by changing the PMT voltages for each channel. For example, if the FL1 fluorochrome emitted 20% as many photons in FL2 as in FL1, but the FL2 channel was set to produce a twofold higher signal from each photon, you would likely need 40% compensation. This is why it is important to set the PMT voltage before compensation is adjusted, and not change PMT settings after compensation has been set.

A digression on statistics and correct compensation

There are two ways that compensation is set up (see Fig. 7). One choice is to set the median of the compensated population (say FL1 fluorochrome positive FL2 fluorochrome negative) equal to the (FL2) median of the negative cells. This method appeals to the purists and makes considerable sense. However, as discussed below, this leaves some of the compensated cells with FL2 values higher than the FL2 values for the negative cells. The other method is to set the compensated population so that no cells are higher than any of the cells in the negative population. This is appealing from a data visualizing standpoint, now all of the FL2 can be assigned to the FL2 fluorochrome.

Why are these two methods different? The answer involves the variance in the signal that you expect when cells are emitting light. Even in a perfect cytometer the properties of random sampling dictate that the variance in the number of photons detected will be at least the square root of the number of photons. Also, the counting noise in the two channels is independent even if it is

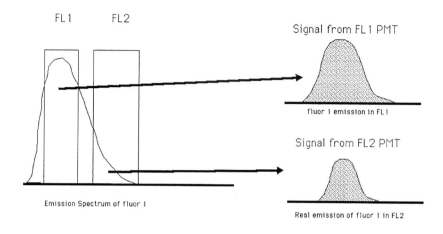

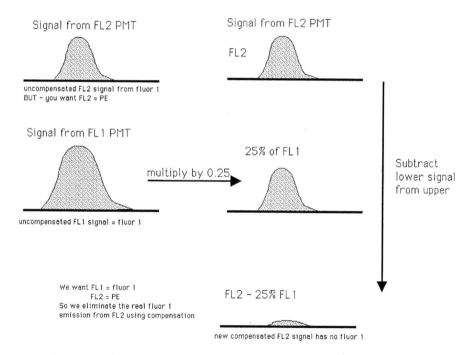

Fig. 6. Analog compensation as used on the FACSCalibur. In this example the signal from the FL1 PMT is multiplied by 0.25 and then substracted from the signal from the FL2 PMT. This gives the observed FL2 = FL2 – 25% FL1. Almost all of the signal from the fluor assigned to FL1 (a hypothetical fluor 1) has been eliminated from the FL2 channel, even though there is emission from fluor 1 in that wavelength band.

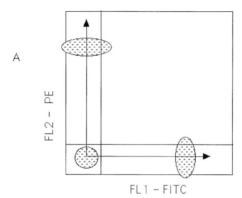

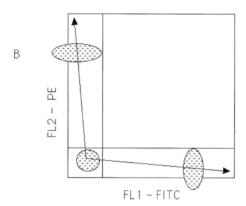

Fig. 7. Two methods for setting up compensation. (*A*) Medians of the populations are aligned. Some events are below zero, and some compensated events are still in the positive quadrants. (*B*) No compensated events are above the negatives. Larger numbers of events are below zero and the medians are no longer aligned.

from the same fluorochrome (it is a quantum effect, each photon has a given probability of being either FL1 or FL2). Why does this suggest that brighter cells in FL1 should appear noisier in FL2 after compensation? Let's assume that we have set up the instrument so that negative cells emit 10 ± 3 (3 = square root of 10) photons in both FL1 and FL2. Let's also assume that with these settings the fluorescein FL2 signal is 25% of the FL1 signal. For a cell that emits 10,000 photons of fluorescein fluorescence in the FL1 channel on average (and neglecting the 10 photon background) we will get

$$FL1 = 10,000 \pm 100 \quad FL2 = 2,500 \pm 50$$

Now, when compensating, we take 25% of FL1 and subtract it from FL2.

$$FL2 \text{ (compensated)} = 10 \pm 75 \text{ photons}$$

This variance (up to 75 photons) is now much higher than for the negatives. To get that highest value down to 13 (10 + 3) we have to get the median down to -62. Yes, that is a negative value and it looks like it will take an extra 0.62% compensation to get the mean this low (total comp 25.6%). Since the system doesn't report negative values, all of the events below zero will get a zero value. This is why the median is suggested rather than the mean. Even if the median is placed at 10 a large fraction of the events will be below zero and the mean will be higher because all of the negative values that should have been added in were moved up to zero. Now for an even worse situation, let's suppose the cell has 100 times as much fluorescein.

FL1 = 1,000,000 ± 1000 FL2 = 250,000 ± 500

By the same argument FL2 (comped) = 10 ± 1500.

Notice that this is 20 times noisier than for the dimmer cells. Also, notice that the extra 0.62% compensation picked from the last example now gives -6200 ± 1500 and is pretty severely overcompensated. If instead the extra 0.15% needed here was used on the previous example we would have obtained -5 ± 75 which would allow events to reach levels of 70. The lower absolute error is more likely to be absorbed in system noise, and it is thus usually safer to set compensation on the brightest population that is likely to be encountered on the day's run. While this argument is presented for an ideal instrument, it is also true in practice. In the real world both system noise and biological variation in dimly fluorescent positive cells can make them unreliable for setup and both undercompensation and overcompensation can occur.

Order of events

While it is easier to think about compensation occurring after the intensity measurement is obtained, the FACSCalibur and FACS Vantage actually put the signal through the compensation circuits before it is log amplified and digitized. This subtraction of analog signals can be thought of as compensating the pulse at every point, not just at the peak. Compensating the analog signals has one main advantage, the compensation occurs before the threshold decision. This means that threshold decisions are made on data that looks like the data which the user is observing to set the threshold discrimination level. In systems in which compensation is done later (on digital values for example) the threshold circuits see uncompensated data. This can complicate matters if fluorescence thresholding is used. For example, a PE-positive cell will emit considerable FL3 just because the PE has red fluorescence (FL3) in its emission spectrum. If the FL3 channel is used for a threshold (using PerCP) and the PE signal is not compensated, PE-positive PerCP-negative events may trigger the system. This is not a problem if these "false" triggers are infrequent, they can usually be removed in post-acquisition analysis. However, if the fluorescence threshold was used to decrease the analysis rate and the PE-reagent is a marker of unwanted cells, this situation could lead to an unexpectedly high threshold rate.

Compensating before thresholding and digitizing has one main drawback in multilaser systems. It is difficult to insure that the pulse shape from cells are the same when they originate from different lasers. It is also difficult to insure that the pulses are aligned properly for subtraction when they originate from different lasers if the alignment of the lasers and the sheath velocity are not predictable. In the fixed alignment, fixed fluidics system of the FACSCalibur the travel time between the two beams is more predictable. Both optical and electronic components are used to align and shape the pulses in order to provide interbeam compensation on this instrument.

Coordinating signals from different beams

In a multilaser instrument, light pulses from a cell occur at different times as the cell moves from one laser intercept to the other. Different strategies are used to correlate the signals from the two lasers in order to collect all of the data from the cell into a single event. In the FACSCalibur the cell enters the red beam first, and the emission excited by this beam is electronically delayed so that it arrives at the signal processing electronics at the same time as the pulses created as the cell is in the blue beam. In the FACS Vantage, the cell enters the primary beam (usually blue) first and the system waits a fixed time interval before digitizing the pulses coming from detectors devoted to the secondary lasers. It is easier to implement (analog) interbeam compensation using the first strategy, but the second strategy is more flexible and is more frequently used when several parameters can be generated from the secondary sources.

Pulse Processing

In some instances additional information about the exact shape of the pulse being measured is helpful. For example, as particles get large compared to the 20 µm beam spot in the FACSCalibur, the pulse height is not as close to a measure of total fluorescence. The pulse area may be more reliable. Also, if aggregates present in the sample can corrupt the data (doublets can be mistaken for cells with twice the resting DNA content, for example) then measuring the pulse width may also provide information. Anomalous broad pulses are often from doublets. The FACSCalibur and the FACS Vantage have pulse processing circuits either standard or as an option that allow the width and area of pulses to be measured. In addition, the pulse processor on the FACS Vantage can produce the ratio of two signals as an output. This is most useful for dyes that emit in two wavelength bands depending on the physical state of the dye. Calcium and pH dependent dyes are most often used in this way.

Noise and autofluorescence

All measurement systems are vulnerable to noise at some level. Flow cytometers can experience stray light, thermal emission from the PMTs, and electronic noise from various components. Noise can manifest itself as higher than expected signals for dim particles or broader than expected variations in signal intensity. On the FACSCalibur and FACS Vantage, considerable design effort has gone into reducing the sources of noise. For most immunophenotyping experiments, the limit for detection of antigen expressed at low levels is actually the autofluorescence of the particle being examined rather than instrument noise. Autofluorescence is the intrinsic emission of the particle even in the absence of the stain. This can be the equivalent of several thousand fluorescein molecules, but is usually lower at longer wavelengths. A cell population that can be separated from baseline in a given parameter is probably actually fluorescing in that channel although some noise may be mixed in with the signal.

Sorting

Both the FACSCalibur and the FACS Vantage are capable of sorting cells of interest. They use very different mechanisms to do so. The FACS Vantage uses the more familiar droplet mechanism, while the FACSCalibur uses a catcher tube method.

Droplet sorting is accomplished by breaking the sample stream into a series of drops. Drops that contain cells of interest are charged, and the charged droplets are deflected by an electric field and land in a well-defined target location. This target can be a test tube for bulk sorting, or a single well of a multiwell plate for cloning experiments. Other types of devices are sorted into for specialized applications, for example agar plates for sorted bacteria.

Droplets are formed by vibrating the sample stream. The sheath and core stream leave a nozzle opening and go past the laser interception point. The nozzle is vibrated to cause the stream to break into droplets, but this breakup does not occur until past the measurement region. Where this breakoff into droplets occurs can be altered to some extent by the pressure pushing the stream through the nozzle and by the frequency and intensity of the oscillator that vibrates the nozzle. Some adjustment is often necessary to obtain a stable breakoff, but this is often minimal if most of these factors are not varied from experiment to experiment.

The hole through which the stream exits the nozzle can have different diameters (50 to 400 µm). In nozzles with large holes the sheath stream is larger, so more energy is required to produce a breakoff into droplets. For large holes (200 µm or more) the MacroSort upgrade may be required to provide enough energy. Also, the diameter of the stream as it breaks into droplets restricts the diameter of the droplets that form. With a larger diameter, fewer drops can be formed per second so the drive frequency is lower for larger nozzles.

When a cell of interest is identified in the sample stream the sort mechanism is activated. A charge is placed on the stream just as the droplet of interest breaks off. This timing is very important for successful sorting. The charged droplet is deflected by an electrical field across two plates that lie on either side of the stream and lands in the appropriate target. The amplitude of the charge placed on the droplet, the magnitude of the field across the plates, and the velocity of the droplet determine how far the droplet is deflected.

The drop-drive frequency determines how many droplets are formed each second. It is also possible to set how many drops are charged during each sort event. The product of the time/drop and the drops/sort give the time/sort, often referred to as the sort envelope. Think of this as the period of time around the cell of interest that is examined for interfering particles. For example a drop drive frequency of 33 kHz and a 3 drop sort gives 100 μsec/sort. If another event occurs within 50 μsecs before or after the target cell it will fall into the same sort envelope and a decision must be made whether to sort the target cell or not. For maximum yield, you want to sort the target no matter what else will come along. For maximum purity and sort count accuracy you want to only sort when you know that nothing else could be included. The different sort modes on the FACS Vantage allow trading off purity for higher yield.

In many cases it is preferable not to sort if contaminating cells are present in the sort envelope. A decision is made to not sort the envelope under these conditions. Note that these "no sort" decisions are not what is reported on the abort counters, although some people call them aborted sorts. The abort counters on the FACS Vantage and on the FACSCalibur report the types of aborts discussed in the Threshold section.

There are two types of sort decision hardware available on the FACS Vantage. The built-in hardware relies on comparison of the analog level of each parameter to decide whether the cell is of interest. This system can only give a "too high", "too low", or "OK" output for the parameter in use. Any or all of the parameters can be used in this way. This amounts to a multi-dimensional box with sides at the high and low levels for each parameter. In any two-dimensional display of this n-dimensional gate you find a two-dimensional rectangular sort gate. Using the analog signal can give very rapid decisions, but has limited flexibility. The Sort Enhancement Module (SEM) makes sort decisions based on the digitized data. A collection of irregular, multisided regions in several parameters can be combined into a sort gate. This type of decision can only be made when the digital data is available, and so is slower than the analog method.

Sorting on the FACSCalibur is done using a catcher tube. A tube allows fluid to flow, and rests in the sheath flow. As a cell of interest moves by, the tube swings out and the sample stream now flows through the tube. The tube then swings back into the resting position. This in-and-out cycle can be thought of as the sort envelope, and the FACSCalibur then looks like a FACS Vantage with a 400 μsec sort envelope. Of course the continual flow of sheath fluid into the sorted sample when in the resting position also complicates recovery of cells. The cells must be recovered from the relatively dilute sample. This can be done by centrifugation, but Becton Dickinson also provides a Concentrator module that continuously removes fluid from the sorted fraction. In the Concentrator, the sort stream is delivered to a filter with pores smaller than the sorted cells. This filter resides in a pressurized chamber, and the pressure forces fluid through the pores. At the end of the sort the sample is available in a relatively small volume. The filtration process has been shown to damage some types of live cells, and is most useful for sorting fixed samples for morphological verification or for further analysis with other methods like PCR or in situ hybridization.

Summary

The FACSCalibur and FACS Vantage are two examples of the tradeoffs in cytometer design. The high flexibility of the FACS Vantage makes it a powerful research tool, but requires considerable knowledge and attention to detail from the operator. The FACSCalibur still provides considerable research capability but limits some of the flexibility in order to provide simpler operation. The instruments have similarities and differences that have been addressed here, but also have details of operation that can't all be addressed in this chapter. The instrument User's Guides should provide additional detail that may be of use in designing and interpreting experiments.

How Flow Cytometers Work – and Don't Work

HOWARD SHAPIRO

Introduction

If you're new to flow cytometry, and/or use one of the modern benchtop flow cytometers which don't require, or even allow, you to mess with the guts of the instrument, you may not care much about how flow cytometers work. A lot of highly competent molecular biologists, immunologists, etc., look at the flow cytometer as a tool, and just want to see samples go in and pretty pictures and numbers come out. This works remarkably well if you're doing the same analyses and analyzing the same samples as everybody else is, but, sooner or later, you may want to do something new and different, and knowing some of the inner details will almost certainly help you design better experiments which will get you the most information for the least effort. You won't get many of the details from these few pages, but I'll try to point you in the right direction; I have written about the subject at much greater length, if you're interested in pursuing things further.

If you're more familiar with flow cytometry and have some acquaintance with my principal *oeuvre*, bear with me anyway, because, since I'm describing how flow cytometers work for what seems to me like the gazillionth time, I thought I'd at least try to rethink the subject and challenge a few cherished assumptions. Also, I thought it might be useful to talk about how flow cytometers don't work, i.e., what can go wrong; I haven't approached the subject from that angle before.

What is a Flow Cytometer?

A **flow cytometer** is a device which passes a stream of cells, ideally in single file, through one or more regions in which physical or chemical measurements are made of the cells. In the context of this volume, we are dealing with instruments which make optical measurements; in particular, they measure incident light scattered by the cells, and fluorescence in one or more wavelength regions emitted from cell-associated dyes and/or from cellular constituents. A **cell sorter**, or **flow sorter**, is a flow cytometer to which additional electronic and mechanical components have been added to permit cells with measurement values in a range preset by the user to be diverted from the stream and collected for further study.

Cellular Parameters and Multiparameter Analysis

Flow cytometers can be used to measure a wide range of cellular characteristics, or **parameters. Intrinsic** parameters, such as cell size and cytoplasmic granularity, may be estimated from light scattering behavior, requiring no reagents and no pretreatment of the cells. **Extrinsic** parameters, i.e., those which require reagents (sometimes called **probes**) for measurement, are classifiable as **structural** (e.g., DNA content, total protein, number of binding sites for a particular antibody) or **functional** (e.g., cytoplasmic Ca++ concentration, intracellular pH, membrane potential). Modern flow cytometry, more often than not, involves **multiparameter analysis,** i.e., the measurement of several parameters for each cell analyzed. This allows a combination of electronic hardware and computer software to be used to identify discrete subpopulations in a mixed cell sample, and, if necessary, to provide a quantitative or qualitative analysis of one or more parameters in each subpopulation. The capacity for multiparameter analysis was largely realized by the marriage of microcomputers to flow cytometers in the 1980's, and is probably the principal basis for the continued utility of flow cytometry in the many disciplines which consider the normal and abnormal behavior of differentiated cells, including cell and developmental biology, genetics, hematology, immunology, and oncology. Flow cytometric analysis and cell sorting have also been used to study and isolate subcellular particles, such as chromosomes, chloro-plasts, and mitochondria, and are becoming more widely used in microbiology, parasitology, and marine biology.

Flow Cytometers

Flow Cytometer Performance Characteristics: Precision and Sensitivity

The performance of flow cytometers is generally described in terms of **precision** and **sensitivity**. Precision is typically ex-pressed in terms of the **coefficient of variation (CV)** of a mea-surement made on a population of nearly identical particles. These can be uniformly sized and fluorescent dye-labeled plastic beads, or diploid cell nuclei stained with a dye which binds stoi-chiometrically to DNA. The mean and standard deviation for the population are calculated; the CV, expressed as a percentage, is 100 times the standard deviation divided by the mean. An instru-ment in good alignment should be able to produce CV's no high-er than 3%.

Sensitivity is generally expressed as the smallest number of molecules of a fluorescent dye or labeled antibody detectable above background noise. This can conveniently be estimated using mixtures of beads directly labeled with different amounts of dye; beads are also available with defined antibody-binding capacities. The mixture usually includes a blank, which does not contain dye or bind antibody. The fluorescence signal from a blank bead will have some finite value, usually deter-mined by the amount of stray light reaching the detectors and the level of noise in the cytometer's electronics. A compar-ison between signal intensities from blank and stained beads will allow the fluorescence signal from the blank to be defined in terms of the numbers of molecule-equivalents of dye or antibody which that signal would represent. The background fluorescence level in many modern instruments is equivalent to fewer than 1,000 molecules of fluorescein or phycoerythrin; it should gen-erally be possible to detect fewer than 10,000 molecules of cell-bound fluorescent antibody above background, even in most older cytometers.

In ancient times, immunologists, who cared about sensitivity but not precision, never ran calibration beads, and might have gotten CV's of 10% if they had run them, while people who wanted to analyze DNA content, who cared about precision but not sensitivity, had no interest in trying either to improve or quantify the antibody detection capabilities of their instruments, which were usually pretty poor because of the DNA dyes that accumulated in the plastic tubing and other parts of the fluidics, which tended to wash off and impart fluorescence to unstained cells.

It is still possible to collect data from a flow cytometer which isn't working very well; making instruments easier to use may actually have made it easier to do so. To avoid the "garbage in, garbage out" problem, precision and sensitivity of a flow cytometer should be estimated, using beads, as part of the daily setup routine.

Flow Cytometers: Structure and Function

What I'll try to do here is first describe what happens when things are working properly, and then consider what can go wrong.

The most common configuration for a fluorescence flow cytometer is described as **orthogonal**, because the axis of the illuminating beam, the direction in which cells travel, and the axis of the lens which collects fluorescence emitted from cells are mutually perpendicular. Lets start with the illuminating beam.

Illuminating the Cells: The Observation Point

Most fluorescence flow cytometers operate at only one illumination wavelength; blue-green light at 488 nm is provided by an argon ion laser, and the beam is focused by simple lenses to a spot, typically 80-100 micrometers wide and 5-20 micrometers high, centered on the cell stream. This intersection defines the **observation point**, or **interrogation point**, of the cytometer. The interrogation point may be enclosed in a round or flat-sided quartz chamber, as is typical in benchtop instruments; cells may also be observed in a stream in air, as is done in medium- and high-speed cell sorters. Flow cytometer illumination is relatively idiotproof as long as your laser is working, and it is relatively simple, although not generally inexpensive, to add additional illuminating beams, which may either converge on a single interrogation point or be focused on additional interrogation points downstream. WHAT GOES WRONG? Lasers lose power as they age; this can result in decreased sensitivity, and, as power drops further and/or the laser becomes noisy, in decreased precision.

Forming the Cell Stream

The cells in a sample are forced into what should be a single file stream ten micrometers or so in diameter by being injected into the center of a larger flowing **"sheath"** stream of water or saline; the sample or **"core"** and sheath streams may be driven either by gas pressure or by pumps. When gas (air or nitrogen) pressure is used, one makes the sample and/or sheath vessels into the functional equivalent of the old-fashioned wash bottles some of you may remember from high school chemistry, where you blow into the bottle, forcing liquid out. In the cytometer, air or gas at a constant, regulated pressure provides the motive force, and the flow rate is relatively constant. Under ideal conditions, flow is laminar, keeping the velocity of cells fairly uniform and maintaining the position of the core stream containing the cells in the center of the sheath.

If one puts nearly identical particles, e.g., the fluorescent beads used for instrument calibration, through the flow cytometer, all of the particles encounter the same illumination profile and spend the same time in the observation region, and thus yield nearly identical signals. WHAT GOES WRONG? The presence of cell aggregates, large pieces of debris, and/or gas bubbles in the flowing stream creates turbulence, which results in cells (or beads) being distributed over a larger portion of the stream and traveling at different velocities; under these conditions, nearly identical particles will not produce nearly identical signals, and precision will be lost.

Orthogonal Light Collection (Fluorescence and Large Angle Scatter)

A single high-power microscope lens, or a lens with similar optical characteristics, is used to collect light scattered orthogonal to the illuminating beam and fluorescence emitted from cells. **Optical filters** and **dichroic mirrors** or **dichroics** (i.e., mirrors which reflect some wavelengths and transmit others), similar to those used in fluorescence microscopes, are used to divide the collected light into different spectral regions, and to divert light from each region to a separate **photomultiplier tube (PMT)**. PMT's are used to detect orthogonal scatter and fluorescence signals because these signals are typically of relatively low intensity, and the built-in amplification mechanism of PMT's provides increased sensitivity. Most modern flow cytometers measure fluorescence in three or four spectral regions, i.e., green (about 530 nm), yellow (about 575 nm) or yellow and orange (about 610 nm), and red (about 670 nm). WHAT GOES WRONG? In ancient times, colored glass filters, which were used to separate fluorescence from light at the excitation wavelength, often added spurious contributions to the fluorescence signal due to excitation of the dye in the filter. Today, most of the filters used are interference filters, which do not fluoresce, but which may degrade with age, changing their light transmission characteristics. This is usually not a problem in a new instrument, but becomes a concern for filters more than a few years old.

Orthogonal scatter signals, also called 90 degree, large angle, or side scatter signals, are influenced by cytoplasmic granularity and by irregularity of the cell surface. Among peripheral blood leukocytes, granulocytes have the largest orthogonal scatter signals, and lymphocytes the smallest, with monocytes falling in the intermediate range.

Fluorescence measurements are made to detect specific fluorescent dyes or other labels in or on cells. Immunofluorescence measurements probably account for the bulk of flow cytometer use, and instruments now measure as few as several hundred molecules of a cell surface or intracellular antigen, using monoclonal antibodies directly labeled with a phycobiliprotein such as phycoerythrin. Fluorescent reagents can also be used to measure cellular DNA, RNA, and protein content, intracellular pH and calcium concentration, cytoplasmic and mitochondrial membrane potential, among other parameters, and it has recently become possible to detect specific nucleic acid sequences using appropriate fluorescent labeled probes.

Forward Scatter Signals: Collection and Detection

The **forward scatter** signal, i.e., light scattered at small angles to the illuminating beam by a cell, is collected by a low-power microscope lens and detected by a **photodiode**; a mechanical obscuration in the light path prevents the illuminating beam itself from reaching the detector. The small-angle or forward light scatter signal is primarily dependent on cell size and refractive index; within a single cell type, larger cells generally have larger forward scatter signals, but the relation of size to signal amplitude is complex. In a preparation of unfixed cells, dead cells which have lost membrane integrity typically show lower forward scatter signals than live cells which have not; this permits discrimination between the two classes of cells provided the size distribution of the overall population is reasonably uniform, as is the case, for example, in unstimulated peripheral blood lymphocytes. WHAT GOES WRONG? The characteristics of light scattered at small angles to the beam may change fairly dramatically with small changes in the angle over which light is collected; this makes forward scatter collection optics much more likely to slip out of alignment than orthogonal collection optics. Because forward scatter signals from eukaryotic cells are typically of reasonably high intensity, a photodiode, which unlike a PMT, has no built-in amplification, is used for detection; a moderate amount of amplification is provided by the electronics. However, when one is analyzing small particles such as bacteria, signals from which are typically around 1/1000 the intensity of those from eukaryotic cells, there may not be enough gain in the amplifier to get signals into an appreciable portion of the measurement scale. Manufacturers of most instruments do not currently offer the option of substituting a PMT for the photodiode detector, but may change their behavior in response to the demands of the increasing market represented by potential users in microbiology.

Front End Electronics: Is a Cell There Yet?

The detectors in a flow cytometer are turned on as long as the instrument is running, and, because they respond to stray light and also generate some **dark current** even when not exposed to light, they are always producing some output in the form of electrical current, which is converted to voltage by the preamplifiers. The **baseline** or background signal from a preamplifier also includes some contribution from the internal electrical noise of the electronics. This signal fluctuates about some value greater than zero, because there is always some noise. The constant (DC) component of the baseline signal is usually subtracted from the signal by a **baseline restoration** circuit in the preamplifier electronics, leaving an output voltage which fluctuates around zero volts, or ground. As a cell passes through the illuminating beam, the light scattered by, and the fluorescence emitted from the cell produce signals at the appropriate detectors and preamplifiers; in order for the cell to be detectable, i.e., for the instrument to "know it's there", at least one of the preamplifier signals must reach a **threshold** value substantially above the level of the normal fluctuations in preamplifier output.

The "front end" electronics of a flow cytometer contain circuitry which compares the value of a **trigger signal** selected by the user to a threshold value, also set by the user. Once the trigger signal rises above the threshold value, the front end electronics initiates the first of several processes needed to collect the data. Some newer instruments allow more than one signal at a time to serve as a trigger signal, the reason for which I will shortly explain. WHAT GOES WRONG? In most cases, at least one of the measured signals will be suitable as a trigger signal. In the typical case in which immunofluorescence measurements are being made on a reasonably clean preparation of eukaryotic cells, the forward scatter signal is used for triggering; all of the cells, whether or not they bear bound fluorescent antibody, will produce relatively strong forward scatter signals. The presence of large amounts of debris in this sample will pose problems, because the electronics used for triggering respond best when the threshold value is set in a range where there are few signals, and the debris will yield a continuum of signal values ranging from the level of background fluctuations all the way up to the threshold level. In another circumstance, the analysis of leukocytes in unlysed whole blood, the use of a forward scatter signal for triggering is impractical, because there are approximately 1,000 erythrocytes and 100 platelets accompanying each leukocyte present in the sample. In this case, the addition of a fluorescent dye which stains DNA will produce strong signals from leukocytes, and little or no signals from erythrocytes and platelets, which contain little or no DNA, providing a good trigger signal. In analyses of relatively small particles, e.g., bacteria, both scatter and fluorescence signals may be fairly close to the level of background fluctuations, and the use of both signals for triggering will improve discrimination, because it is less likely that both scatter and fluorescence signals will simultaneously fluctuate above threshold than that one or the other signal will. This makes the capacity to use multiple trigger signals useful in an instrument.

Pulse Characteristics: Peak Detection, Integration, and Pulse Width

In order to utilize flow cytometric data, the measured parameter values represented by voltages at the preamplifier outputs must be converted to digital form and transmitted to and stored in a computer. However, the signal pulses produced by cells' passage through the observation point are typically of only a few microseconds' duration, and the **analog-to-digital (A-D) converter** circuits typically used in the data capture process cannot respond in this short a time. Instead, analog pulse processing circuits with their own short-term memory are used to hold the desired signal voltages. An analog **peak detector** stores the highest voltage presented to it; an analog **integrator** computes the integral or area of a pulse, and a **pulse width measurement circuit** produces an output voltage proportional to the length of time for which a signal value remains above a threshold value. All of these circuits have to be reset, typically to an output of zero, in time to respond to each new cell's passage through the system; digital logic signals which do this are generated in the front end electronics.

Typically, an **analog delay line**, which reproduces its input at its output after a delay of a few microseconds, is used at the inputs to integrators and pulse width measurement circuits. The delay is necessary because the front end electronics must "see" the rising edge of a pulse before generating the reset signals. A peak detector can "catch up" to the maximum value of the pulse, but integrators and pulse width measurement circuits would miss a significant part of the signal if it were not delayed. WHAT GOES WRONG? The biggest problem with peak detectors and integrators is that their performance degrades considerably at small signal values. This places some constraints on the ways in which an instrument can solve two of the thornier problems encountered in flow cytometry, i.e., logarithmic transformation of the data and fluorescence compensation, discussed below.

Logarithmic Amplification and Alternatives

The signal values encountered in flow cytometry encompass a wide dynamic range; this is particularly true in the case of immunofluorescence measurements, where some cells may carry only a few thousand binding sites for an antibody, while other cells in this same kind of sample may bear hundreds of thousands of binding sites. It is therefore convenient to display flow cytometric data on a logarithmic scale. In principle, one could digitize signal values first, and then use a table of logarithms to convert the scale from linear to logarithmic. However, there are several problems with this. The first relates to the dynamic range of analog-to-digital conversion. Until recently, it was common practice to digitize signals to 10 bits' precision; i.e., voltages nominally between ground and 10 volts would be represented by numbers from 0 to 1023. The smallest change in signal representable in this fashion is 1/1023 of 10 volts, or just under 10 mV. However, the data may span a range of 1 mV to 10 V, which is sometimes described as four decades, with the highest decade corresponding to voltages between 1 and 10 V, or digitized values from 103 to 1023, the next highest corresponding to voltages between 100 mV and 1 V, or digitized values from 11 to 102, and the next to voltages between 10 mV and 100 mV, or digitized values between 2 and 10. These leaves only digitized values of 0 and 1 to represent the voltages between 1 mV and 10 mV in the lowest decade, with the result that data in this decade will look extremely ratty on a logarithmic scale, and data in the next highest decade won't be too pretty, either.

Things improve if a higher precision A-D converter is used; 16 is about the minimum number of bits needed for the data to be presentable, and 18 or 20 bits would be better. With a 16-bit converter, voltages between 0 and 10 V are converted to numbers between 0 and 65535, and the smallest voltage difference which can be represented is 1/65536 of 10 V, or about 153 u[micro]volts. The 1-10 V decade corresponds to channels 6554-65535, 100 mV-1 V to channels 656-6553, 10 mV-100 mV to channels 66-655, and 1-10 mV to channels 7-65. One can convert 16-bit data to a fairly respectable looking 4 decade logarithmic scale with 256 channels (i.e., an 8-bit log scale); things get a little motheaten in the bottom half of the bottom decade (1-3 mV), but that's down around where the noise level is, so it doesn't matter much. The major objection in the past to digitizing to 16 or more bits' precision was the high cost of high precision A-D converters. These are now relatively inexpensive, bringing to the fore a second problem, which is that peak detectors and integrators are generally not well behaved when signal values are in the range of millivolts to tens of millivolts. There are ways around this, which have to date (mid-1997) been implemented in one commercial instrument (the Coulter XL) and in flow cytometers built in my lab and at Los Alamos.

All other commercial and laboratory-built instruments do conversion from linear to logarithmic scales using **logarithmic amplifiers**, or **log amps**, which are analog circuits with output signals proportional (in theory) to the logarithm of the input signal. WHAT GOES WRONG? In principle, everything should be fine; put a 10 V pulse into a log amp and you get a 10 V pulse out, put a 1 V pulse in and get a 7.5 V pulse out, a 100 mV pulse becomes a 5 V pulse, a 10 mV pulse becomes a 2.5 V pulse, and a 1 mV pulse yields an output of 0 V, or ground. Also, in principle, the logarithmic relationship should be preserved across the range of intermediate values. In practice, things don't work that way; real log amps deviate, often substantially, from the ideal, especially when input signals get down in the range of a few millivolts, and the deviations can be unpredictable and may change with time. Log amps also make it necessary to use additional electronic circuitry for fluorescence compensation, which opens up a whole new can of worms.

Fluorescence Compensation: Why and How

Producing antibody labels excitable at 488 nm with emission in the green, yellow, orange, and red spectral regions has been possible principally because nature provided phycoerythrin (PE), an algal photosynthetic pigment with an emission maximum in the yellow, around 575 nm. The organic chemists had already come up with fluorescein (FITC, an abbreviation which represents the isothiocyanate derivative used to bind the dye to antibodies) which has green emission, with a peak near 525 nm), and were able to covalently bond two other dyes to PE, producing **tandem conjugate** molecules with orange and red emission. The tandem conjugate of PE and Texas Red (PE-TR) has an emission maximum around 610 nm (orange); the conjugate of PE and Cy5 (PE-Cy5) emits maximally in the red at about 670 nm. In an ideal world, one would be able to set up a fluorescence measurement system in which the green channel measured only FITC, the yellow channel only PE, the orange channel only PE-TR, and the red channel only PE-Cy5. As you should have come to expect from reading this far, it isn't an ideal world. The green channel measures mostly FITC fluorescence, but picks up a small amount of PE fluorescence as well; the yellow channel measures mostly PE but also responds to FITC and PE-TR, and so on.

Fluorescence compensation is a process in which the real green, yellow, orange, and red fluorescence signals are manipulated arithmetically to produce signals more nearly proportional to FITC, PE, PE-TR, and PE-Cy5 fluorescence. This is typically accomplished by subtracting a small fraction of the PE signal from the FITC signal, subtracting some of the FITC and some of the PE-TR signal from the PE signal, and so on. The required arithmetic produces new signals from linear combinations of the old signals. In order for this to work, one has to have linear signals to start with; the game can't be played with the output signals of log amps. In instruments with log amps, it is necessary to build analog electronic fluorescence compensation circuits, which are placed between the preamplifier outputs and the inputs to the log amps; the amount of a given signal which is subtracted is typically set by the operator interactively and intuitively, using either a knob or a computer pointing device. The compensation circuit itself is similar to what is used in tone control and equalization circuits in stereo equipment. WHAT GOES

WRONG? As the number of colors goes up, the number of adjustments needed for compensation and the resulting complexity of the circuit increase, almost inevitably increasing the level of electronic noise at the log amp input. Analog compensation circuitry for 4-color fluorescence typically restricts the available dynamic range to three decades or so, simply because of the higher noise level. Even if there were no noise, however, it would be difficult to adjust compensation empirically because one would have to twiddle twelve knobs to get everything right. At its base, n-color fluorescence compensation involves the solution of n equations in n unknowns, which is relatively simple algebra in the computer age; one can obtain the coefficients of the equations from measurements of samples stained with each of the antibodies; these measurements, however, must be on a linear scale.

In a system with high-resolution A-D conversion, fluorescence compensation and subsequent transformation of data from a linear to a logarithmic scale can be done with a digital computer, providing algebraically correct compensation and logarithmic values much closer to ideal than could be obtained using log amps; this is clearly the way in which things will be done in the future. As a matter of fact, it is now possible, and should shortly be economically feasible, to replace peak detectors, integrators, etc. with special-purpose signal processing computers, one per measurement channel, each of which will be able to derive pulse properties by analyzing data from a high-speed, high-precision A-D converter sampling the preamplifier output signal a few million times a second. However, more fundamental problems with fluorescence compensation, electronic or computational, remain.

The compensation process is based on the assumptions that the fluorescence signals are dominated by fluorescence from the probes involved, and that the contributions from a fixed amount of any probe to any given fluorescence signal remain constant. The first assumption tends to break down for cells bearing relatively small amounts of label or exhibiting relatively high levels of autofluorescence; the circuitry or algorithm ends up inappropriately subtracting one portion of an autofluorescence signal from another, often resulting in negative values of the compensated signal. These are usually adjusted to zero, or positive values near zero, either electronically or computationally, because one cannot apply the logarithmic transformation to negative values. The displayed data tend to look uglier than they would have had compensation not been applied; one can achieve a cosmetic fix by adding random noise to the data, but I believe this is counter-productive.

The assumption that the fraction of fluorescence contributed by any given probe to any given signal remains constant can break down whenever probe molecules are close enough to one another for nonresonant energy transfer to occur; in this circumstance, emission from the probe with the shorter wavelength emission peak is decreased, and emission from the probe with the longer wavelength emission peak is increased. In many flow cytometers, e.g., the B-D FACScan, no provision is made in the compensation circuitry to subtract red from green or green from red fluorescence, since, in typical situations, there is little spectral overlap between red- and green-emitting fluorescent antibody labels. However, when, for example, a green-emitting fluorescein-labeled antibody to a nuclear antigen is used in conjunction with a red-emitting nuclear DNA stain such as propidium or 7-aminoactinomycin D, there may be substantial energy transfer from the fluorescein to the DNA stain; this is difficult to compensate for according to the classical model, and impossible to compensate for in a FACScan or other instrument which simply hasn't got the right hardware.

Data Acquisition and Analysis Systems

Almost all flow cytometers now incorporate data acquisition and analysis systems based on personal computers, compatible with either Intel's hardware and Microsoft's operating systems or Apples Macintosh hardware and software. Once signals from a cell have been captured in digital form, compensated or log transformed as necessary, they are typically added to a data file, stored on a hard disk. The existence of a more or less standard file format, the Flow Cytometry Standard or FCS file format, allows manufacturers and third party developers to produce data analysis software which, at least in principle, works with data from any instrument. (See Larry Seamer's chapter on Flow Cytometry Standard in Section 2 Chapter 3, of this manual).

Flow Cytometric Data Analysis: Gating is Central

In the early days of flow cytometry, instruments did not have computers attached, and couldn't measure more than two or three parameters. If you wanted to know the distribution of DNA content in a subpopulation of cells bearing a particular surface antigen, and could only measure forward scatter and one fluorescence signal, you would have to stain cells with fluorescent antibody, sort out the fluorescent cells, restain them with a DNA stain, and rerun them through the instrument. Pulse height analyzers, really hardwired special purpose computers originally developed for nuclear physics, were used to record distributions of a single parameter. Dot plots of two parameters could be accumulated on the screen of a storage oscilloscope and photographed.

The pulse height analyzer itself had the capacity to define a **region of interest (ROI)**, or **gate**, i.e., a measurement region smaller than the entire range of signals, to which analysis might be restricted. In the cell sorter itself, electronics were used for the same purpose, allowing the user to select only those cells within a desired range of parameter values for sorting. The input to one analyzer could be governed by a gate set on another. In two-parameter instruments, it was generally possible to set upper and lower bounds on both parameters, defining the sort region, or sort gate, as a rectangular area in the overall measurement space. Fluorescence compensation came into use primarily because, when two-color staining became possible, uncompensated data points from cells didn't conveniently fall into rectangular regions.

Once two-color fluorescence measurements were available, the concept of **gated analysis**, which is central to modern flow cytometry, began to catch hold. It was no longer necessary to sort cells to determine the distribution of DNA content in a subpopulation defined by immunofluorescence; the input to a pulse height analyzer collecting a distribution of fluorescence from a DNA dye could be gated by the output of another analyzer operating on the immunofluorescence signal. The pulse height analyzers, or separate external counters, could be used to define the total number of cells analyzed and the number falling within a particular gate.

Although minicomputer-based data analysis systems for flow cytometers became available during the 1970's, it was not until the mid-1980's that a microcomputer and associated software became an integral part of a typical flow cytometer; without the computer, it would have been nearly impossible to utilize the five-parameter measurement capabilities of the instrument. The computer eliminated the need for pulse height analyzers and counters, and made it possible, using a mouse or other pointing device, to define elliptical, polygonal, or freeform gate regions as well as the rectangular gates to which analysis had previously been restricted.

A typical gated analysis is exemplified by the procedure used for determining the percentage of CD4-positive T-lymphocytes in peripheral blood. A small volume of blood is incubated with a mixture of labeled antibodies against CD45, the common leukocyte antigen, CD3, the T-cell receptor, and CD4. A polygonal gate is then drawn on a two-dimensional display of CD45 vs. orthogonal scatter, surrounding the cluster of lymphocytes identifiable by their low orthogonal scatter and high CD45 values. Next, four rectangular gates, arranged as quadrants, are drawn on a display of CD3 vs. CD4, on which only data points from the lymphocyte gate are plotted; the quadrants define regions containing cells bearing CD3, CD4, both, and neither. The computer is then used to generate counts of cells bearing CD3 and CD4 and of all cells in the lymphocyte gate, from which the percentage of CD4-positive T-cells is calculated. Other types of **quadrant statistics** are also commonly sought in flow cytometric analysis; one of the reasons fluorescence compensation is critical is that without proper compensation, cell populations don't fit into rectangular quadrants. However, rectangular quadrants are a vestige of the pre-computer era. When computers are used for data analysis, there is no reason why quadrants with sloping sides can't be defined just as easily and serve just as well, making compensation less critical. The principal problem with this approach at present is that few software packages provide the option.

Histogram Comparison: The Happy Median

I could say (and have said) a lot more about data analysis, but I'll close with a strong opinion about comparing histograms. People who are concerned about DNA content distributions collect data on linear scales and apply mathematical models which allow them to use classical statistical methods to compare histograms. People who collect distributions of immunofluorescence, or of almost anything besides DNA, work on a logarithmic scale, and wonder about whether to use mean, median, or modal values and what to do about standard deviations or other measures of distribution width. Problems arise because calculation of means and standard deviations requires converting log scale data back to linear. The easiest way to deal with the data is to use medians and percentiles, which can be found without converting all channel values from log to linear. Medians and percentile points (5th and 95th, 25th and 75th, etc.) are also what statisticians refer to as "robust statistics", meaning that they retain their utility even when distributions deviate substantially from the normal distribution, as distributions of immunofluorescence and most other cytometric parameters typically do. Finally, it is possible to plot distributions in a minimalized form, showing the median, 25th and 75th, and 5th and 95th percentile points, and allowing a large number of distributions to be printed in a single figure for easy comparison by visual inspection.

What's Left for the Future?

Nothing succeeds like success. A few people have progressed beyond four- and five-color fluorescence measurement to eight or more colors. This is basically impossible to do without a multibeam flow cytometer. If you do immunofluorescence measurements, the first beam you add to 488 nm is typically yellow (usually a big dye laser, but He-Ne lasers can be made to emit in this region) or red (He-Ne or diode in smaller systems, krypton in bigger ones), allowing you to take advantage of allophycocyanin and its tandem conjugates, Cy5, etc. If you do DNA measurements, you want to add UV excitation, because DNA dyes such as DAPI, which generally gives the most precise measurements, and Hoechst 33342, which is the only dye which stains DNA stoichiometrically in live cells, are UV-excited and blue fluorescent. UV is also essential if you want to measure calcium using indo-1. Unfortunately, the candidate UV light sources now available all have their deficiencies. UV argon or krypton ion lasers are very expensive, water-cooled, and consume a lot of power, while He-Cd lasers, which cost less and draw less power, often become noisy, leading to decreased measurement precision. Mercury or xenon arc lamps are good UV light sources, but cannot be installed on most laser source instruments. However, it now appears that solid-state UV lasers suitable for flow cytometry will become available within a few years.

One might ask whether it might not be easier to synthesize DNA-specific dyes, usable in live cells, which could be excited at 488 nm or by red lasers, than to add UV lasers to more instruments. The answer appears to be no; good chemists have been trying to do that for a long time without success, although somebody might still get lucky. Which leads me to the bottom line: flow cytometers generally work pretty well; they may be somewhat limited by the innate characteristics of the reagents used, and much more severely limited by the characters who misuse the reagents. I hope the rest of this book keeps you out of trouble from that direction.

Flow Cytometry Standard (FCS)
Data File Format

LARRY SEAMER

What is a data file standard?

First, what is a data file? A flow cytometry data file is the compilation of measurements from the acquisition of a single sample
on a flow cytometer. A data files standard is essentially a list of
rules describing how the computer data file should be organized.
When adhered to, the rules yield a defined and uniform computer file structure. The letters 'FCS' refer specifically to the Flow
Cytometry Standard format, which is a comprehensive data-file
structure with a 16 year history of use in flow cytometry.

FCS version 1.0 was proposed in 1984 by Robert Murphy and
Thomas Chused in a three-page article in *Cytometry*.[1]. FCS1.0
established the basic four-part flow cytometry data file still in
use today (described in more detail below). The standard was
revised and expanded in 1990 (to FCS 2.0) as the Society for Analytical Cytology (now called the International Society for Analytical Cytology, or ISAC) adopted FCS as its official data file
standard[2]. Because of the increased complexity of the experiments, advances in computer and communications technology,
FCS3.0 has recently been proposed and is currently available
through the ISAC office as a 30 page document or on the
ISAC WWW home-page[3]. Most (if not all) manufacturers
have voluntarily embraced and routinely use the FCS standard
file structure in one of its three forms. This demonstrates that
even though the file format has been standardized, FCS is flexible
enough to address the varying needs of individual manufacturers and to accommodate many types of data files.

Why do we need a standard file format?

Before 1984 each type of instrument, whether commercial or 'home-built', collected unique data files. Therefore, it was very difficult, if not impossible, to analyze data from one instrument on an instrument of a different make. This caused great difficulty in sharing data or performing uniform analyses. Also, many different proprietary formats limited the kind of analysis available to those that were bundled with their flow cytometer. The FCS data file standard has allowed flow cytometry data to be shared among different types of computers, running different software and has greatly facilitated the growth of the third-party software industry, providing the end users with much greater data analysis flexibility.

What is the FCS file structure?

FCS files are divided in four major segments; HEADER, TEXT, DATA and ANALYSIS. These segments can be arranged in any order, with the exception of the HEADER. All FCS standard data files must begin with the HEADER segment. The HEADER first shows the FCS version number followed by a string of numbers giving the byte-offsets to the other three segments of the file. In other words, the HEADER segment points to the beginning and end of each of the other segments, allowing analysis software to find all segments in the file. A typical HEADER looks like this:

FCS3.0 128 512 513 9520 9521 9600

FCS3.0' is the version number. The first byte offset numbers always points to the TEXT segment of the file. Therefore, in this example the TEXT segment begins at byte 128 and ends at byte 512. The next pair of numbers point to the beginning and end of the DATA segment. And, the third pair of numbers point to the beginning and end of the ANALYSIS segment. Other pairs of numbers may follow these required pairs, if the instrument manufacturer wants to include other information in the file.

As noted above, the second major segment of an FCS file is the TEXT segment. The TEXT segment is a list of descriptors called Keywords and their associated Values. Keyword-Value pairs provide information about the identity of the file, how the data was collected, instrument and gate set-up and any other information the instrument manufacturer wants to include. Some of this information is required by the FCS standard. Other defined Keyword-Value pairs are optional and still others are created by instrument manufacturers to address their specific needs. The Keywords and Values are separated by a text character of the manufacturers choosing called a delimiter character. A portion of a TEXT segment could look like this:

/$PAR/4/$P1N/FALS/ Lab Supervisor/ Dr. Somebody/

In this example the '/' character is the delimiter. The first Keyword shown is $PAR (a required Keyword). Its value, 4, gives the number of parameters collected during sample acquisition. The next Keyword is $P1N. Its Value, FALS, is the name of parameter number 1. The '$' character indicates that these Keywords are defined in the FCS standard document. The last Keyword in this example is 'Lab Supervisor' and its Value is 'Dr. Somebody'. This Keyword is one created by the instrument manufacturer. Therefore, it does not begin with the '$' character. For a complete list of defined Keywords consult the references at the end of this document or the FCS3.0 document on the ISAC Web site.

The next major segment, DATA, is usually the largest part of any FCS file. Data can be stored in one of three modes; uncorrelated, correlated and list. Uncorrelated is simply a single parameter histogram. Correlated is the data representing a multiparameter histogram. When graphed, these are usually represented as dot-plots, contour-plots or cloud-plots (for three-parameter data). And lastly, list mode data is usually a string of integers representing the measured intensity of each event collected (usually cells) on each parameter acquired, stored in the order the events were acquired. Therefore, once a list-mode file is collected, that sample can be "re-acquired" from the computer file. List mode data retains the maximum flexibility, allowing the user to change dot-plot parameters, move or add gates and even perform mathematical operations on the data such as adjusting compensation or creating calculated parameters.

To make matters a little more confusing, any of these three modes can be stored in one of several formats; binary or ASCII (text), integers or decimal numbers. However, binary integers is by far the most common data type used. Binary data is relatively compact and easy for computers to read. However, it is not readable or displayable by text readers such as computer screens or printers. If an FCS file is read with a text reader, the data portion is that part of the file that appears as strange characters, symbols and assorted beeps.

The final major file segment is the ANALYSIS segment. ANALYSIS is an optional file segment that is often added to the file after data acquisition. It is structured similar to the TEXT segment using Keyword-Value pairs to store the results of data analysis. Examples would be; the results of mean fluorescence channel calculations, percent positive or cell-cycle analysis.

Why is this important to me?

Most flow cytometer operators are not computer programmers. So, how does the FCS standard help us? An understanding of the basic file structure is useful when trouble shooting data analysis problems. Files can be corrupted and data deleted. Comparing the reported size of the various file segments with their actual size can often reveal such problems. With the widespread use of third-party analysis programs that list or edit Keywords and their associated Values, it is necessary to understand the nature and meaning of the information found in the TEXT portion of all FCS files. Also, because instrument to instrument file variations exist, it is possible, in multi-instrument labs, to determine which instrument collected any given file by examining the file structure. However, one of the most useful aspects of the FCS standard is that all of the descriptive information regarding any file is in plain text, readable by many common text reading programs. This allows users to review archived files, obtaining such information as patient names, sample numbers and date the file was acquired, without the need to use cumbersome analysis software. This is more powerful than might be immediately apparent. Computer programs exist that index text information in a file. Therefore, a large number of archived files can be indexed and later searched very quickly for files of interest. For example, if a user wanted to re analyze phenotype data gathered from only Balb/c mice over the past year, the Keyword Value

"Balb/c" could be searched and within seconds all pertinent files located.

The future

Because the FCS standard is both simple and flexible it has gained wide acceptance and is now implemented in some form by most instrument manufacturers. However, as computational power, experimental design and instrument technology increase in complexity, the need to store large amounts of data will continue to soar. For example, new digital data acquisition hardware, such as the DiDac system currently in use by the National Flow Cytometry Resource at Los Alamos, will have the ability to store individual pulse shapes as a parameter. Such a parameter will include many individual integers describing a pulse. In other words, one of the measured parameters for each event in a list-mode file may itself be a single parameter histogram. Even without pulse-shape data, it is now common for a single data file to be more than 1 megabyte in length, big enough to quickly fill the largest hard-drive. With this glut of data comes the need to organize it in a way that is easily accessible and widely understandable. The FCS standard must continue to evolve to accommodate such growth in information gathering.

Pressures are mounting to integrate flow cytometry data into larger files of related information. For example, a flow cytometry list-mode or histogram file could be one element in a pathology report file that may also include MRI or x-ray images, clinical laboratory data, etc. This means that the entire FCS structure described above may be only a small part of some as yet undetermined mega-file structure.

References

1. Murphy RF, Chused TM: A proposal for a flow cytometric data file standard. Cytometry 1984; 5:553-555.
2. Dean PN, Bagwell CB, Lindmo T, et al. Data File Standard for Flow Cytometry. Cytometry 1990; 11:323-332.
3. http://nucleus.immunol.washington.edu/ISAC.html

Options for Data Display

SUSAN DEMAGGIO

There are several means of displaying the data collected by flow cytometry including single parameter histograms, two parameter dot plots, contour and density plots, 3D histograms (isometric plots), Cloud displays, kinetic plots, and overlays. A sample and brief description of each type of display is helpful in deciding how to collect flow data and how best to interpret the results.

The first and most commonly used display is the single parameter histogram. It displays only one of the parameters collected by flow cytometry, either forward scatter, 90° side scatter, a fluorescence signal for one of the 2, 3, or more colors available on your instrument. It displays the intensity of the signal on the x axis in 256 or 1024 channels, in linear or 3 or 4 decade logarithmic scale, depending upon the instrument capabilities. The number of events in each of the channels is displayed on the y axis. In general, the scatter parameters are collected in linear mode, and im-

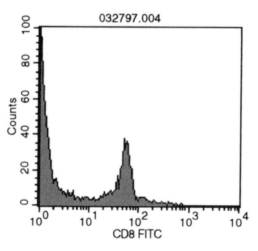

Fig. 1. A single parameter logarithmic histogram

Fig. 2. A DNA histogram of area under the FL2 peak

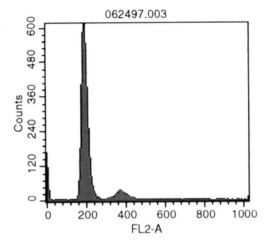

muno-fluorescence height in log mode. Fig. 1 is sample of a typical single parameter histogram of FL1 or FITC fluorescence.

Single parameter histograms are also used for the collection of fluorescence for DNA analysis. The area under the peak of the FL2 (for PI staining) or FL3 (for 7-AAD staining) parameter is usually collected in linear mode for this analysis. An example of FL2-A (area) histogram is displayed in Fig. 2.

The measurement of DNA cell cycle componants can be analyzed by modeling programs which fit the best gaussian distri-

Fig. 3. Modfit analysis of DNA stained with PI

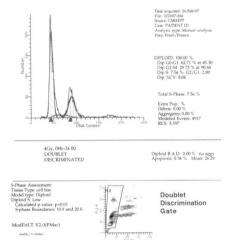

bution curve to each peak, G_0-G_1, G_2-M, and then calculate the resulting S-phase. Fig. 3 demonstrates such a modeling of a normal DNA cell cycle file using the Verity Software House program Modfit LT.

Two parameter histograms can be used for the display of any two parameters, but is very commonly used for two displays in particular. The first type is a scatter or dot plot of the forward and side scatter indicating the size and complexity of the cell surface and cytoplasm. The second type is the display of cells stained with two different fluorescently conjugated antibodies or probes. Fig. 4 displays a scatter plot. The x axis here is the forward scatter and the y axis, the side or 90° light scatter, although some instruments display them in the opposite way. After you gain experience viewing this type of histogram, you will be able to set the photomultiplier tube (PMT) high voltage settings to position the populations of cells in your sample at certain places on the dot plot consistently and recognize divergent sized cell populations simply from their scatter pattern.

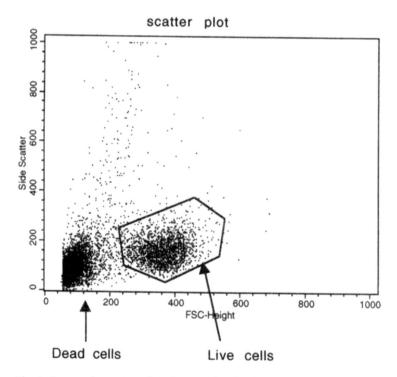

Fig. 4. Scatter plot or Dot plot of Forward and side scatter

The next few plots display data derived from the simultaneous analysis of 2 stains or probes. Fig. 5 displays the data of a cell population stained with two fluorescent antibodies which are mutually exclusive; that is they stain with one or the other of the antibodies, not both. In Fig. 6, there is a distinct population of cells which stains with both antibodies. Analysis of two parameter histograms is very helpful in the detection of cells with multiple epitopes and can assist in the division of subsets of cells such as T cells. Many times there will be not only a negatived subset, but also single staining subsets, double staining subsets, intermediate or dim staining subsets of either antibody. Smears or multiple binding affinities of the antibodies to the cells may indicate cells in transition or development. These cells may demonstrate varing numbers of binding sites to a particular anti-

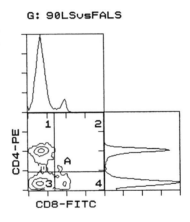

Fig. 5. Two parameter fluorescence staining of CD4 & CD8 cells in Spleen. Gated data collected from a 90°SC + FSC dot plot

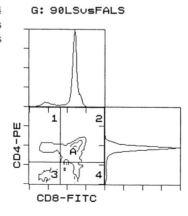

Fig. 6. Two parameter staining of CD4 and CD8 in Thymus. Tissue differences can account for double staining of cells

body. All these conditions may be important to the researcher. Be sure you have titrated the antibody well, so you can interpret all of the population subsets accurately. ●●●(See Sec. 3, Chap 2 for a discussion on titration of antibodies). Compensation setting is also important for an accurate interpretation of results (See Sec. 2 Chap 7 for a discussion on compensation).●●●

Data from 2 parameters can also be displayed in density or contour plots. Samples of data from 2 parameters are displayed in Fig. 7 utilizing first a density plot and then a contour plot. Settings for the histogram can be adjusted to show more or fewer levels of density and therefore more or less of the cells. See your instrument software manuals for a complete description of linear and log density and probability options for data display.

Data can also be displayed in 3 dimensions to give a feeling for the number of cells displaying the characteristics of two color staining or surface complexity and size. The histogram in Fig. 8 shows an oblique view of two colors of staining. This appears like a mountain range and is sometimes referred to as an isometric plot.

Three dimensional data can also display the staining of three fluorescent parameters or the display of two fluorescent antibodies and the size of the cells utilizing forward scatter. This type of display is called a cloud histogram and although it gives you a new view of the data, it is difficult to do any quantitation with this analysis, and is not found often in publications.

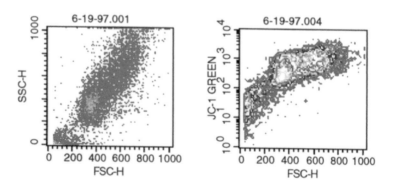

Fig. 7. A Density plot & Contour Plot

Fig. 8. A 3D plot (isometric) of 2 color antibody staining

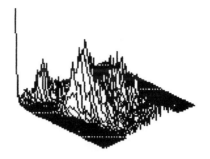

The overlay of several single parameter histograms can give you a good comparison of separate samples stained with the same antibody. Fig. 9 shows the FL1 staining data of two samples of cells collected under different conditions. You can compare the staining of the cells over time with this overlay.

Statistical analysis of data can also be done on single or double parameter histograms. Fig. 10 displays an example of a histogram marker and the statistical analysis of the cells within the marker.

The multiple data generated by a quadrant statistical analysis of two parameter histogram shows you how much more information is gleaned from simultaneously staining a sample of data. Fig. 11a and b each shows the region statistics for a single parameter analysis for first the green analysis and then the red analysis separately. The same sample analyzed utilizing a quadrant

Fig. 9. An overlay of two samples stained with FITC

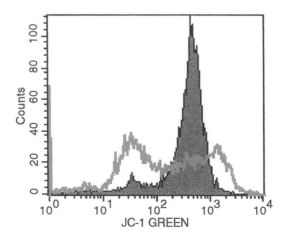

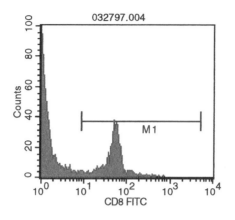

Histogram Statistics

File: 032797.004   Log Data Units: Linear Values
Sample ID: A Lymph node   CD4/CD8   Acquisition Date: 27-Mar-97
Gate: No Gate   Gated Events: 10000
Total Events: 10000   X Parameter: FL1-H CD8 FITC (Log)

Events	% Gated	% Total	Mean	CV	Peak Ch
10000	100.00	100.00	12.61	229.63	1
2078	20.78	20.78	55.74	73.56	47

Fig. 10. A Single parameter marker and statistics

statistical analysis, Figure 11c, shows much more information, including the simultaneous staining of some of the green cells within the red positive region. Utilizing this form of data analysis, we can see a population of cells stained simultaneously with T4 and T8, one stained separately with T4, and one with T8, as well as cells stained with neither.

A last type of histogram shows the ratio of parameters as observed over a time period. Such a histrogram might be used for the ratioing of Calcium influx or pH. And example of this histrogram is shown in Fig. 12, page 70.

The skillful use of histograms can enhance your data and the inefficient use of histograms can waste time and precious antibodies. More information can be learned from one sample tube stained with three antibodies than with 4 separate tubes stained individually. Interactions between antibodies and the population dynamics can not be understood from single stained samples. The power of the assay is multiplied by the number of colors

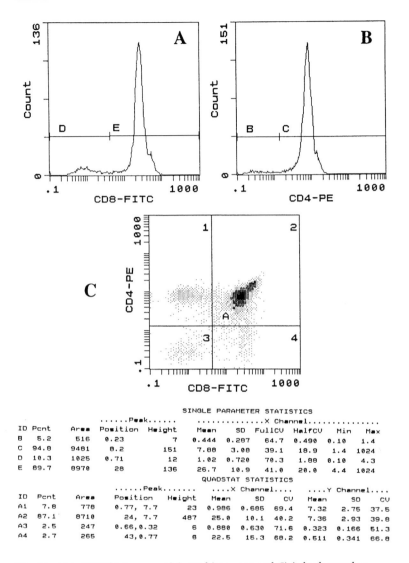

Fig. 11. A) is FITC histogram, B) is PE histogram and C) is both together

assayed togetehr. For example, with 1 color analysis, a minimum of 2 cell populations, positive and negative, and possibly a third dim positive population, can be observed. When 2 colors are analyzed together, a minimum of 4 populations are observed. When 3 colors are used simultaneously, 12 or more populations are observed.

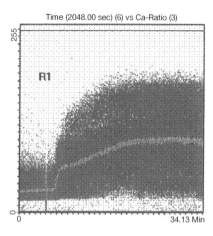

Fig. 12. The ratio of Indo-1 excitation at two wavelengths over time

The combinations of antibodies are also important for optimum analysis. Certain antibody combinations work better with certain dyes and the skillful design of panels of antibodies is often critical to diagnosis of disease. The choice of fluorochrome, antibodies, and filter configuration are all critical to accurate analysis (See Sec. x Chap c on spectral analysis). Refer to the most recent publications for a discussion of the choice of antibody combinations which better evaluate the particular disease or tissue of your interest. An excellent resource for the combination of antibodies is also the manufacturers of the antibodies you are using. Technical advice on the interaction of antibodies and proper combinations of them is available from any reputable antibody manufacturer.

Guidelines to Improve Flow Cytometry Data Display and Interpretation

WENDY SCHOBER-DITMORE

Applications of flow cytometry continue to increase in a variety of cellular and molecular research. Flow cytometry is also a recognized tool in clinical programs such as cancer, gene therapy and transplantation. While investigators in these fields seek to use the advantages of FCM to further their work, without good background information about FCM, the experiments conducted may be poorly controlled or be misinterpreted. In addition, this FCM data submitted in manuscripts for peer review may be misrepresented and published if reviewers are not aware of basic FCM principles. This results in further misinterpretation by the novice readers and a lack of reproducibility of quality FCM data.

There are a number of common mistakes that FCM users make which can end up published thus perpetuating or confounding the problem. A description of these problems will be provided as well as guidelines which will correct mistakes. Investigators using Flow cytometry should keep these points in mind to maintain high standards in their experiments and represent quality information to the field of flow cytometry.

Instrumentation set-up

Routine instrumentation QC is important and has been described elsewhere.[1] For this discussion the focus is on attention to instrumentation details for the particular experiment or application. One of the basic features of instrument set-up is the type of emission filters for fluorescence used for each application. The filter sets used in the instrument with a particular flow cytometry experiment are key to the resultant data. It is therefore important for reproducibility purposes to include filter information in protocols of specific applications.

In addition, sometimes instrumentation is upgraded or customized to accommodate current technology or experiments. When data from these instruments are displayed, it is important to explain the instrumentation differences from commercially available equipment.

Compensation

An example of the ever expanding uses of flow cytometry involve staining one population with multiple fluorochromes. To properly control these experiments, cells must be stained and analyzed with each of the single fluorochromes separately. Using the single stained controls, electronic compensation for spectral overlap is accomplished. This process of compensation, which is necessary to return accurate results from multicolor data, is complicated and further expanded upon in Section 2 Chapter 7. In addition to using single stained cells, cells stained with combinations of two or three of the fluorochromes are necessary to verify compensation is set correctly and assure that event percentages are consistent between control combinations and the sample stained with the complete set of multiple fluorochromes. It is important for investigators using multi-color analysis to acknowledge that these important controls have been run and state how much compensation was ultimately needed.

Recently, there have become commercially available products, either beads or fixed cells, which are marketed as compensation controls. The manufacturer's directions for use are intended to cover a wide range of flow cytometry applications. Careful evaluation and description by the investigator as to how these products are used to set and control compensation of any one particular applications is necessary.

Experimental Controls

Because FCM is measuring a change of biological properties, a baseline of those properties for the experimental particle type must be first established. This is the negative control. For example, a sample without any fluorochrome added, or a fluorochrome reagent that will not bind, or cells not exposed to an agonist gives background characteristics and serve for comparison to the experimental.

A positive control is equally important, although it is sometimes more difficult to obtain. Often the antibody or probe and it's specificity is the subject of the experiment, so no known biological positive exists. However, with careful experiment planning, a similar biological positive control is likely to be found.

Comparing the experimental sample to the appropriate controls is essential, but it alone will not always give confidence to the FCM data. It is important to represent and critique flow cytometry data within reference to the biological system being studied and to confirm with other biological evidence.

Mean Fluorescent Intensities

Analysis of fluorescence on any flow cytometer will return a variety of data including an average fluorescent intensity over a selected distribution. This average is the Mean Fluorescent Intensity (MFI). The correct way to compare and quantitate the amount of fluorescence from a distinct population is by using this MFI value. Too often, intensity differences are expressed in the literature using "+" and "-" or "dim" and "bright." This designation is vague and impossible for the reader to use or reproduce or compare similar data. Within each experiment analyzed on one particular instrument, the MFI values from that data reveal direct quantitative information. To compare MFI' s instrument to instrument there are bead standards commercially available which when analyzed along side the experimental samples return specific numbers of fluorochrome binding sites from which numbers of antibody binding sites can be derived. Follow the manufacturer's recommendation for use of these beads.

Scale Designations

Another common source of misinterpretation of flow cytometry data is faulty use of the x- and y- axis scales of the graphic displays. Data from events analyzed on any instrument is processed either linearly or logarithmically and then plotted against event count as single or double parameter information in the form of histograms. It is very important when presenting these histograms to display the data using the entire instrument scales. Cutting off parts of the axis fails to represent the full range of the instrument intensity measure, and is easily misinterpreted by the novice.

It is also imperative that the axes from the histograms be labeled with proper tic marks reminding the viewer as to whether the data was processed linearly or logarithmically. Some of the histograms from the older instrumentation print out linear tic marks with data that has been logarithmically processed. It is important that the investigator understand and clarify the axes in the display to reveal the proper scale.

Interpretation

Flow cytometry is a method to measure changes of biological properties, on a large number of particles, rapidly, after a variety of manipulations. It integrates data from each particle, and compiles all the particles data to reveal a biological trend. Any of the manipulations (stimulating, staining, washing, etc.) can artificially alter the baseline biological properties. Therefore, when interpreting or displaying flow cytometry data reflecting these biological changes, it is important to evaluate the entire set of data. Failure to include quantities and characteristics of the normal, unaffected, or negative events and controls beside those defined as abnormal, affected or positive introduces doubt about the significance of the biological change which is being studied.

When comparing the negative events to be expected positive on any instrument remember to acquire sufficient number of events to achieve statistical significance. Other biological experimental systems rely on specimen repetitions to assure quality of measure, but because flow cytometry can examine each particle individually, this repetition is inherent, so to obtain statistical credibility, enough events must be analyzed. Depending on the histogram total event quantities and distributions being studied, the number of events necessary will vary. There are a number of biostatistical data analysis techniques that can be referenced and used to confirm significance.[2,3] One possible general rule when choosing total event count, may be to consider the number of parameters being studied and increase the total event count accordingly. For example, on single color data collect a minimum of 10,000 total events, for two color, 20,000 events, three color, 50,000, 100k for four color, etc., although this alone will not guarantee statistical significance.

A common recurring problem in the literature is small differences in percents being described as a valid change without considering the quality or quantity of events these percentages represent. Be especially aware that comparing small percentage differences may or may not be truly statistically different, depending on the characteristics and number of events those percentage represent. If possible, confirm small changes with other biological evidence.

References

1. Hurley A. Safety Procedures and Quality Control. In: Robinson JP, Ed. Current Protocols in Cytometry. NY: Wiley & Sons, 1997: 3.1.1 - 3.2.4.
2. Watson JV. Flow Cytometry Data Analysis: Basic Concepts and Statistics. Cambridge University Press 1992.
3. Zar J., ED. Biostatistical Analysis, 2nd Edition. Prentice-Hall 1974.

Data Management and Storage

DAVID M. CODER

Introduction

The need for proper data management

As flow cytometric assays involve more parameters, the data files generated grow in size and number, and the time required to analyze them grows proportionately. This means that the users of a flow cytometry facility need to have access to their data in a timely fashion whether the data are from today, last month or last year.

Depending on the type of lab, there may also be legal requirements for the storage of data files and related data for specified periods of time. It then falls on the manager of the facility to provide the user's data when needed. How a system of data management is implemented will depend on a variety of factors, thus generalizations defining the "best" method are difficult if not impossible. This chapter is a guide to strategies for planning successful data management, and describes some solutions that can provide adequate service for users. Solutions include an overview of the means and methods-what are the system parts and how they fit together, as well as some examples of data distribution and storage systems.

The search for an appropriate data management solution

There are several questions that the manager of flow facility must address in deciding how best to provide optimal data management. These include:

- What to transfer?

- Who needs access?

- How to send it?

- Where to store it?

- How to get it back again?

What to transfer? Types of data

On the face of it, the answer to the first question may seem obvious, but one should really think of the kinds of reports that are generated in the lab to decide what to keep and how to retrieve it. That is, a clinical laboratory will have very different reporting requirements, than the biotechnology company or the basic research lab in a university.

There are two types of data to consider: **raw data** and **processed data**. The former includes list mode files and histogram files, the latter includes images, and text. The reason to distinguish the two is that there is much time invested in producing the latter, and adequate storage of processed data should be provided. (Just who provides it is an issue to be dealt with latter.) Raw data files should include the list mode data. From this any data subset (such as histograms) can be generated. Some laboratories may produce and save only histograms as the sample is run. Although this may be adequate in some instances, it is very easy to have inadequately gated or annotated histograms. Moreover, these may be irreproducible if lost. For this reason, list mode data files should be maintained. This is especially important if there are legal ramifications-propriety value or medical regulatory requirements.

Processed data may include extracts from the data both as images of histograms (multiply gated and annotated), as well as population statistics and text commentary. Reproducing these may be very time consuming, this it may be advantageous to store these as well. In any event one should keep in mind the frequency of transcription errors.

You should enter data only once:
Data entered repeatedly are prone to error.

Another "data" set that you should copy is the various instrument setup routines and user's profiles. Much time has been invested in such files and if lost, it could make the lab work much more slowly. If disaster strikes, then restoring these files will get the lab back to normal much faster.

Who needs access?

At a basic level, obviously the person who produced the data needs access to list mode files in order to analyze them. For other kinds of data-statistical data or reports-access will depend on who owns the data or who should have access to it. In a clinical setting, this might include laboratory managers, data entry personnel, and various clinicians on the service including the physician who ordered the test. Where the data have propriety value, access should be limited and this becomes a major criterion in the establishment of the data management system. Clearly, the complexity of a system can make it cumbersome and inefficient. Simplicity of operation with adequate safeguards is sometimes a difficult balance to maintain.

How to send data

Transfers of data from the FCM to users may be accomplished via several methods. These include:

- removable disks
- serial: RS232C
- parallel: IEEE488
- Ethernet

DISK TRANSFERS

For small labs, were the number of users is small and the data management requirements are modest, providing data storage on removable disks may be an adequate solution. This will depend, of course, on how the data are analyzed. If all data analysis is done on the flow cytometer computer, then the solution conceivably could work. If data are done elsewhere, then adequate compatibility among the disk formats must be provided.

Compatibility could be a hardware or software solution. For a hardware solution, the disk is simply shuttled between the flow cytometer computer and the user's computer. This will work where both are the same computers and the same driver exists on both computers, or if the computers are dissimilar, the disk can be read on the user's computer. The combinations are summarized in the table below.

Flow Cytometer Computer	User Computer	
	PC	Macintosh
PC		Can read disk
Macintosh	May read disk	
HP	Software available	Software available

For example, the Macintosh can generally read disks for Iomega (Zip, Bernoulli) drives that are formatted on a PC; the converse, however, is not true. For either computer, there is software available that allow either a PC or Macintosh to read 3.5" floppy discs from a HP. The Bernoulli drives on the HP cannot be read on a PC or Mac[DC1].

DIRECT CABLE TRANSFERS BETWEEN COMPUTERS

The next most efficient systems involve connecting two computers together via cables. Most computers have a serial port by which data can be transferred via a null modem cable. Because this is very slow it is not recommended. (AppleTalk is a serial transfer method but will be discussed below in networking.)

Transfers via parallel ports can be done from HP computers to PCs via a HP-IB (IEEE 488) connection. The speed can be adequate, but it is not nearly as fast as a good network. For the HP310 computers, you'll need an IEEE 488 card for the PC (about $500), and software which can be acquired via the Internet (see Eric Martz' collection of on-line software at http://www.bio.umass.edu/mcbfacs/flowcat.html). The software includes instructions for setup of both HP and PC.

If the HP is the more recent HP340 system, there is a similar software package available from Verity software for HP-IB transfers.

NETWORK DATA MANAGEMENT

A chief goal for a flow cytometry facility of any size is getting the raw data off the flow cytometer computer and to a central archive for access by the users, and the most efficient way to do this is via a network. Although Ethernet-based systems are today dominant and continuing to displace alternative networks, AppleTalk for the Macintosh is about the only alternative to consider if the lab has only Macintosh computers. Existing networks may include NetWare (Novell), Vines (Banyan), 3COM, LANtastic (Artisoft), OS/2 (IBM), or some flavor of unix. You may have to integrate the flow cytometry lab into this environment. If you are fortunate, an existing network will be Ethernet-based. If none exists, then Ethernet for even a small lab will be the best choice. Whatever the system, a good data management strategy will still include considering the following:

Network criteria

- data flow
- speed
- convenience
- security; fault tolerance
- off-site data transfer compatibility

- analysis/reporting software

- prints

- network manager

- user training

- growth

- implementation

Data flow: how will the data be sent to an archive, and how will it be distributed to the users? *Speed:* how rapidly do the users need access to current and archived data? Are there any bottle necks that could impede flow?

Convenience: how will users access the data? What kinds of computers will they have and how will they connect to the computer that stores the data?

Security: Must the data be protected from any intrusion from the outside? Also, what is the requirement for resistance to failure, that is, what are the requirements for fault tolerance if part of the network, a server, or a disk drive fails?

Off-site data transfer compatibility: If you are transferring data to remote sites, how will remote networks and computers perform with your network?

Analysis/reporting software: will you also supply analysis and reporting software from the network server? *Prints:* will you provide a network printer?

Network manager: who will manage the network day-to-day? *User training:* who will train users in the use of the network and its components?

Growth: as the needs for network service increase, how will you plan for future expansion?

Implementation: who will purchase and setup all the parts of the network?

Answers to these questions will help you determine the scale of you needs, and the amount of work involved in providing network service. If you think you need a network to connect multiple computers, and have no or little experience with networks, then you will need to find a network consultant. Most large institutions have a person whose job it is to provide this service. Find them.

What's in a Network?

Networks consist of both hardware and software parts. The hardware components consist of:

- Computers; one may be a file server

- Network controller cards

- Cabling

- Routers, hubs

Each computer must have a network controller card to connect via cabling to other computers (several may be connected directly), or to hubs and routers for a network of any size. Controller cards may be of variety of types, although Ethernet cards or adapters are by far the most common for new computers. Many current Macintosh computers, for example, have an Ethernet adapter as a standard connection. Ethernet cards for PCs cost between $50 and $100. One or several computers may act as fileservers. That is, the server contains a high capacity disk drive (3GB or larger) and provides a central repository for data. For an Ethernet-based network, each computer is connected to the network from the network adapter via cabling to an Ethernet port. The two common types of cabling are: 10BaseT (RJ45 connector; twisted-pair cable), or 10Base2 (BNC connector; RJ60 coaxial cable.) The latter permits direct connections among computers as long as the end computers are terminated with a 50-Ohm terminator. The cabling, however, is more awkward, costly, and more prone to problems. In contrast the most common cabling, 10BaseT, makes a separate connection for each computer to the network and computers cannot be connected directly to each other. One way to get around this is to use a mini-hub. These relatively cheap devices (Asante Friendly Net is one example, http://www.asante.com) connect several computers to a single Ethernet network connection, or if a stand-alone network, allow several computers to be connected together.

Network software mediates connection of the computer to the network, and provides programs to access files on other computers and to print using networked printers. There may be several types of network software running concurrently over the same kind of hardware. (There are various "layers" of networking software, which may accommodate a variety of file types and transfer methods. Fortunately, the arcane business of network software an issue that you need not be concerned with in most instances.) Much of the interface software comes with the computer as part of the basic operating system, or is available for little or no cost over the Internet. The same is true for the network software-it comes with the operating system in Mac OS and Windows 95; both are the dominant desktop computer operating systems.

NETWORK EXAMPLES

Ethernet-based networks are the most common ways that desktop computers talk to one another and the world at large. Despite the past prevalence of Novell NetWare in the PC field, other network software is far more common on PCs. Fortunately, several types of network software can run at the same time over an Ethernet-based network. If starting a network today, most likely you'll using PCs (undoubtedly running Windows 98 if the computer was purchased within the year) or Macs (running OS 9 if purchased recently.) A server computer may be running Mac OS, Windows NT, or unix; individual users' computers ("clients") will most likely be running Mac OS or Windows 98. To determine the compatibility of computers on the network, you'll have to consider the computer that runs the flow cytometer as well as the user's computers. The following table shows the compatibility among various operating systems.

Compatibility Among Network Operating Systems

Flow Cytometer Computer	Server (and/or) User Computer	
	PC	Macintosh
PC	Windows 95, NT; TCP/IP	Windows NT, TCP/IP
Macintosh	Windows NT, TCP/IP	EtherTalk, AppleTalk, TCP/IP
HP	TCP/IP	TCP/IP

What the network does: file access, file transfer, printing, software access

Depending on complexity of the network and the needs of the user group, the network may serve a variety of functions. The most basic is file transfer from the flow cytometer computer to the users. This will most likely be done via a network server which acts as the central repository of data. The most convenient (from the user's point of view) is to have the server disk act just like another disk on the user's computer. This is the case for most of the combinations of PC's and Mac's listed in the table above. For heterogeneous combinations of computers, Windows NT servers can provide file service for both PCs and Macs. Alternatively, there is third party software that mediates connections among Macs and any Windows-based network; DAVE from Thursby Software Systems (*http://www.thursby.com/*), is one example. If the Mac is the network server, AppleShare for Windows permits Windows-based PCs to access Mac disks and printers.

A less desirable, but workable solution is to provide file access by copying the files from the server onto the local user's computer. Probably the best solution is via standard ftp (file transfer protocol) software that runs on the client computer. The server is setup as an ftp server, and the user's computer can copy files for remote analysis. There are two very good ftp client programs, and both are free for governmental and/or nonprofit and academic use: Fetch (*http://www.dartmouth.edu/pages/softdev/fetch.html*) for the Mac, and WS_FTP LE (*http://www.ipswitch.com*) for PCs.

Where to store it

DATA STORAGE CRITERIA

In planning a data archive and distribution system, you need to consider the following:

- What to store?

- How long to store it?

- How accessible should it be for users?

The first issue was addressed above when discussing general network planning. You must decide if you will be providing access to and archiving raw data, or various forms of processed data for users. The latter will most likely be smaller in size and perhaps important since much work has gone into the process of analyzing the data and producing prints. Alternatively, the responsibility for all but the raw data may be defined as the responsibility of the users. At the extreme, you may provide no archiving service, but provide only the raw data as they are produced and require that each user provide their own storage.

In providing basic file service it's useful to distinguish two types data storage: primary storage and secondary storage. Primary storage is for current data. This means that access to it is frequent and must be fast; this also means that it's more expensive. Secondary storage is for the data archive. Access to it is infrequent and will be slower; this means it will be less expensive.

CHOICES FOR STORAGE

The medium works us over completely.
– Marshall McLuhan

- Primary: hard disk

- Secondary: removable hard disk; optical; magneto/optical disks; tape

The obvious choice for primary storage is, of course, a hard disk. Storage capacities are now large (you should consider nothing smaller than 10GB), speeds are fast, and the prices are relatively modest. The 10GB disk recommended will cost less than $200 as of today (7 November 1998), and will only get cheaper. One reason that hard disks provide for fast access to files is that they are a random-access medium. That is, files on the disk can be read in any order, in contrast to slower media like tape, files can be read only serially.

There are a variety of choices for secondary storage. Your decision of which to use will be based on the convenience of storing and retrieving data, cost of the medium, and speed of access. The following table gives a comparison of some current choices.

The choices for storage media continue to expand. Capacities have increased and the prices have generally decreased. Also, the overall speed of reading and writing data has increased greatly. For large archives that will be accessed only occasionally, tape is

Mass Data Storage Alternatives

Drive Type (Manufacturer)	Capacity	Read/Write Speed	Drive cost/ media cost (per MB)	Advantage	Disadvantage
FLOPPY DISK REPLACEMENTS					
Zip Drive (Iomega)	250MB	slow to moderate; parallel port version slower	$150, $17	Cheap; easy to use; in widespread use.	Less capacity than other solutions; relatively high cost per megabyte.
3.5-inch magnetic cartridge			(7 cents)		
REMOVABLE DISKS					
3.5-inch MO	640MB (320MB per side)	fast	$330, $26	Moderately good capacity and very good speed; rugged media with long shelf life; backward compatibility	use is decreasing in competition with removable magnetic drives
3.5-inch magneto-optical cartridge			(4 cents)		
(Fujitsu, Olympus, Pinnacle)					
5.25-inch MO	5.2GB	fast	$1,600, $90	Good capacity and very good speed; rugged media with long shelf life.	relatively expensive drive cost; use is decreasing in competition with removable magnetic drives
5.25-inch magneto-optical cartridge	(1.3GB per side)		(1.7 cents)		

Mass Data Storage Alternatives (Continued)

Drive Type (Manufacturer)	Capacity	Read/Write Speed	Drive cost/ media cost (per MB)	Advantage	Disadvantage
(HP, Maxoptix, Sony)					
Jaz Drive (Iomega)*	2GB	fast	$200, $90	Good capacity and very good speed.	somewhat high price per megabyte
5.25-inch magnetic cartridge			(4.5 cents)		
Apex (Pinnacle Micro)	4.6GB	very fast	$1,300, $85	Excellent capacity and speed; rugged media with long shelf life.	Delayed introduction raised concerns about product viability; not compatible with legacy MO formats.
3.5-inch magnetic cartridge	(2.3GB per side)		(1.8 cents)		
Compact disk-recordable	650MB	very slow to write; moderate to read	$250, $1	Lets you create disks for widespread distribution; low cost per megabyte.	Slow; creating disks can be difficult; read problems in some CD-ROM drives
5.25-inch optical disk			(0.2 cents)		
(HP, Sony, Yahama)					

Mass Data Storage Alternatives (Continued)

Drive Type (Manufacturer)	Capacity	Read/Write Speed	Drive cost/ media cost (per MB)	Advantage	Disadvantage
OTHER OPTIONS					
DAT	74GB	slow to write; slow to moderate to read	$1,500, $30	Good capacity; low cost per megabyte.	Relatively slow due to sequential nature of tape
3.5-inch magnetic tape			(0.1 cents)		
(Sony, HP, Exabyte)					
Travan 5	20GB	slow to write; slow to moderate to read	$260, $40	Excellent capacity; low cost per megabyte.	Relatively slow due to sequential nature of tape
3.5-inch magnetic tape			(0.1 cents)		
(HP, Sony)					
DVD-RAM*	5.2GB	slow to write; moderate to read	$700, 40	Good capacity.	Available in June of 1997 or later.
5.25-inch optical disk	(2.6GB per side)		(0.7 cents)		
(Hitachi, Panasonic, Toshiba)					

hard medium to beat. The capacities are very high (20GB per tape), the costs are low, but the speed and convenience of restoring data will be slow. The TRAVAN tape medium has replaced quarter inch tape cassettes (QIC) as a medium for storing large archives. If speed of restoration is important, then recordable CD's are an option. The medium has increased in speed and ease of use over the past year, and the cost is relatively small. If considering a CD-R for archives, software that permits multi-session recording is far easier to use. Unlike music CDs, however, the surface is susceptible to damage. Removable magnetic disks (e.g., Jaz) provide nearly the same performance as a hard fixed disk at a relatively cheap price. They have largely replaced the magneto-optical disks that were the best choice a few years ago. For personal use, the super floppy disks (e.g., Zip) provide a useful solution with capacities of about 250MB per disk and a relatively low cost for the drive. Another consideration in choosing backup systems is the software for making copies and restoring the data. Most will provide scheduled backups making the formation of the archive and automatic process. You should also consider the ease of retrieving single files or multiple files from the stored data.

Archiving Strategy: How to store it, how to get it back again

Once you've decided on what to backup and how to do it, you need to devise a plan to ensure that all data are saved and plan for disasters as if they were going to happen. Naturally, you'll store the data files produced. You may be storing processed data as well as instrument setup files and settings. In general, you should make a copy as soon as a file is created or changed. For data files, this might occur immediately after a user finishes collecting data, or at some convenient interval during the day. Once data files are on the server, they should copied to the archive on a daily basis. Most backup software will automate this task for you via a script. Simply stating the time at which you want the backups to occur, defining the disk or tape that will hold the archive, and placing the right disk or tape in the appropriate drive will help protect against data loss.

You should have more than one set of copies. Some formal procedures for backups of data require that you have separate tapes for each day, and backup the entire system at some interval. Given that you will always be replacing data on the server, it is wise to have at the very least two sets of archives to guard against the failure of a tape or a disk. It is also wise to store them is separate locations protected from hazards. Providing more than two sets of archives will depend on institutional or licensing requirements for data storage.

FUTURE: Nothing but Net?

Trying to predict the future of computer technology is difficult at best. At the least I can say with certainty that the computer you buy next month will be faster, have more storage capacity, and cost proportionately less. Beyond that it's guesswork.

Having said that, there are some developments that look probable. The most obvious is the predominance of the Internet. Much of current software development is oriented toward working with an internet browser. The good thing is that the applications may gain a degree of independence of the operating system, and using an internet browser provides for a common interface among applications. Also, sending and receiving files is further simplified because working via the internet assumes a common network operating system, and a defined set of file types. On the downside, this means that computer on your desktop must be made to conform. If hardware is not up the task, then you need to upgrade or replace it. New software may further push up the costs. (I don't expect the current trend in nearly free Internet software to continue.)

As far as storage options are concerned, we can only expect to see capacity and speed increase and the cost per unit of storage decrease. The CD-ROM will probably disappear to be replaced by the DVD. Variously named digital virtual disk or digital video disk (depending on the application), this disk may eventually replace CD-ROM, music CDs, and video tape. DVD now consists of serval formats. They include: DVD-Video, DVD-ROM and DVD-RAM, a rewriteable version of the computer format and DVD-EW. DVD-RAM (restores) could be very useful for storage given the 2.6 GB per side capacity of the disks. Storage space on the disk should incease to 17 GB in the year 2002.

Although a network may be costly to implement remember:

The most expensive part of your network is the information stored.

Coloring Up:
A Guide to Spectral Compensation

ROCHELLE A. DIAMOND

Spectral overlap is one of the hurdles that faces a flow cytometer user when considering more than one color analysis. It grows increasingly complex when trying to use multiple colors and more than one excitation source.

The first consideration should be the choice of fluorochromes. You should try to choose fluorochromes that minimally overlap in their emission spectrum given your excitation and quantum efficiency restrictions. Consult the fluorescence spectra of your fluorochromes and examine not only the maxima of emission but also the breadth of the extended long wavelength tails that are responsible for the spillover of fluorescence from the shorter wavelength dyes into the detector channels of the longer wavelength dyes.

Next consider your choice of spectral filters wisely to pick up the most emission range (central wavelength and optical bandwidth) with the least overlap even if it means cutting off the peak maxima as long it still allows sufficient detection. Optical filters are used to separate the spectral bands (colors) from your fluorochromes for measurement by detectors. The efficacy of the filters vary depending on the wavelengths they separate.

There are several types of optical filters. Interference filters reflect and pass specific wavelengths of light and come in several flavors- short pass, long pass, and band pass. When splitting the spectrum into two wavelength ranges the filter is set at a 45° angle to the incident light and is known as a "dicroic" (meaning 2 color) mirror. This is used for steering light by reflection and transmission. "Long pass" filters pass light of a certain wavelength and longer and, if it is a dicroic mirror, reflect shorter wavelengths than the "cut off wavelength", specified by the half maximal wavelength of transmission. "Short pass" filters pass light of a certain wavelength or shorter and, if it is a dicroic mirror, reflect light longer than the "cut off wavelength". "Band pass" filters transmit light of a very narrow range specified by the center of the spectral band and the width at half maximal transmission. For instance to measure fluorescein we commonly use a 530/30 band pass filter which is centered on 530 nm with a spectral width of 30 nm. The filters on hand, provided standardly by the manufacturer, usually dictate the choice of optical filtration sets. However, custom filters can be purchased from companies like Omega Optical (check out their web site at http://www.omegafilters.com for a good glossary of terms and spectra specifications of various filters and fluorochromes). If you are in doubt about the spectral transmission curves of your filters consult the manufacturers or use a scanning spectrometer to generate curves, but be careful to position the filters in the correct orientation and angle.

Many filters (interference) are made by thin layer vacuum deposition of rare dielectric earths and coatings on glass or quartz that vary for the specifications called out. The manufacturers can help you decide the center and width of the spectral band selection for your purposes to optimize the signal to background given the requirements for filter manufacture. The expense of the filter is directly proportional to the number of coatings creating interferences and blocking "cavities" on the filter. Once you have made a choice, make sure that you orient your filters correctly for proper use. Dicroic filters are aligned at a 45° angle for proper light passage. Blocking and bandpass filters are usually aligned in a perpendicular direction to the incoming light. The coatings are also directional for filters and the markings should be oriented correctly for light passage usually toward the laser. Look for markings on the filter frame or consult the manufacturer. Beware of pinhole leaks in filters, which may occur over time and let light leak in.

The last consideration and most essential for measurements of antigen density and dim cells in multicolor flow cytometry is fluorescence compensation. Simply put, fluorescence compensation is the procedure used to compensate for the resulting spectral overlap from the tails of emission spectra of your fluorochromes that escape removal by your optical filtration system which inappropriately fall into the detectors measuring the other fluorochromes. Compensation can be implemented in two ways. Most commercial cytometers employ hardware analog circuitry, which you adjust by potentiometers to add and subtract a certain percentage of signal from one detector to the signal of another detector. Newer machines and some stand alone software packages have programs that compensate "off-line" using a matrix of algebraic equations relating the measured values of fluorescence in your control samples to determine the proper compensation values. This can be quite useful for uncompensated or undercompensated samples, but it is a wonderful tool for use with multiple laser excitation where a dye may be excited by both lasers. This type of cross-laser compensation cannot be easily done with analog compensation.

How is compensation determined? It is a given that the spectrum of a fluorochrome will remain the same as long as the solvents and microenvironment remain the same. Therefore the emission spectrum coming from the same fluorochrome will always vary proportionally according to the filters that are used for its collection. If you collect light from a single fluorochrome through your filter set arrangement then you will see how much light is transmitted by that fluorochrome through all the filters going to all of your detectors. That amount of transmission will always be proportional to the intensity of the fluorochrome from one detector to another. From this you can calculate the compensation value as the ratio of change in one signal to the change in the other signal via a simple set of algebraic equations.[1,2,3]

In practice, you simply need to make a brightly stained single color control sample for each fluorochrome that you will be measuring along with one unstained control. Run these sample controls individually. Create bivariate plots from your fluorochrome signal intensities measured from the appropriate detector vs each signal in the inappropriate detector. Subtract the signal produced from your fluorochrome in the inappropriate detectors using either the analog compensation hardware on your cytometer or a software program for compensation. Using the negative unstained cell population control as a benchmark, move the compensated cell population so that it lines up with the median of the negative cell population in your bivariate plots. When running, for example, your fluorescein sample uncompensated you will notice that there is a nice bright signal in the appropriate detector but in addition smaller significant positive signals are picked up in the other detectors that were set up for PE and PE-Cy5 tandem conjugate signals (see figure 1 panel 1, A-

C). These smaller signals need to be subtracted away. The next control sample to be run is the PE sample which gives a lovely bright signal in its detector but has inappropriate signal overlap with the fluorescein and the PE-Cy5 detector. These two signals need to be subtracted away. Finally, the PE-Cy5 sample is run and its signal is measured in the appropriate detector, but it has a crossover in the PE detector which needs to be subtracted out. Thus all control samples provide measurable inappropriate signals which we use to set the compensation values to be subtracted out of the inappropriate detectors. It is important that a control sample also contain negative cells because the amount of signal to subtract will be measured against the center of the median of the signal from this negative cell population. Stained cells or calibration beads are routinely used to make these subtractions but you should keep in mind that fluorochromes (isomers) may vary from one manufacturer to another in slight but significant ways. We always use a control of the fluorochrome employed to measure the experimental samples just to be safe. For two-color analysis, compensation matrix ratios must be performed for one fluorochrome into the other fluorochrome's detector and vice versa, because both will spillover to one extent or another. For three color multiparameter analyses, the signals from all three detectors need to be compensated just as for the two-color analysis, i.e., a matrix of one-color controls is performed to compensate signal cross-over on each of the detectors. An example of the three-color compensation is presented in figure 1D - before compensation and after compensation. The same is true of four, five, and n color analysis. At this point, however, computer compensation is the way to go because the number of analog compensation boards would overrun your cabinet.

In summary consider the following for establishing good and true signals for your fluorochromes:

- Choose your fluorochomes according to their spectrum and quantum yield.

- Choose your filter sets to minimize cross-talk of the spectral tails for your fluorochromes.

- Properly compensate by using bright singly stained and unstained control cell samples, preferably with the same autofluorescence backgrounds.

Single Color Controls for Compensation

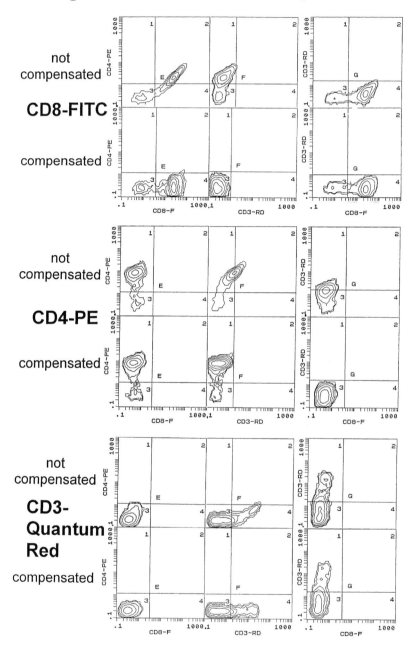

Three Color Analysis

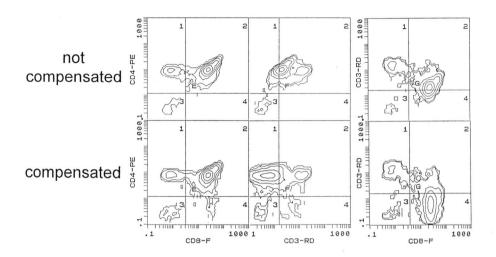

- Set your detector voltages high enough so that your unstained cells are off the axes of your histograms to insure good negative median values to match the positive cell populations after compensation.

- Align the centers of the negative cell populations with the positive populations to match median fluorescences in bivariate plots of singly stained cell populations versus the other fluorochromes' detectors.

- Remember if you are either under or over compensated you will have a difficult time distinguishing dimly fluorescent cells in your sample population from negative cells.

References

1. http://cmgm.stanford.edu/~roederer/compensation/GenComp.html
2. Shapiro HM. Practical Flow Cytometry, 3rd Edition. Wiley-Liss, New York. 1995 pp214-215.
3. Bagwell CB, Adams EG. Fluorescence spectral overlap compensation for any number of flow cytometry parameters. In: Ann NY Acad Sci Landay AL, Ault KA, Bauer KD et al (eds); 677:167-18, 1993.

Quality Control Guidelines for Research Flow Cytometry

ROCHELLE A. DIAMOND

Introduction

Accurate and reproducible test results are integral to all laboratories. Quality control methods identify and minimize sources of variation in instruments and reagents. In clinical laboratories, quality control methods require rigorous documentation, certification, and monitoring for reproducibility and accuracy. This entails validation of chain of custody information (sample source, collection procedures, transportation and storage of viable samples), documented sample preparation procedures, staining conditions, reagent performance, instrument reproducibility, validated measuring conditions, meaningful data analysis, and comparative reporting of data. The bottom line for clinical cytometrists is to provide valid comparisons of clinically relevant parameters within a sample, between samples, between laboratories, over time. For an in depth review of these issues see reference one[1] which provides a good bibliography on quality control for clinical applications.

Research laboratories, unlike clinical laboratories, are not officially regulated or monitored. The responsibility falls on the shoulders of the investigators themselves and/or the flow cytometry core facility operators. The guidelines presented here are addressed solely to these research oriented laboratories and are in no way intended to give advice on clinical quality control. Research investigators, however, can certainly take the clinical guidelines set forth by regulatory agencies, consensus committees, and quality control conferences and workshops as good procedural advice in order to maintain their experimental reproducibility over time.[2]

Determining flow cytometric instrument accuracy has been relegated in many research laboratories to optimizing instrument performance and then monitoring over time to identify trends and drifts which may need attention for preventive maintenance or repair service. There are presently no certifiable standards for monitoring absolute accuracy, although there is some movement in that direction. There are, however, many commercially available control, calibration, and standardization materials that can be used to monitor the instrument on a daily basis. These consist of a variety of cell sized plastic dye encapsulated beads or fixed cell based particles which ideally give stable values over an extended length of time (months). They come in many spectral flavors and size ranges and should be selected to match the parameters of the reagents that you will be measuring. There are particles that are generally used for setting up and monitoring instrument performance because of their broad spectral range and uniformity in fluorescence peaks and scatter profile. The first step for open optical systems that are operator aligned is to optimize the instrument for optical and electronic performance. Using these beads, optical alignment is performed which involves adjusting the components of the system for maximal signal intensity, as measured by the mean channel number, and minimal variability, as measured by the standard deviation or coefficient of variation (CV) of the bead peaks registered for each detector. This may mean aligning lasers, optical filters, focussing lenses, flow cells, adjusting sheath streams, flow rates, photodetectors, and other conditions that may influence instrument performance.

Following optical alignment, initial values must be established for the calibration particles so that they can be monitored regularly. A quality control log should be started and the values recorded daily and actively followed. There are two general procedures that can be used to monitor daily optical performance (see protocols following this text). The first protocol measures instrument performance by monitoring the reproducibility of the particle mean intensities and CVs under specified instrument conditions on a day to day basis. The second protocol monitors the reproducibility of the instrument settings needed to achieve specified mean channel numbers for the alignment particles. This second protocol gives an indication of instrument drift over time. In either protocol, a range of acceptable variation is established which defines acceptable instrument performance. For fixed optical systems, such as the FACScan, the manufacturer has specifications for certain alignment products tied to specific instrument parameters and settings that define acceptability. For operator aligned optical systems, a range of acceptability can be defined by running the protocol of choice with selected calibration particles under the same defined conditions for a repetitive number of times (>20) over time (>5 days), while recording all pertinent parameters for your specific application. The range is therefore established on the observed parameter (two mean standard deviations.[1,2,3] Be sure to test and record new lots of calibration beads in parallel with the old lot using optimized instrument settings before switching lots. Some systems come with special software or suggested protocols for instrument set-up and monitoring. Commercial programs like QC Tracker (Phoenix Flow Systems) are available for monitoring this kind of information and keeping validation records. You can also create one yourself using a spreadsheet such as Excel to plot out weekly and monthly quality control data. Good record keeping and consistent monitoring of acceptable established performance ranges are essential to maintain control of instrument performance. Trends and drifts tracked graphically can give instantaneous warning of instrument problems that need to be investigated. This can minimize down time and help to maintain high quality data output.

Quality control is also used to establish conditions relevant to specific sample measurement protocols. Standard biological controls should be chosen to be as close to the unknown sample as possible in both scatter characteristics (size of the cells) and cell type. There are many variables in sample preparation. Positive and negative sample controls should be prepared simultaneously with the experimental sample. This kind of biological quality control establishes reagent quality and protocol validity for the assay system. The positive control should be designed to verify reagent and sample specificity and spectral compensation (see chapter this section on spectral compensation). The negative control should be designed to establish background information and report false positives (see Alice Givens chapter in section 3 on sample staining). If possible, controls for dim versus negative samples should be used to monitor sensitivity and autofluorescence. It is important to examine the literature and references provided by the various authors throughout this book to set up meaningful controls for your experiments. Clinical researchers should be aware of the literature for various consensus conferences, workshops and committee publications[3] such as the DNA Cytometry Consensus Conference[4], the International Leukocyte Typing Workshops,[5] and CDC MMWR recommendations and reports.[6]

Outline

Building a system for optimizing and monitoring quality control is individual to the laboratory and the instrumentation on hand. Still, common threads hold from lab to lab, which you can use as quality control guidelines for research flow cytometry:

- You should understand your instrument's operation, alignment, and optimization practices (consult the operator's manual and training course guide for recommended specifications and procedures).
- Minimum internal laboratory quality control procedures are important and should be clearly documented utilizing standard reagents.
- Choose relevant instrument parameters and establish initial ranges for instrument settings like photomultiplier tube voltage settings, compensation values and CV's of bead peaks.
- Monitor the above instrument conditions after alignment on a daily basis and keep a log or at least good records of the data. Tabulate or chart means and standard deviations for your parameters to generate data that makes trends and shifts easy to see. Progressive drift away from mean values or a sharp change from the values should be a cause for further evaluation, adjustment, or service call. If these procedures fail to produce expected values, common sense should prevail and the experiment delayed until the cause is tracked down and remedied.
- Use relevant biological controls to monitor your specific assay system for your instrument parameters. Establish an expected range of values for your control samples.
- Regularly review your quality control data on known control samples. Set cut-offs for acceptability or rejection of quality control data. Establish an overall monitoring frequency to evaluate the quality control system.
- Monitor your instruments before and after service calls. Document and monitor QC data for variability over the life-

time of your instrument. If you follow these guidelines you can have confidence in the reliability and reproducibility of the data generated by your flow cytometer.

Subprotocol 1
Quality Control for Standard Channel Settings

▓▓ Materials

Equipment
- Flow Cytometer

Solutions
- DNA-Check Fluorospheres - Coulter Cytometry part number 6603488

▓▓ Procedure

Clean instrument as needed

1. Clean instrument according to manufacturer's recommendations.

Align optics

2. Set up your instrument with optimal standard voltages and gains in linear mode. Create histograms for all parameters that you will be measuring.

Note: The following should be done daily after the instrument is warmed up.

Run Fluorospheres

3. Run sample of alignment calibration beads. Be sure that the beads will fluoresce for all parameters to be measured. They should have reasonable scatter and fluorescence peaks. Adjust sample flow for 60-100 beads per second flow rate and let stabilize.

Note: We use DNA-check beads from Coulter Cytometry. Record lot number. Compare any new lot of beads with the current one in use before using routinely.

4. Align the optics so that the beads are optimally displayed for the standard voltages on your fluorescence detectors and for standard gains on both of your scatter detectors according to your manufacturer's instructions.   **Adjust optics**

Note: Adjust for the brightest and tightest peaks possible for scatter signals and fluorescence signals according to manufacturer instructions (some machines cannot be aligned by the user/ operator – ask for service if not meeting specifications).

5. Place the alignment /calibration beads in a specific channel for each detector that is standard on a daily basis by adjusting the flow cytometer voltages as necessary.   **Assign beads to particular channel numbers**

6. Acquire data for the beads at these settings.   **Acquire data for beads**

7. Record photomultiplier voltages and positions for each peak – mean, median, and CV   **Record voltages and bead statistics**

8. Monitor the daily standardizations to identify any changes in performance by plotting the values over time.   **Monitor daily and plot over time**

9. Compensate for spectral overlap daily with single color biological samples labeled with fluorochromes of interest respectively. Monitor changes for fluorochrome usage.   **Compensate for spectral overlap and record settings**

10. Run relevant QC biological control samples that are negative, dim, and bright to determine sensitivity for each parameter to monitor.   **Establish minimum expected range of biological control values**

11. Establish minimum acceptable distances between negative and dim peaks. Record values. Monitor for acceptability before performing subsequent assays.   **Monitor and document QC data for variability over time**

Subprotocol 2
Quality Control for Specified Voltage Settings

▦▦ Materials

Equipment
– Flow Cytometer

Solutions
– DNA-Check Fluorospheres - Coulter Cytometry part number 6603488

▦▦ Procedure

Clean instrument as needed

1. Clean instrument according to manufacturer's recommendations.

Align optics

2. Set up your instrument with optimal standard voltages and gains in linear mode. Create histograms for all parameters that you will be measuring.

Note: The following should be done daily after the instrument is warmed up.

Run Fluorospheres

3. Run sample of alignment calibration beads. Be sure that the beads will fluoresce for all parameters to be measured. They should have reasonable scatter and fluorescence peaks. Adjust sample flow for 60-100 beads per second flow rate and let stabilize.

Note: We use DNA-check beads from Coulter Cytometry. Record lot number. Compare any new lot of beads with the current one in use before using routinely.

Adjust optics

4. Align the optics so that the beads are optimally displayed for the standard voltages on your fluorescence detectors and for standard gains on both of your scatter detectors according to your manufacturer's instructions.

Note: Adjust for the brightest and tightest peaks possible for scatter signals and fluorescence signals according to manufacturer instructions (some machines cannot be aligned by the user/operator – ask for service if not meeting specifications).

5. Acquire 10,000 events.

Acquire bead information

6. Analyze the histograms for half peak coefficient of variation (CV) and mean fluorescence of the bead peaks.

Analyze histo-gram statistics

7. If CV's are greater than 2.0 then readjust the optical alignment until the values are under 2.0. The bead peaks should ideally be close to midrange of the histogram.

8. Record CV and mean values with the instrument settings in your records.

Record values and plot for variation

References

1. Muirhead KA. "Quality Control for Clinical Flow Cytometry." In: Bauer KD, Duque RE, Shankey TV. *Clinical Flow Cytometry Principles and Application.* Williams and Wilkins 1992; pages 177-199.
2. Hurley, AA. "Quality Control in Phenotypic Analysis by Flow Cytometry" In: *Current Protocols in Cytometry.* 1997; John Wiley & sons, Inc. pages 6.1.1 to 6.1.4.
3. National committee for Clinical Laboratory Standards. Internal Quality Control testing: principles and definitions; approved guideline. Villanova, PA, 1991 NCCLS Document C24-A.
4. "DNA Cytometry Consensus Conference". *Cytometry* 1993; 14:471-500.
5. *Leukocyte Typing V* 1995; Oxford University Press, Oxford.
6. 1997 Revised Guidelines for Performing CD4+ T-Cell Determinations in Persons Infected with Human Immunodeficiency Virus (HIV). *MMWR* January 10, 1997; 46(RR-2):12. U.S. Department of Health and Human Services Public Health Service Centers for Disease Control and Prevention (CDC), Atlanta Georgia 30333.

Section 3

Sample Preparation and Cell Surface Staining

ROCHELLE A. DIAMOND

Now that we are familiar with the instrumentation, how do we go about preparing our biological samples to utilize all this technology? Rochelle Diamond *et al* present a discussion on various methodologies to make the all-important single cell mono-dispersed samples. They provide protocols for cleaning up debris, getting rid of dead cells, and ways to enrich for the investigator's particular population of interest. Alice Givans follows with an in-depth description of the fundamentals for cell surface staining with antibodies. This is the bedrock of much flow cytometry, as we know it. Andrew Beavis then takes us to another level of staining with two chapters on complex multicolor flow cytometry using different techniques to reach the same goal. David McFarland and Gary Durack give us an entirely different perspective on measuring exterior proteins by providing a means of encapsulating cells to entrap and measure their secreted products.

Cell Preparation and Enrichment for FCM Analysis and Cell Sorting

ROCHELLE A. DIAMOND, HUA WANG, FEI CHEN, AND MINDY WILKE-DOUGLAS

Introduction

Flow cytometry and cell sorting are really tools for answering biological questions. Some of these questions can be answered directly by analyzing cell populations that reside within heterogeneous starting material. This kind of analysis is very powerful, in many cases quick, and does not require cell separation. Good statistics can be generated on as few as several hundred cells even if they represent less than 0.1% of the starting population as long as they define a unique population. Detection and analysis of such a minor population is always much easier than attempting to physically isolate and purify such cells without significant cell loss.

Many other types of biological questions require physical isolation and purification of the cells for further characterization. The success of this kind of experiment and the protocols which one utilizes to gain access to these cells is strongly depend on the end use for the cells. In particular, one needs to know the relative cell numbers that will be required post purification to perform the rest of the experiment. In addition, the percentage of the cells of interest residing in the starting heterogeneous population has an enormous influence on the strategy. For instance, if only a small number of cells or single cells are needed to clone long term cultures or perform rtPCR assays, then one could go directly to the sorter and purify out the interesting cells with one or two passes through the machine. If on the other hand, one needed several million cells from a starting population that contains less than 1% of the cells of interest, then given the common commercial sorting capacity of 5000-10,000 cells per second, it would take hours of precious sorting time (approximately 6 hours) to accomplish such a task. A better strategy to approach this kind of problem is to enrich for the population of interest and then let the sorter do the purification step.[1]

The first step to isolation is to obtain cellular material and process it into a single cell suspension followed by pre-enrichment by one or more of a variety of techniques, then purification by flow cytometry and electronic cell sorting.

Cell Preparation – Mechanical or Biochemical Dispersion

Preparation of cells for flow cytometry and cell sorting will vary depending on the source and specific requirements for each cell type. The one absolute requirement is that the cell suspension be monodispersed at the end of preparation. For peripheral blood or suspension cell culture this is relatively easy. For cell masses such as organs, tissues, or tumors, gentle mechanical force or biochemical dispersion is a must. Care should be taken to dissociate tissue samples quickly after dissection, as cells may become anoxic or begin to autolyse. Cell integrity is essential for the quality of any protocol. If storage is necessary, no more than 24 hours in an unfixed state is recommended. Organs such as spleen or thymus can be gently minced, teased, pushed through stainless steel wire mesh, or rubbed between frosted glass slides. Bone marrow may be pushed through 25g needles. Vigorous pipetting may disperse some cultured monolayer cells, while others require chelation agents such as EDTA with or without enzyme treatments such as trypsin or collagenase.[2,3] Tumors and cells that form the parenchyma of organs may need to be dispersed with cocktails of combining agents such as trypsin, pronase, and papain to break intercellular linkages, together with agents such as collagenase, dispase, hyaluronidase, and elastase to break up stromal components.[4] Matrices of time, temperature, and concentration are usually performed to optimize with the preparation conditions. Enzyme matrices are usually more reproducible than mechanical means for setting up standard conditions, although they tend to change from lot to lot and therefore should be lot tested. In most of cases it is advisable to use DNase (20-100µg/ml) to keep the cells from aggregating due to the leakage of the sticky DNA from dead cells. Of course if you are planning to analyze DNA content, this should be kept in mind so that there is enough EDTA in your post dispersion buffers to inactivate any

DNase which remains after washing the cells. Careful planning is necessary if surface molecules are to be measured. Trypsin, papain, and pronase are all proteases, which can clip off or destroy cell surface molecules and receptors or remove some of their antigenic determinants. A pilot experiment is warranted in such cases to determine if the dispersive agent affects the outcome of the experiment. Thus a cell suspension protocol needs to be created and optimized for the particular cell type of interest.

Once the cells are dispersed they must stay dispersed. This is important because clumping or clotting can ruin an analysis or sort by clogging lines and flow tips. Aggregation can also artificially skew results by selectively sequestering cells. Several things can be done to avoid clumping- a) use a buffer that is Ca^{2+} and Mg^{2+} free unless an enzyme or measurement depends on these ions; b) use DNase in your sample buffer; c) work at concentrations that minimize clumping; d) work at temperatures that are compatible for your cells (some cell cultures prefer room temperature and others clump at $37°C$); e) use anti-clotting agents such as heparin in whole blood samples. It's always best to check your dispersion technique with a microscope to evaluate aggregation and viability before starting the staining or fixation procedure.

Experimental design is an important key to a successful outcome. In setting up your experimental protocol the choice of reagents can be crucial. Antibodies or chemical probes that you may want to use to define your cells of interest might perturb the cellular function that you want to study. Many cell surface molecules that define phenotype for example act as receptors for signals that modulate cell behavior. The binding of antibodies to these molecules may inhibit or enhance normal cellular responses to subsequent stimulation or perturb the measurement of physiological data. If you are planning to perform cell assays post-sorting; you should give this serious consideration. When monitoring cellular response it is important to include appropriate controls of stained versus unstained cells as well as stained but unseparated cells to rule out such effects. Culturing your sorted cells for a day or two usually sheds the antibody complex and may be useful prior to measuring response capability, but this in and of itself may produce artifacts. The way out of this dilemma is to use negative selection to isolate the cells of interest.

Using complementary sets of staining reagents, one may be able to label and eliminate undesired subsets of cells from the starting population while collecting the negative unlabeled desired cells.

Pre-enrichment for analysis and sorting

Pre-enrichment for a rare cell sort is advantageous for several reasons. Pre-enrichment increases discrimination of a population and thereby increases purity and sort efficiency. By reducing sort time, researchers have time to go back to the bench and do more experiments with his/her precious cells, not to mention that it is easier on the pocketbook. There are as many ways of accomplishing pre-enrichment as there are particulars that define the type of cell that you are interested in enriching. Cells can be defined and fractionated by their size, buoyant density, molecules on their cell surface, molecules not on their cell surface, their viability, or other functional properties. Some of the ways to use these properties for preparative purposes will be highlighted in the following protocols.

Antibody and lectin panning

Antibodies to cell surface molecules have allowed researchers to physically separate cells by several means. One method is to coat petri dishes with the antibody of interest and allow the cells which bind the antibody to adhere to the surfaces of the plates. This is known as antibody panning.[5,6] Along similar lines, carbohydrate sequences that end in certain sugars are expressed on specific cells. This is also used to stick cells to plates by coating the plate with lectins (plant molecules that specifically bind sugar moieties). In both plate fractionation methods, the cells expressing cell surface molecules that bind to either the antibodies or lectins are stuck to the plate and the cells devoid of these molecules are poured and washed off the plates. In the case of antibody panning, the positive cells can be scraped off the plates for further use or incubated overnight in media to release the antibodies from the cell surface. The lectin binding cells can be released by competition with the free sugar and then washed off the plate for further use. Figure 1 shows the populations that can be enriched for by panning mouse thymocytes on peanut agglutinin (PNA) plates. Note that thymocytes double positive for CD4 and CD8 are enriched on the PNA+ plate.

Depletion by complement elimination

Another way to use antibodies for depletion is to utilize the ability of the immune system to recognize antibody bound to cells and specifically lyse those cells. [7,8] This can be done by adding a group of serum proteins known as complement that participate in an enzymatic cascade to generate a cytolytic attack complex which destroys the antibody labeled cells. This methodology is accomplished by first binding the cells to be depleted with an antibody which recognizes a cell surface molecule. The antibody must be capable of fixing complement. After incubation, the complement containing serum (prescreened for low toxicity and high specificity) is added and incubated at 37 °C to create the complex that lyses the cells. The live cells are then washed out of the complement and counted for recovery. Debris and dead cells are usually removed after this treatment. Bulk preparation of enriched cells may be made in this way. One drawback of this method is that the incubation at 37 °C in serum proteins may perturb the cells and induce artifacts into the experiment. A good mock complement control should always be included in the experimental design. It is important that the antibody isotype used be a good complement fixer like IgM. The complement is commercially available as baby rabbit serum (Pel Freeze, Cedarlane) or guinea pig serum. Do not use adult rabbit serum, as this can be highly toxic to cells. Some antisera may have anti-complement antibodies and should therefore only be used in a multiple step procedure that removes the excess antibody before adding complement.

Separating with super para-magnetic beads (MACS)

If one's needs exceed the number of cells that can be sorted on any one day, then magnetic bead selection, known as MACS (Miltenyi Biotec, Auburn CA), may be the way to go for bulk separation and enrichment. [9,10] This method also separates cells by their surface molecule expression. The cell surface molecules are targeted with antibodies that are either directly coupled to super-paramagnetic microbeads or indirectly labeled with a primary antibody followed by a secondary antibody that is coupled to the microbeads. Alternatively, a biotin-to-avidin microbead system can also be employed. The microbeads are biodegradable and degrade in cell culture. They are extremely tiny (around 50 nm in diameter) compared to other magnetic beads and do not appear to interfere with cellular function. They are invisible to FACS and do not change the scatter properties of cells, which makes them ideal.

Cells are labeled using the MACS microbeads in exactly the same way as for FACS, i.e. make a single cell suspension and bind with antibodies. As a matter of fact, the magnetic labeling reagents can be spiked with fluorescent antibody so that the cells can be followed for quality control by FACS. After the magnetic labeling, the cells are passed through a separation column, which is located within a strong permanent magnet. The column matrix serves to create a high gradient magnetic field, which sequesters the magnetically labeled cells and retains them on the column while the unlabeled cells pass right through. After washing to remove any remaining antibody negative cells, the column is removed from the magnet and the positive cells are eluted. Subsequent cell staining for quality control can be performed by FACS.

Two types of matrices for separation columns are available. The positive selection matrix column is composed of a spherical type matrix with a ferromagnetic core. There is also a column made with a ferromagnetic fibrous matrix, which is designed for depletions. All column matrices are coated with a cell friendly plastic allowing fast and gentle cell selections. Although the column matrices are designed for positive selections (enrichments) or negative selections (depletions), the columns can be optimized for efficient recovery of both fractions. Both types of columns are available in various sizes. The column size is selected based on the number of magnetically labeled cells expected to be retained on the column matrix in addition to the total number of cells to be loaded. Columns are available for retention capacities ranging from 10^7 to 10^9 cells and total load capacities of 2×10^8 to 2×10^{10} cells. Columns are packaged sterile and ready to use.

Extremely rare cells can be highly enriched with this methodology.[11] Even cells with low surface expression can be separated from those with high surface expression by titrating the antibody system for appropriate binding, as demonstrated in the following protocol for heat stable antigen titration.

Removal of dead cells, red cells, and debris

Dead cells and debris can increase sorting time and disturb scatter patterns and staining profiles. Their elimination is therefore important. Dead cells often take up labels nonspecifically due to the DNA they release and therefore have an increased autofluoresence. Released DNA can entrap live cells into aggregates, which will be filtered or gated out of the picture, or worse, may clog the cytometer/sorter. Dead cell removal prior to staining is especially important because by the time you get to the sorter, it may be too late to get rid of interfering cells that may overlap or obscure the population you are trying to sort. This can be accomplished by using various ficoll type support gradient media, which requires a one step centrifugation.[12] These are commercially available (Accurate, Nycomed), fairly inexpensive, and quick. The only things to keep in mind are 1) density changes with temperature, so don't expect the same separation at 0°C as you get at room temperature; 2) different cell types have different buoyant densities, so they require different density gradient medias; 3) the gradient media needs to be thoroughly washed away prior to subsequent steps to maintain cell health and prevent cell loss.

Red blood cells may also be depleted from cell preparations using gradient media. Formulations vary and you should pick one to suit your species and is compatible with your cell type. Another way to get rid of unwanted red blood cells is to lyse them specifically with hypotonic shock, tris-buffered ammonium chloride, or hemolytic Gey's solution.[7]

Not only can you physically deplete, but you can also use the cytometer to gate out dead cells and debris with your flow cytometer hardware by setting thresholds on raw data prior to computerization so that the cells are never counted in the initial data. Essentially one is blindfolding the computer from seeing the small particles. This can be used to eliminate machine noise as noted in section 2 of this book. You should not rely on hardware gating to eliminate dead cells and debris. The old saying still applies- "garbage in, garbage out". The dead cells can overwhelm your analysis without viability discrimination or elimination in one form or another.

Subprotocol 1
Enrichment by Panning on Antibody Coated Plates

▓▓ Materials

Solutions

Complement lysis buffer (CLB)
- 25 mM Hepes buffer pH 7.4
- 1 mM NaN_3
- 2 mg/ml BSA fraction V
- 50 µg/ml DNase
- medium of choice (RPMI 1640 for lymphocytes)

Preparation
- Dilute purified goat anti-rat IgG (H+L) in PBS to 100 µg/ml.

▓▓ Procedure

This protocol is for enrichment by depletion, although it may also be used for selecting positive cells if you use a ten-fold dilution of the goat anti-rat IgG to coat the plates in order to facilitate removal of the cells from the plate.

Prepare reagents

1. Prepare reagents
   Complement lysis buffer (CLB)

Prepare antibody coated plates

2. Add 4 mls of dilute goat anti-rat antibody to 100 mm polystyrene bacteriological petri dish.

You can also use 150 mm plates polystyrene plates, but you need to increase the volume of antibody solution to 10 ml.

3. Swirl to coat plate until the plate is evenly coated all around and the liquid loses surface tension.

4. Incubate one hour at room temperature or overnight at 4°C.

5. Pour off anti-rat antibody.

6. Wash plate 4 times with PBS and store with 4 mls of PBS to keep the surface wetted.

7. Harvest mouse thymi.

Use 4-6 week old mice. We typically harvest 2-3 x 10^8 thymocytes per thymus.

Prepare mouse thymocytes

8. Prepare single cell suspension in CLB by mincing and pushing thymocytes through stainless steel wire mesh fabric.

9. Centrifuge at 300 x g for 10 minutes at 4°C.

10. Resuspend cells in 5 mls CLB and count on a hemacytometer. Aliquot 5 x 10^6 cells for control staining

11. Centrifuge as above. Resuspend cells in 1 x 10^7/ ml of rat anti-mouse CD25 (clone 7D4, ATCC) hybridoma supernatant.

Label thymocytes with rat anti-mouse CD25

You can use affinity purified, or ascites as well as hybidoma cell supernatant. Titrate for antibody binding by FACS prior to the experiment using dilutions of CD25 and revealing with a fluorescent goat anti-rat IgG secondary antibody.

12. Incubate 40 minutes on ice.

13. Centrifuge as above and resuspend in 4 x 10^7 cells/ml in CLB.

14. Add 5 mls of cells to each plate. Swirl gently.

If you use 150 mm plates you can add 10 mls of cells.

Plate out cells on antibody coated plate.

15. Incubate at room temperature for 30 minutes.

16. Gently swirl and pour off cells into a second coated plate.

17. Incubate cells at room temperature for 30 minutes.

Harvest depleted cells and plate a second time

18. Gently swirl and pour off into a third coated plate.

19. Incubate cells at room temperature for 30 minutes.

Harvest depleted cells and plate a third time

Harvest depleted cells

20. Swirl and pour off cells into a 50 ml polypropylene centrifuge tube.

Wash cells

21. Add CLB to fill tube and centrifuge as above.

Count cells

22. Resuspend cells in 5 mls of CLB and count cells on a hemacytometer for cell number and viability.

23. Aliquot 1×10^6 cells for quality control by FACS

Stain for Quality Control

24. Stain aliquots of fractionated and unfractionated thymocytes for quality control.

We use pretitrated CD8 FITC, CD4 Cychrome and CD25 PE (PharMingen), in a one step staining for quality control staining of mouse thymocytes.

Subprotocol 2
Procedure for Lectin Panning with Peanut Agglutinin (PNA)

▓▓ Materials

Solutions

- 50mM Tris-HCl with 15M NaCl, pH9.5 (Tris/NaCL-9.5)
- Phosphate buffered saline with 0.6mM $MgSO_4$ and 1.3 mM $CaCl_2$, pH 7.3 (PBS+)
- 1% fetal calf serum in PBS+
- 5% fetal calf serum in PBS+
- *1 mg/ml Peanut Agglutinin (PNA)* (Vector Laboratory) dissolved in Tris/NaCl-9.5 – *Wear gloves to handle PNA with care. PNA is toxic.*
- 0.2M D(+)-Galactose in PBS(+)Preparation
- Hank's Balanced Salt Solution without phenol red containing 2 mg/ml Bovine Serum Albumin Fr. V (HBSS/BSA)

▓▓ Procedure

Prepare reagents for panning

1. Prepare reagents for panning.

2. Dilute out PNA to 10μg/ml in Tris/NaCl-9.5 (25 mls is enough to cover six non-tissue culture petri dishes at 4 mls per plate).

Prepare PNA plates

Note: *Wear gloves and handle PNA with care. PNA is toxic.*
10 μg/ml PNA gives better results for PNA (+) fraction purity. If you need the PNA (-) fraction purer, use 40 μg/ml.

Note: This procedure can be used with any lectin, but you should titrate the amount of lectin needed for your purposes. For best results, make plates the day before. This allows the FBS proteins to bind up non-specific binding sites on the plastic. If you are in a hurry, leave the FBS on the plate at least one hour. PNA working solution can be reused several times before disposal.

3. Add 4 mls 10μg/ml PNA to each plate. Swirl to coat plate until the plate is evenly coated all around and loses surface tension.

Bind PNA to plates

4. Allow to sit on plate for 1 hour at room temperature.

5. Pipet solution off and store at 4°C for reuse.

6. Wash PNA plates three times each with 4 mls of PBS (+).

Wash off excess PNA

7. Wash plate once with 1% FBS in PBS (+).

8. Add 4 mls 1% FBS in PBS(+) and leave overnight at 4°C.

Bind up non-specific sites on plate

9. Harvest mouse thymi.

Prepare mouse thymocytes

Note: Use 4-6 week old mice. We typically harvest 2-3 x 10^8 thymocytes per thymus. CLB is complement lysis buffer (see complement elimination protocol for recipe).

10. Prepare single cell suspension in CLB by mincing and pushing thymocytes through stainless steel wire mesh fabric.

11. Centrifuge at 300 x g for 10 minutes at 4°C.

12. Resuspend cells in 5 mls CLB and count on a hemacytometer. Aliquot 5 million cells for control staining.

Count cells

13. Centrifuge as above, resuspend cells at 1 x 10^7/ml in 5% FBS in PBS+.

Ready PNA plates

14. Pour off 1% FBS/PBS+ from PNA plates.

Plate thymocytes on PNA

15. Add 4 mls of cells (4×10^7 cells) to the PNA plates.

Note: Plates must be cold and on a flat, level surface so that the cells do not accumulate and pile on top of each area in one area of the plate.

Allow cells to bind

16. Incubate plates on a flat and level surface for 70 minutes at 4°C.

17. Swirl and tap at least 2 times gently at 15 minute intervals to assure that all cells have an opportunity to sit on the plate surface and attach.

Harvest PNA- fraction

18. After incubation gently swirl plates and collect the non-adherent cells (PNA- fraction).

Wash and combine PNA-fractions

19. Wash plates gently 3 times with 1%FBS in PBS+ and add to the PNA (-) fraction.

Note: The trick here is to be gentle enough not to disturb bound cells, but strong enough to make sure the unbound cells wash off cleanly.

Discard marginal cells

20. Wash plates vigorously 2 times with 1% FBS in PBS+ and discard the washes.

Add competitive sugar to remove bound cells

21. Add 5 ml of 0.2M D- (+) galactose in PBS+ to the plates.

Incubate to allow competition

22. Let plates stand at room temperature for 10 minutes.

Harvest PNA+ fraction

23. Swirl roughly and squirt the cells off the plate by pipetting against the plate to remove the PNA (+) cell fraction.

24. Collect the cells into a 50 ml polypropylene centrifuge tube.

Wash and combine PNA+ fractions

25. Wash the plates twice more with 5 ml of galactose using vigorous pipetting and combine with the PNA (-) cell fraction.

26. Add PBS+ to fill both fractions' tubes. Mix well. Centrifuge fractions as above.   **Wash cell fractions**

27. Wash fractions once more in PBS+.

28. Resuspend each fraction in 5 ml HBSS/BSA.

29. Aliquot cells from each fraction for quality control staining on FACS.

30. Stain unfractionated and PNA panned fraction aliquots with cocktail to evaluate separation.   **Stain for Quality Control Analysis on FACS**

Note: We use pretitrated CD8 FITC and CD4 PE (PharMingen) in a one step staining for quality control staining of mouse thymocytes.

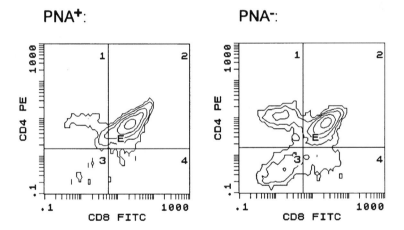

PNA⁺:   **PNA⁻:**

Fig. 1. Thymocytes from $C_{57}Bl_6$ mice (4-6 week old) were panned for binding to the lectin, peanut agglutinin as described in the protocol. The separate harvested fractions were stained with CD8-fitc and CD4-PE to evaluate the separation purity. This is a common method to enrich for double positive thymocytes. The figure represents a typical fractionation pattern for these cells. Note the preponderance of double positive cells and the decrease of single positive cells in PNA⁺ fraction as compared to the PNA⁻ fraction and the unseparated pattern seen in Figure 3 later in this chapter. The PNA⁻ fraction is very difficult to obtain cleanly because a good fraction of the cells sit on top of the layer of cells that are binding to the plastic and never have the chance to bind properly. Thus this fraction is usually contaminated with cells that have the PNA binding capacity. Only through multiple rounds of panning will the PNA⁻ fraction obtain a degree of purity

Subprotocol 3
Procedure for Complement Mediated Cell Lysis

▓▓ Materials

Equipment

Circulating 37 water bath
Centrifuge

Solutions

Complement lysis buffer (CLB)
- 25 mM Hepes buffer pH 7.4
- 1 mM NaN$_3$
- 2 mg/ml BSA fraction V
- 50 µg/ml DNase
- medium of choice (RPMI 1640 for lymphocytes)

▓▓ Procedure

Prepare reagents

1. Prepare complement lysis buffer

2. Thaw pretitrated baby rabbit complement on ice. Dilute to 1:10 in CLB (2x stock).

Prepare cells

3. Make 2 fold concentrated, monodispersed, cell preparations in CLB at 1×10^7 cells/ml.

Note: Make one cell sample for the elimination and a smaller one for a mock elimination to check for cytotoxicity and to provide positive staining controls for quality control analysis.

Bind antibody to cells

4. Add antibody specific for cell surface antigen to the cell sample to be eliminated.

Note: The antibody should be pretitrated in a pilot experiment and evaluated by FACS staining if possible. As a rule use the antibody neat if you are using a hybridoma supernatant or use the concentration that you use for indirect FACS staining if you are using ascites, concentrated, or affinity purified antibody.

The azide in the CLB prevents the cell surface molecule from modulating off the cell when the antibody is bound.

5. Add CLB to the mock elimination sample.

6. Incubate on ice 5-10 minutes.

7. Add an equal volume of prewarmed 2x concentrated complement to the labeled and mock cell suspensions. Mix well.

Add complement

Note: Complement should be pretested for cytotoxicity and titrated against a known antibody prior to use in an experiment. We use baby rabbit complement (Pelfreeze). Adult rabbit complement is usually cytotoxic. Cedarlane offers cytotoxicity pretested complement. Complement can also be used from guinea pig.

8. Incubate at 37°C in a circulating water bath for 1 hour.

Incubate

9. Centrifuge cells at 300 x g for 10 minutes. Resuspend cells and wash in a full tube of CLB. Centrifuge again and resuspend in 5 mls of CLB.

Wash

10. Count cells on hemacytometer with eosin or trypan blue for viability.

Check cell viability

11. Eliminate dead cells using a step gradient cushion. (see step gradient protocol)

Remove dead cells

12. Check for quality control by FACS staining to see if you have eliminated the population that you labeled. Use appropriate fluorochrome conjugated second antibody to detect cells bound but not lysed. It is always good to include a sample stained with directly conjugated antibodies to other markers for the cells that you wanted to eliminate as a check on the antibody/complement elimination efficiency.

FACS Analysis

Subprotocol 4
Enrichment of lin⁻ Cells from Fetal Liver by MACS

▓ ▓ Materials

Equipment

Miltenyi Biotec VarioMACS Magnet

Solutions

PBS with 2% Fetal Bovine Serum (FBS)

1 x Hank's Balanced Salt Solution (HBSS/BSA) – column and staining buffer consists of:
- 50 ml 10 x HBSS with azide
- 1.25 g Bovine Serum Albumen (frac.V)
- 5 ml 1M Hepes buffer pH 7.0
- Distilled water to 500 ml.

10 x HBSS with azide (1L) consists of:
- 4.0 g KCL
- 0.6 g KH_2PO_4
- 80 g NaCl
- 10 g glucose
- 0.475 g Na_2HPO_4
- 3.25 g $NaHCO_3$
- 3.25 g NaN_3 adjusted to pH 7.6
- filter through 0.22 micron sterile filter.
  You may substitute PBS/BSA pH 7.2 if your cells prefer it.

Preparation

MACS lin⁻ staining cocktail consists of biotin labeled antibodies for:
- Ter119 (PharMingen)
- B220 (Caltag Laboratories)
- GR-1 (PharMingen)
- CD3ε (PharMingen)
- CD8 (PharMingen)
  All antibodies are pretitrated for saturation binding and diluted in HBSS/BSA staining buffer for a 10 x stock. (Most of these are used at 1:40 dilution)

FACS staining cocktail for stem cells consists of the appropriately titered antibodies:
- Sca-1 FITC conjugate (PharMingen)
- CD117 PE conjugate (PharMingen)
- Streptavidin-Red670 (Life Technologies)

▨ ▨ Procedure

1. Prewash VS+ column with 3 mls sterile HBSS/BSA. Chill at 4°C.

   Prepare separation column and buffer

Note: Columns are sterile and coated with a protective polymer that washes off readily in column buffer.
Column buffer/staining buffer is HBSS/BSA with sodium azide or PBS/BSA pH 7.2.

Note: The magnet should be located in a cold room or in a chromatography cabinet at 4°C. There are several variations of magnets and adaptors for the columns depending on the size and magnetic field required. Check the magnet manual for column sizes, adaptors and set up.

2. Harvest day 14 mouse fetal liver.

   Prepare fetal liver cells

3. Homogenize liver in 40 ml PBS with 2% FBS by dissociation between frosted glass slides.

4. Filter cells through 50 micron stainless steel wire mesh fabric (Tetko). Collect filtrate in 50 ml polypropylene centrifuge tube.

5. Centrifuge cells at 300 x g for 10 minutes at 4°C. Discard supernatant.

6. Resuspend cell pellet in 40 ml PBS with 2% FBS.

7. Filter through 30 micron nytex nylon mesh fabric (Tetko) to remove dead cell clumps.

Note: Make sure cells are in a mono-dispersed suspension for good staining and column action. Clumps will clog column and dead cells will be sticky and take up antibody and magnetic beads non-specifically and trap live negative cells on the column.

8. Count cells with eosin-y or trypan blue on a hemacytometer.

9. Transfer 1×10^8 cells to a 3 ml glass conical staining tube. Centrifuge 300 x g for 10 minutes at 4°C.

Block cells for Fc receptor binding

10. Resuspend cell pellet in 1 ml neat rat antimouse Fc receptor monoclonal antibody supernatant (clone 2.4G2 from ATCC). Incubate on ice for 5 minutes. (Be sure to reserve on ice several million cells for control stains).

Note: This antibody is also sold as Fc Block by PharMingen. If you are planning to use an anti-rat secondary antibody on the cells then this is not recommended. Instead substitute mouse serum or mouse IgG.

Bind lin⁻-antibody cocktail to cells

11. Add 100µl volume of a biotin conjugated cocktail of lineage negative antibodies which is 10 times concentrated. (If you normally use 1:40 antibody dilution then use 1:4).

Note: All antibodies are pretitrated for saturation binding and diluted in HBSS/BSA staining buffer for a 10 x stock. (Most of these are used at 1:40 dilution).

12. Mix well and incubate on ice for 40 minutes.

Wash cells

13. Fill tube to dilute with HBSS/BSA. Mix by vortexing.

14. Centrifuge at 300 x g for 10 minutes at 4°C.

15. Resuspend cell pellet in 0.9 ml HBSS/BSA.

Label with magnetic bead conjugated streptavidin

16. Add 0.1 ml magnetic beads conjugated with streptavidin (Miltenyi Biotec, Auburn CA).

Note: Be sure to follow the package insert instructions for cell number, and volumes for your particular usage as different types of beads may vary.

17. Incubate for 15 minutes on ice.

18. Place VS+ separation column in magnet.

Separate cells on matrix column

19. Load cells onto top of column.

Harvest lineage negative cells

20. Collect effluent.

21. Wash column 3 times with 3 ml of HBSS/BSA and add to collected effluent.

22. Count collected effluent cells with eosin-y or trypan blue on hemacytometer.

23. Centrifuge cells 300 x g for 10 minutes at 4°C. Discard the supernatant. Resuspend in FACS staining cocktail with enough volume appropriate to stain the number of cells in step 22 for stem cells.

Stain cells for sorting stem cells on FACS

Note: Antibodies are diluted in HBSS/BSA.

Note: Be sure to stain control cells (unseparated cells for each single color to set up FACS machine and compensation.

24. Incubate at room temperature in the dark for 25 minutes.

25. Dilute cells to wash with 1 ml HBSS/BSA. Add 0.3 ml FBS to the bottom of the tube to create a cushion gradient and centrifuge the cells through the FBS at 300 x g for 10 minutes at 4°C. (Do not pour off.)

Wash cells

26. Resuspend the cells in 0.5 ml of HBSS/BSA and filter through 30 micron nytex nylon fabric mesh to get rid of cell aggregates.

27. Set up FACS for 488nm Argon excitation. Collect FITC, PE, and Red670 spectrum parameters.

28. Run single color stained cells separately to set scatter settings for viable cells and set backgrounds for each color in the first decade. Adjust compensation percentages for the 3 color channels.

Run controls on FACS

29. Set up 2D dot plot histograms for FITC versus PE, FITC versus Red670, and PE versus Red670. Adjust compensation percentages for all 3 channels such that the signals are orthogonal to each other as you run all three singly stained control samples individually.

30. Run 3 color sample analysis for 50,000 cells. Gate on viable cells by scatter and gate away any Red670 cells.

31. Set sort gates for fetal liver stem cells on Sca-1$^+$-FITC, CD117 $^+$-PE, Streptavidin-Red670 $^-$ (lin$^-$-biotin) cells.

Sort cells

Note: The stem cells should stain positive for Sca-1 and CD117 (c-kit receptor) and negatively for any biotin labeled lineage negative cells that were not taken out from the magnetic bead separation.

BEFORE MACS: AFTER MACS:

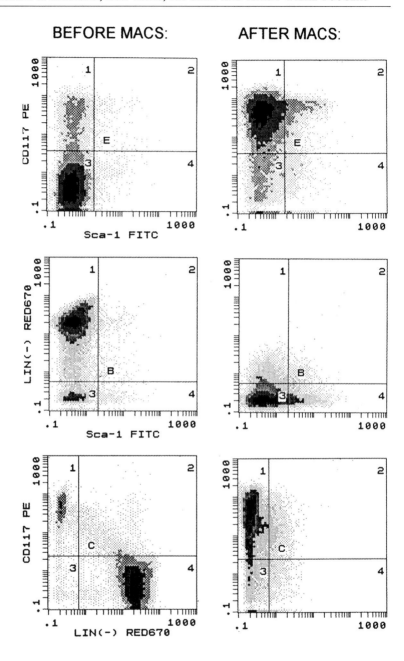

◀ **Fig. 2.** Murine day 14 fetal liver cells were harvested from $C_{57}Bl_6$ mice. They were fractionated by staining with a mature hematopoetic lineage cocktail of biotinylated antibodies prior to binding with avidin-labeled magnetic beads and separating on a magnetic bead column magnet as described in the protocol. This protocol is commonly used to obtain hematopoetic stem cells from fetal liver and also adult bone marrow. This figure is an example of the enrichment for these stem cells as evaluated in the post fractionation plots by the disappearance of lineage cocktail markers (lin(-)) and the increased presence of the stem cell markers CD117 (c-kit) and Sca-1 (stem cell antigen)

32. Collect into chilled FBS coated tubes.

33. Rerun an aliquot of sorted cells for purity.

Subprotocol 5
Procedure for MACS Separation of Antigen^{low} cells from Antigen^{hi} Cells

▨▨ Materials

Equipment

Miltenyi Biotec VarioMACS Magnet.

Solutions

CLB (see page 118)

1 x Hank's Balanced Salt Solution (HBSS/BSA) – column and staining buffer consists of:
- 50 ml 10 x HBSS with azide
- 1.25 g Bovine Serum Albumen (frac.V)
- 5 ml 1M Hepes buffer pH 7.0
- Distilled water to 500 ml

10 x HBSS with azide (1L) consists of:
- 4.0 g KCL
- 0.6 g KH_2PO_4
- 80 g NaCl
- 10 g glucose
- 0.475 g Na_2HPO_4

- 3.25 g $NaHCO_3$
- 3.25 g NaN_3 adjusted to pH 7.6
- filter through 0.22 micron sterile filter.
  You may substitute PBS/BSA pH 7.2 if your cells prefer it.

▨ ▨ Procedure

Prepare Reagents and VS+ separation columns

1. Prewash 6 each VS+ separation columns with 3 mls of HBSS/BSA. Chill at 4°C.

Note: You may substitute PBS/BSA pH7.2 if your cells prefer it.

PREPARE THYMOCYTES

2. Dissect and harvest thymus from 4-6 week old mice.

Note: Up to 10 normal thymi may be prepared in batch using this preparation protocol.

3. Place in 10 ml CLB with 20µg/ml DNAse.

4. Mince thymus and push cells through stainless steel wire mesh fabric to release thymocytes into 100mm petri dish.

5. Collect cells by quantitative transfer into a 50 ml conical polypropylene centifuge tube using 40 ml of CLB.

6. Centrifuge 300 x g for 10 min at 4°C.

7. Resuspend in 5 ml CLB and count cells on hemacytometer using eosin-y or trypan blue for viability and cell number.

8. Add 8 x 10^7 cells/ml to each of six 50ml centifuge tubes.

Label cells with antibody dilutions

9. Add an equal volume of antibody dilution to the appropriate pairs of tubes so that the final concentrations of antibody are 1:5, 1:20, and 1:100.

Note: The concentration of antibody that is used for cell labeling is important to separate low antigen expressing cells from high antigen expressing cells. For positive selection, lower concentrations of antibody yield better purity, but poor recovery of the cells. Higher concentrations give better recovery but poorer purity. For negative selection higher concentrations of antibody yield better purity, but poor recovery. Lower concentrations give better recovery but poorer purity. By titrating the labeling antibody with at least 3 dilutions you should be able to determine how much antibody to use. In this sample experiment we are titrating anti-mouse heat stable antigen hybridoma supernatant (M1/69 from ATCC).

Make 2 x concentrated antibody dilutions so that equal volume to cell volume will give appropriate dilution.

10. Incubate cells at 4°C for 30 minutes.

11. Centrifuge at 300 x g for 10 minutes at 4°C.

12. Resuspend cells in 1 ml HBSS/BSA . Centrifuge as above.

13. Resuspend cells in 90 µl or 95µl of HBSS/BSA per 10^7 cells.

14. Add either 5 or 10 µl MACS Microbead labeled secondary antibody (anti rat IgG).

Magnetic bead labeling

Note: The amount of magnetic beads bound to the labeled cells may also affect yield. In this case we are also testing a two fold dilution of microbeads – 5 & 10 µl.

Note: The normal recommendation for anti rat IgG microbeads is a 1:5 dilution, but in this case we are titrating antigen^{low} cells so we need to be in a range where the low cells do not bind.

15. Mix well. Incubate at 4°C for 15 minutes in the refrigerator.

16. Wash twice by diluting with 3 ml buffer and centrifuging as above.

17. Resuspend cells in 3 mls ice cold HBSS/BSA. Filter through 30 micron nytex swiss nylon mesh fabric.

Note: Be sure not to have any clumps of cells that may clog up the column and ruin yield.

18. Count cells (N_{total}).

19. Insert prechilled, prewashed VS+ column into magnet and ready collection tube.

20. Pipet dilution sample onto column.

Load cells on column

21. Allow sample to run into the column while collecting efflu-ent. Wash column 3 times with 1 ml HBSS/BSA. Collect all effluent as the negative fraction.

Collect antigen^{low} fraction

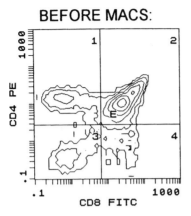

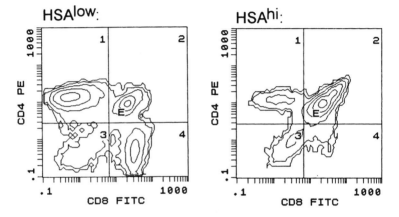

Fig. 3. The object of this enrichment is to isolate a population of cells that express a low amount of an antigen, in this case a heat stable antigen low (HSA^{low})cells. Adult (4-6 week old) thymocytes were harvested from C$_{57}$Bl$_6$ mice, labeled with a titrated amount of antibody, then reacted with anti-rat IgG tagged magnetic beads and separated over a magnetic bead column magnet as described. The resulting cells were evaluated by staining with subpopulation markers, in this case CD4PE and CD8FITC. Note the increase of single positive markers and the decrease in the double positive population in the HSA^{low} separated population. The majority of cells in the unfractionated adult thymus are HSA^{hi} which is why the HSA^{hi} cells greatly resemble the unfractionated population

22. Wash column 3 times with 3 ml HBSS/BSA. Discard.

23. Remove column from the magnet. Place column over a new collection tube. Apply 5 ml buffer to the column and flush out cells with the column plunger. Collect these cells as the positive fraction.

Collect Antigen^{high} fraction

24. Spin down all cell fractions as above. Resuspend in 2 ml and count cells (N $_{fraction}$).

Centrifuge cells

25. Calculate recovery as follows:
R=(N$_{fraction}$) / (N$_{total}$) where:
N$_{fraction}$ is the number of cells in each fraction
N$_{total}$ is the number of cells applied to column

Calculate recovery

26. Aliquot 1×10^6 cells from each fraction to stain with fluorescent labeled antibodies to check for cell purity by FACS.

Check cell purity by FACS

Note: For this heat stable antigen separation we use CD8 FITC and CD4 PE (PharMingen) to evaluate the enrichment. Antibodies are diluted in HBSS/BSA.

Subprotocol 6
Procedure for Splenic T-Cell Enrichment by MACS

▓▓ Materials

Equipment

Miltenyi Biotec VarioMACS Magnet.

Solutions

Complement Lysis Buffer (CLB, see page 118)

$1 \times$ Hank's Balanced Salt Solution (HBSS/BSA) – column and staining buffer consists of:
- 50 ml 10 x HBSS with azide
- 1.25 g Bovine Serum Albumen (frac.V)
- 5 ml 1M Hepes buffer pH 7.0
- Distilled water to 500 ml.

10 x HBSS with azide (1L) consists of:
- 4.0 g KCL
- 0.6 g KH_2PO_4
- 80 g NaCl
- 10 g glucose
- 0.475 g Na_2HPO_4
- 3.25 g $NaHCO_3$
- 3.25 g NaN_3 adjusted to pH 7.6
- filter through 0.22 micron sterile filter.
  You may substitute PBS/BSA pH 7.2 if your cells prefer it.

Preparation

MACS antibody cocktail consists of:
- Ter119 biotin conjugate (PharMingen, San Diego). Tags erythrocyte lineages.
- B220 biotin conjugate (Caltag). Tags B cells.
- GR-1 biotin conjugate (PharMingen). Tags granulocytes and macrophages. Antibodies are diluted in HBSS/BSA.

Stain for T-cell enrichment with the following FACS antibody cocktail:
- B220 APC conjugate (PharMingen)
- CD3 PE conjugate (PharMingen)

▧ ▧ Procedure

Prepare Reagents and VS+ separation columns

1. Prewash a VS+ separation column with 3 mls of HBSS/BSA. Chill at 4°C.

   Note: You may also use PBS with BSA at pH 7.2 if your cells prefer in place of HBSS.

Dissect spleens

2. Dissect and harvest spleen from at least 2 month old mice.

3. Place in 10ml CLB with 20 µg/ml DNase.

Harvest spleen cells

4. Mince spleens and push cells through stainless steel wire mesh fabric to release lymphocytes into 100mm petri dish.

5. Collect cells by quantitative transfer into a 50 ml conical polypropylene centrifuge tube using 40 ml of CLB.

6. Centrifuge 300 x g for 10 min at 4°C.

7. Resuspend in 5 ml CLB, filter through 30 micron nytex nylon mesh fabric.

Note: Be sure not to have any clumps of cells that may clog up the column and ruin yield.

8. Count cells on hemacytometer using eosin-y or trypan blue for viability and cell number.   Count cells

9. Add 1×10^8 cells to 15 ml conical centifuge tube. Centrifuge as above.

10. Resuspend cells in 1 ml of anti-mouse Fc Receptor hybridoma supernatant (2.4G2, ATCC). Incubate 10 minutes on ice.   Block Fc receptors on B cells

11. Add MACS antibody cocktail. Mix well.   Label cells with antibody

Note: In this case we are attempting to label all cells except T cells with biotin antibodies to cell surface molecules.

12. Incubate on ice for 40 minutes.   Incubate to bind antibody

13. Wash cells by filling tube to dilute cells with HBSS/BSA and centrifuge at 300 x g for 10 minutes at 4°C.   Wash cells

14. Resuspend cells in 0.9 ml HBSS/BSA.

15. Add 100µl MACS Microbead labeled Streptavidin (Miltenyi Biotec, Auburn CA).   Magnetic bead labeling

Note: Streptavidin MACS microbeads bind to the biotinylated antibodies.

16. Mix well. Incubate at 4°C for 15 minutes.   Incubate sample

17. Filter through 30 micron nytex swiss nylon mesh fabric.

Note: Be sure not to have any clumps of cells that may clog up the column and ruin yield.

18. Insert prechilled, prewashed VS+ column into magnet and ready collection tube.   Ready column

Load cells on column and collect T-cell fraction

19. Pipet sample onto column.

20. Allow sample to run into the column while collecting effluent. Wash column 3 times with 3ml HBSS/BSA. Collect all
effluent as the negative fraction.

Note: This is known as depletion or negative selection.

Check cell purity by FACS

21. Aliquot 1 x 10^6 cells from each fraction including unfractionated to stain with fluorescent labeled antibodies to check
for cell purity by FACS. Be sure to include single color controls to set compensations for the FACS.

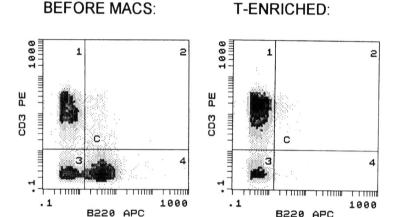

Fig. 4. The object of this enrichment is to purify splenic T cells away from splenic B cells. Adult murine spleen was harvested and the cells labeled with a B cell
and macrophage specific cocktail of biotinylated antibodies. The cells were then
reacted with Streptavidin-magnetic beads and run over a magnetic bead column
as described. The resulting T-enriched fraction was stained with CD3-PE (a T-
cell specific marker) and anti-B220-APC (a B-cell specific marker). Note the disappearance of the B cells and the increase in the T-cell staining pattern from the
T-enriched sample

Subprotocol 7
Procedure for Red Blood Cell Lysis

▓▓ Materials

Solutions

Stock Solutions for RBC lysis buffer:
- 0.16 M NH_4Cl - dissolve 83g in 1L H_2O.
- 0.17M Tris-HCl pH 7.65 - dissolve 20.6g Tris base in 900 ml H_2O. Adjust pH to 7.65 with HCl and bring to 1 Liter.

Working solution: Mix 90 ml of 0.16 M NH_4Cl with 10 ml of 0.17 M Tris-HCl. Adjust to pH 7.2 with HCl
  Growth media RPMI 1640 with 5% FBS

▓▓ Procedure

1. Prepare stock and working solutions.

Prepare solutions

Note: There are several kinds of red blood cell lysis buffers that you can purchase or make yourself. This protocol is an example of the use of one such buffer.

2. Collect cells and centrifuge at 300 x g for 10 minutes at 4°C.

3. Resuspend cells (0.1 ml packed cells) in 1 ml Tris-NH_4Cl working solution.

Treat cells with lysis buffer

4. Incubate cells at room temperature for 2-3 minutes while pipetting up and down.

5. Underlay cells with 0.5 ml bovine serum.

Remove cells from buffer

6. Centrifuge cells through the serum at 300 x g for 10 minutes at 4°C.

Note: The serum provides a separation barrier between the cells and the lysis buffer and stops the reaction.

7. Repeat steps 3-6 if pellet is still red.

8. Wash once with growth medium. Resuspend cells in 1 ml medium and count with eosin-y or trypan blue on a hemacytometer.

Count cells for viability and cell number

Subprotocol 8
Procedure for Step Gradient Separation for Viable Cells

▓▓ Materials

Solutions

Growth media RPMI 1640 with 5% FBS

Step Gradient Medium

Note: Lympholyte-M (Cedarlane) for mouse lymphocytes.

Note: Ficoll-Hypaque (Pharmacia) for human lymphocytes.

Note: Check literature for recommended gradient media for your specific cell type.

Note: Warm step gradient media and medium of choice to room temperature. Remember that density changes with temperature.

▓▓ Procedure

Prepare cells
1. Prepare monodispersed cell suspension in a 15 ml polystyrene or polybutadiene (Nunc) conical centrifuge tube.

2. Count cells on a hemacytometer with eosin-Y or trypan blue for viability and cell number.

Note: For larger numbers of cells use a 50 ml tube. We find that it is harder to harvest cells from the larger tubes however and separation is not as good.

3. Centrifuge at 300 x g for 10 minutes.

4. Resuspend in 5 ml of room temperature growth medium between 0.5–10 x 10^7 cells per ml.

Load step gradient
5. Underlayer cells with 5 mls of room temperature step gradient medium.

Centrifuge
6. Centrifuge at 300 x g for 10 minutes.

Separate live cells from dead
7. Carefully remove live cells laying on the interface between the gradient medium and the growth medium to a fresh tube. Discard gradient tube. Try to keep the amount of gradient media to a minimum when you harvest your cells.

8. Wash cells with by filling up the tube room temperature growth medium. Mix well.

Wash away gradient medium

Note: Be sure to mix washes well to dilute gradient medium because the gradient media will tend to settle at the bottom and stay behind with the cells. Most cells do not tolerate the gradient medium for extended lengths of time.

9. Centrifuge 300 x g for 10 minutes.

10. Repeat steps 8 & 9 two times.

11. Count cells on a hemacytometer with eosin-y or trypan blue to check efficiency of dead cell depletion.

References

1. Rothenberg EV. Cell separation and analysis: a strategic overview. In: Diamond RA, Rothenberg EV, eds. Methods: A Companion to Methods in Enzymology 1991; 2(3):168-172.
2. Fresney RI. In: Fresney RI, ed. Culture of Animal Cells: Manual of Basic Techniques. Alan Liss Inc., 1983.
3. Morasca L, Erba E. In: Fresney RI, ed. Animal Cell Culture, a Practical Approach. IRL Press, 1986:137.
4. Pretlow TG, Pretlow TP. Sedimentation for the separation of cells. In: Diamond RA, Rothenberg EV, eds. Methods: A Companion to Methods in Enzymology 1991; 2(3):184-185.
5. Wysoki WL, Sato VL. "Panning" for lymphocytes: a method for cell selection. Proc Natl Acad Sci 1978; 75:2844-2848.
6. Mage MG. In vitro assays for mouse B and T cell function. In: Coligan JE, Kruisbeck AM, Marguilies DH et al. eds. Current Protocols in Immunology 1991:1:3.52-3.56
7. Mishell BB, Shiigei SM. In: Mishell BB, Shiigei SM eds. Selected Methods in Cellular Immunology. Freeman 1980:173-234.
8. Hathcock KS. T dell depletion by cytotoxic elimination In: Coligan JE, Kruisbeck AM, Marguilies DH et al. eds. Current Protocols in Immunology 1991:1:3.41-3.43
9. Miltenyi S, Pfluger E. High gradient cell sorting. In Radbruch A, ed. Flow Cytometry and Cell Sorting. Springer Verlag, 1992:141-152.
10. Miltenyi S, Muller W, Weichel W. et al. High gradient cell separation with MACS. Cytometry 1990; 11:231-238.
11. Radbruch A, Rechtenwald D. Detection and isolation of rare cells. Current Opin Immun 1995; 7:250-273.
12. Boyum A, Berg T, Blomhoff R. In: Richwood D, ed. Iodinated Densisty Gradient Media: A Practical Approach. IRL Press 1983:156.

The Basics of Staining for Cell Surface Proteins

ALICE L. GIVAN

Introduction

A flow cytometer detects fluorescent light given off by cells. Although flow cytometry can be used to analyze the intrinsic fluorescence of cells ("autofluorescence"), cells differ from each other in many ways that do not result in autofluorescent variation. For this reason, staining of cells is usually required to make biochemical variation "visible" to the photodetectors of a flow cytometer. By making use of fluorescent stains that are specific for different molecules, we can use a flow cytometer to assay a large array of different cell constituents.

Antibodies represent what are arguably the major staining agents for flow cytometric analysis. In this chapter, we will describe what is currently the major application for antibody staining: phenotyping cells according to their expression of surface membrane proteins. Mice (or other animals) can be encouraged to produce mouse (or other) antibodies against foreign proteins. The binding of these antibodies to their reciprocal molecules ("antigens") is strong and effectively irreversible. As a result of monoclonal antibody technology, antibodies are now available that bind specifically to many cellular proteins. Many cell proteins for which monoclonal antibodies are available have been given a "CD" number, so that there are now over a hundred proteins on leukocytes with their own CD numbers (x); each of these proteins can be stained specifically by its reciprocal anti-CD(x) antibody. Although antibodies are not in themselves fluorescent, they can be co-valently linked or "conjugated" to any of a number of differently-colored fluorescent molecules ("fluorochromes") thereby becoming specific fluorescent stains for a large number of cellular proteins. Using a fluorochrome-conjugated antibody to stain the particular cellular protein to which that antibody binds is known as "direct staining" (Fig. 1).

Because antibodies are a class of proteins that bind to other proteins, it is, indeed, possible to produce antibodies that bind to other antibodies. For example, goats can produce antibodies that bind to mouse antibodies. With a so-called "indirect staining" technique, a cell can be stained with a colorless monoclonal antibody (called the "primary antibody") against one of its proteins and then, in a second step, the protein can be visualized by using another antibody (the "secondary antibody") that is specific for the primary antibody and that has been conjugated to a fluorochrome. For example, a human cell can be stained with a mouse monoclonal antibody against a specific human protein and then this staining can be visualized with a fluorochrome-conjugated goat antibody that binds to the constant region of any mouse antibody (called a "goat-anti-mouse secondary antibody" Fig. 2).

There are many advantages in this indirect staining method: It is often less expensive because labs that don't want to get involved in conjugating fluorochromes to antibodies can purchase one fluorochrome-conjugated secondary antibody and use it with all their home-made primary antibodies. Indirect staining makes comparison of the surface expression of different proteins on cells somewhat easier because the visualizing antibody used

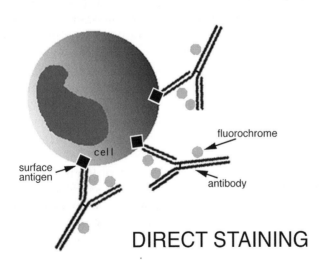

DIRECT STAINING

Fig. 1

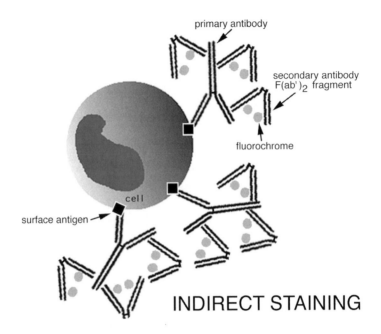

Fig. 2

for a panel of primary antibodies is identical, with a uniform fluorochrome/antibody (F/P) ratio. Indirect staining is flexible, making it easy to change the color of the staining reaction – a selection of secondary antibodies conjugated to different fluorochromes can be part of a laboratory's reagent arsenal. Most importantly, indirect staining increases fluorescence intensity because the secondary antibody is usually polyclonal, binding to multiple sites on the primary antibody; the second step in the reaction therefore involves an amplification of the signal. This can be a critical point in the detection of weakly-expressed proteins.

For many purposes, the advantages of the indirect staining method may, however, be outweighed by its disadvantages. Because the indirect method involves a second procedural step, it doubles the time that it takes to stain cells. In addition, the use of a secondary antibody increases the potential for non-specific staining and therefore the need for extra controls. The most important disadvantage of the indirect staining technique is that it is practically incompatible with multicolor analysis, because each secondary antibody may have trouble knowing to which primary antibody it is supposed to bind. It is possible to find secondary antibodies with tight specificities for certain classes of primary antibodies; it is also possible to stain cells first with a two-step indirect staining procedure with one fluorochrome, followed by blocking of the bound secondary antibody with non-immune serum, followed then by staining with a directly-conjugated monoclonal antibody of a second color. But these techniques are cumbersome and fraught with pitfalls of cross-reactivity to sabotage the unwary (and occasionally even the sophisticated) scientist. For the most part, multicolor experiments demand the direct staining method.

A variation on the indirect staining technique, but with some of the advantages of both the direct and indirect methods, makes use of a primary antibody conjugated to biotin, followed by avidin or streptavidin conjugated to a fluorochrome. Staining is achieved by virtue of the specific binding reaction between biotin and avidin. A biotinylated primary antibody can be used to stain cells simultaneously with one or more different directly fluorochrome-conjugated antibodies. The following avidin-fluorochrome reagent will react only with the biotinylated primary antibody.

Materials

EQUIPMENT

- Microcentrifuge or centrifuge with plate carriers
- Pipetter (Gilson-type :20, 200, and 1000 µl) with tips
- Repeating pipette, if available
- Multichannel pipette (if using plates)
- Suction line with trap
- Eppendorf tubes or 96-well round-bottomed plates

SOLUTIONS

2% formaldehyde

Formaldehyde crosslinks proteins, thereby fixing them so that they remain stable for long periods of time. Laboratory "formalin" contains methanol as well as formaldehyde and should not be used for flow cytometry because the methanol increases background fluorescence. Formaldehyde is formed when p-formaldehyde crystals are dissolved in solution. Therefore a solution of p-formaldehyde will contain pure formaldehyde that is suitable for fixation of cells for flow cytometry. It is also possible to buy EM-grade pure (methanol-free) formaldehyde in solution and this also is suitable. Cells stained for surface antigens and then fixed with formaldehyde at 1% (final concentration) should be stored for at least 24 hours (in the dark and at 4°C) and then read on a flow cytometer within 2 weeks (although most cells with most stains will probably last much longer).

Combine
- 20 gm of p-formaldehyde
- 700 ml of phosphate-buffered saline (PBS)

in a 1 liter flask. Stir in a fume hood until all the solid dissolves. This will take several hours and can be left overnight. After the solution is clear, correct the pH to 7.3 and make the volume to 1 liter with PBS. Store in a tightly-covered bottle at 4°C. This solution has indefinite stability; some labs make it fresh weekly and others use it for months with no problem. To be safe, make it fresh monthly.

Phosphate-buffered saline (PBS)

Although PBS can be bought as a mix of crystals or as a concentrated solution, making it up from scratch is not difficult.

Weigh
- NaCl 8.0g (138 mM)
- KCl 0.2g (2.6 mM)

- KH_2PO_4 (anhydrous) 0.2 g (1.5 mM)

- Na_2HPO_4 (anhydrous) 0.9 g (6.3 mM)

and add crystals to 700 ml of distilled water. Stir until dissolved. Adjust pH to 7.3 and make final volume to 1 liter. Filter through an 0.2 micron sterilizing filter into sterile bottles and store at 4°C.

Staining/Washing (S/W) Buffer

Some labs stain and wash cells in medium (minus phenol red and biotin) supplemented with 0.1% azide. Alternatively, PBS (or other balanced salt solution compatible with your cells) supplemented with 1% bovine serum albumin (BSA) and 0.1% azide is adequate to satisfy most cells throughout the staining and washing procedure. The BSA goes part way to making the cells think that they are back home on the farm and the azide is a metabolic poison that prevents capping and internalization of surface proteins and inhibits microbial growth. This buffer can also be used for diluting antibodies.

To make S/W Buffer, add
- 1 gram of sodium azide (NaN_3)

- 10 grams of bovine serum albumin

to 700 ml of PBS. Stir gently to avoid foaming of the protein. When dissolved, check that the pH is 7.3. Make the final volume to 1 liter. Filter through an 0.2 micron sterilizing filter into sterile bottles and store at 4°C.

Blocking Solution

Blocking Solution is important in the staining procedure because it fills the Fc-receptors on cells, thereby preventing irrelevant binding of whole antibodies to these receptors. Human cells should be blocked with human immunoglobulin.

To make Blocking Solution for human cells, weigh 120 mg of human γ-globulin (Sigma) and make up to 10 mls with S/W Buffer. Stir very gently at room temperature until dissolved. Filter through an 0.2 micron sterilizing filter into vials and store frozen.

Titered Antibodies

Antibodies should be titered by halving dilutions. The concentration of each antibody added is that required to give saturating fluorescence intensity when 5-20 µl is diluted to a final volume of 60 µl.

Procedure

Add anti-body(ies) to tubes/wells

1. Label Eppendorf tubes or template for microtiter plate wells. Add directly conjugated antibody (ies) (for direct staining procedure) or primary unconjugated antibody (for indirect staining procedure) to tubes or wells. Antibodies should have been titered by halving dilutions. The volume of each antibody added is that required to give saturating fluorescence intensity in a final volume of 60µl.

Add buffer to each tube/well (if necessary, to make 20 µl volume)

2. Cold S/W buffer is added to bring the total antibody volume to 20 µl. Cover tubes or plates and keep on ice and in dark until cells are ready.

Suspend cells (10^7 to 10^8/ml)

3. Suspend cells in cold S/W buffer to a final concentration of 10^7 to 10^8 per ml.

Note: You need 20 µl of cells for each tube in your staining panel – plus an extra 50 µl for pipetting margin.

Add blocking solution to cell suspension

4. Mix cell suspension with an equal volume of cold blocking solution.

Add 40 µl of cells/block to each tube or well

5. Add 40 µl of cells with block to each Eppendorf tube or well (already containing the antibody (ies)).

Note: Use an automatic pipetter for repeating additions (or use a multichannel pipette for wells if you have enough cells to give you sufficient volume). Be careful not to drag antibody from one well to another.

Swirl plate or vortex tubes

6. Swirl plate or vortex tubes to make sure that antibodies and cells are well mixed.

7. Incubate cells covered, in the dark, at 4°C for 30 minutes.

Incubate cells, for 30 minutes

Note: During incubation, use this free time to prepare flow cytometer sample tubes, with 150 µl of cold 2% formaldehyde solution in each. Use test tubes that are appropriate to the sample intake on your flow cytometer. Make one tube for each sample in your experiment.

8. This is the wash step that is repeated three times. Using an automatic or multichannel pipette, add cold S/W buffer – 100 µl into each well or 1 ml into each Eppendorf tube.

*Add S/W buffer

9. Swirl plate or vortex tubes to resuspend cells.

Resuspend cells

10. Centrifuge at about 1000 x g for 2 seconds (or just long and fast enough to lightly pellet cells).

Centrifuge

Note: If using Eppendorf tubes and an angled rotor, place tubes with lid hinges toward the outside.

11. Use a glass Pasteur pipette attached to a suction line (with a trap containing 10% chlorox bleach) to remove supernatant from the cell pellet.

**Remove supernatant

Note: If using Eppendorf tubes, the pellet will be on the hinge side of the tube. Be careful to leave the pellet behind (!).

12. Repeat washing of cells (from step 8 to step 11 two more times (total of three centrifugations).

Repeat from * to ** twice more

13. *If using an indirect procedure, skip to step 13 x.*If using a direct staining procedure, you are almost done now. Add 150 µl of cold S/W buffer to each tube/well. Swirl or vortex to resuspend cells completely.

If direct staining, resuspend cells in 150 µl S/W buffer

14. Use a pipette to remove cell suspension from each tube/well and pipette into one of the prepared test tubes containing 150 µl of cold 2% formaldehyde.

Add cells to equal volume of formaldehyde solution

Store for at least 24 hrs	**15.** Store covered, in the cold and dark, for at least 24 hours to give FSC, SSC, and fluorescence intensities time to stabilize after fixation.
Read on flow cytometer	**16.** If necessary, roughly dilute each sample with cold S/W buffer from a squeeze bottle so that cells go through the cytometer at 100-1000 cells per second. That's it!
If indirect staining, add 50 µl of secondary antibody to each cell pellet and resuspend cells	**13. x** *If using an indirect staining procedure,* add 50 µl of fluorochrome-conjugated secondary antibody to each cell pellet and swirl or vortex to resuspend. **Note:** The secondary antibody should be at a concentration that produces saturating fluorescence intensity, but low levels of background.
Incubate for 30 minutes	**14 x.** Incubate covered, in the cold and dark, for 30 minutes.
*Add S/W buffer	**15 x.** Using an automatic or multichannel pipette, add cold S/W buffer – 100 µl into each well or 1 ml into each Eppendorf tube.
Resuspend cells	**16 x.** Swirl plate or votex tubes.
Centrifuge	**17 x.** Centrifuge at about 1000 x g for 2 seconds (or just long and fast enough to lightly pellet cells). **Note:** If using Eppendorf tubes, and an angled rotor, place tubes with lid hinges toward the outside.
**Remove supernatant	**18 x.** Use a glass Pasteur pipette attached to a suction line (with a trap containing 10% chlorox bleach) to remove supernatant. **Note:** If using Eppendorf tubes, the cell pellet will be on the hinge side of the tube.
Repeat from * to ** twice or more	**19 x.** Repeat washing of cells two more times (total of three centrifugations for the primary antibody step and now three centrifugations for the secondary antibody step).
Resuspend cells in 150 µl S/W buffer	**20 x.** Add 150 µl of cold S/W buffer to each tube/well. Swirl or vortex to resuspend cells.

21 x. Use a pipette to remove cell suspension from each tube/ well and pipette into one of the prepared test tubes containing 150 μl of cold 2% formaldehyde.

Add cells to equal volume of formaldehyde solution

22 x. Store covered, in the cold and dark, for at least 24 hours to give FFC, SSC, and fluorescence intensities time to stabilize after fixation.

Store for at least 24 hrs

23 x. If necessary, roughly dilute each sample with cold S/W buffer from a squeeze bottle so that cells go through the cytometer at 100-1000 cells per second. That's it!

Read on flow cytometer

CRITICAL ASPECTS OF STAINING

A typical experiment for flow cytometric analysis involves a series of samples, each of which contains cells that have been stained for one (single color) or more (multicolor) membrane proteins. Often, the cells may be the same in all samples; each sample then represents the analysis of those cells for different combinations of proteins. Alternatively, the cells may differ in each sample – and all types of cells may be analyzed for expression of the same protein(s). Whatever the overall purpose of the analysis, there are certain general aspects of staining that need to be understood in order to plan an effective experiment.

Cell Concentration/Cell Number

For *each* sample, you will need between 10^5 and 10^6 cells. If you are new to flow cytometry, use the higher number of cells – to give yourself a margin for error (you always lose more cells than you expect during the staining and washing procedures). Trying to stretch too few cells into too many sample tubes in order to maximize results from a single experiment almost always leads to no results at all (it's happened to all of us).

After staining and washing, resuspend the cell pellet in 0.3-1.0 ml, according to the configuration of the sample uptake on the cytometer. Use the smallest volume you can; the suspension can always be diluted with buffer later if cells are flowing too fast. When at the cytometer, adjust this final concentration of cells with more buffer if necessary, so that cells flow through the cytometer at about 100-1000 cells per second. Slower than 100 cells per second gets very boring. Faster than 1000 cells per second may lead to problems because the computer may not keep up or because two cells close together in the laser beam may be seen by the cytometer as a single "event." Constraints on adjusting cell concentration come from the total number of cells that are available and the minimum volume that your cytometer can handle. In general, you want to have cells at a high concentration so that you can use the lowest pressure possible to push them through the cytometer (but still achieve a flow rate close to 1000 cells per second). Pushing the cells too hard to achieve this high flow rate has deleterious effects on intensity resolution (by widening the sample core in the flow stream).

If you are setting up a new procedure or are needing to adjust instrument settings, it is especially helpful to have a few extra control samples with more cells in a larger volume. Running these cells initially, before you start collecting data, will give you time to adjust FSC and SSC parameters so that all cells are on scale, to make sure that background fluorescence is at an appropriately low position on the fluorescence scale, and to check out compensation if necessary for multicolor analysis.

Tubes vs Plates

Cells can be stained in any container for which you have an appropriate centrifuge. You can stain in the test tubes that will eventually fit on your cytometer, although the large bottom diameter of these tubes will make it difficult to get consistent results with a small volume of staining solution. Eppendorf tubes are classical favorites because of their conical shape. Staining can take place in a small volume in an Eppendorf tube and then the suspension can be diluted to a full 1 ml for each washing step.

If you are staining more than a few samples and have a centrifuge with plate adaptors, consider using 96-well, round-bottomed microtiter plates. While the dilution factor in the washing step is not as effective in these plates (because the maximum volume is 200 μl in each well), the benefits that derive from not having to organize, label, and manipulate dozens of test tubes make plate-staining highly attractive. Photocopy a template to use for identifying wells; plan the layout of samples in the plate with some thought in order to make addition of reagents easy. It usually works to add the antibody or antibodies to wells in advance while cells are being prepared, keep the plates covered in the dark at 4°C until the cells are ready, and then use a multichannel or repeating pipette to add aliquots of the cell suspension to the rows of wells. Care needs to be taken not to drag antibody from one well to the next in this procedure. Although it takes a bit of practice to become confident about staining and centrifugation in plates, most people become converts very quickly.

Centrifugation and Washing

The centrifugation step pellets cells from the staining suspension, thus permitting removal of the supernatant fluid, and the washing of unbound antibody from the cells. The unbound antibody is removed by dilution in the sequential washing steps. Exact advice about centrifugation speeds is not possible, as different centrifuges take more or less time to reach full speed. In general, you want to spin cells down hard enough that the supernatant fluid can be removed with little loss of cells, but not so hard that the cells are difficult to resuspend. If staining in Eppendorf tubes, a high speed microfuge with an angled rotor can be used; 1-2 seconds is long enough to pellet most cells well. If the Eppendorf lids are lined up in the rotor so that the hinges all point outward, then the cell pellets (even if not visible) will all be at the hinge side of the tube after spinning. The supernatant fluid can then be removed with confidence by inserting a glass disposable ("Pasteur") pipette into the other (non-hinge) edge of the tube. The disposable pipette should be hooked up to suction such that the aspirated fluid will go into a trap containing bleach. After addition of 1 ml of S/W buffer, the cells can be resuspended on a vortex mixer before spinning again.

If staining in microtiter plate wells, use a centrifuge with plate adaptors. At about 1,500 x g, cells will pellet well in about 2-3 seconds. Removing supernatant fluid after centrifugation can be done by flicking the plates and then blotting once on clean paper towels. More securely, the supernatant can be removed from each well with a glass disposable ("Pasteur") pipette hooked up to suction and a trap. Wash buffer can be added with a multi-channel pipette – a squirt from this pipette is enough to resuspend a cell pellet if it has not been centrifuged too hard.

Specificity and Non-Specificity

In order to draw conclusions about particular proteins on cells, we have to be sure that the stains that we use are specific in their binding to the proteins in which we are interested. Lack of specificity may occur as a result of several factors: Firstly, an antibody may bind to a common epitope on several different proteins. Secondly, different sorts of cells contain the same proteins, so a staining reaction may not be as specific as one would like in classifying different cell types. These first two types of problems require knowledge of the appropriate reagents for answering particular questions. Two other types of specificity problems are more difficult to control. Antibodies bind to many cell types by their non-specific (Fc) ends. Monocytes, in particular, profes-

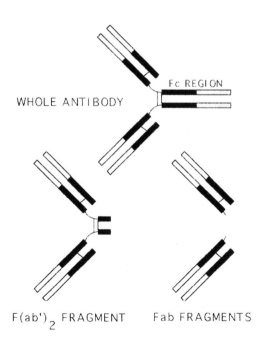

Fig. 3

sionally bind many antibodies through their Fc-receptors. The use of Fab or F(ab')$_2$ fragments (antibodies without their Fc ends) is recommended whenever possible and helps to alleviate this problem (Fig. 3). Blocking the Fc-receptors (see below) also helps. Finally, dead cells, with compromised membrane integrity, tend to be sticky and to bind all sorts of reagents with careless abandon. The dead cell problem can be helped by careful cell preparation or by "gating out" dead cells from analysis – either through propidium iodide staining to positively identify dead cells by their membrane permeability or, less securely, through avoiding cells with low FSC intensity (cells that are dead usually show low FSC signals probably because the refractive index of their cytoplasm is similar to that of the surrounding medium). In the final analysis, however, in order to permit conclusions from fluorescence staining about the classification or chemical constituents of a cell, flow experiments must make use of careful controls.

Controls

In an experiment with indirect staining, it is very necessary to stain cells with the secondary antibody alone to control for non-specific binding of this polyclonal antibody to dead or sticky cells. This is the so-called "secondary control" that marks the level above which fluorescence intensity can be considered specific. Comparing the intensity of cells stained with the secondary antibody alone to the background autofluorescence of unstained cells is often diagnostic of problems with the secondary reagent. Ideally, the secondary control should be no brighter than the unstained cells. It is also absolutely necessary to have singly-stained cells to set the spectral cross-talk compensation for multicolor experiments. There is considerable debate, however, about the question of appropriate negative controls when staining cells with a direct protocol. If positive cells are much brighter than unstained (autofluorescent) cells and if the experiment contains some samples where no cells stain, then, arguably, the unstained cells within the stained samples are, in themselves, controls for non-specific staining. Most people, however, feel happier about ruling out problems with dead cells or Fc-receptor binding if their experiment contains samples stained with so-called isotype control antibodies. Isotype control antibodies are antibodies with no specificity for the cells in question but with all the non-specific characteristics of the antibodies used in the experiment. Since antibodies come in different classes ("isotypes") with different binding characteristics, an experiment should probably have an isotype control antibody for each class of antibody used for staining. Such antibodies are available from manufacturers of flow cytometric reagents. Sceptics will wonder about our ability to know the protein concentration of every antibody that we are using and to know the number of fluorochrome molecules conjugated to each of these antibodies. Compromises may be necessary, since each stain in a panel of many stains may be of a different isotype, at a different concentration, and with a different F/P ratio – and will therefore require its own control. There are no perfect answers here; the requirement for controls is often determined by the nature of the experiment and needs to be considered carefully in relation to the stringency of the questions being asked of the resulting data.

Blocking

One important way to minimize non-specific staining is by the use of a so-called blocking reagent. A blocking reagent contains a high concentration of immunoglobulin that will bind to the Fc-receptors on cells like monocytes, thereby blocking the non-specific binding of the staining antibody reagents to these receptors. The blocking reagent should be immunoglobulin from the species whose cells you are staining. For human cells, use the γ-globulin fraction from human serum at a final concentration of 4 mg/ml. Alternatively, one can purchase specific antibodies directed against the Fc receptors of the cells in question; care needs to be taken here as there are different classes of Fc receptors and, in addition, not all antibodies against Fc receptors block the receptor binding site.

Titration and Proportionality

Flow cytometry has the potential for providing quantitative information about the relative expression of membrane proteins on different cells. For example, we may want to determine if cells, when stimulated, express more of a given protein than they do under resting conditions. In order to be able to draw quantitative conclusions from flow cytometric data, we have to be sure that the fluorescence intensity of a stained cell is directly and linearly proportional to the protein we are measuring. This involves using concentrations of reagents that are not limiting, so that the staining intensity of a cell will be related to the amount of the protein on a cell and not to the number of cells present nor to pipetting inaccuracies in the exact amount of reagent used. Determining the amount of stain to use involves titrating each antibody reagent to find a concentration that approaches saturation within the time of the reaction. With low affinity/avidity antibodies, strict saturation may not be possible in short staining times at reasonable antibody concentrations; therefore use of such antibodies makes conclusions about relative intensities difficult.

To determine a saturating concentration, run a titration curve by using halving dilutions of antibody, covering a range above and below that suggested by the manufacturer (if any). Make this 1:2 dilution series (using 10 halving dilutions) of both your antibody of interest and also of an irrelevant antibody matched by concentration and isotype. Then stain cells with the standard staining protocol, using both the irrelevant antibody and the antibody of interest from each of the 10 dilutions. As your final working concentration, pick a dilution of stain well on the intensity plateau for the stained cells, where small changes in stain concentration have little effect on cell fluorescence intensity. Very high antibody concentration is not, however, a panacea: cells often stain less well at very high antibody concentrations than they do at lower concentrations. Additionally, as can be seen from the cells stained with the control antibody, even negative controls often begin to stain non-specifically at high antibody levels. The important thing here is to maximize the signal-to-noise ratio: use a concentration on the plateau that gives maximal staining with the positive antibody but little staining with the control antibody. Fig. 4 and Fig. 5.

With indirect staining, titration with halving dilutions is necessary for both the primary and the secondary antibody. Although it may seem as if there are too many unknowns in this equation, it is possible to guess at a reasonable concentration of one primary antibody, titrate halving dilutions of the secondary antibody with that concentration of the primary antibody, and then, using the concentration of secondary antibody determined to be optimal, titrate all primary antibodies in turn. Because non-specific staining will increase at high antibody concentrations, when titrating the secondary antibody it is particularly important to compare the staining of cells with and without primary antibody present. Cells stained with the secondary antibody, but without any primary antibody, are known as a "secondary control." As with titrating directly-conjugated antibodies, you want to use a level of secondary antibody that maximizes the signal-to-noise ratio, that is, gives a saturating signal when the primary antibody is present but little signal when the primary antibody is absent.

At staining concentrations that reach saturation, the amount of the stain present is hardly ever limiting: that is, there is usually enough antibody around to stain all the relevant cell proteins on all the cells without significantly lowering the concentration of free antibody (see Kantor and Roederer[1] for examples of this calculation). The *number* of cells is not, therefore, critical to the

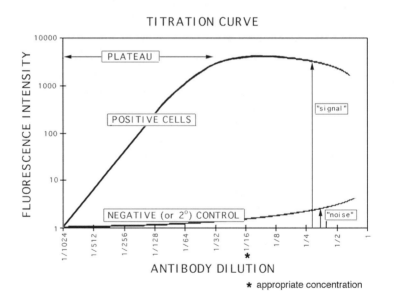

Fig. 4

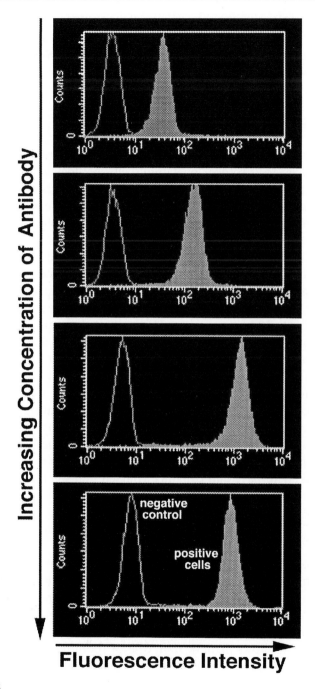

Fig. 5

staining intensity nor is the *volume* of the suspension. We simply need a high enough *concentration* of antibody so that effective equilibrium between bound and unbound antibody will be achieved in a reasonable staining time.

Staining Volume (and Expense!)

Many staining reagents, particularly monoclonal antibodies, are expensive. Any protocol for staining cells should have expense as one of the criteria for experimental design. Although the number of cells for an experiment is constrained by both experimental and flow cytometer design and the concentration of stain used is constrained by a desire to achieve near saturation (see above), the staining volume is under user control. The smaller the staining volume, the less stain will be required to achieve saturating concentration. Although many companies selling staining reagents suggest staining in 100 µl volume, it usually works to reduce this to 50 µl or even less. This reduces the requirement for stain by 2- or more-fold. The only considerations here are that you don't go to such a small volume that pipetting becomes inaccurate, nor that the amount of stain becomes limiting (that is, that there isn't enough stain to go around to all the receptors on all the cells present). With 50 µl of stain at a saturating concentration, 10^6 cells with antigens within the usual range of density will not use up a significant fraction of most antibodies. Check this out by using 2-fold more and 2-fold fewer cells than you are ever likely to have and convince yourself that the fluorescence intensity of the cells in all cases is the same.

Time and Temperature

Antibody concentration, temperature, and time interact in the staining reaction: higher antibody concentrations reach effective staining saturation more quickly than lower concentrations and any concentration reaches saturation faster at a higher temperature. Even though azide is a metabolic poison (and some labs find that it allows them to stain cells at room temperature, rapidly and with impunity), most labs stain with all reagents/solutions at 4°C even in the presence of azide to be sure of minimizing capping / internalization/ miscellaneous loss of surface-bound antibodies. At 4°C, 30 minutes is a reasonable staining time. Antibody titrations can all be done at 4°C and for 30 minutes. After titration to determine a saturating antibody concentration, check out the effect of time and use a time that is long enough that a bit more time or a bit less time makes no difference to the results.

Stability

Once you have stained cells, it is necessary to know how long the cells with the fluorochrome on them will remain as you want them. Cells die with time, antibodies are internalized gradually at warm temperatures, fluorochromes become bleached in the light, and microbes will grow everywhere eventually. The answer to all of these problems is to keep cells (before and after staining) scrupulously at 4°C and in the dark, use a metabolic poison like azide in all buffers if you are not concerned with recovering cell function, and, finally, fix the cells with formaldehyde after staining if you are not going to analyze them within several hours on the flow cytometer. Although 1% formaldehyde fixation does change the FSC and SSC of some cells and may increase their background fluorescence and decrease their staining fluorescence just a bit, it provides great cell stability after staining and also a large margin of safety when dealing with potentially pathogenic samples. A good, general procedure for antibody-stained cells is to fix them in an equal volume of 2% formaldehyde after staining, let them sit at least 24 hours, (covered, in the cold and dark) to reach final FSC, SSC, and fluorescence levels, and then read them on the flow cytometer within 2 weeks (although most cells will last much longer). Check out your own cells with your own stains before deciding on the length of time that you are willing to risk this. But fixation after staining is definitely the way to go if possible – giving great flexibility in planning time on that cytometer. As a reminder, you cannot fix cells before flow if you want to use a membrane-impermeant stain like propidium iodide to check cells for viability or to gate out dead cells in flow analysis. All cells become dead and take up the dye after formaldehyde fixation.

Safety

Fixation of cells after staining and before running them through a flow cytometer is highly recommended if there are any questions about the pathogenicity of the cells involved. Since there are real concerns about the potential pathogenicity of all living cells, fixation after staining is always recommended unless the cells will be sorted and cultured afterward or unless you need to look at membrane integrity of the cells as a measure of viability.

Fluorochromes for Single and Multiparameter Analysis

The early experiments in flow cytometry made use of antibodies conjugated to fluorescein, with sequential samples stained with different antibodies, all conjugated to this same fluorochrome. Within a short period of time, however, instruments were developed with the capacity for multicolor or multiparameter analysis. Multiparameter analysis permits efficient use of small numbers of cells and, more importantly, leads to far more information than can be obtained from single color sequential samples. Because it is difficult to prevent cross-reactivity between multiple secondary antibodies and multiple primary antibodies, multicolor analysis usually makes use of directly-conjugated antibodies (direct staining) or perhaps in combination with an antibody conjugated to biotin (followed by streptavidin conjugated to a fluorochrome).

Various fluorochromes can be conjugated to antibody reagents. Depending on the number of lasers available, the color of their excitation light, and the number of photomultiplier tubes available for analysis of emitted fluorescent light, different combinations of fluorochromes are possible. The overall strategy in picking fluorochrome combinations is that the fluorochromes must all be able to absorb light from the lasers present and must emit light at different wavelengths from each other so that their fluorescence signals can be distinguished. The farther apart their fluorescence peaks, the easier it will be to resolve their signals without recourse to excessive compensation.

There are different strategies for allocation of antibody/fluorochrome combinations. In general, the brightest fluorochrome should be assigned to the antibody binding to the protein with lowest expression. In many situations, phycoerythrin will give the brightest staining. If expression is very low, it may, however, be necessary to test antibodies conjugated to a range of different fluorochromes as conjugation procedures are variable and it is not always possible to predict brightness from photochemical principles alone.

READINGS

Although there are many books giving individual adaptations of methods for staining cells, I have found two articles particularly helpful for discussion of some of the general principles involved. These articles would certainly serve as useful supplements to the topics discussed here.[1,2]

FLUOROCHROME	ABS.	EM.	LASER
fluorescein (FITC)	490	515	argon(488)
phycoerythrin (PE)	480/565	578	argon(488)
PE-Cy5 (tandem)	480/565/650	670	argon(488)
PE-Texas Red (tandem)	480/565	615	argon(488)
PerCP	488	677	argon(488)
allophycocyanin (APC)	650	660	HeNe or krypton
AMCA	350	445	argon(UV)
Cascade Blue	375/400	423	argon(UV)
Cy3	512/552	565/615	krypton
Cy5	625-650	670	HeNe or krypton

Acknowledgements

I am grateful to Allan Munck for a very helpful discussion of binding kinetics. Paul Guyre, dependable as always, came through with that cartoon of an antibody (whole and fragments). I would also like to thank Michael White for introducing me to the staining of cells for flow cytometry a long time ago and Kathleen Wardwell for generously offering sound tips and wise advice in my current life. Both Kathy and Mike will no doubt hear echos of their own voices in much that I have written here.

References

1. Kantor AB, Roederer M. FACS analysis of lymphocytes. In: Weir DM, et al eds. The Handbook of Experimental Immunology. 5th edition. 1997:49.1-49.13.
2. Stewart CC, Stewart SJ. Cell preparation for the identification of leukocytes. In: Darzynkiewicz, Z et al eds. Flow Cytometry, 2nd edition. Methods in Cell Biology. 1994: 41: 39-60.

Phenotypic Analysis Using Five-Colour Immunofluorescence and Flow Cytometry

ANDY BEAVIS

Introduction

The correlated data obtained from multicolour, multiparameter flow cytometric analysis provides an accurate method to characterize subpopulations of cells. Multicolour immunofluorescence is also applicable where cell numbers are very low but there is a requirement for the determination of many specific cell surface markers, such as fine needle aspirates or bone marrow specimens. Commercial flow cytometers (both the sorters and the benchtop analysers) are available with two or more lasers which permits the use of multiple excitation wavelengths and a wider range of fluorochromes.

Materials

Equipment

Flow Cytometer equipped with 2-3 lasers with appropriate filter configuration for the excitation and emission of the dyes chosen

Solutions

- Lysing buffer: 0.17M Ammonium Chloride, 0.16M Tris-HCl, pH 7.0
- Staining Buffer PBS, 2% FCS, 0.1% azide
- Blocking buffer, Staining buffers / Antibody solutions as manufacturer indicates

▓ Procedure

<table>
<tr>
<td>

Staining with biotinylated and directly-conjugated antibodies

</td>
<td>

1. Prepare single cell suspensions for cell types of interest. For tissue samples such as spleen and thymus, dissociate tissues with forceps and filter.

2. Remove contaminating RBC using lysing buffer (Amm. Chloride, Tris). Resuspend cells in lysing buffer (1-5 ml), incubate for 5 min at r.t. and centrifuge at 200 g, 5 min.

</td>
</tr>
</table>

Note: Lyse RBC with buffer containing 0.16M Tris-HCl, 0.17M NH_4Cl, pH 7.0.

3. Discard supernatant, resuspend cells in 10 ml cold PBS and centrifuge as previous.

4. Wash cells X2 in staining buffer (PBS, 2% FCS, 0.1% azide), enumerate and resuspend at 1×10^7/ml.

5. Aliquot 100 µl cells/well into a 96-well plate (V-bottomed, polypropylene) and add 5 µl of a blocking antibody. Mix and incubate for 5 min at 4°C.

Note: The blocking antibody is used to minimize non-specific binding (e.g. an anti-Fc receptor antibody or an IgG of the same species of cell to be used). When using indirect immunofluorescence and second antibodies, ensure that the anti-Ig second antibody will not bind to the blocking antibody used in the previous step.

6. Add biotinylated antibody and directly-conjugated antibodies (10-20 µl of each), mix gently and incubate for 30 min at 4°C.

Note: Titrate the antibodies to be used for optimal staining. An example panel of antibodies used for 5-colour immunofluorescence analysis of murine spleen cells: IAd-FITC, L3T4-PE, B220-RED613, LYT2-APC, Thy 1.2 - biotin /streptavidin-CASCADE BLUE.

7. Add 100 µl staining buffer to each well, mix and centrifuge. Discard the supernatant with a rapid inversion of the plate.

8. Resuspend the cells in 100 µl staining buffer, add 10 µl of the streptavidin- CASCADE BLUE and incubate for 15 min at 4°C.

Note: Titrate the Streptavidin-CASCADE BLUE for optimal staining of the biotinylated primary antibody with minimal non-specific staining.

9. Wash X2 as previous, resuspend in 250 µl staining buffer and transfer cells to tubes for flow cytometric analysis. Keep tubes on ice and protected from light.

10. Optimize laser alignment and instrument performance according to standard laboratory procedures.

Flow Cytometric Analysis

11. Obtain forward scatter (FSC) and side scatter (SSC) signals. Identify population of intact cells of interest with a wide gate.

Note: The FSC and SSC signals are collected from the 488 nm laser using the 488 / 10 BP filters. Use a wide gate to include all cells and use subsequent phenotypic analysis to accurately quantitate subpopulations of cells.

12. Adjust PMT voltages for each channel (log) to position autofluorescence signals on scale.

13. Check fluorescence of each single-colour control separately to ensure that positive signals remain on scale. Adjust compensations as needed to correct for spectral overlap.

14. Check the 5-colour sample with all compensations set (Fig. 1).

15. Acquire listmode data for ALL samples at these instrument settings: unstained, isotype control, single-colour controls and five-colour samples.

Note: Collect enough events for accurate analysis of the subpopulation of cells present at the lowest frequency. e.g. for a 2% subpopulation collect 100,000 cells giving 2000 cells for analysis.

16. Define the lymphocyte population gate from a FSC/SSC dotplot. Define a series of dual-parameter dot-plots to display the possible 2-colour combinations for the multicolour sample.

Data Analysis

17. Establish quadrants to define the negative (autofluorescence) cells for the unstained sample in the lower left quadrant. Determine background binding of isotype controls and readjust quadrants to define cells as negative.

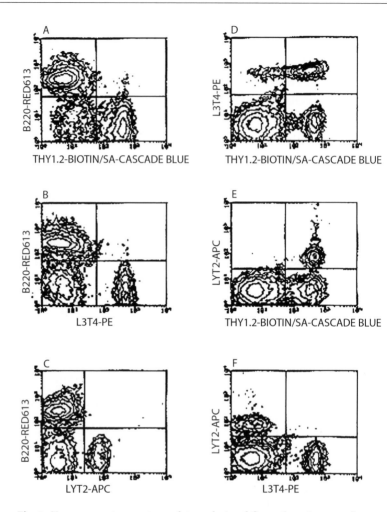

Fig. 1. Two-parameter contour plot analysis of five-colour immunofluorescence data. Murine spleen cells were stained with monoclonal antibodies to B-cell, T-cell and T-cell subset antigens and the activation marker IA-d. Fluorescence data was collected on a log scale with 1024 resolution using a standard FACS Vantage (Becton Dickinson Immunocytometry Systems, San Jose, CA). Single cells were gated from a 2-D plot of forward and side scatter (not shown) and analysed on 2-D contour plots: **A:** Thy1.2biotin/SA-Cascade Blue vs B220-RED613; **B:** L3T4-PE vs B220-RED613; **C:** LYT2-APC vs B220-RED613; **D:** Thy1.2biotin/SA-Cascade Blue vs L3T4-PE; **E:** Thy1.2biotin/SA-Cascade Blue vs LYT2-APC; **F:** L3T4-PE vs LYT2-APC. "Reprinted from Cytometry, volume 15(4), Andrew J. Beavis and Kenneth J. Pennline. Simultaneous Measurement of Five Cell Surface Antigens by Five-Colour Immunofluorescence. Pages 371-376, 1994 with kind permission of Wiley-Liss, Inc. 605 Third Avenue, New York, NY 10158-0012, USA"

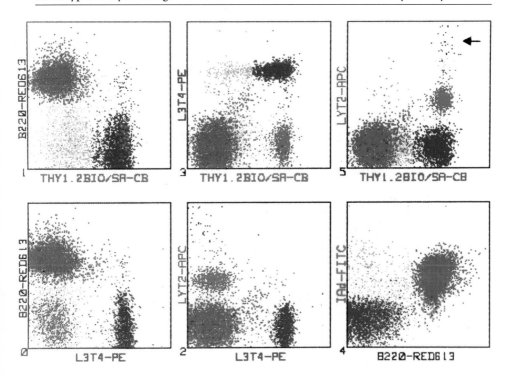

Fig. 2. Phenotypic analysis of murine spleen cells by five-colour immunofluorescence and flow cytometry. The listmode data was analysed using Paint-A-Gate Plus (Becton Dickinson Immunocytometry Systems, San Jose, CA) and the phenotype of each subpopulation assigned a colour. The position of each subpopulation on a particular two-dimensional dotplot can then be identified by the specific colour of the dots: Thy1.2(+)/Lyt2(+)/L3T4(-) = VIOLET; Thy1.2(+)/Lyt2(-)/L3T4(+) = BLUE; Thy1.2(-)/Lyt2(+)/L3T4(-) = CYAN; Thy1.2(-)/Lyt2(-)/L3T4(+) = YELLOW; Thy1.2(+)/B220(+)/Lyt2(+) = BLACK; B220(+)/Thy1.2(-)/IAd(-) = GREEN; B220(+)/Thy1.2(-)/IAd(+) = RED. "Reprinted from Cytometry, volume 15(4), Andrew J. Beavis and Kenneth J. Pennline. Simultaneous Measurement of Five Cell Surface Antigens by Five-Colour Immunofluorescence. Pages 371-376, 1994 with kind permission of Wiley-Liss, Inc. 605 Third Avenue, New York, NY 10158-0012, USA"

18. Quantitate phenotypes for the multicolour samples using the adjusted quadrants.

19. A useful analysis strategy may also be to identify subpopulations from a 2-d dot-plot of SSC and phenotypic marker. Apply Boolean gate logic to further quantitate subpopulations of cells. It can be useful to apply this method of analysis to eliminate a specific cell type from the quantitation of other specific subpopulations. For example, the inclusion of a monocyte marker in the five-colour panel can be used to GATE OUT the monocytes from any lymphocyte subset analysis.

Note: The T-cells can be identified from a 2D-plot of SSC and THY 1.2. Gate the THY 1.2 (+) cells onto a plot of L3T4 and Lyt2 to further quantitate the T-cell subpopulations. This prevents the exclusion of T-cells from the analysis due to atypical light scatter properties such as for activated or large-granular lymphocytes.

20. Do Cluster Analysis

Note: Cluster analysis can also be a useful technique for analysis of multicolour data. An example of this is shown in Fig. 2 , where specific subpopulations of cells of a given phenotype (defined by positive or negative fluorescence for each of the five monoclonal antibodies) are assigned a specific colour. This permits the identification of any subset of cells on any 2-D dotplot by the location of the coloured dots.

Use of Indirect and Direct Antibody Conjugates

It is preferable to use directly-conjugated monoclonal antibodies for multicolour immunofluorescence as this reduces the number of wash steps preserving cell number and minimizes background binding and non-specific fluorescence. If indirect antibodies need to be used, a biotinylated antibody and streptavidin-fluorochrome are preferable. When using unconjugated antibodies and second step, anti-Ig antibodies, it is necessary to perform the two-step, indirect binding first before the addition of the directly-conjugated reagents to prevent indiscriminate binding of the anti-Ig secondary antibody to all primary antibodies.

Use of primary antibodies of a different isotype/subclass or species than other primary antibodies in the protocol can minimize this problem.

Fluorochrome Combinations

The use of specific fluorochromes is determined primarily by the available excitation wavelengths, optical filters and the number of detectors on the flow cytometer. The choice of fluorochromes that may be used in combination is determined by the ability to spectrally resolve their respective emissions for accurate quantitation of each fluorescent colour. The initial five-colour immunofluorescence protocol utilizes commercially-available fluorochromes and was developed for use on a standard FACS Vantage equipped with 3 lasers (blue 488 nm, UV 351-364 nm, red 633 nm or 647 nm) and five fluorescence PMTs (see Beavis and Pennline, Cytometry 15(4), 1994). The 488 nm line was used to excite FITC, PE and RED613 (PE-Texas Red tandem), the UV line to excite Cascade Blue and the red line (633 nm HeNe or 647 nm krypton) to excite APC.

The second protocol utilizes a novel tandem dye combining APC and CY7 (named ALLO-7, see Beavis and Pennline, Cytometry 24, pp 390-394, 1996) that can be excited by a red laser line, with resonance energy transfer from the APC to CY7, producing an emission at 780 nm. This unique tandem dye permits dual-colour immunofluorescence analysis using a 633 nm HeNe or a 647 nm krypton laser and can be used for two-laser, five-colour immunofluorescence analysis with an argon laser (488 nm) and FITC, PE and RED613.

To further optimize the data collection, it is useful to match fluorochrome-antibody combinations such that the brighter fluorochromes (higher emission intensities) are conjugated to antibodies used to detect antigens expressed at low levels or to antibodies with lower affinity of binding to the ligand.

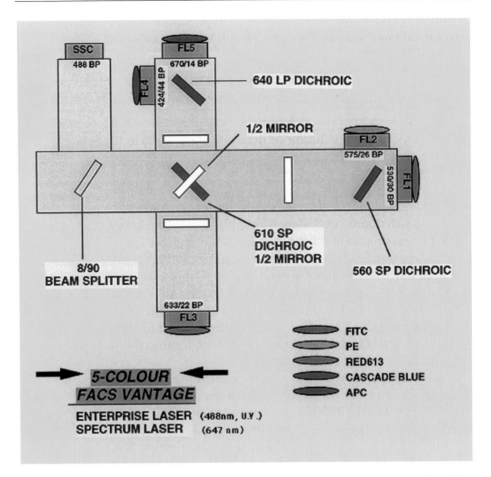

Fig. 3. Diagrammatic representation of the optical configuration of the FACS Vantage for collection of seven parameter, five-colour immunofluorescence data. This configuration uses 3 laser lines (488 nm and 351-364 nm from the Coherent Enterprise and 647 nm from Coherent Spectrum) with the 488 nm line as the primary beam for excitation of FITC, PE and RED613 and the UV and 647 nm lines as secondary beams delayed by 15 usec for excitation of Cascade Blue and APC respectively. The forward scatter (FSC) and side scatter (SSC) signals are collected from the 488 nm beam with a FSC diode (not shown) and the SSC PMT. "Reprinted from BioTechniques, volume 21 No. 3 (September), Andrew J. Beavis and Kenneth J. Pennline. Detection of Cell Surface Antigens Using Antibody-Conjugated Fluorospheres (ACF): Application for Six-Color Immunofluorescence. Pages 498-503, 1996 with kind permission of Eaton Publishing Co. 154 East Central Street, Natick, MA 01760, USA"

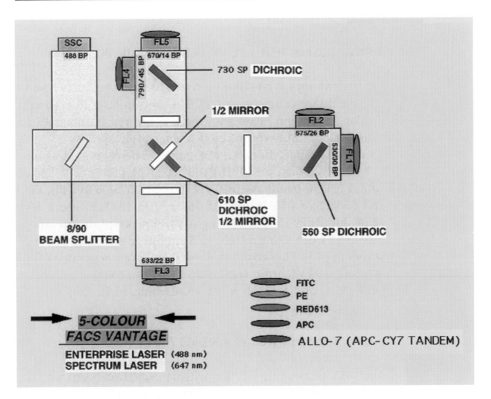

Fig. 4. Diagrammatic representation of the modified optical configuration of the FACS Vantage for collection of seven parameter, five-colour immunofluorescence data using ALLO-7. This configuration uses 2 laser lines (488 nm from the Coherent Enterprise and 647 nm from Coherent Spectrum) with the 488 nm line as the primary beam for excitation of FITC, PE and RED613 and the 647 nm line as a secondary beam delayed by 15 usec for excitation of ALLO-7 and APC respectively. The APC and ALLO-7 fluorescence emissions are separated with a 730 shortpass dichroic mirror and collected with a 670/14 bandpass (APC, FL5) and a 790/45 bandpass (ALLO-7, FL4) and provides for the ability to collect two colours from the red laser line for immunophenotyping. The forward scatter (FSC) and side scatter (SSC) signals are collected from the 488 nm beam with a FSC diode (not shown) and the SSC PMT. This is a modification of an original figure "reprinted from BioTechniques, volume 21 No. 3 (September), Andrew J. Beavis and Kenneth J. Pennline. Detection of Cell Surface Antigens Using Antibody-Conjugated Fluorospheres (ACF): Application for Six-Color Immunofluorescence. Pages 498-503, 1996 with kind permission of Eaton Publishing Co. 154 East Central Street, Natick, MA 01760, USA"

Optical Configurations

3-lasers, 5-colours (FITC / PE / RED613 / CASCADE BLUE / APC)

A standard optical configuration was used to spectrally resolve the five fluorescence emissions. The 488 nm laser line was the primary excitation beam with the UV and red beams (633 nm HeNe or 647 nm krypton) coincident and delayed by 15 to 20 usec. The primary fluorescence signals were separated using a 610 shortpass dichroic (SPDi) and 560 SPDi filters and collected with a 530/30 bandpass (BP) (FITC, FL1), 575/26 BP (PE, FL2) and 630/22 or 610/20 BP (RED613, FL3). [NOTE: use a 630/22 BP for RED613 when using a 647 nm krypton line and a 610/20 BP when using a HeNe 633 nm line to eliminate red laser scatter in FL3]. The secondary laser signals were separated with a 640 longpass dichroic (LPDi) and collected with a 670/14 BP (APC, FL4) and a 424/44 BP (Cascade Blue, FL5). (Fig. 3).

2-lasers, 5-colours (FITC / PE / RED613 / APC / ALLO-7)

The 488 nm laser line was the primary excitation beam with the red beam delayed 15 to 20 usec. The primary fluorescence signals were separated as described above. The secondary laser signals were separated with a 730 SPDi and collected with a 790/45 BP (ALLO-7, FL4) and a 670/14 BP (APC, FL5). The 730 SPDi and the 790/45 filters were custom manufactured by Omega Optical, Brattleboro, VT. (Fig. 4).

References

Beavis AJ and Pennline KJ. Simultaneous measurement of five cell surface antigens by five-colour immunofluorescence. Cytometry 15: 371-376; 1994.

Beavis AJ and Pennline KJ. Allo-7: A new fluorescent tandem dye for use in flow cytometry. Cytometry 24: 390-394; 1996.

Beavis AJ and Pennline KJ. Detection of cell surface antigens using antibody-conjugated fluorospheres (ACF): Application for Six-Colour immunofluorescence. BioTechniques 21: 498-503; 1996.

Detection of Cell Surface Antigens Using Antibody-conjugated Fluorospheres and Flow Cytometry

ANDY BEAVIS

Introduction

Fluorescent particles provide a versatile alternative to conventional antibody-fluorochrome systems for the detection of cell surface antigens. The fluorescent particles typically contain relatively large amounts of fluorescent material producing a high emission intensity which facilitates analysis of cellular antigens expressed at low levels. Reactive particles can easily be tagged with antibodies or other ligands for analysis of a wide range of cellular receptors and antigens. Furthermore, particles are available with unique excitation and emission properties which can be spectrally resolved from those of conventional fluorochromes expanding the multicolour capabilities. This protocol illustrates the use of antibody-conjugated fluorospheres (ACF) for detection of cell surface antigens that may be used in combination with multicolour applications using standard flow cytometers.

Materials

Equipment

Flow cytometer equipped with 1, 2 or 3 lasers, one of which is capable of red or far red emission wavelength.

Solutions

– fluorospheres antibody buffer - 1.55 ml phosphate buffer (0.1M, pH 7.0).

- PBS containing 10% FCS or 2% BSA
- RBC lysing buffer - 0.16M Tris-HCl, 0.17M NH$_4$Cl, pH 7.0.
- staining buffer (PBS, 2% FCS, 0.1% azide)

Procedure

Attaching Antibody to Fluorespheres

1. Vortex fluorospheres to disaggregate any clumps. Sonicate briefly if necessary.

2. Aliquot 250 μl of 0.45 μm SKY BLUE fluorospheres (1%wt/vol) into a polypropylene tube. Add 200 μl antibody (1mg/ml) and 1.55 ml phosphate buffer (0.1M, pH 7.0).

Note: Polypropylene tubes prevent adherence of particles to the sides of the tubes. The v-bottomed tubes (e.g. Nunc vials) produce a tight pellet of particles and allow complete removal of antibody and washing solutions

3. Aliquot 300 μl of 2.08 μm SKY BLUE fluorospheres (1%wt/vol) into a polypropylene tube. Add 100 μl antibody (1mg/ml) and 1.6 ml phosphate buffer (0.1M, pH 7.0).

Note: Adjust the amount of antibody incubated with particles of different diameters as the available particle surface area will be different. Also incubate *control fluorospheres* in buffer alone.

4. Vortex the mixtures and incubate for 1 hr at r.t.

5. Centrifuge (3000 x g, 15 min), discard the supernatant and wash the fluorospheres twice in 4 ml of PBS.

6. Resuspend the fluorospheres in PBS containing 10% FCS or 2% BSA and incubate for 1 hr at r.t.

Note: The proteins block any unoccupied "sites' on the surface of the particles.

7. Wash the fluorospheres X2 in PBS and resuspend in 4 ml PBS. Store at 4°C in the dark.

Note: Treat the antibody-conjugated particles just as you would a directly-conjugated monoclonal antibody.

8. Prepare single cell suspensions for cell types of interest. For tissue samples such as spleen and thymus, dissociate tissues with forceps and filter.

9. Remove contaminating RBC using lysing buffer. Resuspend cells in lysing buffer (1-5 ml), incubate for 5 min at r.t. and centrifuge at 250 x g, 5 min.

Staining of cells with antibody-conjugated fluorospheres

Note: RBC lysing buffer contains 0.16M Tris-HCl, 0.17M NH₄Cl, pH 7.0.

10. Discard supernatant, resuspend cells in 10 ml cold PBS and centrifuge as previous.

11. Wash cells X2 in staining buffer (PBS, 2% FCS, 0.1% azide), enumerate and resuspend at 1×10^7/ml.

12. Aliquot 100 µl cells/well into a 96-well plate (V-bottomed, polypropylene) and add 5 µl of a blocking antibody. Mix and incubate for 5 min at 4°C.

Note: The blocking antibody is used to minimize non-specific binding (e.g. an anti-Fc receptor antibody or an IgG of the same species of cell to be used).

Note: When using indirect immunofluorescence and second antibodies, ensure that the anti-Ig second antibody will not bind to the blocking antibody used in the previous step.

13. Add the antibody-conjugated fluorosphere or unconjugated, control fluorospheres (10-100 µl of each), mix gently and incubate for 30 min at 4°C.

Note: Titrate the volume / concentration of fluorospheres required for optimal staining.

14. Wash cells X2, transfer to tubes for flow cytometric analysis and resuspend in 500 µl staining buffer.

15. Align flow cytometer as per standard protocol. Set gates and PMT settings. Obtain FSC/SSC signals for unstained cells, gate population of interest and adjust PMT voltages to set autofluorescence signals (log) on scale.

Flow Cytometric Analysis

Note: Collect the SKY BLUE fluorescence with a 680 shortpass dichroic and a 730/30 bandpass filter.

16. Check that the fluorescence for the single-colour ACF positive sample remains on scale. Check positive controls and set compensations as required.

Note: SKY BLUE can be collected simultaneously with APC. There is some spectral overlap between SKY BLUE and APC.

17. Establish a 2d-plot of FSC versus SKY BLUE and SSC versus SKY BLUE (Fig. 1) for gating of the SKY BLUE positive and negative cells.

18. Acquire listmode data for unstained cells, cells with unconjugated particles and ACF samples at these instrument settings.

19. Run a sample of ACF alone and with non-fluorescent beads to check that the positive signal observed is due to fluorescence and not to increased light scatter due to the particle attached to the cell surface.

Note: The particles-cell complex may result in an increase in the SSC signal. Analysis of SKY BLUE and non-fluorescent particles of similar size should show that only the SKY BLUE particles show positive emission in the fluorescence channel.

20. Look at your cells under the microscope to visualize the particles attached to the cells. This will be useful to confirm and interpret what you see from the flow cytometer.

Multicolour Immunofluorescence

21. Stain cells with conventional monoclonal antibody conjugates as described above. Then incubate cells with ACF, wash and analyse for fluorescence as for standard multicolour analysis.

Data Analysis

24. For data analysis, use a large scatter gate as the attachment of the ACF to the cells may increase forward and side scatter signals slightly.

25. A 2D-plot of SSC and ACF fluorescence can be used to identify the ACF(+) cells for gating in or out of further analysis (Fig. 1). Another strategy for multicolour immunofluorescence data analysis is to use cluster analysis as shown in Fig. 2. Here, cells of a given phenotype are assigned a specific colour so that each subset of cells can be identified on any two-parameter dot-plot by the location of the subpopulation dot colour.

Fig. 1. Identification of macrophages in a murine spleen cell sample with CD11b-SKY BLUE ACF. Murine spleen cells were stained with monoclonal antibodies specific for six distinct cell surface antigens and analysed for fluorescence using a FACS Vantage (Becton Dickinson Immunocytometry Systems, San Jose, CA). The CD11b-SKY BLUE(+) and CD11b-SKY BLUE(-) cells can be identified from a single parameter histogram of CD11b-SKY BLUE (R2 and R1 respectively) or from a two-parameter plot of FSC and CD11b-SKY BLUE [CD11b(+) cells gated in R3]. The CD11b positive and negative cells can then be gated onto separate 2D dotplots for further analysis. For this example, the macrophages were gated out to permit five-colour analysis of the lymphocyte subpopulations. "Reprinted from BioTechniques, volume 21 No. 3 (September), Andrew J. Beavis and Kenneth J. Pennline. Detection of Cell Surface Antigens Using Antibody-Conjugated Fluorospheres (ACF): Application for Six-Color Immunofluorescence. Pages 498-503, 1996 with kind permission of Eaton Publishing Co. 154 East Central Street, Natick, MA 01760, USA"

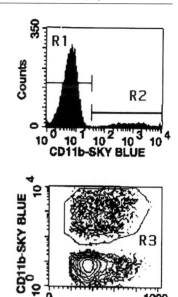

Fluorospheres that are useful for flow cytometry

Typically, monoclonal antibodies are available as conjugates of the more frequently used fluorochromes (FITC, PE, RED613, PE-APC or PE-CY5) or biotinylated for use as indirect reagents. Fluorescent particles with emissions similar to these fluorochromes would be used in place of these. However, the use of particles with unique emissions that can be spectrally resolved from the commonly used fluorochromes and other fluorescent reagents would provide versatility and compatability with existing protocols. The TRANSFLUOSPHERES 720 (TSF720) (Molecular Probes, Eugene, OR) have an excitation maximum near 488 nm with a corresponding emission at 720 nm. This large Stokes shift permits use of the particles with other 488 nm-excited dyes and was used for 5-colour analysis using a single 488nm laser (Beavis and Pennline, unpublished data). The LIGHT YELLOW fluorospheres (Spherotech Inc, Libertyville, IL) have an excitation in the UV region (351-364nm) with an emission around 470nm. The SKY BLUE fluorospheres (Spherotech) can be excited with a red laser (HeNe 633nm, krypton 647 nm) with a bright emission at 730 nm that is spectrally resolved from FITC, PE, PE-CY5 and CY5 or APC. The SKY BLUE

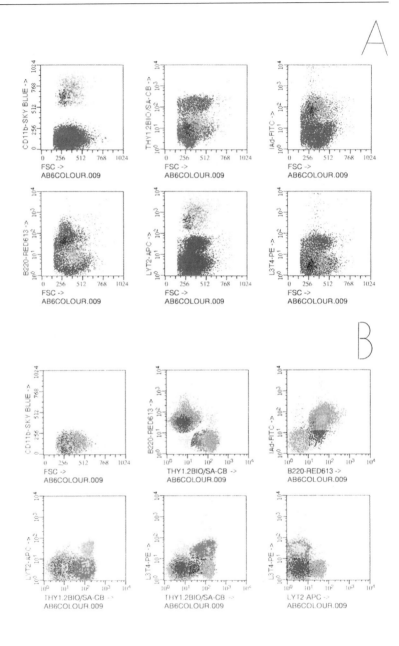

◀ **Fig. 2.** Phenotypic analysis of murine leukocytes by six-colour immunofluorescence. Murine spleen cells were stained with six monoclonal antibodies and analysed for fluorescence using a FACS Vantage (Becton Dickinson Immunocytometry Systems, San Jose, CA) and a custom set of optics. The listmode data was analysed using the ATTRACTORS software (Becton Dickinson Immunocytometry Systems, San Jose, CA) whereby the specific phenotype of a subpopulation is gated and assigned a colour. Each subpopulation may be identified on any two-dimensional dotplot by the position of the coloured dots.
A: Resolution of SKY BLUE (FL6) from the other five fluorescence signals: CD11b(-) = RED; CD11b(+) = BLUE and phenotypic analysis of the CD11b(+) subpopulation of splenocytes: CD11b(+)/Thy1.2(+) = YELLOW; CD11b(+)/B220(+) = GREEN.
B: Phenotypic analysis of the CD11b(-) cells: Thy1.2(+)/B220(-)/L3T4(-)/Lyt2(+)/IAd(-) = PINK; Thy1.2(+)/B220(-)/L3T4(+)/Lyt2(-)/IAd(-) = LIGHT BLUE; Thy1.2(-)/B220(+)/L3T4(-)/Lyt2(-)/IAd(-) = RED; Thy1.2(-)/B220(+)/L3T4(-)/Lyt2(-)/IAd(+) = GREEN; Thy1.2(+)/B220(+)/L3T4(-)/Lyt2(-)/IAd(+) = ORANGE; Thy1.2(-)/B220(-)/L3T4(+)/Lyt2(-)/IAd(+,-) = YELLOW. "Reprinted from BioTechniques, volume 21 No. 3 (September), Andrew J. Beavis and Kenneth J. Pennline. Detection of Cell Surface Antigens Using Antibody-Conjugated Fluorospheres (ACF): Application for Six-Color Immunofluorescence. Pages 498-503, 1996 with kind permission of Eaton Publishing Co. 154 East Central Street, Natick, MA 01760, USA"

fluorospheres therefore provide the ability for dual-colour analysis using a red laser as well as up to 6-colour analysis using a standard FACS Vantage (Beavis and Pennline, BioTechniques 21(3), pp 498-503, 1996).

Conjugation of the Fluorospheres with Monoclonal Antibodies

The fluorescent particles are available with specific chemical moeities for attachment of ligands to the surface. The simplest and easiest conjugation method though is passive adsorption of monoclonal antibodies to the polystyrene fluorospheres. The exact choice of particle (polymer, chemical moeities, fluorochrome, size) will depend upon the application and may require some experimentation to optimize these variables. The particles that were used in these studies were obtained from:

• SPHEROTECH, Libertyville, IL, Tel.(847) 680 8922, Fax (847) 680 8927

- MOLECULAR PROBES, Eugene, OR, Tel.(541) 465 8300, Fax (541) 344 6504. Another useful source for particles, information and advice is PAINLESS PARTICLES[TM], Bangs Laboratories, Carmel, IN Tel. (317) 844 7176, Fax (317) 575 8801.

Optical Configurations

Red Laser, 2-colours (APC / SKY BLUE)

A custom optical configuration was used to spectrally resolve the fluorescence emissions of the APC and SKY BLUE signals. The emissions were separated with a 680 SPDi and collected with a 670/14 BP for APC and a 730/30 BP (Omega Optical, Brattleboro, VT) for SKY BLUE. Signals were well resolved with compensation settings of 5% and 20% required to correct for the spectral overlap of APC into SKY BLUE and SKY BLUE into APC respectively.

3-lasers, 6-colours
(FITC / PE / RED613 / CASCADE BLUE / APC / SKY BLUE)

A standard optical configuration was used to spectrally resolve the five conventional fluorescence emissions and was combined with a custom set of optics for resolution of the SKY BLUE signal (Fig. 3). The 488 nm laser line was the primary excitation beam with the UV and red beams (633 nm HeNe or 647 nm krypton) coincident and delayed by 15 to 20 usec. The primary fluorescence signals were separated using a 610 shortpass dichroic (SPDi) and 560 SPDi filters and collected with a 530/30 bandpass (BP) (FITC, FL1), 575/26 BP (PE, FL2) and 630/22 or 610/20 BP (RED613, FL3). [NOTE: use a 630/22 BP for RED613 when using a 647 nm krypton line and a 610/20 BP when using a HeNe 633 nm line to eliminate red laser scatter in FL3]. The secondary laser signals were separated with a 640 longpass dichroic (LPDi) and collected with a 670/14 BP (APC, FL4) and a 424/44 BP (Cascade Blue, FL5). The SKY BLUE fluorescence (FL6) was collected in place of side scatter using a 680 SPDi, a 730/30 BP and a red-sensitive PMT.

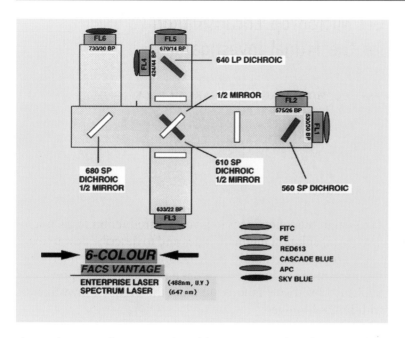

Fig. 3. Diagrammatic representation of the custom optical configuration of the FACS Vantage for collection of seven parameter, six-colour immunofluorescence data. This configuration uses 3 laser lines (488 nm and 351-364 nm from the Coherent Enterprise and 647 nm from Coherent Spectrum) with the 488 nm line as the primary beam for excitation of FITC, PE and RED613 and the UV and 647 nm lines as secondary beams delayed by 15 usec for excitation of Cascade Blue (UV) and APC (647 nm) and SKY BLUE (647 nm). The SKY BLUE signal was collected in place of side scatter (SSC) using a 680 shortpass dichroic half-mirror, a 730/30 bandpass filter and a red sensitive PMT. The forward scatter (FSC) signal was collected from the 488 nm beam with a FSC diode (not shown). "Reprinted from BioTechniques, volume 21 No. 3 (September), Andrew J. Beavis and Kenneth J. Pennline. Detection of Cell Surface Antigens Using Antibody-Conjugated Fluorospheres (ACF): Application for Six-Color Immunofluorescence. Pages 498-503, 1996 with kind permission of Eaton Publishing Co. 154 East Central Street, Natick, MA 01760, USA"

This work was performed in collaboration with Dr. Kenneth Pennline, Director of Biopharmaceutical Support Services, Cytometry Associates, Brentwood, TN.

References

Beavis AJ and KJ. BioTechniques 21: 498-503; 1996

Gel Microdrop Encapsulation for the Frugal Investigator

DAVID C. MCFARLAND AND GARY DURACK

Introduction

The goal of the following is to give an introduction to gel micro-drop encapsulation methodologies and to describe how to make microdrops using, mainly, basic laboratory equipment. The encapsulation procedure described is a modification of previously published protocols tailored to meet the needs of investigators with limited budgets.

Background

Microdrop encapsulation is a method for isolating single cells in agarose microenvironments. Briefly, a cell suspension is mixed with an agarose solution and then added to stirring oil. The emulsion is chilled to gel the agarose and the microdrops are harvested.[1]

Poisson statistics predicts the cell distribution among micro-drops. The two main variables are the size of the microdrops and the cell concentration. In most cases, the number of cells desired per microdrop is one. Proper values for both variables must be chosen to give the desired distribution. To this end, it is important to start with a **single cell suspension**. For a more in depth discussion, with formulae, see references 1 and 2.

Rapid diffusion of small molecules (<250 kDa) through the agarose matrix is achieved if microdrops are sufficiently small. Microdrop diameter should be only 5-10 times the diameter of the encapsulated cell to maintain unihibited exchange of molecules between the cell and the microdrop exterior.[1] Encapsulated cells should remain viable in a normal culture environment if this condition is met. This rapid diffusion also allows labeling of cells with antibodies (Ab), after encapsulation, with very little non-specific retention of Ab in the gel matrix. The microdrops may be analyzed by **flow cytometry** and sorted with a **cell-sorter** in the same manner as free cells.

Proliferative cells will grow out of microdrops after prolonged incubation[1]and/or addition of agarase.[3] In our hands, three Chinese hamster ovary cell lines tested broke free of the microdrops and grew normally after a week in culture, without the addition of agarase (McFarland and Durack, unpublished observation).

Subprotocol 1
Agarose Preparation

Making a 4% agarose solution seems straightforward. However, a solution of 4% agarose is fairly saturated and the agarose does not readily dissolve at this concentration. A quick microwave method, used routinely for making 1-2% agarose, and an autoclave method were both attempted to no avail. Therefore, an easy method for making 4% agarose is also included in the protocol section.

▓▓ Materials

Solutions

25 ml of a 4% agarose solution consists of 1 g of agarose in 25 ml water.

▓▓ Procedure

1. Measure 1 g of agarose on weighing paper or boat and set aside. **Weigh agarose**

2. Place 1 inch stir bar in 100 ml beaker and tare. **Tare vessel**

Note: The vessel is tared so that the solution can be quantified by weight.

3. Add about 30 ml deionized water to beaker and place on hotplate on "high". **Boil**

Note: The volume does not need to be monitored closely since the solution will be quantified by weight.

4. Heat until boiling while stirring.

Add agarose 5. Add a small amount of agarose to boiling water.

Note: Wait for the water to boil before adding any agarose. Do not add a bolus amount or it will take an enormous amount of time to dissolve.

6. Continue to stir and boil until initial agarose has dissolved.

7. Continue to add agarose in small amounts. Wait for it to dissolve before adding more.

Add deionized water 8. Add deionized water as deemed necessary.

Note: Add more dH_2O as it boils off. The more volume, the easier the agarose will go into solution. Boil it down after everything has dissolved.

Quantify solution 9. Weigh vessel. Adjust to 26 g.

Note: 25 ml of a 4% agarose solution consists of 1 g of agarose in 25 ml water. Ignoring the volume contribution of the agarose, 25 ml of 4% agarose should weigh 26 g. (1 g agarose + 25 ml water at 1 g/ml).

Sterile filtration 10. Filter through 0.22 μm filter directly into sterile 1.5 ml microfuge tubes, placing about 1 ml in each tube.

Note: Under a sterile hood, use a 0.22 μm luer lock filter (Sterivex-GS, Millipore) with a 30 cc syringe and dispense directly into 1.5 ml lock-cap microfuge tubes that have been previously autoclaved.

Agarose storage 11. Label tubes and place at 4°C until needed.

Note: This should be enough agarose to do about 75 experiments.

Subprotocol 2
Gel Microdrop Encapsulation

Microdrop-making apparatus

A relatively inexpensive microdrop-making apparatus can be constructed with a fast stirplate (Thermix stirrer model 120S, Fisher), a 100 ml titration beaker, a smooth 1 inch stir bar, a ring stand with clamps, a small lab jack and a 1 ml threaded plunger syringe (Hamilton). Water baths at 37° C and 70° C are also required.

To apply Poisson statistics effectively, the microdrops need to be fairly uniform in size. The threaded plunger syringe, along with the lab jack and ringstand, are very helpful in this regard. In addition, the tip on an 18 gauge needle may be ground down to a blunt end in order to get proper dispensation from the syringe into the oil. This has the added benefit of eliminating the safety hazard of the sharp needle. If uniform microdrops are not important for the application of interest, a regular syringe or even a Pasteur pipette may be used instead of the threaded plunger syringe.

Alternate microdrop formation method

Microdrops can also be formed by adding a cell suspension to oil and vortexing.[4,11] The stirring method was attempted first and gave satisfactory results. The vortex method was not investigated extensively, but the microdrops formed seemed to be less uniform in size. Once again, the method of choice depends on the demands of the application.

▪▪ Materials

Equipment

- microdrop-making apparatus
- Cell sorter

Solutions

4% agarose solution.

▓ ▓ Procedure

Assemble microdrop-making apparatus

1. Set up microdrop-making apparatus as described in the text and pictured in fig. 1.

Melt agarose

2. Melt 1 ml aliquot of 4% agarose in 70°C water bath (about 5 minutes).

3. Cool agarose in 37°C water bath.

Bring oil to proper temperature

4. Place 18 ml of DMPS in sterile titration beaker with 1" stir bar. Cover with parafilm and place in 37°C water bath.

Note: Dimethylpolysiloxane (DMPS, Sigma) is one of several emulsifiers mentioned in the references listed.

Make cell suspension

5. Count cells and suspend 4.5×10^6 cells in 0.3 ml medium and place in 37°C water bath.

Note: Cells may be surface labeled before encapsulation, if desired. (This facilitates discrimination of empty and occupied microdrops.)

Microdrop-Making Apparatus

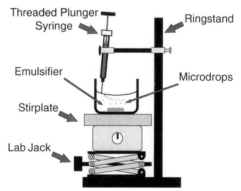

Fig. 1

6. Add 0.3 ml 4% agarose solution to cell suspension to get a final agarose concentration of 2%.   **Add agarose**

Note: The agarose solution is very viscous. Aspirate and dispense slowly to prevent soiling of pipetman.

7. Vortex until homogeneous; replace in water bath.

8. Place beaker with oil on stirplate and begin stirring.   **Place beaker on stirplate**

9. Load syringe with agarose-cell suspension.   **Load syringe**

10. Lower syringe into the beaker.   **Adjust syringe**

11. Adust syringe so that the needle is near the surface of the stirring oil but not dipping into the oil (Fig. 2).

Note: The cell suspension/agarose solution should be sheared from the tip of the needle in a fine stream as it exits. The needle should not be dipped into the oil to dispense.

12. Begin adding the agarose solution to the oil. Add the solution evenly over about a minute.   **Making the microdrops**

Note: As the agarose solution is added, the height of the syringe may need to be adjusted to keep it out of the oil. The lab jack is very useful for this fine adjustment.

Proper Dispensation into Emulsifier

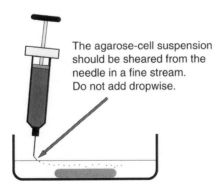

The agarose-cell suspension should be sheared from the needle in a fine stream. Do not add dropwise.

Fig. 2

Note: The stirring speed as well as the rate of agarose addition can affect microdrop size. Experiment to find the speeds that give the drop size needed.

Note: If done properly, the oil will become cloudy as the emulsion is formed. If individual drops are large enough to be seen, they are probably too big for flow cytometry.

Gelling the agarose

13. Place the emulsion in an ice bath and stir slowy for 12 minutes to solidify the microdrops.

Washing the microdrops

14. Place emulsion on 6 ml layers of cold medium in two 50 ml conical centrifuge tubes. Centrifuge at 1300g for 12 minutes at 4°C.

Note: If the cells/microdrops are to be manipulated further, leave in medium. If they are ready to run them on the flow cytometer, wash with PBS instead.

15. Remove pellets by aspirating with a 9" Pasteur pipette. Pool drops and place in 50 ml conical tube. Fill tube to 50 ml with cold medium and spin again as above.

Note: Do not try to aspirate the oil off the top. Just stick the pipette down through the oil layer and aspirate the aqueous phase. However, try to minimize the amout of oil transfered.

16. Aspirate off excess medium. Resuspend the pellet in a few ml of medium and transfer to a 15 ml conical tube. Fill tube to 15 ml with cold medium and spin again.

17. Resuspend pellet in 1-2 ml medium and divide among 4 microfuge tubes. Spin at 2500 g for 3 minutes and resuspend in 200 µl medium. Place samples on ice.

Microscopy

18. Place a few microliters of your microdrop suspension on a microscope slide and cover with a cover slip. Visualize with a light/fluorescence microscope.

Note: Visualizing the microdrops by light microscopy before continuing is highly recommended to make sure that the expected results were obtained. Empty versus occupied microdrops(singly or multiply) should be easily distinguished. Fluorescence microscopy is also useful if the cells were labeled before encapsulation.

19. Use a hemacytometer to estimate the ratio of occupied/ empty drops. (Optional)

Note: The cell density used in this protocol should give approximately 10% *singly-occupied* microdrops if the microdrops are about 5 times cell diameter. Very few drops should contain multiple cells and the remainder should be empty. If this is not the case, then either a *single cell suspension* of the proper density was not obtained or the microdrops formed are not the proper size.

20. Add a pre-titrated amount of Ab to microdrop suspension. Vortex gently for 20 seconds. Place on ice.

Stain encapsulated cells

Note: As mentioned above, Ab should pass freely through the gel matrix, so staining of encapsualted cells is possible. However, the microdrops may settle out quickly, so longer incubation times and frequent resuspension may be necessary.

21. Stain for one hour, resuspending every 5 minutes by gently vortexing for 5 seconds.

22. Wash microdrops in PBS. Resuspend in PBS and place on ice until needed.

23. Dilute all samples 1:4 with PBS, to prevent clogging of the sampling system.

Flow Cytometry

24. Make the necessary modifications to the flow cytometer

Note: Some modifications of the flow cytometer may need to be made when running gel microdrops. A few suggestions are to use at least a 100 μm flow cell, a larger gauge sample insertion needle and sample tubing and higher sheath pressure.
FALS vs Log 90 has not been very useful to identify populations of interest (ie occupied microdrops). The gel matrix seems to mask the light scatter properties of cells to some extent. Cells surface labeled with a PE conjugate allowed for gating on Log 90 vs PE to identify occupied microdrops. In addition, by triggering on PE, signals from empty drops were ignored.

25. Analyze samples.

26. Run a sample of 10% bleach for 2 minutes.

Clean the flow cytometer

27. Run a sample of 70% ethanol for 2 minutes.

Note: A normal cleaning procedure of 10% bleach followed by 70% ethanol seems adequate to clear out agarose debris.

Applications

Microdrop encapsulation has already been employed in various experimental applications.

Cell contact inhibition assay

Microdrop encapsulation inhibits cell-to-cell contact. Investigation of processes that may require cell-to-cell contact (eg T-cell activation) can be scrutinized.[1,4,5]

Microdrop in situ hybridization (MISH)

Stabilizing chromosomes in gel microdrops allows detection of hybridization by flow cytometry.[6]

Growth assay

Single cells are encapsulated in gel microdrops, allowed to proliferate, and then analyzed by flow cytometry. Slow growing cells can be manipulated and sequestered by cell-sorting, even if they represent a rare population.[1,7,8]

Cytotoxicity assay

Cytotoxicity of a toxin is determined by inhibition of proliferation of encapsulated cells using a growth assay similar to above.[1,8,9]

Secretion-capture assay

This is an elegant assay for protein secretion from single cells. An Ab specific for the protein of interest is incorporated into the gel matrix via a biotin-avidin linkage after cells are encapsulated. The cells are allowed to secrete for a period of time and then a fluoresceinated reporter Ab is used to label captured secreted protein. The microdrops are analyzed by flow cytometry and fluorescence is correlated with captured protein.[1,3,10,11]

References

1. Weaver JC, Bliss JG, Harrison Giet al. Microdrop technology: A general method for separating cells by function and composition. Methods: A Companion to Methods in Enzymology. 2. 1991; 3:234-247.
2. Weaver JC. Gel microdroplets for microbial measurement and screening: Basic principles. Biotechnology and Bioengineering Symp. No. 17., 1986.
3. Gray F, Kenney JS, Dunne JF. Secretion capture and report web: Use of affinity derivatized agarose microdroplets for the selection of hybridoma cells. J. of Immunological Methods. 1995; 182:155-163.
4. Su MW-C, Walden PR, Eisen HN. Cognate peptide-induced destruction of CD8+ cytotoxic T lymphocytes is due to fratricide. J. of Immunology 1993; 151:2:658-667.
5. LaSalle JM, Toneguzzo F, Saadeh M, et al. T-cell presentation of antigen requires cell-to-cell contact for proliferation and anergy induction. J. of Immunology 1993; 151:2:649-657.
6. Nguyen B-T, Lazzari K, Abebe J, et al. In situ hybridization to chromosomes stabilized in gel microdrops. Cytometry 1995; 21:111-119.
7. Gift EA, Park HJ, Paradis GA, et al. FACS-based isolation of slowly growing cells: Double encapsulation of yeast in gel microdrops. Nature Biotechnology 1996; 14:884-887.
8. Ryan C, Nguyen B-T, Sullivan, SJ. Rapid assay for mycobacterial growth and antibiotic susceptibility using gel microdrop encapsulation. J. of Clinical Microbiology 1995; 33:1720-1726.
9. Goguen B, Kedersha N. Clonogenic cytotoxicity testing by microdrop encapsulation. Nature 1993; 363:189-190.
10. Kenney JS, Gray F, Ancel M-H, Dunne JF. Production of monoclonal antibodies using a secretion capture report web. Biotechnology 1995; 13:787-790.
11. Powell KT and Weaver JC. 1990. Gel microdroplets and flow cytometry: Rapid determination of antibody secretion by individual cells within a cell population. Biotechnology 8:333-337.

Section 4

Reporter Genes and Cell Trackers

ROCHELLE A. DIAMOND

This section describes and gives procedures for one of the most cutting edge and exciting technologies in biology today. Reporter genes, enhancer-traps, gene expression systems, no matter what you call them, are all answering many biological questions at the forefront of developmental, cell, and molecular biology. The power of flow cytometry in conjunction with cell sorting enhances these areas of investigations by providing a means to evaluate, isolate, and purify cells transfected with these genes and DNA expression regulators. Steven Kain presents a thorough discussion of the green fluorescent reporter system and its many variants. Steven Fiering describes the FACS-Gal system, which uses the lac-Z gene to produce β – galactosidase enzyme that cleaves a non-fluorescent substrate into a fluorescent reporter product. Poulos et al. follow with a procedure for using a bicistronic vector system for following gene expression. Raghu et al. provide a new cell surface receptor reporter system using an antibody for the expressed nerve growth factor receptor to evaluate transfected cells. In addition, the new Miltenyi Biotec system has two different vectors for truncated cell surface molecules which are targeted for enrichment by antibodies conjugated to super-paramagnetic beads.

Following cells by reporter genes has mostly been employed for isolation and purification, although they can be used for cell tracking. Other kinds of cell trackers have come into vogue recently to follow migrating cells, measure their lifespan, and view their offspring both *in vivo* and *in vitro*. Andrew Beavis and Poon et al. provide us with protocols and information on the lipophilic membrane dyes of the PKH family. Ken Ault and Cathy Knowles provide an alternative methodology for cell tracking by *in vivo* biotinylation.

Flow Cytometric Analysis of GFP Expression in Mammalian Cells

STEVEN R. KAIN

Introduction

In the jellyfish *Aequorea victoria*, light is produced when energy is transferred from the Ca^{2+}-activated photoprotein aequorin to green fluorescent protein or GFP (Fig. 1).[1-3] This process occurs in specialized photogenic cells located at the base of jellyfish umbrella, where each protein is found at very high concentrations. The cloning of the wild-type GFP gene (wt GFP)[4,5] and its subsequent expression in heterologous systems[6,7] has established GFP an important reporter protein for the analysis of gene expression and protein localization in a wide variety of experimental designs. When expressed in either eukaryotic or prokaryotic cells and illuminated by blue or UV light, wt GFP emits a bright green fluorescent signal which is easily detected by fluorescence microscopy, flow cytometry, or other fluorescence imaging techniques. GFP fluorescence is species-independent and does not require additional cofactors, substrates, or gene products - the protein is "naturally fluorescent". Moreover, detection of GFP and its variants can be performed in living samples, and is ideally suited to real time analysis of molecular events.

GFP, like the chemical fluorophore fluorescein, has a fluorescence quantum yield (QY) of about 70-80%,[8] although the extinction coefficient (Em) for wt GFP is much lower (Table 1). Nevertheless, in fluorescence microscopy, wt GFP has been found to give greater sensitivity and resolution than staining with fluorescently tagged antibodies.[7] GFP has the advantage of being both resistant to photobleaching, and of avoiding background caused by nonspecific binding of primary and secondary antibodies.[7] Although binding of multiple antibody molecules to a single target offers a potential amplification not available for

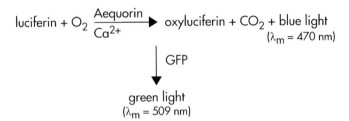

luciferin + O_2 $\xrightarrow[Ca^{2+}]{Aequorin}$ oxyluciferin + CO_2 + blue light

(λ_m = 470 nm)

↓ GFP

green light
(λ_m = 509 nm)

Fig. 1. Energy transfer reactions occurring in the photocytes of *Aequorea victoria* . In the jellyfish this process involves the interplay of two photoproteins - *Aequorin* and GFP. When used as a genetic reporter GFP is excited directly with either UV or blue light, thereby bypassing the *Aequorin* catalyzed reaction

GFP, this is offset because neither labeling of the antibody nor binding to the target is 100% efficient.

In addition to expression of GFP alone, GFP and it's variants have also been used extensively to express chimeric genes. In many cases, hybrid genes encoding either N- or C-terminal fu-

Table 1. Fluorescence properties of wild type GFP and selected GFP variants.

Variant	Mutations	Excitation max. (nm)	Emission max. (nm)	Em	QY	Comments
Wild type	none	395 (470)	509	21.000 (7,150)	0.77	
S65T	S65T	489	511	39,200	0.66	Single 489 exc. peak; brighter fluorescence
GFP mut1 EGFP	F64L S65T	488	507	~250,000	~0.7	Single 488 exc. peak; brighter fluorescence
"cycle 3"	F99S M153T V163A	395	509	N/D	N/D	Single 395 exc. peak; brighter fluorescence
P4-3	Y66H Y145F	381	445	14,000	0.38	Blue emission
EBFP	F64L S65T Y66H Y154F	380	440	37,000	0.20	Blue emission; resistance to photobleaching
10C	T203Y S65G V68L S72A	513	527	36,500	0.63	Yellow-green emission; brighter fluorescence

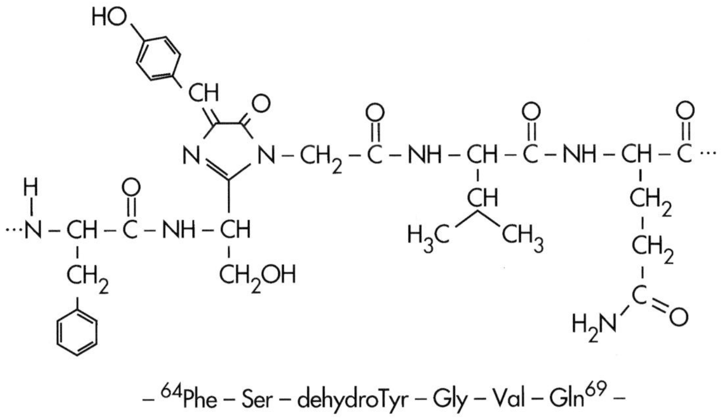

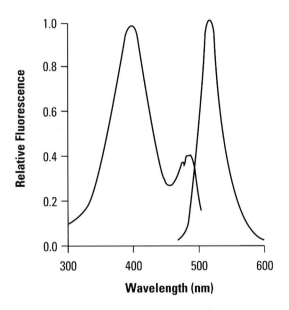

$- ^{64}Phe - Ser - dehydroTyr - Gly - Val - Gln^{69} -$

Fig. 2. Chemical structure of the GFP chromophore containing the Ser-Tyr-Gly cyclic tripeptide. The figure shows a hexapeptide found in GFP which has the identical absorption spectra as full-length GFP, but is completely non-fluorescent

Fig. 3. Fluorescence excitation and emission spectra of wild type GFP

sions to GFP retain the normal biological activity of the heterologous partner, as well as maintaining fluorescent properties similar to native GFP.[7,9,10,11] The use of GFP and its variants in this capacity provides a "fluorescent tag" on the protein, which allows for *in vivo* localization of the fusion protein. GFP fusions can provide enhanced sensitivity and resolution in comparison to standard antibody staining techniques,[7] and the GFP tag eliminates the need for fixation, cell permeabilization, and antibody incubation steps normally required when using antibodies tagged with chemical fluorophores. Lastly, use of the GFP tag permits real-time kinetic studies of protein localization and trafficking. [9,7,12-14]

The GFP chromophore

A remarkable feature of GFP is that the protein is naturally fluorescent, encoding the fluorescent moiety within the primary amino acid sequence. The GFP chromophore consists of a cyclic tripeptide derived from Ser-Tyr-Gly at positions 65-67 in the protein (Fig. 2),[15]. GFP is 238 amino acids in length, and there have been no reports of significant truncations in GFP (other than a few amino acids from the C-terminus) which retain fluorescence. The crystal structures of GFP and a variant termed S65T (see discussion below) have revealed a tightly packed "β-can" structure enclosing an α-helix containing the chromophore.[16,17] This structure provides the proper environment for the chromophore to fluoresce by excluding solvent and oxygen. Nascent GFP is not fluorescent, since chromophore formation occurs post-translationally with a significant lag time.[18] The chromophore is formed by a cyclization reaction and an oxidation step that requires molecular oxygen.[19,20] These steps are either autocatalytic or use factors which are ubiquitous, since fluorescent GFP forms in a broad range of cells and organisms. Chromophore formation may be the rate-limiting step in generating the fluorescent protein, especially if oxygen is limiting.[19] The wt GFP absorbs UV and blue light with a maximum peak of absorbance at 395 nm and a minor peak at 470 nm and emits green light maximally at 509 nm (Fig. 3).[3,6]

Subprotocol 1
DNA transfection Procedure I by CalPhos Maximizer transfection

▪▪ Materials

Equipment

There are a variety of commercial instruments that one can use for   **Cytometers**
detection of GFP and GFP variants. Published flow cytometry data
has been reported for the following instruments from Becton Dickinson, Inc.: FACScan, FACStar^{Plus}, FACS Calibur, and FACS Vantage. There have also been reports using the EPICS Elite-ESP
cytometer from Coulter Corporation. Most instruments used
for detection of GFP and the red-shifted GFP variants are equipped
with argon ion lasers tuned to 488 nm excitation. The EPICS Elite-
ESP instrument equipped with a krypton ion mixed-gas laser has
been used to detect the blue emission variant P4[31].

Solutions

- Cell culture medium (e.g., Dulbecco's modified Eagle med-   **Materials**
  ium [DMEM] or another appropriate growth medium for   **required**
  mammalian cells in culture)
- Fetal bovine serum, newborn calf serum, or equivalent, to
  supplement the growth medium

Phosphate buffered saline (PBS) (pH 7.4)

	Final concentration	To prepare 2 L
Na_2HPO_4	58 mM	16.5 g
NaH_2PO_4	17 mM	4.1 g
NaCl	68 mM	8.0 g

Dissolve components in 1.8 L of ddH$_2$O. Adjust to pH 7.4 with
0.1 N NaOH. Add ddH$_2$O to a final volume of 2 L. Store at
room temperature.
- 1 x Trypsin/EDTA (Life Technologies #25300-054)

Preparation

B. CalPhos Maximizer transfection

Any calcium phosphate transfection procedure may be used, but we recommend using the CalPhos Maximizer Transfection Kit for the highest transfection efficiencies with an incubation of only 23 hours. This protocol is designed for use with adherent cultures growing in 10-cm tissue-culture plates. If you are using plates, wells, or flasks of a different size, adjust the components in proportion to the surface area of the container you are using.

For each transfection, prepare solution A and solution B in separate sterile tubes.

Solution A: add components in the following order:
- 1020 µg GFP plasmid DNA
- Sterile H_2O
- 40 µl CalPhos Maximizer
- 62 µl 2 M Calcium Solution
- 0.5 ml Total Volume

Solution B
- 0.5 ml 2 x Hepes-buffered saline (HBS)

Note: To reduce variability when transfecting multiple plates with the same plasmid DNA, prepare master solutions A and B sufficient for all plates.

▓ ▓ Procedure

CalPhos Maximizer transfection

Note: All steps of the following protocol should be performed in a sterile tissue culture hood

Prepare the cells

1. Plate the cells the day before the transfection experiment. The cells should be 50-80% confluent the day of transfection. Generally, we plate $1-2 \times 10^6$ cells/10-cm plate.

Note: Any calcium phosphate transfection procedure may be used, but we recommend using the CalPhos Maximizer Transfection Kit for the highest transfection efficiencies with an incubation of only 2-3 hours. This protocol is designed for use with adherent cultures growing in 10 cm tissue-culture plates. If you are using plates, wells, or flasks of a different size, adjust the components in proportion to the surface area of the container you are using.

2. 0.5-3 hr prior to transfection, replace culture medium on plates to be transfected with 9 ml of fresh culture medium per 10 cm plate.

3. For each transfection, prepare solution A and solution B in separate sterile tubes.

Prepare transfection reagent solutions

Note: To reduce variability when transfecting multiple plates with the same plasmid DNA, prepare master solutions A and B sufficient for all plates.

4. Carefully add solution A dropwise to solution B while gently vortexing solution B. (Alternatively, blow bubbles into solution B with a 1-ml sterile pipette and an autopipettor while slowly adding solution A dropwise.)

5. Incubate the transfection solution at room temperature for 5-20 min

6. Briefly vortex transfection solution and then add solution dropwise to culture plate medium until 1 ml of transfection solution per 10-cm plate has been added.

Add transfection solution to medium

7. Gently agitate by moving the plates back and forth to distribute transfection solution evenly.

8. Incubate plates at $37°C$ for 2-6 hr in a CO_2 incubator.

Incubate

9. Remove calcium phosphate-containing medium by aspiration, and replace it with fresh complete growth medium.

Replace transfection medium with growth medium

10. Return plates to CO_2 incubator.

Incubate

11. Prepare cells for flow cytometry 24-72 hrs posttransfection.

Prepare for Flow analysis

Skip to flow cytometry procedure.

Subprotocol 2
DNA transfection for Flow Cytometric Analysis of GFP Expression in Mammalian Cells: Liposome-mediated transfection

▉ ▉ Materials

Equipment

Cytometers   There are a variety of commercial instruments that one can use for detection of GFP and GFP variants. Published flow cytometry data has been reported for the following instruments from Becton Dickinson, Inc.: FACScan, FACStar[Plus], FACS Calibur, and FACS Vantage. There have also been reports using the EPICS Elite-ESP cytometer from Coulter Corporation. Most instruments used for detection of GFP and the red-shifted GFP variants are equipped with argon ion lasers tuned to 488 nm excitation. The EPICS Elite-ESP instrument equipped with a krypton ion mixed-gas laser has been used to detect the blue emission variant P4[31].

Solutions

Materials   – Cell culture medium (e.g., Dulbecco's modified Eagle med-
required   ium [DMEM] or another appropriate growth medium for
                 mammalian cells in culture)
             – Fetal bovine serum, newborn calf serum, or equivalent, to
                 supplement the growth medium

Phosphate buffered saline (PBS) (pH 7.4)

	Final concentration	To prepare 2 L
Na_2HPO_4	58 mM	16.5 g
NaH_2PO_4	17 mM	4.1 g
NaCl	68 mM	8.0 g

Dissolve components in 1.8 L of ddH$_2$O. Adjust to pH 7.4 with 0.1 N NaOH. Add ddH$_2$O to a final volume of 2 L. Store at room temperature.

- Transfection liposome reagent.
- 1 x Trypsin/EDTA (Life Technologies #25300-054).

Preparation

Prepare the followin solutions:

Solution A: add components in the following order:
- Dilute 2.0-4.0 µg of the GFP plasmid into 200 µl of serum-free medium.

Solution B:
- Dilute 4-24 µg of liposome reagent (2-6 x the amount of plasmid DNA) into 200 µl of serum-free medium

▣▣ Procedure

Note: There are number of suppliers for liposome transfection reagents. We have had the best results for transfection of GFP vectors using Lipofectin or Lipofectamine (both from Life Technologies, Inc.) and CLONfectin (CLONTECH Laboratories, Inc.). The method described below is compatible with each of these reagents.

Liposome-mediated transfection

1. In a 60-mm tissue culture plate, seed 2 x 105 cells in the appropriate growth media supplemented with serum.

Prepare Cells

2. Incubate at 37°C in a CO_2 incubator for 18-24 hr, or until cells reach a density of 40-80%.

3. Prepare the liposome reagent solutions found in the flow chart.

Prepare the transfection reagent solutions

4. Combine the two solutions, mix gently, and incubate at room temperature for 10-15 min.

5. Wash cells once with serum-free medium.

Wash cells

6. Add 1.8 ml of serum-free medium to the transfection mixture, and mix gently.

Overlay mixture onto the cells and incubate

7. Overlay mixture onto the cells and incubate at 37°C for 1-12 hr in a CO_2 incubator.

Note: Do not add antibacterial or antifungal agents to the medium during transfection.

Replace the transfection mixture with complete medium

8. Replace the transfection mixture with complete medium containing serum, and incubate at 37°C in a CO_2 incubator.

Prepare cells for flow cytometry

9. Prepare cells for flow cytometry 24-72 hrs posttransfection.

Skip to Flow Cytometry Procedure

Subprotocol 3
DNA transfection for Flow Cytometric Analysis of GFP Expression in Mammalian Cells By Electroporation

▨▨ Materials

Equipment

Cytometers

There are a variety of commercial instruments that one can use for detection of GFP and GFP variants. Published flow cytometry data has been reported for the following instruments from Becton Dickinson, Inc.: FACScan, FACStar[Plus], FACS Calibur, and FACS Vantage. There have also been reports using the EPICS Elite-ESP cytometer from Coulter Corporation. Most instruments used for detection of GFP and the red-shifted GFP variants are equipped with argon ion lasers tuned to 488 nm excitation. The EPICS Elite-ESP instrument equipped with a krypton ion mixed-gas laser has been used to detect the blue emission variant P4[31].

Solutions

- Cell culture medium (e.g., Dulbecco's modified Eagle medium [DMEM] or another appropriate growth medium for mammalian cells in culture)
- 1 x Trypsin/EDTA (Life Technologies #25300-054)
- Fetal bovine serum, newborn calf serum, or equivalent, to supplement the growth medium

Phosphate buffered saline (PBS) (pH 7.4)

	Final concentration	To prepare 2 L
Na_2HPO_4	58 mM	16.5 g
NaH_2PO_4	17 mM	4.1 g
NaCl	68 mM	8.0 g

Dissolve components in 1.8 L of ddH$_2$O. Adjust to pH 7.4 with 0.1 N NaOH. Add ddH$_2$O to a final volume of 2 L. Store at room temperature.

Procedure

1. Ethanol precipitate 20-40 µg of the GFP plasmid. Resuspend the DNA in 20 µl of PBS.

2. Wash cells once with PBS. Harvest cells with 1 x trypsin/EDTA or dilute PBS-EDTA solution.

3. Resuspend cells in complete medium containing serum at a density of 4 x 10^6 cells/ml. Ensure that a uniform single-cell suspension is produced, with no clumps of cells.

4. Add the GFP plasmid DNA to 0.4 ml of the cell suspension, mix well, and incubate at room temperature for 10 min.

5. Transfer the mixture to a 0.4-cm gap electroporation cuvette.

6. Electroporate at 950 µF, 0.22 kV/cm (t = 20-30 ms).

Incubate   7. Incubate the electroporated mixture at room temperature for 10-20 min, and plate cells in complete medium containing serum in a 60-mm plate.

8. Return plates to CO_2 incubator.

Prepare cells for flow cytometry   9. Prepare cells for flow cytometry 24-72 hrs posttransfection.

Continue with Flow cytometry Procedure

Subprotocol 4
Flow Cytometric Analysis of GFP Expression in Mammalian Cells

Materials

Equipment

Cytometers   There are a variety of commercial instruments that one can use for detection of GFP and GFP variants. Published flow cytometry data has been reported for the following instruments from Becton Dickinson, Inc.: FACScan, FACStar[Plus], FACS Calibur, and FACS Vantage. There have also been reports using the EPICS Elite-ESP cytometer from Coulter Corporation. Most instruments used for detection of GFP and the red-shifted GFP variants are equipped with argon ion lasers tuned to 488 nm excitation. The EPICS Elite-ESP instrument equipped with a krypton ion mixed-gas laser has been used to detect the blue emission variant P4[31].

Procedure

Preparation of cells   1. Harvest cells at the desired posttransfection interval. If cells are adherent, use 1X trypsin/EDTA or 5 mM EDTA in PBS to remove cells from the plate.

2. Centrifuge cells at 200 x g for 10 minutes.

3. Resuspend cells gently in PBS at 106-107 cells/ml in a total volume of 0.5-1.0 ml making sure the cells are monodispersed and free of clumps. Cells can be stained at this point for surface markers prior to analyzing on FACS.

4. Add propidium iodide to a final concentration of 5.0 ug/ml as a viability marker (see Section 5, Chapter 2).   **Stain with PI**

5. Set up flow cytometer according to the GFP variant which was transfected.   **Set up flow cytometer**

6. Analyze samples immediately, or store on ice for 1-2 hours. Run mock transfected (cells incubated with transfection reagents but no vector) sample first to establish baselines and autofluorescence pattern. Place the negative baseline fluorescence within the first decade of the log scale.   **Analyse**

7. Use forward angle light scatter vs propidium iodide to gate on the viable cells. Exclude debris, aggregates, and all positive propidium iodide cells (dead or dying cells) from the rest of the analysis.

8. Collect log green fluorescence for wtGFP, EGFP, S65T, cycle 3, or 10C variants. Collect log blue fluorescence for P4-3 or EBFP variants.

9. Plot gated viable cell's fluorescence vs scatter (or other surface color markers).

GFP: GAIN OF FUNCTION VARIANTS

Red-shifted GFPs

A number of "red-shifted" mutants of wt GFP have been described, most of which contain one or more amino acid substitutions in the chromophore region of the protein. The red-shifted terminology refers to the position of the major fluorescence excitation peak, which is shifted for each of these variants towards the red, from 395 nm in wt GFP, to 488-490 nm. The emission spectra for such variants is largely unchanged, and these mutants still generate green light with a wavelength maxima of approximately 507-511 nm. The major excitation peak of the red-shifted variants encompasses the excitation wavelength of commonly used fluorescence filter sets, so the resulting signal is much brighter relative to wt GFP. Similarly, the argon ion laser used in most flow cytometers has a major line at 488 nm, so excitation of the red-shifted GFPs is more efficient than excitation of wt GFP. In practical terms, this means the detection limits are considerably lower with the red-shifted variants. The red-shifted excitation spectra also allows such variants to be used in combination with wt GFP in double-labeling studies. Although the peak positions in the emission spectra of the red-shifted GFP variants are virtually identical, double-labeling can be achieved by selective excitation of wt GFP and red-shifted GFP.[21,22]

The two most commonly used red-shifted GFP variants are S65T[18,23] which contains a Ser-65 to Thr substitution in the chromophore, and GFPmut1 or EGFP[24,25] which contains the same S65T change plus a Phe-64 to Leu mutation (Table 1). GFPmut1 and EGFP have identical amino acid sequences, but the EGFP coding sequence has been further modified with 190 silent base changes to contain codons preferentially found in highly expressed human proteins[26]. The "humanized" backbone used in EGFP contributes to efficient expression of this variant in mammalian cells, and subsequently yields very bright fluorescence (Fig. 4). Based on spectral analysis of equal amounts of soluble protein, EGFP fluoresces approximately 35-fold more intensely than wt GFP when excited at 488 nm,[24] due to an increase in its extinction coefficient (Em). Studies with wt GFP expressed in HeLa cells[27] have shown that the cytoplasmic concentration must be greater than ~1.0 µM to obtain a signal that is twice the autofluorescence. This threshold for detection is lower with the red-shifted variants: ~200 nM for S65T and ~30 nM for EGFP. In addition to improved sensitivity of detection, other advantages of the EGFP and S65T variants include: (1) improved solubility; (2) more efficient protein folding; (3) faster chromophore oxidation to form the fluorescent form of protein; and (4) reduced rates of photobleaching.

UV-optimized GFP

The "cycle 3" GFP variant (commercially available as GFPuv from CLONTECH) is optimized for maximal fluorescence when excited by UV light (360-400 nm) and for higher bacterial

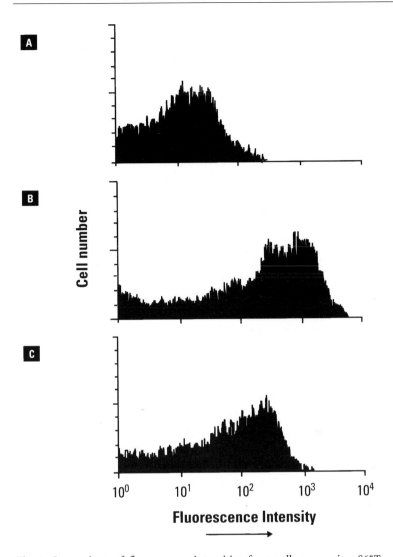

Fig. 4. Comparison of fluorescence intensities from cells expressing S65T, EGFP, and another commercially available GFP reporter protein. 293T cells were transfected separately with the expression vector pS65T-C1 (**Panel A**), pEGFP-C1 (**Panel B**), and another commercially available GFP reporter (**Panel C**). Transfections were performed using Lipofectamine (Life Technologies), the cells harvested 48 hours posttransfection, and prepared for flow cytometry by suspension in PBS containing 0.5 µg/ml propidium iodide (PI). Cytometric analysis was performed using a FACScan Cytometer (Becton Dickinson) equipped with a 488 nm argon ion laser. The FL1 channel was used to monitor green fluorescence, and the FL3 channel used to identify PI red fluorescence. The results indicate that a significantly brighter signal can be obtained using the EGFP variant for transient expression studies

expression. The cycle 3 gene is a synthetic GFP gene in which five rarely used Arg codons from the wt gene were replaced by codons preferred in *E. coli*. Consequently, this variant is expressed very efficiently in *E. coli*, but also functions quite well in mammalian systems. The cycle 3 variant is ideal for applications in which GFP is excited with UV light such as macroscopic imaging using a hand-held light source (transgenic organisms, yeast or bacteria colonies, screening studies, etc.), or in flow cytometry applications using krypton ion lasers. The cycle 3 variant was developed using an *in vitro* DNA shuffling technique to introduce point mutations throughout the GFP coding sequence.[28] Colonies were visually screened using a long-wave UV lamp (365 nm), and the brightest clones were pooled and subjected to another round of DNA shuffling. The cycle 3 variant emerged as the brightest GFP mutant after three rounds of shuffling and selection, hence the name "cycle 3". This variant contains three amino acid substitutions (Phe-99 to Ser, Met-153 to Thr, and Val-163 to Ala), none of which alter the chromophore sequence (Table 1). These amino acid changes make *E. coli* expressing the cycle 3 variant fluoresce 18 times brighter than wt GFP.[28]

The development of EGFP and cycle 3 (GFPuv) may allow these variants to be used in combination for double-labeling experiments by selective excitation at 488 nm and 395 nm, respectively. Some promising applications that use this dual reporter combination are 1) Analysis of multiple cell populations in a mixed culture; 2) monitoring gene expression from two different promoters in the same cell, tissue, or organism; 3) monitoring the localization of two different protein fusions in the same cell, tissue, or organism; and 4) Fluorescence activated cell sorting (FACS) of mixed cell populations (e.g., cells expressing GFPuv, EGFP, and nonfluorescent cells).

Blue and yellow fluorescent proteins

The properties of the red-shifted and UV-optimized variants largely overcome the limitations of wt GFP for single reporter studies. However, one important feature of wt GFP and each of these variants remains unchanged – they all produce green fluorescence. As discussed above, several combinations of GFP variants can be used for dual-reporter studies by employing selective excitation conditions, but this process is complicated in microscopy as the image collected from each reporter is green. Dual-color images must be generated by pseudocoloring techniques,[18,22] or by depicting separate images for each variant. Moreover, the utility of green emitting reporters is limited in both microscopy and flow cytometry in cases where cellular green autofluorescence is a concern, or when the reporter is used in conjunction with chemical fluorophores such as fluorescein. For each of these reasons, it is desirable to have emission variants of GFP capable of producing distinct colors.

The first such examples of useful variants are the blue emission mutants of GFP, each of which contain an invariant Tyr-66 to His mutation in the GFP chromophore.[19] The initial variant of this type referred to as P4 contains the single point mutation Tyr-66 to His , and yields a cobalt blue signal, but only dim fluorescence. Despite the weak signal of this variant, one investigator has demonstrated the simultaneous detection of P4 and wt GFP in cotransfected 293 cells by flow cytometry[31]. An improvement to P4 was termed P4-3, which contains an additional Tyr-145 to Phe substitution. The P4-3 double mutant has a major shift in both the excitation an emission maxima, with values of 381 nm and 445 nm, respectively. The P4-3 mutant is approximately two-fold brighter than P4, primarily due to a higher Em value which is similar to that of wt GFP (Table 1). The P4-3 mutant has recently been used in conjunction with wt GFP in co-transfection studies to simultaneously visualize mitochondria and the nucleus in the same living mammalian cell.[29] Such studies are facilitated by the fact that wt GFP and P4-3 excite with UV light, and allow the real time analysis of mitochondrial and nuclear dynamics in proliferating cells.

In addition to weak fluorescence, another important disadvantage of the P4-3 variant is rapid photobleaching. In experiments with conventional epifluorescence microscopy we observe complete loss of the P4-3 signal in transfected mammalian cells within a few seconds. Moreover, all previous reports with blue emission variants such as P4-3 have used GFP genes containing the wild type jellyfish codons. As stated above for EGFP, the use of a humanized GFP gene is certain to improve the mammalian cell expression of blue emission variants as well. To address these needs, we have recently developed a novel humanized blue emission variant termed EBFP, which contains four mutations: Phe-64 to Leu, Ser-65 to Thr, Tyr-66 to His, and Tyr-145 to Phe (Table 1). This variant contains the same 190 silent mutations found in EGFP, and consequently yields more efficient expression in mammalian cells. The Em of EBFP is 31,000 cm-1 M-1 for 380 nm excitation, leading to a fluorescent signal that is markedly brighter than P4-3. The fluorescence excitation and emission maxima of EBFP are similar to other blue emission variants at 380 nm and 440 nm, respectively (Fig. 5). The EBFP reporter works quite well for flow cytometry applications (Fig. 6), and is also suitable for dual-reporter studies using cotranfection with EGFP (data not shown). Lastly, while detailed studies are yet to be completed, our results with EBFP transfected CHO-K1 cells indicate this variant is resistant to photobleaching for at least several minutes. Because of the brighter signal, photostability, and elevated expression of EBFP this mutant is likely to be the blue variant of choice for most applications in mammalian cells.

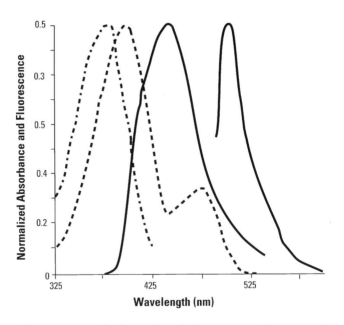

Fig. 5. Fluorescence excitation and emission spectra of EBFP. Excitation (dashed lines) and emission (solid lines) spectra of wt GFP (black) and EBFP (gray). The emission data were obtained with excitation at 390 nm

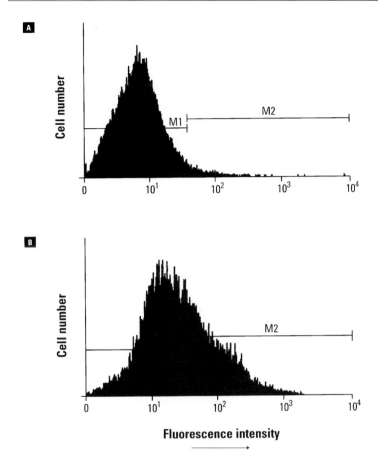

Fig. 6. Analysis of transfected cells expressing EBFP. Untransfected cells (**Panel A**) and cells transfected with the vector pEBFP-C1 (**Panel B**) were analyzed by flow cytometry. 293 EBNA cells were transfected separately with each vector using Lipofectamine (Life Technologies), the cells harvested 48 hours posttransfection, and prepared for flow cytometry. Cytometric analysis was performed using a FACStar PLUS Cytometer (Becton Dickinson) equipped with an argon-ion laser tuned to 488 nm with 300 mW power. UV fluorescence was detected through a 5 Watt argon-ion pulse plasma tube with 45-20 amps discharge current, and 380 nm with 200 mW power, 424 DF 44 filter. Photomultiplier pulses were amplified logarithmically. The fraction of cells showing elevated blue fluorescence (monitored in the FL4 channel) was ~12-fold higher in **Panel B** than in the untransfected cells (**Panel A**)

The analysis of the crystal structure for S65T has revealed a critical association between Thr-203 and the phenolic group on Tyr-66 in the GFP chromophore.[16] Therefore, a targeted mutagenic strategy was used to replace Thr-203 with several other residues in hope of yielding GFP mutants with desirable spectral properties. One such mutant, termed "10C" gave dramatic redshifts in both the excitation and emission maxima at 513 nm and 527 nm, respectively.[16] The 10C variant contains a total of four mutations: Thr-203 to Tyr, Ser-65 to Gly, Val-68 to Leu, and Ser-72 to Ala (Table 1). The Val-68 to Leu and Ser-72 to Ala mutations improve the folding of this variant at 37 °C, but are not significantly responsible for the shift in excitation or emission maxima.[24] This variant has Em and QY values similar to those for S65T, and therefore yields a bright fluorescent signal.

As indicated by the emission maxima for 10C, this variant produces a yellow-green fluorescence. It seems feasible to design microscopy and flow cytometry filter sets compatible with dual-reporter applications using 10C and several other GFP variants. In fact, it now appears possible to perform 3-color detection using the appropriate filter combination and reporter constructs expressing 10C, EBFP and either EGFP or S65T. The combination of these spectral variants opens a wide range of applications such as the simultaneous analysis of different promoter elements, multiparameter flow cytometry, intracellular localization of several different proteins, and real time analysis of protein-protein interactions[18]. Given the wealth of possibilities provided by currently availiable and yet to come variants, the future certainly appears bright for GFP technology.

DNA transfections PROTOCOLS

Appropriate GFP vectors may be transfected into mammalian cells by a variety of techniques. These methods include calcium phosphate, liposome-mediated transfection, and electroporation. The efficiency of DNA uptake and expression is largely dependent on the host cell line being transfected. Different cell lines may vary by several orders of magnitude in their ability to take up and express exogenous DNA. Moreover, a transfection method that works well for one host cell line may be inferior for another. Therefore, when working with a cell line for the first

time, it is advisable to compare the efficiencies of several transfection protocols. This can be accomplished using a GFP vector
which has a strong viral promoter (such as CMV) to provide
high-level expression in a wide variety of cell types. After a method of transfection is chosen, it must be optimized for parameters
such as cell density, the amount and purity of DNA, media conditions, transfection time, and the post-transfection interval
prior to GFP detection. Once optimized, these parameters should
be kept constant in order to obtain reproducible results.

The transfection protocols have been used for transient expression of GFP and GFP-variants in several cell lines. GFP expression can typically be detected by flow cytometry 24-72 hours
after transfection by each method, depending on the host cell
line used.

Troubleshooting guide and helpful hints

General considerations for mammalian cell expression of GFP
- Variability in the intensity of GFP fluorescence has been
  noted. This may be due in part to the relatively slow formation
  of the GFP chromophore and the requirement for molecular
  oxygen[19]. This problem is largely circumvented by use of the
  red-shifted GFP variants S65T and EGFP which form a functional chromophore more readily in mammalian cells.

- The slow rate of chromophore formation and the apparent
  stability of wt GFP may preclude the use of GFP as a reporter
  to monitor fast changes in promoter activity[19,20]. This limitation is again reduced by use of S65T or EGFP, which acquires
  fluorescence faster than wt GFP[19,24].

- Some people have put GFP expression constructs into their
  system and failed to detect fluorescence. There can be numerous reasons for failure, including use of an inappropriate filter
  set, expression of GFP below the limit of detection, and failure
  of GFP to form a functional chromophore. GFP Monoclonal
  or Polyclonal Antibodies may be used to confirm GFP expression in these cases by performing Western blots or immuno-
  precipitations. In this manner, the researcher can discriminate between problems asssociated with GFP gene expression
  versus detection of GFP fluorescence. The commercial anti-

bodies listed in Appendix I are suited to detection of all GFP variants and fusion proteins.

- If you have not been able to detect wt GFP in a mammalian expression system, try using an EGFP expression vector instead. The brighter fluorescence and improved translation should make detection of EGFP expression easier than expression of wt GFP.

- Have you verified your GFP plasmid construct and concentration with a restriction digest? Verify that all subcloning steps have been done correctly, keeping in mind specific restriction sites in the vectors may be inactivated by methylation. GFP-sequencing primers should be used to verify sequence junctions for N- and C-terminal protein fusion constructs.
    - Is the vector compatible with your cell type? Has the promoter been used previously for transient expression in your cells?
    - Do you have another assay to estimate transfection efficiency? This can be accomplished by transfection with a second reporter plasmid which contains the same promoter element, for example a vector expressing the *lacZ* gene, and employing standard assays or stains to detect β-galactosidase activity.

- Fusion of a protein to wt GFP may cause a redshift in the absorption spectrum. Excitation at longer wavelengths (450-490 nm) may give improved fluorescence.

- Exciting GFP intensely for extended periods may generate free radicals that are toxic to the cell. This problem can be minimized by excitation at 450-490 nm. Such toxicity problems are not of major concern in flow cytometry applications with GFP, as the time interval used for excitation is too brief to yield significant free radical production.

Requirements for GFP chromophore formation
- Formation of the GFP chromophore appears to be temperature sensitive. In some cases, *E. coli,* yeast, and mammalian cells expressing GFP have shown stronger fluorescence when grown at lower temperatures[19,32]. Hence, incubation at a lower temperature may increase the fluorescence signal. If the cell type is amenable, culture temperatures of 32-35 °C may yield a marginal increase in signal. However, the improved folding

properties of EGFP largely overcome this problem, yielding bright fluorescence at 37 °C.

- GFP chromophore formation requires molecular oxygen[19,20]; therefore, cells must be grown under aerobic conditions.

Photobleaching of the chromophore
- Photobleaching is primarily a concern with fluorescence microscopy applications of GFP, and is not a major problem is flow cytometry again due to the relatively brief period of fluorophore excitation.

- Excite at 488 nm for wt GFP and the red-shifted GFP variants S65T and EGFP. Excitation at the 395-nm peak for wt GFP may result in rapid loss of signal. Photobleaching with blue-emission variants such as EBFP is unavoidable, but can be minimized by using the longest possible wavelength to excite the fluorophore.

Autofluorescence
- Some samples may have a significant background autofluorescence, e.g., worm guts[6,27]. A bandpass emission filter may make the autofluorescence appear the same color as GFP; using a long-pass emission filter may allow the color of the GFP and autofluorescence to be distinguished.

- Most autofluorescence in mammalian cells is due to flavin coenzymes[27] (FAD and FMN) which have absorption/emission = 450/515 nm. These values are very similar to those for wt GFP (470 nm excitation peak) and the red-shifted GFP variants, so autofluorescence may interfere with the GFP signal. The use of DAPI filters may make this autofluorescence appear blue while the GFP signal remains green.

- Flow cytometry of transiently transfected 293 cells with S65T and wt GFP has shown that the relative intensity of autofluorescence is least with excitation at 407 nm relative to UV (351 nm) or 488 nm[30]. These results suggest that GFPuv may be a good alternative when autofluorescence is excessive using blue light excitation. Excitation at 407 nm also yielded relatively mild blue autofluorescence (440-475 nm), indicating these conditions should be effective for the detection of EBFP and other blue emission variants.

- Some cell types can produce blue autofluorescece due to mitochondrially bound NADH which has an absorption maximum of ~365 nm[30]. This problem is minimized with excitatory light around 488 nm, as opposed to UV excitation[27], and is therefore not a major concern when using S65T or EGFP. When using EBFP, such autofluorescence is minimized by excitation above 380 nm.

- For mammalian cells, autofluorescence can increase with time in culture. For example, when CHO or SCI cells were removed from frozen stocks and reintroduced into culture, the observed autofluorescence (emission at 520 nm) increased with time until a plateau was reached around 48 hours[33]. Therefore, in most cases it is preferable to work with freshly plated cells.

- Always use a mock-transfected control or cells transfected with a promoterless vector such as pEGFP-1 or pGFP-1 to gauge the extent of autofluorescence.

References

1. Shimomura O, Johnson FH, Saiga Y. Extraction, purification and properties of aequorin, a bioluminescent protein from the luminous hydromedusan, *Aequorea*. J. Cell. Comp. Physiol. 1962; 59: 223227.
2. Morin JG, Hastings JW. Energy transfer in a bioluminescent system. J. Cell. Physiol. 1971 ;77: 313318.
3. Ward WW, Cody CW, Hart RC., et al. Spectrophotometric identity of the energy transfer chromophores in *Renilla* and *Aequorea* green-fluorescent proteins. Photochem. Photobiol. 1980; 31: 611615.
4. Prasher DC, Eckenrode VK, Ward WW, et al. Primary structure of the *Aequorea victoria* green fluorescent protein. Gene 1992;111:229233.
5. Inouye S, Tsuji FI. *Aequorea* green fluorescent protein: Expression of the gene and fluorescent characteristics of the recombinant protein. FEBS Letters 1994; 341: 277280.
6. Chalfie M, Tu Y, Euskirchen G, et al. Green fluorescent protein as a marker for gene expression. Science 1994; 263: 802805.
7. Wang S, Hazelrigg T. Implications for bcd mRNA localization from spatial distribution of exu protein in *Drosophila* oogenesis. Nature 1994; 369: 400403.
8. Ward WW. Properties of the Coelenterate green-fluorescent proteins. In: DeLuca, M. & McElroy, W. D., eds. Bioluminescence and Chemiluminescence: Basic Chemistry and Analytical applications. New York; 1981: 235242.

9. Flach J, Bossie M, Vogel J, et al. A yeast RNA-binding protein shuttles between the nucleus and the cytoplasm. Mol Cell Biol 1994; 14: 83998407.

10. Marshall J, Molloy R, Moss GW, et al. The jellyfish green fluorescent protein: a new tool for studying ion channel expression and function. Neuron 1995; 14: 211215.

11. Stearns T. The green revolution. Curr Biol 1995; 5: 262264.

12. Carey KL, Richards SA, Lounsbury, et al. Evidence using a green fluorescent protein-glucocorticoid receptor that the RAN/TC4 GTPase mediates an essential function independent of nuclear protein import. J Cell Biol 1996; 133: 985996.

13. Náray-Fejes-Tóth A, Fejes-Tóth G. Subcellular localization of the type 2 11β-hydroxysteroid dehydrogenase. J Biol Chem 1996; 271: 1543615442.

14. Kaether C, Gerdes H-H. Visualization of protein transport along the secretory pathway using green fluorescent protein. FEBS Letters 1995; 369: 267271.

15. Cody CW, Prasher DC, Westler WM, et al. Chemical structure of the hexapeptide chromophore of *Aequorea* green-fluorescent protein. Biochemistry 1993; 32: 12121218.

16. Ormö M, Cubitt AB, Kallio K, et al. Crystal structure of the *Aequorea victoria* green fluorescent protein. Science 1996; 273: 13921395.

17. Yang F, Moss LG, Phillips GN. The molecular structure of green fluorescent protein. Nature Biotechnol 1996; 14: 12461251.

18. Heim R, Tsien RY. Engineering green fluorescent protein for improved brightness, longer wavelengths and fluorescence resonance energy transfer. Curr Biol 1996; 6: 178182.

19. Heim R, Prasher DC, Tsien RY. Wavelength mutations and posttranslational autoxidation of green fluorescent protein. Proc Natl Acad Sci USA 1994; 91: 1250112504.

20. Davis DF, Ward WW, Cutler MW. Posttranslational chromophore formation in recombinant GFP from *E. coli* requires oxygen. Proceedings of the 8th international symposium on bioluminescence and chemiluminescence. Neuron. 1995 Feb, 14; 2: 211-5.

21. Kain SR, Adams M, Kondepudi, et al. The green fluorescent protein as a reporter of gene expression and protein localization. BioTechniques 1995; 19: 650-655.

22. Yang T T, Kain SR, Kitts P, et al. Dual color microscopic imagery of cells expressing the green fluorescent protein and a red-shifted variant. Gene 1996; 173: 19-23.

23. Heim R, Cubitt AB Tsien RY. Improved green fluorescence. Nature 1995; 373: 663664.

24. Cormack BP, Valdivia R, Falkow S. FACS-optimized mutants of the green fluorescent protein (GFP). Gene 1996; 173: 3338.24.

25. Yang TT, Cheng L, Kain SR. Optimized codon usage and chromophore mutations provide enhanced sensitivity with the green fluorescent protein. Nucleic Acids Res. 1996; 24: 4592-4593.

26. Haas J, Park E-C, Seed B. Codon usage limitation in the expression of HIV-1 envelope glycoprotein. Curr Biol 1996; 6: 315324.

27. Niswender KD, Blackman SM, Rohde L, et al. Quantitative imaging of green fluorescent protein in cultured cells: comparison of microscopic techniques, use in fusion proteins and detection limits. J Microbiol 1995; 180: 109116.

28. Crameri A, Whitehorn EA, Tate E, et al. Improved green fluorescent protein by molecular evolution using DNA shuffling. Nature Biotechnol. 1996; 14: 315319.

29. Rizzuto R, Brini M, De Giorgi F, et al. Double labelling of subcellular structures with organelle-targeted GFP mutants *in vivo*. Curr Biol 1996; 6: 183188.

30. Ropp JD, Donahue CJ, Wolfgang-Kimball D, et al. *Aequorea* green fluorescent protein analysis by flow cytometry. Cytometry 1995; 21: 309-317.

31. Ropp, J.D., Donahue, C.J., Wolfgang-Kimball, D. et al. *Aequorea* green fluorescent protein: simultaneous analysis of wild-type and blue-fluorescing mutant by flow cytometry. Cytometry 1996; 24: 284-288.

32. Lim CR, Kimata Y, Oka M, et al. Thermosensitivity of green fluorescent protein fluorescence utilized to reveal novel nuclear-like compartments in a mutant nucleosporin Nsp1. J Biochem 1995; 118: 13-17.

33. Aubin JE. Autofluorescence of viable cultured mammalian cells. J Histochem Cytochem 1979; 27: 36-43.

34. Lybarger L, Dempsey D, Franek KJ, et al. Rapid generation and flow cytometric analysis of stable GFP-expressing cells. Cytometry 1996; 25: 211-220.

Suppliers

CLONTECH LABORATORIES, INC.

Appendices

Products available from CLONTECH Laboratories, Inc.

Product	Cat. #
GFP Variant Vectors	
pEGFP Vector	
pEGFP-1 Promoter Reporter Vector	6086-1
pEGFP N-Terminal Protein Fusion Vectors	many
pEGFP C-Terminal Protein Fusion Vectors	many
phGFP-S65T Humanized GFP Vector	6088-1
pEBFP Vector	6068-1
pEBFP N-Terminal Protein Fusion Vectors	many
pEBFP C-Terminal Protein Fusion Vectors	many
pGFPuv Vector	6079-1
Wild-Type GFP Vectors	
pGFP Vector	6097-1
pGFP-1 Promoter Reporter Vector	6090-1
pGFP N-Terminal Protein Fusion Vectors	many
pGFP C-Terminal Protein Fusion Vectors	many
p35S GFP Plant Expression Vector	6098-1
Other GFP Products	
GFP-N Sequencing Primers	many
GFP-C Sequencing Primers	many
GFP Monoclonal Antibody	8362-1
GFP Polyclonal Antibody (IgG Fraction)	8363-1, -2

Fluorescent Proteins Newsgroup

A newsgroup for the discussion of fluorescent proteins has been created within the bionet hierarchy of Newsgroups. This newsgroup is intended to provide a forum for discussion of bioluminescence, to promote further development of reporter proteins obtained from bioluminescent organisms (e.g., GFP, luciferases, and *Aequorin*), and to facilitate their application to interesting biological questions. (A full copy of the newsgroup charter can be found in the BIOSCI archives.)
We hope you will find this newsgroup useful and encourage you to participate in the discussions.

Discussion Leaders: Steve Kain & Paul Kitts, CLONTECH Laboratories, Inc. *Administration:* BIOSCI International Newsgroups for Biology *Newsgroup Name:* bionet.molbio.proteins.-fluorescent *To subscribe:* If you use USENET news, you can participate in this newsgroup using your newsreader.
You can also access the newsgroup on the World Wide Web using the URL http://www.bio.net/hypermail/FLUORESCENT-PROTEINS/

To receive "The BIOSCI electronic newsgroup information sheet" which describes the BIOSCI newsgroups and gives instructions on how to subscribe via e-mail:

If you are located in the Americas or the Pacific Rim

- Send a mail message to the Internet address:
  biosci-server@net.bio.net

- Leave the subject line of the message blank and enter the following line in the mail message: info usinfo. This message will be automatically read by the computer and you will be sent the latest copy of the information sheet.

If you are located in Europe, Africa, or central Asia

- Send a mail message to the Internet address:
  biosci-server@net.bio.net

- Leave the subject line of the message blank and enter the following line in the mail message: info ukinfo. This message will be automatically read by the computer and you will be sent the latest copy of the information sheet.

FACS-Gal: Flow Cytometric Assay of β-galactosidase in Viable Cells

STEVEN FIERING

Introduction

FACS-Gal is a system that enables the fluorescence activated cell sorter (FACS) to sensitively assay expression of the E. Coli lacZ (β-galactosidase, β-gal) reporter gene, and sort viable cells based on the levels of expression of this enzyme.[1,2] The system depends on the enzymatic conversion of the nonfluorescent β-gal substrate fluorescein di-β-galactopyranoside (FDG), into the fluorescent molecule fluorescein. FACS-Gal effectively combines a selectable marker with a reporter gene and in the combination produces novel experimental possibilities. The system is an approach to applying the analytical and sorting capabilities of the FACS to solving problems in molecular biology. FACS-Gal and its variants have been used with cells from a variety of organisms including E. Coli,[3] yeast,[3] Drosophila,[4] transgenic mice,[5-7] and most frequently with mammalian cell lines.[8-13] For a more extensive list of references using the FACS-Gal assay see.[14] This review focuses on the technical basics of FACS-GAL, how to most productively use the system, and the strengths and weaknesses of this system in comparison to other presently available alternatives.

Reporter Genes

The analysis of cis-regulatory DNA elements in eucaryotic cells has depended on the use of reporter genes. By comparing gene expression of two constructs that differ in regulatory sequences but express the same mRNA encoding the reporter gene, an investigator can attribute expression differences to differences in regulatory sequences. Reporter genes incorporate the following characteristics: (*i*) they are not expressed in most cells so the assay will have a very low background; (*ii*) they can be assayed with precise quantitation across a broad range of activities and the assay is easy and very sensitive; (*iii*) the expression of the reporter gene in cells does not have a biological influence on the cell. The requirement for sensitivity has made enzymes the predominant choice, so far, for reporter genes.

Most reporter gene assays have measured the expression of the reporter gene in bulk biochemical lysates produced from a cell population and therefore information regarding cell-to-cell variation of expression is not available. In contrast, FACS-Gal is a cell by cell assay system that allows analysis of the variation of expression within a cell population. The ability of the FACS to rapidly analyze large numbers of cells provides accurate statistical information about the variation of reporter gene expression in the assayed populations.

Selectable Markers

Selectable markers are used to select for (or against) rare cells that express (or don't express) the marker gene. Most markers can only be selected in one direction, that is, for expressing or for non-expressing cells. Positive selection is selection for expression, for example the commonly used neomycin phosphotransferase (neo) gene that confers resistance to the drug G418 in mammalian cells. Such a marker will rescue expressing cells from a toxic drug. Non-expressing cells or cells expressing insufficient quantities of the marker protein are killed by the drug and therefore are unavailable for further experimentation. There are also negative selectable markers, like the HSV-TK gene, which mediate the death of expressing cells in the presence of a drug. Once again there is a threshold for activity and the expressing cells are unavailable for further analysis. It is occasionally possible, but certainly very difficult, to titrate the selecting drug in order to select for varying levels of expression among cells expressing these drug-based selectable markers. Some selectable markers can be selected both positively and negatively in succession or vice versa. The best example of this is the HPRT gene. An endogenous gene, HPRT expression can be selected negatively with 6-thioguanine and positively with HAT medium (a closely related bacterial gene, GPT, acts similarly). A novel fusion protein has been developed that combines positive and negative selectable markers.[15] The HYTK gene expresses a protein fusion of the positive selectable marker, Hygromycin resistance, with the negative selection marker herpes simplex thymidine kinase. Once again, though, after a given selection only one population is alive for further analysis when using a positive-negative drug selection.

Using FACS-Gal, expression of β-gal can be selected positively and negatively with the sorting capability of the FACS. One advantage as compared to other selectable markers is that the level of enzyme activity can be used to make the selection, there is no practical threshold level of selection. This is particularly true since the assay is sensitive enough to recognize cells with as little as 5 active β-gal enzyme molecules.[2] Furthermore, in the same FACS sort one can select for cells with different levels of enzyme activity. Therefore, the cells below or above some specific level of enzyme activity can be separated and utilized for further analysis as opposed to a drug selection in which cells either survive or are killed by the selection.

Lac-Z Gene

The lacZ gene has a long history of utility as a reporter gene. It has a variety of assays for its activity in multicell lysates. When FACS-Gal was being developed lacZ was the only commonly used reporter gene that had a single cell assay for its activity, the x-gal assay. The x-gal assay is performed on fixed nonviable cells and is only coarsely quantitative. Nevertheless, it has made the lacZ gene the reporter of choice for expression studies of regulatory elements in whole organisms. The historical use of lacZ, its presence in many strains of transgenic mice and the ability to correlate the FACS-Gal assay with other bulk lysate and single cell assays caused us to focus our efforts on lacZ. In addition, we knew FDG was a fluorogenic substrate for β-gal [16] and could be easily assayed by a standard flow cytometer.

Sensitivity

Enzymes are favored as reporter genes since the detection of their protein products is often exquisitely sensitive. This is the case for lacZ when using the FACS-Gal assay. Studies have shown that FACS-Gal can detect less than 5 molecules of active β-gal in mammalian cells (as compared to a few hundred active molecules required for detection by a x-gal assay).[2] Since β-gal functions as a tetramer this means less than 20 monomeric β-gal subunits can be detected. This can be compared to the requirement of expression of many hundreds of cell surface molecules for detection of surface antigens by antibody staining. However, the question of how many molecules are required for detection is really only one part of answering the question of comparative sensitivity of assays for various reporter genes. Since reporter genes use functional protein production to assay transcription, the efficiency of the transcriptional and posttranscriptional processing of the messenger RNA as well as the efficiency of translation and posttranslational modification of the protein will affect the assay. These variables have not been studied in regard to lacZ (or other reporter genes) and therefore it is quite difficult to compare different reporter genes in regard to overall sensitivity for detecting transcribed message.

FACS-Gal Used to Analyze Transgenic Mice

LacZ has been favored as a reporter gene in transgenic mouse experiments in particular to determine the tissue and developmental specificity of a set of transcriptional regulatory elements. LacZ is expressed well in the embryo and the use of the x-gal assay has provided a reliable semiquantitative assay for lacZ expression in whole mount embryos, sections of tissues, or individual cells. The FACS-Gal assay works well with any type of mouse cells that can be isolated as single cell suspensions. In addition, the compatibility of FACS-Gal with analysis of cell surface antigens of hematopoietic cells has made it possible to correlate reporter gene expression with cell type.[5] However, unfortunately expression of lacZ in hematopoietic cells of adults has not been as dependable as expression in the embryo. There have been examples of expression of lacZ in adult hematopoietic cells,[5-7] but in many other instances transgenic lacZ constructs have not expressed. The basis for this unreliable expression in the hematopoietic system is speculated to be due to a tendency of the lacZ gene to be a target of position effect variegation but evidence for this is anecdotal, (see ref. 14 for a more detailed discussion of this phenomena).

How does FACS-Gal work?

The FACS-gal assay treats each cell like an individual reaction vessel. Substrate (FDG) is loaded into each cell by exposing the cells to the substrate in hypotonic medium. Following this short (1-3 minute) loading period, during which the cells are at 37°C, near normal tonicity is restored and the cells are chilled by the addition of a 10-fold excess of ice-cold isotonic medium. Chilling the cells changes the cell membrane so that the substrate, FDG, and the product, fluorescein, are kept within the cell. The cells are kept chilled during the following fluorescence development period in which the enzyme activity is producing fluorescein from the non-fluorescent substrate FDG. Fluorescence is then assayed on the flow cytometer using the normal set-up for fluorescein analysis.

The level of fluorescence produced in a given cell is proportional to time and enzyme concentration. The rate of fluorescence increase is linear with respect to time but a logarithmic function for lacZ enzyme activity per cell.[2] Fluorescence level is approximately the square (actually 1.8 power) of the enzyme concentration (fluorescence = enzyme concentration $^{(1.8)}$ X time). This logarithmic function makes it more complicated to interpret the fluorescence data quantitatively, and lowers the range of activity observable from any particular sample to 10-20 fold.

A given FACS-Gal analysis identifies three general cell types based on β-gal expression- non-expressing cells, cells that have exhausted the substrate during incubation on ice, and cells that have β-gal activity but have not yet exhausted the substrate. On a typical four-decade log plot of fluorescence, the first decade is β-gal negative cells. The negative cells are clearly separated by 1-2 decades from the β-gal positive cells which have exhausted their substrate with the positive cells that still have substrate located between them (see fig. 2B top panel).

The cells that have exhausted their substrate form a tight peak in the higher fluorescence levels. The loading system is not able to introduce concentrations of substrate significantly above the Michaelis-Menton concentration needed to make the rate of enzyme activity independent of substrate concentration. Therefore the ability of the assay to quantitatively compare enzyme levels between cells requires that the cells have similar concentrations of substrate when the reaction starts. Fortunately, the loading technique fulfills this requirement as shown by fig. 1.[2]

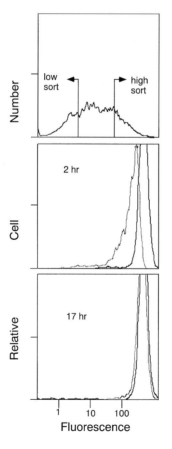

Fig. 1 The hypotonic treatment introduces uniform concentrations of FDG into a population of cells. BW5147.28 lacZ (+) cells (1) were loaded with FDG, and immediately after loading 100,000 each of the 25% most fluorescent and 25% least fluorescent cells were sorted into ice cold media. The cells were kept on ice and reanalyzed periodically. The middle panel shows the fluorescence distribution of the sorted populations 2 hours after substrate loading. The bottom panel shows the fluorescence of these populations 17 hr after loading (low sort is the broken line and high sort is the solid line). With enough time to hydrolyze the substrate, the two histograms are almost identical, demonstrating that the intracellular FDG concentration is uniform and does not contribute to the observed differences in measured ß-gal activity

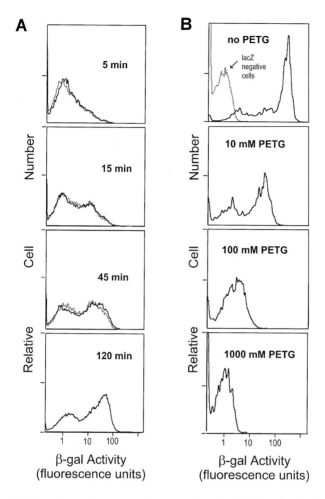

Fig. 2. PETG can be used to stop or slow the hydrolysis of FDG by ß-gal. **A:** Jurkat NFATZ human T lymphocytes with an inducible lacZ construct were stimulated to produce ß-gal as discussed elsewhere (11). A single sample of cells was loaded with FDG and incubated in ice. At the indicated timepoints, the cells' fluorescence was assayed on the FACS; simultaneously an aliquot was removed and PETG was added to 1mM. At the final timepoint (120 min.), the aliquots previously treated with PETG were assayed. The histograms displayed demonstrate that the PETG-treated aliquots (broken lines) assayed at 120 minutes after loading, compared with the assays done at the time of PETG addition (solid lines) have not increased in fluorescence since the addition of PETG. **B:** BW5147.42 lacZ (+) cells (1) were loaded with FDG in the presence of the indicated concentrations of PETG. The hypotonic treatment was terminated by the addition of 10 x ice cold buffer containing PETG at the same concentrations as were used during the loading. After 30 min. of incubation on ice the cells were analyzed by FACS. Displayed are histograms of that analysis. The broken line in the top panel shows the fluorescence of lacZ (-) BW5147 cells loaded with FDG without any PETG

Subprotocol 1
FACS-Gal Assay in Viable Cells

▓▓ Materials

Solutions

- DMSO at 200mM final FDG substrate at 2mM in 99% ddH$_2$O and 1% DMSO
- 1 µM propidium iodide (PI) from a 100 x stock

▓▓ Procedure

Prepare the FDG substrate

1. Dissolve the dry substrate in DMSO at 200 mM.

Note: FDG is available from Molecular Probes (Pitchford, OR) or Sigma Chemical (St. Louis, MO). It should dissolve quickly and completely.

2. Dilute the stock 100 X 37°C water. Transfer to a 37°C bath.

Note: Upon addition of the water, the FDG will initially precipitate. This is the final substrate at 2 mM in 99% ddH$_2$O and 1% DMSO which is very close to the solubility of FDG in water. Thus the substrate will need to be warmed at 37°C, occasionally vortexed and periodically checked for loss of visible precipitate. The final solution should be clear and colorless or slightly yellow.

Freeze FDG substrate

3. Make 1 ml or larger aliquots of 2 mM FDG substrate and freeze.

Note: The 2mM FDG substrate is very stable at temperatures up to 65 °C . Repeated freeze/thaw cycles do not cause appreciable deterioration, so aliquots can be large and repeatedly thawed, redissolved, used and refrozen. Freezing or refrigerating the substrate will cause it to reprecipitate, mandating redissolution as above.

Prepare cell samples

4. Set up 6 types of sample cells:
- lacZ(-) cells
- lacZ(+) cells

- lacZ(-) + lacZ(+) at a 10:1 ratio
- lacZ(-) + lacZ(+) at a 1:1 ratio
- lacZ(-) + lacZ(+) at a 1:10 ratio
- lacZ (+) cells that will not receive substrate

Note: For best results it is important to do this pilot experiment using the cell types of interest to you. It is important to validate the system using lacZ(-) cells and lac Z(+) cells. If you do not have a lacZ expressing cell line it can be produced by transfecting your cells with a β-geo expression construct (17).

Note: The best results are obtained with healthy dividing cells.

5. Centrifuge cells at 200 x g for 10 min. Resuspend in medium at 10^6 cells per ml.

Note: The medium can be any isotonic medium that optimally preserves the health of the particular cell line. Normally the cells standard growth medium is used. This will also be used later to end the loading procedure.

6. Dissolve an aliquot of 2mM FDG stock by maintaining at 37 °C in a water bath.

7. Place a measured volume of cells (50 µl) into a 12 x 75 polystyrene culture tube (FACS tube) and equilibrate in a 37 °C water bath for a few minutes.

8. Add an equal volume (50 µl) of prewarmed 2mM FDG to the cells.

Load cells by hypotonic shock

Note: If sterility is desired for a sterile sort then use the culture tube caps, otherwise they can be discarded.

9. Mix briefly by tapping the tube or pipeting the mixture and replace in the 37 °C bath.

10. After 1-3 minutes stop the loading and restore the cells to near normal tonicity by the addition of at least 10 x ice cold media and place tube on ice.

11. Incubate for 2 hours on ice.

Fluorescence development

Note: Duration of incubation on ice varies by experimental design. Generally to reveal all lacZ(+) cells but those with the least amount of enzyme we incubate for 2 hours.

12. Add 1 µM propidium iodide (PI) from a 100 x stock to the cells.

Note: Dead cells should be excluded from the analysis based on the uptake of propidium iodide (PI).

FACS analysis and sorting

13. Set up flow cytometer with the standard filters for propidium iodide (>620 nm) and fluorescein (515-545nm).

14. Run lacZ(-) cells first to establish the baseline for fluorescein background in the first decade of the log scale.

Note: For optimal sensitivity and separation of the positive and negative cells the appropriate compensation adjustment should be made so that when looking at β-gal positive cells in a 2 D fluorescein (FL1) vs. PI (FL2) plot the population is not on a slant with brighter cells for fluorescein also brighter for PI.

15. Collect histograms of PI vs forward scatter. Set up gate to exclude PI (+) cells from the analysis histograms.

16. Run lac Z(+) cell next to adjust compensation for any spill over in the fluorescence channels.

17. Set up histograms for log of fluorescein fluorescence vs. cell number.

18. Collect data from all of the samples (10,000 cells for gated histogram).

19. Perform sort.

Note: For analysis the cells do not have to be kept on ice while they are on the machine. For sorting the cells must be kept iced to maintain standard sample conditions over the long term.

Subprotocol 2
FACS-Gal Assay With PETG to slow fluorescence development

▦▦ Materials

Solutions

- PETG
- 10μM PETG (add 1μl of a stock of PETG that has been diluted to 500 μM in isotonic media)
- 100 μM PETG (add 1μl of a stock of PETG that has been diluted to 5mM in isotonic media)
- 2mM FDG stock
- propidium iodide (PI) 100 x stock

▦▦ Procedure

Note: The concentrations of PETG described here, 10μM and 100μM are for illustration. Experimentation with the cells of interest will reveal the best PETG concentration to appropriately slow the reaction. It will likely fall somewhere between 10μM and 100μM.

1. Follow steps 1-5 from the FACS-Gal assay protocol.

2. Dissolve the PETG in isotonic medium to a concentration of 50mM.

   Prepare the β-gal inhibitor PETG.

Note: PETG stock solution may be stored frozen until ready to use.

3. Pre-label tubes for time points.

4. Dissolve aliquots of 2mM FDG stock by maintaining at 37 °C in a water bath.

5. Add a measured volume of cells (50μl) into a 12 x 75 polystyrene culture tube and equilibrate in a 37°C water bath for a few minutes.

Loading the cells with PETG

6. Bring the warming cells to either 10μM PETG (add 1μl of a stock of PETG that has been diluted to 500 μM in isotonic media) or 100 μM PETG (add 1μl of a stock of PETG that has been diluted to 5mM in isotonic media) mix briefly and continue to warm for 2 minutes.

Note: PETG is used in the FDG loading to prevent cells with very high levels of lac-Z from exhausting the FDG substrate during loading. A very low concentration of PETG, depending on the severity of the problem (10-100 μM), is included in the loading reaction to slow the lacZ activity proportionally in each cell. The exact concentration should be determined for every cell type.

7. Add PETG to an aliquot of your ice cold isotonic diluent to bring the diluent to either 10μM PETG or 100μM PETG.

Note: The diluent should have the same PETG concentration as the cells while they were being warmed in step 6. The size of the aliqout is related to the number of reactions that are being treated this way.

8. Add an equal volume (50μl) of prewarmed 2mM FDG to the cells. Mix briefly by tapping the cells or pipeting the mixture and replace in the 37°C bath.

9. After 1-3 minutes stop the loading reaction and restore the cells to near normal tonicity by the addition of at least 10 x ice-cold isotonic media and place on ice.

Note: The added ice-cold media should contain the same concentration of PETG as the cells were incubated in step 6.

10. Incubate on ice for an appropriate time to get the range of fluorescence appropriate for the experiment.

11. Add 1 μM propidium iodide (PI) from a 100 x stock to the cells.

Note: Dead cells should be excluded from the analysis based on the uptake of propidium iodide (PI).

12. Set up flow cytometer with the standard filters for propidium iodide (620nm) and for fluorescein (515-545 nm).

13. Run LacZ(-) cells first to establish the baseline for fluorescein background in the first decade of the log scale.   **FACS analysis**

14. Collect histograms of PI vs forward scatter. Set up gate to exclude PI (+) cells from the analysis histograms.

15. Run Lac Z(+) cell next to adjust compensation for any spill over in the fluorescence channels.

Note: For optimal sensitivity and separation of the positive and negative cells the appropriate compensation adjustment should be made so that when looking at β-gal positive cells in a 2 D fluorescein (FL1) vs. PI (FL2) plot the population is not on a slant with brighter cells for fluorescein also brighter for PI.

16. Set up histograms for log of fluorescein fluorescence vs. cell number.

Note: For analysis the cells do not have to be kept on ice while they are on the machine. For sorting the cells must be kept iced to maintain standard sample conditions over the long term.

17. Collect data from all of the samples (10,000 cells for gated histogram). See fig. 2B.

Subprotocol 3
FACS-Gal Assay With PETG to stop fluorescence development

▮▮ Materials

Solutions

- 2mM FDG stock
- PETG stock solution
- Propidium iodide (PI) from a 100x stock

Preparation

PETG is an isotonic medium to a concentration of 50mM.

▓▓ Procedure

Prepare the β-gal inhibitor PETG

1. Follow steps 1-5 from the FACS-Gal assay protocol.

2. Dissolve the PETG in isotonic medium to a concentration of 50mM.

Note: PETG stock solution may be stored frozen until ready to use.

3. Pre-label tubes for time points.

Loading the FDG

4. Dissolve aliquots of 2mM FDG stock by maintaining at 37 °C in a water bath.

5. Place a measured volume of cells (50μl) into a 15 ml tube and equilibrate in a 37 °C water bath for a few minutes.

6. Add an equal volume (50μl) of prewarmed 2mM FDG to the cells. Mix briefly by pipeting the mixture and replace in the 37 °C bath.

Fluorescence development

7. After 1-3 minutes stop the loading and restore the cells to near normal tonicity by the addition of at least 10 x ice cold media and place on ice for 2 hours.

Note: Duration of incubation on ice varies by experimental design. Generally to reveal all LacZ(+) cells but those with the least amount of enzyme we incubate for 2 hours. Cells that have exhausted their FDG substrate during the fluorescence development period on ice pile up as a highly fluorescent peak. The addition of PETG to stop the reaction at various timepoints allows the cells to be assayed and sorted for variations in their β-gal activity before topping out in the high fluorescent peak.

8. Remove aliquots (100 μl) of cells to a 12 x 75mm culture tube on ice at different timepoints after the loading with FDG. Add 2μl PETG (final concentration of 1mM) to the cells and mix briefly to stop the accumulation of fluorescence.

FACS analysis

9. Add 1 μM propidium iodide (PI) from a 100 x stock to the cells in each tube.

Note: Dead cells should be excluded from the analysis based on the uptake of propidium iodide (PI).

10. Set up flow cytometer with the standard filters for propidium iodide (›620nm) and fluorescein (515-545 nm).

11. Run LacZ(-) cells first to establish the baseline for fluorescein background in the first decade of the log scale.

Note: For analysis the cells do not have to be kept on ice while they are on the machine. For sorting the cells must be kept iced to maintain standard sample conditions over the long term.

12. Collect histograms of PI vs forward scatter. Set up gate to exclude PI (+) cells from the analysis histograms.

13. Run Lac Z(+) cell next to adjust compensation for any spill over in the fluorescence channels.

Note: For optimal sensitivity and separation of the positive and negative cells the appropriate compensation adjustment should be made so that when looking at β-gal positive cells in a 2 D fluorescein (FL1) vs. PI (FL2) plot the population is not on a slant with brighter cells for fluorescein also brighter for PI.

14. Set up histograms for log of fluorescein fluorescence vs. cell number.

15. Collect data from all of the samples (10,000 cells for gated histogram). See fig. 2B.

Subprotocol 4
Chloroquine use in FACS-Gal assay to reduce endogenous β-gal activity

▪▪▪ Materials

Solutions

– 200μM and 600μM aliquots of chloroquine in growth media.
– 2mM FDG substrate
– Propidium iodide (PI) from a 100 x stock

Procedure

Note: Chloroquine is a weak base, which is non-ionic at neutral pH and ionizing after entering the lysosomes that concentrate and trap it. The ionization raises the lysosomal pH thereby reducing the activity of endogenous β-gal. This treatment may not be entirely effective and can be toxic enough to cause loss of viability. Several concentrations should be tested for your particular cell type. It is a protocol of last resort when cells in log phase growth still exhibit too high an endogenous β-gal activity.

Prepare FDG reagent

1. Follow steps 1- 5 from the FACS-Gal assay protocol.

Prepare chloroquine

2. Make up 100μl aliquots of 200μM and 600μM aliquots of chloroquine in growth media. Place at 37 °C.

3. Make up several mls of 100μM and 300μM aliquots of chloroquine respectively in growth media for terminating the loading of FDG in step 11. Place on ice.

4. Pre-label tubes for time points.

5. Use lac-Z negative cells in log growth for this chloroquine pilot determination, which is done along side of the normal FACS-Gal assay for viable cells.

6. Place a measured volume of 2 x concentrated cells (25μl) into each of three 12 x 75 mm polystyrene culture tubes and equilibrate in a 37°C water bath for a few minutes.

Pretreat cells with chloroquine

7. Add an equal volume (25μl) either of growth medium, 200μM chloroquine in growth medium (100μM final), or 600μM chloroquine in growth medium (300μM final) to the respective labeled tubes of cells.

8. Incubate for 30 minutes at 37 °C.

9. Dissolve aliquots of 2mM FDG stock by maintaining at 37 °C in a water bath.

Loading the FDG substrate

10. Add an equal volume (50μl) of prewarmed 2mM FDG to the cells. Mix briefly by tapping the tube or pipetting the mixture and replace in the 37 °C bath.

11. After 1-3 minutes step the loading and restore the cells to near normal tonicity by the addition of 10 x ice cold media containing the same final concentration of chloroquine as used in stop 7.

Stop loading reaction

Note: If chloroquine helps to lower the endogenous background activity it should be kept in the assay all the way through the protocol.

12. Incubate on ice for 2 hours.

Fluorescence development

Note: Duration of incubation on ice varies by experimental design. Generally to reveal all lacZ(+) cells but those with the least amount of enzyme we incubate for 2 hours.

13. Add 1 μM propidium iodide (PI) from a 100 x stock to the cells.

Add propidium iodide

Note: Dead cells should be excluded from the analysis based on the uptake of propidium iodide (PI).

14. Set up flow cytometer with the standard filters for propidium iodide (>620 nm).

FACS analysis

15. Run lacZ(-) cells first to establish the baseline for fluorescein background in the first decade of the log scale.

16. Collect histograms of PI vs. forward scatter. Set up gate to exclude PI (+) cells from the analysis histograms.

17. Run lacZ(+) cell next to adjust compensation for any spill over in the fluorescence channels.

Note: For optimal sensitivity and separation of the positive and negative cells the appropriate compensation adjustment should be made so that when looking at β-gal positive cells in a 2 D fluorescein (FL1) vs. PI (FL2) plot the population is not on a slant with brighter cells for fluorescein also brighter for PI.

18. Set up histograms for log of fluorescein fluorescence vs. Cell number.

19. Collect data from all of the samples (10,000 cells for gated histogram). See figure 2B.

Note: For analysis the cells do not have to be kept on ice while they are on the machine. For sorting the cells must be kept iced to maintain standard sample conditions over the long term.

Subprotocol 5
FACS-Gal Assay coanalyis with DNA content using Hoechst 33342

▓▓ Materials

Solutions

- 200µg/ml stock of Hoechst 33342 in growth media.
- 2mM FDG stock.
- Propidium iodide (PI) from a 100 x stock

▓▓ Procedure

Note: The analysis of DNA content in living cells[19] can be performed in combination with the FACS-Gal assay.[11].

Prepare Hoechst 33342

1. Follow steps 1-5 from the FACS-Gal assay protocol.

Note: Make a 250µg/ml stock of Hoechst 33342 in growth media.

Note: Use 5µg/ml Hoechst 33342 (Sigma Chemical, St. Louis MO) in growth media to label cells.

2. Dilute Hoechst 33342 to a concentration of 10µg/ml (2 x stock) in growth media.

3. Make a 1:1 dilution of the 2 x Hoechst 33342 to be used as the cold diluent in step 7.

Incubate cells with Hoechst 33342

4. Add an equal volume of 2 x Hoechst 33342 to cells and allow to incubate at 37°C for 1 hour.

5. Dissolve aliquots of 2mM FDG stock by maintaining at 37 °C in a water bath.

6. Place a measured volume of cells (50µl) into a 12 x 75 mm polystyrene culture tube and equilibrate in a 37°C water bath for a few minutes.

Loading the FDG substrate

7. Add an equal volume (50µl) of prewarmed 2mM FDG and 1 µl Hoechst 33342 undiluted stock to the cells. Mix briefly by tapping the tube or pipeting the mixture and replace in the 37 °C bath.

Note: Cells should be maintained in 5µg/ml Hoechst 33342 at all times during this assay. Different types of cells will try to pump out the Hoechst dye at different rates.

8. After 1-3 minutes stop the loading and restore the cells to near normal tonicity by the addition of at least 10 x ice cold media containing 5µg/ml Hoechst 33342 and place on ice for 2 hours.

Fluorescence development

Note: Duration of incubation on ice varies by experimental design. Generally to reveal all lacZ(+) cells but those with the least amount of enzyme we incubate for 2 hours. The cold diluent must contain the Hoechst dye.

9. Add 1 µM propidium iodide (PI) from a 100 x stock to the cells.

Add propidium iodide

Note: Dead cells should be excluded from the analysis based on the uptake of propidium iodide (PI).

10. Set up flow cytometer with a UV laser exciting at 362nm and a standard Argon laser exciting at 488nm. Set up standard filters for propidium iodide (›620nm), fluorescein (525-540nm), and Hoechst 33342 (440-460nm) .

FACS analysis

11. Run lacZ(-) cells first to establish the baseline for fluorescein background in the first decade of the log scale.

12. Collect histograms of PI vs forward scatter. Setup gate to exclude PI (+) cells from the analysis histograms.

Note: For optimal sensitivity and separation of the positive and negative cells the appropriate compensation adjustment should be made so that when looking at β-gal positive cells in a 2 D fluorescein (FL1) vs. PI (FL2) plot the population is not on a slant with brighter cells for fluorescein also brighter for PI.

13. Run lacZ(+) cells next to adjust compensation for any spill over in the fluorescence channels.

14. Set up histograms for log of fluorescein fluorescence vs. Cell number, Hoechst 33342 vs. cell number, and fluorescein vs Hoechst 33342.

15. Collect data from all of the samples (10,000 cells for gated histograms).

Note: For analysis the cells do not have to be kept on ice while they are on the machine. For sorting the cells must be kept iced to maintain standard sample conditions over the long term.

Troubleshooting

The lacZ negative substrate loaded cells are very much more fluorescent than the unloaded control cells

One possibility is that the FDG is contaminated badly with fluorescein. This would most likely be a manufacturing problem. Is the FDG solution distinctly yellow? Fluorescein is yellow in solution and this would indicate contamination with fluorescein or fluorescein-mono-galactoside. FDG should be colorless or only very faintly yellow-tinged.

Another less likely possibility is that the cell line being assayed has unusually high levels of endogenous lysosomal β-gal activity (see above discussion of endogenous β-gal). If you suspect that this is the case you might try to test this by using a different cell line or possibly mouse thymocytes or splenocytes. We have not found this situation with any of the cell lines we have assayed including, NIH3T3, Jurkat, BW5147, Hela, 293, L cells, Burkitt Lymphoma lines, Abelson virus mouse pre-B lines, K562 and others, but it is always a possibility.

Known negative cells assayed in a mixture with positive cells are more fluorescent than negative cells assayed by themselves

This problem is due to the reaction product fluorescein leaking from the lacZ (+) cells into the lacZ (-) cells. If this is the case, the problem will be more pronounced in the mixtures with a higher percentage of lacZ (+) cells. Cells loaded with FDG by this technique generally have only 5μM FDG intracellularly even though the FDG concentration in the loading reaction is 1 mM, so the loading has no effect on the concentration of FDG outside the cell and loading is not limited by that factor. As mentioned above, the important variable impacting the cell density in the loading reaction is the total amount of β-gal in the loading reaction. During the period of loading when the cells are at 37°C in hypotonic medium, substrate is being introduced into the cells but also product from the enzymatic cleavage of the FDG by β-gal is being produced and leaking from the cells. If there is too much total β-gal activity in the loading reaction the fluorescein will leak from the cells with high β-gal activity into cells with low or no β-gal activity and produce false positives. Extreme examples of this problem can be clearly seen by a strong yellow color of the cell suspension after loading and dilution with ice-cold isotonic media. This problem does not occur often and can be alleviated by reducing the concentration of lacZ enzyme in the reaction. This can most easily be done by using fewer cells but it can also be done by increasing the volume of the loading reaction (but this uses more substrate per assay).

The peak of the lacZ (+) cells is not well resolved from the lacZ (-) cells (even though lacZ- cells are not much more fluorescent than unloaded control cells, see 1 above)

The first possibility is that the cells have not truly exhausted their substrate. The best way to determine this is to leave the cells on ice for quite a few more hours, even overnight, and reassay. If the fluorescence has increased significantly, are the negative and positive peaks now well resolved? If the substrate is exhausted and the positive peak is not much brighter after further incubation then the problem is that not enough substrate has been introduced. Do an experiment in which the cells are hypotonically loaded for more than a minute. We have found that most cell lines can be hypotonically loaded for up to 3 minutes with very little impact on cell viability. Introduction of FDG is roughly linear with respect to time during the first three minutes of the loading reaction as show by figure 3.

Rare false-positive cells

When the FACS-Gal assay is done on cells without lacZ there is "background" of rare cells that are significantly more fluorescent

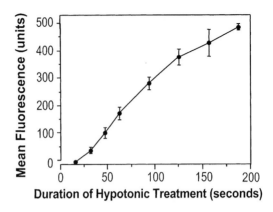

Fig. 3. Duration of the hypotonic treatment influences the amount of FDG introduced into the cell. ß-gal-expressing BW5147.56 cells (1) were hypotonically treated for varying periods with 1 mM FDG in 37 C medium with a tonicity of 50% isotonic. After the hypotonic treatment, the samples were incubated on ice until all intracellular FDG had been hydrolyzed and then the samples were analyzed by FACS. The geometric mean of the fluorescence is plotted against the duration of the hypotonic treatment

than the rest of the population. The frequency of these cells is generally between 0.01% and 1%, varies by cell type and is dependent upon loading of the cells with FDG.[2] It is hypothesized that for some reason more FDG is introduced into the lysosomes of these cells since in many cases, maintaining the cells in logarithmic growth will reduce the incidence of these cells, (however chloroquine treatment did not affect the frequency of these cells, see above).

When these "rare false-positive" cells are sorted from the rest of the population they are found to be viable and most importantly, their progeny is not enriched for cells with this characteristic. Therefore, FACS-Gal can select lacZ positive cells occurring at less than one in a million in a population by performing cycles of FACS-Gal sorting, expansion in culture and resorting until the β-gal expressing cells are enriched above the background of rare false-positive cells. In addition, two fluorescence characteristics of the rare false-positive cells helps to differentiate them from β-gal expressing cells (thus reducing the need for growth and resorting). They are generally not as fluorescent as cells with enough β-gal activity to have hydrolyzed all the substrate they contain, and they tend to be bright for emission in the yellow wavelength (562-588 nm) as well as for the fluorescein (green 515-545 nm) wavelength. Avoiding cells with low green fluorescence and above average yellow fluorescence is the best way to select true β-gal expressing cells when they are infrequent in the population (see fig. 4).

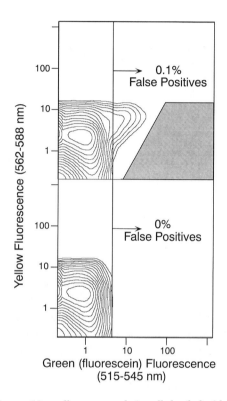

Fig. 4. Rare false-positive cells appear only in cells loaded with FDG, are moderate in fluorescein fluorescence and are above average in yellow fluorescence. The top panel displays a logarithmic contour plot of NIH3T3 cells without lacZ after a standard hypotonic loading of FDG. The bottom panel displays the same type of plot on NIH3T3 cells that were treated identically except that an isotonic FDG solution was used instead of the normal hypotonic FDG solution. The contour plots are 50% logarithmic, a representation that emphasizes areas of the plot with very low density. The shaded area depicts the fluorescence levels from which the highest ratio of infrequent ß-gal expressing cells over false-positive cells can be selected

Comments

Getting Started

The initial experiment to validate the system uses lacZ negative and lacZ positive cells. It should include at least 6 samples of which 5 samples receive the substrate – lacZ negative cells, lacZ positive cells, and 3 mixtures of positive and negative cells in roughly 10:1, 1:1, and 1:10 proportions. The sixth sample does not receive substrate. The cell density is not important in the loading. Generally, 10^6 cells are convenient for an assay, but the density of the cells will not affect the concentration of FDG introduced into the cells. The total amount of β-gal enzyme in the loading reaction can influence the results so one may need to dilute the cells for loading in that case (see troubleshooting). The amount of substrate used is proportional to the volume of the loading reaction; thus one can conserve substrate by using a small volume for the loading reaction (50 μl). The cells should be loaded identically (except for the no substrate control) and allowed to incubate on ice for roughly 2 hours before FACS analysis.

A good assay will have the lacZ negative loaded cells only slightly more fluorescent (less than 2 fold mean fluorescence) than the unloaded cells. The lacZ positive cells will have a distinct peak that is clearly separated from the lacZ negative peak. The mixtures will show well separated negative peaks and positive peaks containing the expected percentages of cells based on the percentage lacZ (+) and lacZ(-) cells used to make the mixture. The peaks for the lacZ(+) and lacZ(-) cells in the mixture should also correspond to the fluorescence levels of the cells assayed separately.

For best results it is important to do this pilot experiment using the cell types of interest to you. If you do not have a lacZ expressing cell line it can be most easily produced by transfecting your cells with a β-geo expression construct.[17] β-geo is a fusion gene that has the activity of β-gal as well as the activity of the drug G418. Cells that are made resistant to G418 by expression of β-geo will be β-gal positive and can be kept that way by selecting with G418. (I have available a PGK-β-geo construct that is expressed in most cells and I will provide it for interested investigators- please E-mail me.)

Use of the β-gal inhibitor, PETG

The use of the competitive inhibitor of β-gal, PETG (phenyl-ethyl-thio-galactoside, Sigma Chemical, St. Louis MO) makes the assay more convenient and comparable between samples when multiple samples are assayed. It also extends the range of β-gal activity that can be analyzed.[2] PETG is a reversible competitive inhibitor of β-gal but it is not able to be hydrolyzed by β-gal so its concentration is stable in the presence of β-gal. PETG can enter cells even when they are chilled (but surprisingly it will not wash out of cells that are colder than room temperature). It is nontoxic, readily available, water soluble and inexpensive.

As discussed above, the cells that have exhausted the substrate pile up in a peak with high fluorescence after incubation on ice. Variation of β-gal activity among cells within this highly fluorescent positive cell peak cannot be determined. To assay the variation within the β-gal expressing population the fluorescence development must be stopped before the cells have exhausted their substrate. To do that, one can remove aliquots at different time points after loading with FDG and stop the accumulation of fluorescence with PETG. Then the multiple timepoints can be analyzed by the FACS following the final timepoint. This makes the assay more convenient, reproducible, and reveals more information regarding the range of activity within a population of cells. It also allows the sorting of the very highest expressing cells, which is a frequently desired objective (fig. 2A).

Occasionally, when cells with very high levels of lacZ activity are assayed, they may exhaust the substrate during the loading period, and therefore cell-to-cell variation of β-gal activity is not revealed. To avoid this problem a low concentration of PETG can be included in the loading reaction and the cold diluent to simply slow the lacZ activity proportionally in each cell. The appropriate concentration of PETG in the loading reaction should be determined for each cell line in which a reduction of activity is desired (fig. 2B).

Endogenous β-gal activity

Eucaryotic cells often have endogenous β-gal enzymatic activity that can cleave FDG into fluorescein. This activity is found in the lysosomes and varies widely among cell types. In mammals, macrophages have the highest levels of endogenous β-gal activity and sperm likely have the lowest. This activity does not cause frequent problems for the FACS-Gal assay since the loading procedure does not introduce significant amounts of substrate into the lysosomes. Occasionally endogenous β-gal will make differentiation of lacZ expressing and non-expressing cells difficult. Maintenance of the cells in optimal logarithmic growth prior to performing a FACS-Gal assay will often greatly reduce endogenous β-gal activity and is a generally recommended practice.

β-gal assays that do not maintain the integrity of the cell membrane, such as the x-gal assay, can avoid the problem of endogenous β-gal activity by performing the assay in a buffer with a neutral pH, since the endogenous β-gal is optimal at a low pH found in the lysosome and is inactive at neutral pH. The FACS-Gal assay is done on intact viable cells and therefore it is more difficult to influence the pH of the lysosome, however it can be done by treating the cells with a weak base such as chloroquine. Chloroquine is non-ionic at neutral pH and ionizes after entering the lysosomes (and is therefore trapped and concentrated in the lysosome). The ionization within the lysosome raises the lysosomal pH and reduces the activity of endogenous lysosomal β-gal thus improving the assay when performed on cells with prohibitively high endogenous β-gal activity. Although chloroquine is the most effective weak base among those tested for the property of lowering endogenous β-gal activity, it is not completely effective and in addition, chloroquine is somewhat toxic to the cells and can impair viability.[2] Therefore chloroquine is not recommended routinely and is best used as a last resort in the rare occasions when logarithmically growing cells still exhibit too much endogenous β-gal activity for the particular FACS-Gal experiment being performed.

Coanalysis with other parameters

FACS-Gal is completely compatible with analysis for cell surface antigens by antibody staining. Simply load the cells and maintain on ice during the incubation, then stain for surface antigens as usual. The only caveat is the cells obviously must be kept cold throughout the staining, including during centrifugation. Most antibody staining protocols, however, are done with cells kept on ice and centrifuged at 4°C to avoid capping the antigen, so this is not usually a problem. The antigen must, however, be revealed with a fluorochrome other than fluorescein since the FACS-Gal assay uses fluorescein.

FACS-Gal with FDG is not compatible with any other assay that must open the cell membrane for its performance. This rules out assays for antibody staining of intracellular antigens or for cell cycle position using the most common techniques for staining cells for DNA content. FACS-Gal can be used with a two-parameter analysis of lacZ activity and DNA content (cell cycle position) by using the vital DNA dye HO33342 (see fig. 5).[11] This dye can enter cells through the membrane and is revealed with an UV excitation laser. This technique is appropriate if cell viability must be maintained, however, the use of HO33342 for cell cycle analysis on viable cells is considerably less reproducible than with DNA dyes that must be used with fixed cells, like propidium iodide. For most experiments, to correlate cell cycle position with gene expression it is best to use a surface antigen as a reporter for gene expression and propidium iodide to assay DNA content in fixed cells.[18]

FACS-Gal substrates other than FDG

The FACS-Gal system using FDG is convenient and sensitive for use with flow cytometry, however the system requires that cells be intact and kept chilled. It is therefore not practical for use with tissue sections, cells adhering to slides or dishes or with situations in which the cells are fixed for analysis. In order to create further possibilities for fluorescent analysis of β-gal activity, alternative substrates have been developed. These substrates have a fatty acid tail attached to the FDG molecule and thus are more lipophilic. FDG with a fatty acid tail (C8FDG or C12FDG, Molecular Probes, Pitchford OR) enters the cell passively and is retained within the cell at 37°C or room temperature. These substrates have been examined for utilization with the FACS but have not found wide utility since the substrate also passively enters lysosomes and is available to endogenous lysosomal β-gal activity, therefore making lacZ negative cells also highly fluorescent. C12FDG has been used successfully in assaying β-gal

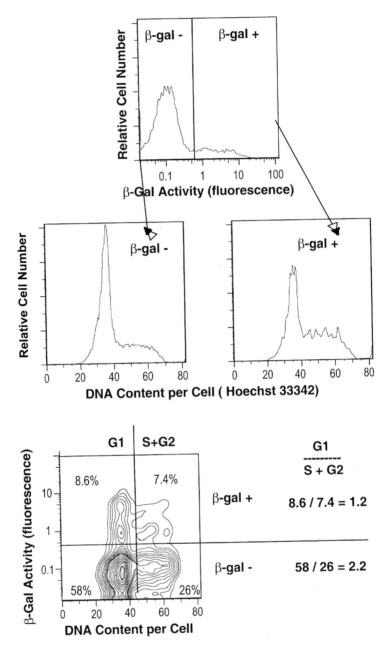

Fig. 5

activity in the sperm of lacZ transgenic mice.[20] Success with sperm cells is likely due to the low level of lysosomal β-gal activity found in sperm.

More recently, another FDG based substrate (5-chloromethylfluorescein di-β-galactopyranoside, CMFDG) has been developed that carries a reactive moiety and complexes with glutathione within cells to cause it to be retained without chilling the cells. This substrate has just become available and its use with the FACS has not been thoroughly characterized.[21] However it may have advantages over FDG for some specialized assays of β-gal by the FACS. This remains to be seen.

Alternative assays to FACS-Gal

Other reporter gene systems are available that maintain cell viability and can be analyzed by the FACS. These have the same analytical and selection advantages as FACS-Gal. What are the comparative advantages of these alternative systems?

Surface antigens can be useful reporter genes as suggested above. One advantage of the surface antigen as a FACS-based reporter is that the cell membrane can be opened to allow correlation with DNA content for cell cycle analysis. A second advantage is that fluorescence is linearly proportional to the quantity of surface molecules, permitting simple quantitation directly from the FACS plots. A third advantage is that many of these antigens are encoded by mammalian genes and are possibly more dependably expressed than the prokaryotic lacZ. The main disadvantage is the reduced sensitivity per expressed molecule since there is no enzyme-mediated amplification of the signal. Generally, many hundred to a few thousand molecules of a surface antigen must be expressed on the cell surface to recognize a cell as expressing. Finally, the use of a surface antigen reporter gene in transgenic mice may lead to unexpected phenotypes since most surface antigens are not biologically neutral. Rather, they are generally a receptor or ligand. This potential problem may become less important, as the detailed function of more surface antigens are elucidated.

Another alternative reporter gene system for FACS analysis of viable cells is the green fluorescent protein (GFP) of Aequoria victoria. This protein is itself fluorescent and therefore requires no substrate, making GFP more convenient and less expensive to use than lacZ. It has been expressed in a wide variety of organisms including transgenic mice and toxicity has not been noted.[22] GFP is rapidly becoming a very widely used reporter gene in a variety of applications. It has not been widely used for FACS until recently (see this section, chapter 1) since for multiple reasons the wild type GFP protein does not have sufficient sensitivity for detection of any but the most highly expressing cells. It has become clear that this molecule can be modified in regard to its stability, excitation spectra, emission spectra and level of fluorescence per molecule. This engineering process has produced variant GFP molecules that show promise as reporter genes for assay by FACS.[22,23] The present variants and expected future variants, which may be even more easily assayed, are likely to provide an important alternative to lacZ as a FACS-based reporter gene system. In addition, other fluorescent molecules found in nature may be identified, cloned and characterized in the future for use as reporter genes.

Future technical development using lacZ, GFP and other reporter genes for usage with the FACS will undoubtedly lead to experiments in which multiple reporter genes are used to provide simultaneous information on multiple experimental parameters in single viable cells. This will permit the examination of a variety of complex transcriptional and regulatory questions that are not presently accessible to experimentation.

Feel free to contact me by email if you have further questions on FACS-Gal. I would like to acknowledge the American society of Hematology and the Burroughs-Wellcome Foundation for grant support and Len Herzenberg, Gary Nolan, Dave Parks, and Mario Roederer for their contributions to the development and characterization of FACS-Gal.

References

1. Nolan GP, Fiering S, Nicolas FF, Herzenberg LA. Fluorescence-activated cell analysis and sorting of viable mammalian cells based on β-D-galactosidase activity after transduction of Escherichia coli lacZ. Proc Natl Acad Science 1988;85:2603-2607.

2. Fiering SN, Roederer M, Nolan GP, et al. Improved FACS-Gal: flow cytometric analysis and sorting of viable eukaryotic cells expressing reporter gene constructs. Cytometry 1991;12:291-301.

3. Nir R, Yisraeli Y, Lamed R, Sahar E. Flow cytometry and sorting of viable bacteria and yeasts according to beta-galactosidase activity. Appl Environ Micro 1990;56:3861-6.

4. Krasnow MA, Cumberledge S, Manning G, et al. Whole animal cell sorting of Drosophila embryos. Science 1991;251:81-5.

5. Zhuang Y, Soriano P, Weintraub H. The helix-loop-helix gene E2A is required for B cell formation. Cell 1994;79:875-884.

6. Robertson G, Garrick D, Wilson M, et al. Age-dependent silencing of globin transgene expression in the mouse. Nucleic Acid Res 1996;24:1465-1471.

7. Zambrowicz B, Imamoto A, Fiering S, et al. Disruption of overlapping transcripts in the ROSA βgeo 26 gene trap strain leads to widespread expression of βgalactosidase in mouse embryos and hematopoietic cells. Proc Nat Acad Science 1997; 94:3789-3794.

8. Yancoupolos GD, Nolan GP, Pollock R, et al. A novel fluorescence-based system for assaying and separating live cells according to VDJ recombinase activity. Mol Cell Biol 1990;10:1697-704.

9. Kerr WG, Nolan GP, Serafini AT, Herzenberg LA. Transcriptionally defective retroviruses containing lacZ for the in situ detection of endogenous genes and developmentally regulated chromatin. Cold Spring Harbor Symp Quant Biol 1989;54 pt 2:767-76.

10. Kartunnen J, Shastri N. Measurement of ligand-induced activation in single viable T cells using the lacZ reporter gene. Proc Nat Acad Science 1991;883972-6.

11. Fiering S, Northrop JP, Nolan GP, et al. Single cell assay of a transcription factor reveals a threshold in transcription activated by signals emanating from the T-cell antigen receptor. Genes Dev 1990;4:1823-1834.

12. Walters MC, Fiering S, Eidemiller J, et al. Enhancers increase the probability but not the level of gene expression. Proc Nat Acad Science 1995;92:7125-7129.

13. Walters MC, Magis W, Fiering S, et al. Transcriptional enhancers act in cis to suppress position-effect variegation. Genes Dev 1996;10:185-195.

14. Roederer M, Fiering S, Herzenberg LA. FACS-Gal: Flow cytometric analysis and sorting of cells expressing reporter gene constructs. Methods: A Companion to Methods in Enzymology 1991;2(3):248-260.

15. Lupton SD, Brunton LL, Kalberg VA, Overell RW. Dominant positive and negative selection using a hygromycin phosphotransferase-thymidine kinase fusion gene. Mol Cell Biol 1991;11:3374-8.

16. Rotman B. Measurement of activity of single molecules of β-D-galactosidase. Proc Nat Acad Science 1961;47:1981-1991.

17. Friedrich G, Soriano P. Promoter traps in embryonic stem cells: a genetic screen to identify and mutate developmental genes in mice. Genes Dev 1991;5:1513-1523.

18. Van den Heuvel S, Harlow E. Distinct roles for cyclin-dependent kinases in cell cycle control. Science 1993;262:2050-2054.

19. Arndt-Jovin DJ, Jovin TM. Analysis and sorting of living cells according to deoxyribonucleic acid content. Jour of Histochem Cytochem 1977;25:585-589.

20. Moynahan ME, Akgun E, Jasin M. A model for testing recombinogenic sequences in the mouse germline. Hum Mol Gen 1996;5:875-886.
21. Poot M, Arttamankul S. Verapamil inhibition of enzymatic product efflux leads to improved detection of β-galactosidase activity in lacZ transfected cells. Cytometry 1997;in press.
22. Zhang G, Gurtu V, Kain SR. An enhanced green fluorescent protein allows sensitive detection of gene transfer in mammalian cells. Biochem Biophy Res Com 1996;227:707-11.
23. Anderson MT, Tjioe IM, Lorincz MC, et al. Simultaneous fluorescence-activated cell sorter analysis of two distinct transcriptional elements with a single cell using engineered green fluorescent proteins. Proc Natl Acad Science 1996;93:8508-8511.

Utilization of a Bicistronic Expression Vector for Analysis of Cell Cycle Kinetics of Cytotoxic and Growth-arrest Genes

NICHOLAS E. POULOS, ERIC STANBRIDGE, AND SUSAN DEMAGGIO

Introduction

Traditional eukaryotic gene expression systems have many shortcomings that make them unsuitable for analysis of cytotoxic and growth-arrest genes. An intestinal alkaline phosphatase bicistronic expression vector was specifically designed to facilitate such studies. Intestinal alkaline phosphatase serves as a marker of cells that have been transfected and therefore must also be co-expressing the gene under study. Using flow cytometry, a trivariate analysis was performed on $p16^{INK4}$ tranfected U87 glioblastoma cells, but other systems could be studied with this technique. This vector has universal applications including: 1) analysis of cytotoxic and growth-inhibitory genes in transient assays; 2) 100% enrichment in gene expression studies, especially in low transfection efficiency experiments; and 3) facilitation of the study of cell cycle kinetics.

The motivation for the design of the human intestinal alkaline phosphatase (IAP) bicistronic expression vector came from a desire to study cytotoxic and growth-arrest genes. Stable transfections are impossible given the nature of these genes which preclude colony formation. Inducible promoter systems regulated by heavy metals or dexamethasone are often "leaky", may have cytotoxicity associated with the inducing heavy metal, require the presence of a functional glucocorticoid receptor, or may suffer from the pleotropic non-specific effects of dexamethasone-mediated gene expression. More sophisticated systems such as the tetracycline system[1] are very cumbersome to use, requiring two stable transfections. Also, because tetracycline is extremely unstable at 37°C in culture media, daily or every other day tetracycline supplementation of media for the duration of the experiment must be employed. Episomal systems such as pCEP4 are another option, but these frequently have cellular toxicity associated with the viral antigen. They also require colony formation, which would be obviated by the constitutive expression of the cytotoxic / growth-arrest genes. Microinjection is very labor intensive and is typically used to study phenomena involving a few hundred cells. Reproducing experiments can be arduous, and drawing conclusions based on such small sample sizes, tenuous. Traditional gene expression techniques are therefore unsuitable for the study of cytotoxic or growth-arrest gees, making functional analysis and characterization of these genes an extremely difficult technical problem.

A system utilizing CD20, a lymphocytic specific surface antigen, developed by Van den Heuvel and Harlow[2] attempts to surmount the aforementioned problems in a unique way. A plasmid expressing CD20 is co-transfected with a plasmid expressing the cDNA of the gene of interest into cells that are CD20 negative, in a transient assay. Flow cytometry is then used to isolate cells expressing the surface antigen and further analyzed for a phenotype associated with the potentially co-transfected gene of interest. A major disadvantage of this system is the potential for false negatives and false positives associated with transfection of only one of the desired constructs.

The IAP bicistronic expression vector represents a major improvement over these conventional systems. Human intestinal alkaline phosphatase is a member of an isoenzyme family. IAP is localized at the extracellular surface of the plasma membrane [3-5] and is primarily expressed at high levels in the apical brush border membranes of enterocytes of the small intestine.[6] It is only rarely expressed on cells grown in culture and, therefore, like CD20, can serve as a surface marker in many transfection experiments. Human IAP was cloned from a HeLa-D98/AH.2 Lambda ZapII cDNA library[7] and a hybridoma that produces a monoclonal antibody; (BD6) that specifically recognizes IAP and not other alkaline phosphatases was developed.[8] These reagents, combined with the cloned internal ribosomal entry site from the encephalomyocarditis virus RNA, [9,10] facilitate the development of the IAP bicistronic expression vector. The IAP bicistronic expression vector produces one transcript with two translation initiation sites, resulting in the simultaneous expression of two proteins. IAP expression serves as a marker of IAP negative cells that have been transfected with the vector and which therefore co-expressed both sites within the bicistronic vector under study. Combined with flow cytometric analysis, this system circumvents the potential problems of false negative and positive inherent in the CD20 system while providing several additional advantages, making it uniquely suited to the study of cytotoxic and growth-arrest genes.

The power of the IAP bicistronic expression vector is that it efficiently identifies transfected cells amid a high background of untransfected cells. There are several parameters that contribute to this success. IAP, unlike most eukaryotic surface antigens, is not ubiquitously expressed, therefore making it an excellent marker of transfected cells. Additionally, it is not hydrolyzed by trypsin, permitting harvesting of cells grown in monolayer for flow cytometric analysis. Finally, a monoclonal antibody is readily available in our laboratory that allows specific identification of cells expressing IAP with minimal background. Because a second monoclonal antibody, anti-BrdUrd B44 (Becton Dickinson), is used in these experiments, anti-IAP was biotinylated. This reaction was successfully performed without modification of functionally important lysine groups involved in antigen binding. A successful blocking reagent, 5% FBS, was then developed to reduce noise from endogenous biotin.

Flow cytometry permits the observation of large numbers of cells and, with the capability of trivariate analysis, allows the cell cycle and S-phase analysis of a specific subpopulation of cells within the cell line or tissue studied. A strategy that would permit the partial denaturation of DNA requisite for anti-BrdUrd binding but which would not denature the surface antigen IAP, thereby destroying the specificity of the system, was needed. The traditional technique of 2N HCl / Triton X-100 DNA denaturation denatured the IAP protein. Furthermore, endonuclease / exonuclease protocols such as BamHI / exonuclease III[11] and DNAse / exonuclease III[12] either denatured IAP or inadequately denatured the DNA, respectively. A crosslinking strategy was developed which covalently linked the terminal amino groups of IAP to those of the BD6 monoclonal antibody, using the homobifunctional NHS-ester crosslinking reagent BS[3]. After the antigen-antibody complex had been formed, 2N HCl / Triton X-100 could then be used to denature DNA, permitting BrdUrd staining while preserving the specificity of IAP staining required for the identification of transfected cells.

The non-trivial obstacle of developing a fluorophore staining protocol that would enable three-color labeling with fluorophores which could be excitable by a 488 nm argon ion laser and which would have a minimum of compensation problems for 1) the cell surface antigen IAP, 2) the nuclear antigen BrdUrd, and 3) the DNA content was accomplished so that a benchtop Becton Dickinson FACScan flow cytometer with conventional optical filters would be sufficient. The method published by Schmid and colleagues[13] was adapted. PE-strepavidin-biotinylated monoclonal antibody BD6 labels IAP, FITC-conjugated monoclonal antibody B44 labels BrdUrd, and DNA content staining is achieved with 7-AAD. Transfected cells may thus be identified based on IAP expression, and further analyzed for DNA content and S-phase labeling.

The utility of the IAP bicistronic expression vector in combination with flow cytometry is impressive. Because of its high specificity and sensitivity, it is indispensable in low transfection efficiency experiments resulting in 100% enrichment in gene expression studies, without the false positives and false negatives associated with the CD20 system. This feature is strikingly illustrated by a p16 experiment[14] where transfection efficiency was estimated to be only 1% after optimization of electroporation parameters by immunofluorescence staining, yet a very convincing G1/S block was observed. Cells expressing cytotoxic and growth-arrest genes therefore may be isolated, studied in transient assays, and functionally characterized. Additionally, time consuming protocols that synchronize cells for cell cycle experiments may possibly be avoided, permitting manipulation of asynchronous populations, as was done in the p16 experiments. This trivariate analysis protocol is an example of the dynamic studies that are possible in cell cycle kinetics. Proliferative indices based on S-phase labeling may be accurately determined at desired time points by pulsing with BrdUrd. Cohorts may also be labeled and their fate ascertained as they progress around the cell cycle. FITC conjugated monoclonal antibodies may also be used to analyze the cell cycle dependent expression of proteins.

In conclusion, a novel IAP bicistronic expression vector was designed, constructed, and successfully tested. This vector, in combination with flow cytometry, has universal applications including: 1) analysis of cytotoxic and growth-inhibitory genes in transient assays; 2) 100 % enrichment in gene expression studies, especially in low transfection efficiency experiments; and 3) facilitation of the study of cell cycle kinetics.

Materials

Solutions

- DMEM with 10% FCS and 2mM L-glutamine without antibiotics
- HEPES-buffered saline pH 7.05
- 2N HCl / 0.5% Triton X-100 in PBS with 5% FBS
- reaction buffer: 1 mM BS^3 in PBS
- 0.1 M $Na_2B_4O_7 \cdot 10H_2O$, pH 8.5.

Equipment

Standard FACScan configuration

Procedure

Transient Assay for Analysis of Cell Cycle Kinetics and Growth Inhibition

1. Cells were cultured in 5% CO_2 at 37° C in DMEM with 10% FCS and 2mM L-glutamine without antibiotics.

   Cell Line & Culture Conditions

   Note: U87 is an established human glioblastoma cell line (1), p16 deleted and IAP negative.

2. pCMV-p16-CITE/IAP Bisctronic expression vector by utilizing multiple cloning site 5' of CITE. PCMV-CITE/IAP functions as control vector.

   Expression Vectors Cloned

3. Resuspend cells in HEPES-buffered saline pH 7.05 at concentration of 2×10^7 cells / ml in a BTX disposable cuvette (4 mm gap).

   Electroporation of cells

4. 40 µg supercoiled plasmid DNA is added.

5. Electroporate cells at 250 volts with a capacitance of 1050 μF, R1(13 ohms), and a pulse length of 8.5 msecs, utilizing a BTX ECM 600 system.

6. Replate under optimal cell growth conditions.

7. Trypsinize cells to collect data points at appropriate time intervals

BrDU Staining

8. Incubate cells with BrDU at a final concentration of 10 μM for 30 minutes at 37° C.

9. Wash cells 2 x in PBS with 0.1% sodium azide.

Note: Sodium azide prevents capping and internalization of antibody.

10. Trypsinize cells, pellet, and fix in 70% EtOH.

Note: Cells can be stored here at -20° C until analysis.

Wash

11. Wash in 1 ml PBS.

Antibody Labeling

12. Label cells in suspension with 100 μl of biotinylated monoclonal antibody BD6 diluted 1:200 in PBS with 5% FBS, incubate 30 minutes at 4° C.

Wash

13. Wash with 1 ml PBS 3 x.

Cross-link IAP to BS³ (bis(sulfosuccinimidyl) suberate)

14. 200 μl of reaction buffer of 1 mM BS³ in PBS for 30 minutes at room temperature.

Quench

15. Add Tris buffer to bring final concentration to 50 mM in reaction buffer, let incubate for 15 minutes.

Wash

16. Wash in PBS.

Denature DNA

17. Resuspend in 1 ml 2N HCl / 0.5% Triton X-100 for 30 minutes at RT.

Neutralize Acid

18. Centrifuge, aspirate supernatant, resuspend in 1 ml of 0.1 M $Na_2B_4O_7$ -10H_2O (pH 8.5).

Wash and Resuspend

19. Wash in 1 ml PBS and resuspend in 60 μl PBS containing 5% FBS.

Note:

20. Label S-phase cells and IAP positive cells with 20 µl of anti-BrDU FITC and 20 µl of streptavidin-PE per 10^6 cells, incubate 30 minutes in the dark at RT.

Antibody Staining

21. Wash in 1 ml PBS.

Wash

22. Resuspend in 1 ml PBS containing 25 µg / ml of 7-AAD, incubate 30 minutes in the dark at room temperature.

DNA Staining

23. Filter cells through 53 m-pore nylon mesh.

Filter

24. Setup flow cytometer to collect the DNA/ 7-AAD in FL3 or far red, BrDU/ FITC in FL1 or green PMT, and IAP$^+$/PE in FL2 or red PMT. Histograms to display are the DNA / FL3 linear signal and the IAP$^+$/PE log signal and dot plots of 7-AAD / FL3 against BrDU / FITC FL1 and 7-AAD area under the peak against width of the peak for double discrimination.

DNA Trivariate Analysis by Flow Cytometry

25. Adjust the high voltages so that the 7-AAD stained G_0-G_1 peak in PMT 3 is set at approximately channel 200, the BrDU /FITC peak of a normal specimen gives a nice horseshoe appearance on the bivariate dot plot and the IAP/PE peak sits in the second decade on the log scale of FL2. (See Sec. 5, Chap 2c.Fig. 2c.2).

26. Acquire data on single color cells to set the photomultiplier tubes and adjust the compensation for each channel against the others. (See Sec. 2, Chap. 7)

27. DNA cell cycle/BrDU sample collection is gated on the top 1-5% of PE positive cells which have passed through the doublet discrimination gate (See Sec. 5, Chap 2c. Fig. 2c.1)

Collect data

Note: Acquire data on a minimum of 10^6 cells to collect 5000 IAP-positive cells within the gate.

28. Analyze DNA Cell cycle using a modeling program. Analysis of % of BrDU positive cells can be calculated using a box analysis to follow cell populations through S-phase of the cell cycle. (See Sec. 5, Chap 2c.Fig.2c.2)

Analyze DNA Cell cycle

Note: This allows an accurate DNA and S-phase analysis of a rare or limited cell population without contamination of information from other cells, simply by the gating process.

References

1. Gossen M, Bujard H. Tight control of gene expression in mammalian cells by tetracycline-responsive promoters. Proc Natl Acad Sci USA 1992; 89:5547-5551.
2. Van den Heuvel S, Harlow E. Distinct roles for cyclin-dependet kinases in cell cycle control. Science 1993; 262:2050-2054.
3. Henthorn PS, Raducha M, Kadfesch T, et al. Sequence and character-ization of the human intestinal alkaline phosphatase gene. J Biol Chem 1988; 263:12011-12019.
4. Millan JL. Oncodevelopmental expression and structure of alkaline phosphatase genes. Anticancer Res 1988; 8:995-1004.
5. Low ML. Biochemistry of the glycosyl-phosphatidylinositol membrane protein anchors. Biochem J 1987; 244:1-13.
6. Alpers DH, Eliakim R, DeSchryber-Kecskemeti K. Secretion of hepatic and intestinal alkaline phosphatase: similarities and differences. Clin Chim Acta 1989; 186:211-224.
7. Latham KM, Stanbridge EJ. Identification of the HeLa tumor-associated antigen, p75/150, as intestinal alkaline phosphatase and evidence for its transcriptional regulation. Proc Natl Acad Sci USA 1990; 87:1263-1267.
8. Bicknell DC, Sutherland DR, Stanbridge EJ, et al. Monoclonal antibo-dies specific for a tumor-associated membrane phosphoprotein in hu-man cell hybrids. Hybridoma 1985; 4:143-152.
9. Parks GD, Duke GM, Palmenberg AC. Encephalomyocarditis virus 3C protease: efficient cell-free expression from clones which link viral 5' noncoding sequences to the P3 region. J Virol 1986; 60:376-384.
10. Elroy-Stein O, Fuerst TR, Moss B. Cap-independent translation of mRNA conferred by encephalomyocarditis virus 5' sequence improves the performance of the vaccinia virus / bacteriophage T7 hybrid expres-sion system. Proc Natl Acad Sci USA 1989; 86:6126-6130.
11. Dolbeare F, Gray JW. Use of restriction endonucleases and exonuclease III to expose halogenated pyrimidines for immunochemical staining. Cytometry 1988; 9:631-635.
12. Takagi S, McFadden ML, Humphreys RE, et al. Detection of 5-bromo-2-deoxyuridine (BrdUrd) incorporation with monoclonal anti-BrdUrd antibody after deoxyribonuclease treatment. Cytometry 1993; 14:640-648.
13. Schmid I, Uittenbogaart CH, Giorgi JV. A gentle fixation and permea-bilization method for combined cell surface and intracellular staining with improved precision in DNA quantification. Cytometry 1991; 12:279-285.
14. Poulos NE, Farmer AA, Chan KWK, et al. Design of a novel bicistronic expression vector with demonstration of a p16^{INK4} induced G1/S block. Cancer Research, 1996 Apr 15; 56(8):1719-23.

NGFR as a Selectable Cell Surface Reporter

M. MUNN, G. RAGHU, S.W. PARK, C-H. PAN, AND A. DAYN

Introduction

To simplify analysis, characterization, and selection of trans-fected/transduced cells by non-cytotoxic methods, several groups have recently used low affinity nerve growth factor receptor or NGFR, as a selectable membrane marker.[1,2] NGFR joins a list of other selectable markers, most notably neomycin resistance gene and green fluorescence protein, GFP. Using a fluorochrome-conjugated monoclonal antibody against NGFR[3], one can easily identify the NGFR+ cell population, which expresses your gene of interest, by flow cytometry. Therefore, NGFR can be incorporated easily into current DNA transfection or retroviral transduction methods targeted against mammalian cell lines and primary cells.

There are other advantages to using NGFR. First is the ability to immediately and rapidly sort for NGFR+ cells by flow cytometry, 3-5 days after transfection/transduction. Therefore, in 3-5 days, one can obtain a pure population of cells containing your gene of interest. In contrast, antibiotic resistance selection (i.e. neomycin) requires at least two weeks. In addition, there are cell toxicity issues involved with selection by neomycin resistance. Even when selection is not required, NGFR provides an immediate assessment of the efficiency of gene transfer on the actual cell population. Historically, a vector construct containing the β-gal reporter gene is transferred into a parallel population of cells, and used to assess efficiency of transfer. NGFR can also be used to study the activity of specific promoters in the target cells.

When NGFR is incorporated into an appropriate vector (i.e. dual promoter or bicistronic vector), NGFR+ cells (which contain your gene of interest) can be easily distinguished from NGFR- cells. We typically see a one to two log separation between

NGFR+ and NGFR- cell populations (Figure 1). Depending on the vector construction and cell type, NGFR expression has been stable for greater than 6 months. Green fluorescence protein is another ideal selectable marker. As the protein itself is fluorescent, therefore obviating an antibody detection step, one can sort out GFP+ cells by flow cytometry. However, in our experience it is much more difficult to distinguish between positive and negative cells with GFP than with NGFR. However, there are humanized and enhanced fluorescent versions of GFP which have not been tested.

NGFR has potential applications in human gene therapy. NGFR is normally expressed on human neurons, and therefore less likely to be antigenic. GFP and neomycin, which are not normally present in humans, have a greater risk of prompting an immune response. NGFR, either the full length or c-terminus truncated versions, has been successfully used with retroviral vectors such as LNCX and LXSN in the transduction of hematopoietic stem cells and peripheral blood lymphocytes. Clinical trials using NGFR as the selectable marker are currently underway. The truncated version is preferred for its non-physiological

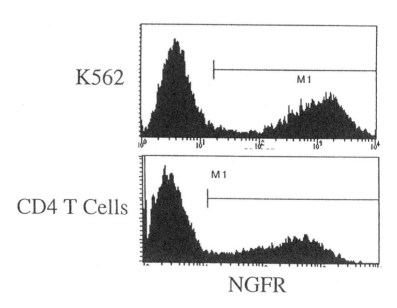

Fig. 1. Two examples of unsorted transduced cells. Gates were set with negative (mock) control cells (not shown)

response. The C-terminus is cut and stops at the transmembrane region.

Subprotocol 1
Simple DNA Transfection of Cell Lines for NGFR Expression Procedure

▪▪ Materials

Equipment

- $37^o/CO_2$ incubator
- Centrifuge
- Hemacytometer
- Flow Cytometer

Solutions

- Growth medium
- 2 x Hank's Balanced Salt Solution (HBSS) can be purchased as 10X and diluted.
- Supercoiled NGFR plasmid (Chromaprobe Inc.) 12-15µg
- Anti-NGFR monoclonal antibody (Chromaprobe, Inc.)
- PBS/EDTA
- PBS with 2% BSA
- $CaCL_2$, 2-2.5M stock solution
- 2.5mM chloroquine stock (100 x)

▪▪ Procedure

1. Grow adherent cell to $2\text{-}5 \times 10^5$

Note: Examples of cell lines are VERO, HELA, NIH3T3.

2. Prewarm cell growth media and 2 x HBSS.

Grow adherent cells of choice

**Prepare trans-
fection DNA/
CaCl$_2$**

3. To a 1.5 ml microcentrifuge tube add 430 µl H$_2$O, 60µl CaCl2, and 12-15 µg NGFR supercoiled plasmid DNA vector or mock DNA control (total volume should be 500µl).

Note: Examples are LNCX and LXSN for hematopoeitic stem cells.

4. To 500µl of 2 x HBSS in a 1.5 ml microcentrifuge tube add the DNA from step3. Mix.

Note: Important to add the DNA to the HBSS, not the other way around.

5. Remove media from the cells.

**Prepare cells
in chloroquin
media**

6. Add 8 – 12 ml cell culture medium containing 1 x chloroquine to the cells.

**Add DNA/
CaCl$_2$**

7. Add the DNA/CaCl2 mixture slowly to each flask (NGFR or mock) of the cells.

Note: CaCl$_2$ is toxic to cells, so add very slowly to the opposite face of the flask to mix first with the medium, then invert.

**Incubate cells
with DNA to
transfect**

8. Place in 37°C incubator (5%CO$_2$).

9. Incubate 5-10 hours.

**Let cells re-
cover**

10. Replace chloroquine medium with 5 ml fresh normal growth medium.

**Analyze Cells
by the fol-
lowing steps:**

11. Analyze cell 24-72 hours after transfection.

Harvest cells

12. Harvest cells with PBS/EDTA.

Count cells

13. Count cells on hemacytometer. Aliquot 1 x 10^6 cells for staining.

14. Centrifuge cells to stain at 500 x g. Discard supernatant.
15. Add 100µl PBS with 2% BSA to cells.

Add antibody

16. Add 20 µl anti-NGFR FITC or PE directly conjugated monoclonal antibody to the cells.

Incubate

17. Incubate on ice for 30 minutes.

18. Add 1 ml cold PBS with 2% BSA to the tube.   **Wash cells**

19. Centrifuge at for 5 minutes at 500 x g to wash cells.   **Centrifuge**

20. Aspirate off the supernatant.

21. Analyze on flow cytometer.   **Analyze on flow cytometer**

22. Set up flow cytometer with for either fluorescein filters or PE filters depending on which fluorochrome you stained with.

Note: You may add 1ug/ml propidium iodide to distinguish and gate out dead cells.

23. Run mock – transfected cells for background control.

24. Use HS294T cells from ATCC as a positive control which expresses NGFR

Note: HS294T cells from ATCC are a neuronal carcinoma cell line expressing NGFR. (These are not a T-cell line.)

**Subprotocol 2
Procedure for Retroviral Transduction of Cell Lines
for Creation of Stable NGFR Expressing Cell Lines**

Materials

Equipment

- Centrifuge with swinging ELISA plate adapter rotor buckets.
- 0.45 micron filter and syringe
- Hemacytometer
- 37°/CO$_2$ incubator
- Parafilm and tape
- Flow Cytometer / cell sorter

Solutions

- Growth medium best for chosen cells.
- 2.5mM chloroquine (100 x) stock solution.
- 2 x Hank's Balanced Salt Solution (HBSS) (can be purchased as 10 x and diluted).

- CaCL$_2$, 2-2.5M stock solution.
- Polybrene 8mg/ml stock solution.
- NGFR DNA.
- Anti-NGFR antibody for flow cytometry and cell sorting (Chromaprobe, Inc.).
- Fetal bovine serum.

Preparation

- Grow PCL (Packaging Cell Line) cells.
- Grow recipient transducible cells
- Prewarm medias and solutions needed for each day

Time Line

1. Wait 3-5 days after transfection for NGFR expression

2. Set up on Tuesday for transduction.

3. Wednesday - spinoculation #1

4. Thursday - spinoculation #2.

5. Friday - spinoculation #3

6. Monday - check for NGFR expression in transduced cell line.

7. Creation of stable cell line takes about 6 months or more.

▨▨ Procedure

Set up cell lines for experiment.

1. Begin to grow PCL cells to split one week in advance. Split cells until you have 16-18 T25 flasks. Begin to grow recipient transducible cell line.

Note: PCL – a retroviral packaging cell line of choice. Transducible cell line.

2. Prewarm growth media and 2XHBSS.

Prepare DNA

3. Label 2 sets of 15 ml tubes and prepare
   - Set A: 2 x HBSS (500 µl each tube)
   - Set B: DNA + CaCL$_2$ + H$_2$O4. Mix set A and set B tubes by pipetting. Use air bubbles to mix.

5. Prepare PCL cells by aspirating media carefully off of cells in the flask.

Note: Example: add 100µl to 10 ml media.

6. Add Chloroquine stock (100 x) to prewarmed growth media (final concentration = 1 x).

7. Add 10 mls of Chloroquine media to each flask.

8. Add DNA solution to the media slowly. Distribute over surface of flask.

Note: CaCl$_2$ is toxic to live cells, so add very slowly.

9. Incubate at 37°C in CO$_2$ incubator for 6-10 hours.

10. Aspirate chloroquine media after incubation and feed cells with fresh 6 ml growth media.

11. Incubate overnight.

12. Prewarm growth media.

13. Collect supernatant from PCL cells into a centrifuge tube.

14. Add a 1:1000 dilution of an 8mg/ml polybrene stock. Mix and filter through 0.45 micron syringe filter. Final dilution is 8µg/ml.

15. Add fresh 6 ml growth media to PCL cells. Optional: check for NGFR expression on PCL cell by flow cytometry (see steps 29,30).

16. Harvest recipient cells.

Note: Almost any cell will work: PBLs, hematopoetic stem cells, HELA, VERO, Jurkat K562, U937 to name a few.

17. Count on a hemacytometer.

18. Spin down cells at 500 x g for 10 min.

19. Aspirate media off of cells.

Side notes (right margin):

Prepare PCL cells for transfection

Add DNA for Transfection to the packaging cells (PCL)

Incubate transfecting cells with DNA

Stop transfection

Incubate cells

Collect viral supernatant at 24, 48, and 72 hours

Put transfected PCL cells back in culture

Prepare recipient transducible cell line

Add virus to cells from step 14

20. Resuspend cells well in the virus-containing PCL/polybrene media.

21. Transfer to T25 flasks.

Spin virus onto cells – Spinoculation #1

22. Parafilm the flask. Tape to a swinging ELISA plate holder centrifuge adapter bucket for centrifugation at 2000 rpm in a Sorvall GLC Tabletop centrifuge for 90 minutes.

23. Transfer to 15 ml tubes. Spin down cells at 500 x g for 10 minutes.

24. Remove supernatant and add 5-10 ml fresh medium, then transfer to a fresh T25 flask.

Incubate cells

25. Incubate overnight in $37°C/CO_2$ incubator.

Infection #2

26. Repeat steps 19-25.

Infection #3

27. Repeat steps 19-23.
28. Transfer cells to T75 flasks. Incubate for ~5 days.

Check packaging cell transfection on flow cytometer

29. Stain 1×10^6 non-tranfected and transfected PCL cells with anti NGFR antibody to check transfection efficiency.

Note: See steps 12-24 from "Simple DNA Transfection of Cell Lines for NGFR Expression" protocol.

Check viral transduction on flow cytometer

30. Stain transduced cells with anti-NGFR antibody to check transduction efficiency.

Note: We usually achieve 60-80% efficiency.

31. If cells are NGFR+ then prepare 1×10^7 to be sorted on cell sorter for enriching or purifying transduced cells.

Sort transducted cells for purity

32. Collect sorted cells sterily in 500μl FBS.

33. Transfer to 15 ml centrifuge tube and centrifuge 300 x g for 10 minutes.

34. Resuspend in media and transfer to flasks for culture.

References

1. Rudoll T, Phillips K, Lee SE et al (1996) High-efficiency retroviral vector mediated gene transfer into human peripheral blood CD4$^+$T lymphocytes. Gene Ther 3:695-705
2. Valtieri M, Schiro R, Chelucci C et al (1994) Efficient transfer of selectable and membrane reporter genes in hematopoietic progenitor and stem cells purified from human peripheral blood. Cancer Res 54:4398-4404
3. Ross AH, Grob P, Bothwell M et al (1984) Characterization of nerve growth factor receptor in neural crest tumors using monoclonal antibodies. Proc Natl Acad Sci 88:6681-6685

Assessment of Cell Migration *in vivo* Using the Flurorescent Tracking Dye PKH26 and Flow Cytometry

ANDY BEAVIS

Introduction

The ability to determine the migration of specific cell subpopulations to certain tissue target sites is important in identifying the role of cell mediated immunity in the pathogenesis of autoimmune disorders. The highly fluorescent PKH tracking dyes produce a stable cell membrane labelling and have been used in combination with flow cytometry and multicolour immunofluorescence to determine the involvement of specific subsets of cells in adoptive transfer of insulin-dependent diabetes mellitus (IDDM) in the non-obese diabetic (NOD) mouse model (see Beavis and Pennline, J. Immunol. Methods, 170, pp 57-65, 1994). Here, the PKH26 tracking dye (Sigma, PKH26-GL) was used to label donor cells and then to follow their migration to various sites in a recipient mouse for up to 28 days following transfer. This was combined with phenotypic analysis using FITC and APC-labelled antibodies.

Labelling of cells with PKH26

The PKH26 is a lipophilic dye that produces a stable labelling of the cell membrane. The labelling process is a simple, rapid method but it is important to optimize dye concentrations and incubation times for each cell type used. It is important to quantitate cell numbers and function pre- and post-labelling to determine that these parameters are not adversely affected.

PKH26 Fluorescence

The PKH26 dye has an excitation maximum at 510 nm and 551 nm and can be excited well at 488 nm facilitating use on benchtop analysers. The fluorescence emission is centred around 567 nm and so can be collected in place of PE. It can be combined with FITC and the red-emission, tandem dyes such as PE-APC or PE-CY5 for use in multicolour analysis on standard benchtop analysers as well as with Cascade Blue and APC or CY5 for use on multi-laser cytometers. There is significant spectral overlap of the emission of PKH26 with that of the tandem dyes when using single laser (488 nm) excitation but this problem can be avoided by use of a 3-colour, dual laser protocol. The following protocol describes such a combination of PKH26 with FITC and APC for 3-colour analysis using dual laser excitation (488 nm blue line and either 633 nm or 647 nm red line).

▒ Materials

Equipment

Flow Cytometer equipped with 2-3 lasers with appropriate filter configuration for the excitation and emission of the dyes chosen.

Solutions

- RBC lysing buffer contains 0.16M Tris-HCl, 0.17M NH_4Cl, pH 7.0.
- PKH26 (Sigma) stock solution (4µM) fresh in Diluent C (Sigma).
- Staining buffer (PBS, 2% FCS, 0.1% azide)

▒ Procedure

1. Sacrifice donor mice by accepted method, remove spleen and tease apart with forceps. Filter disaggregated tissue through a pasteur pipette stuffed with a 2 cm x 2 cm piece of sterile gauze. Collect single-cell filtrate in 15 ml polypropylene tube.

Prepare Single Cell Suspension

2. Wash cells in 10 ml PBS, centrifuge 250 x g, 5 min and discard supernatant. Repeat twice.

3. Lyse RBC with 2 ml lysing buffer for 5 min at r.t. Wash X3 in PBS as previous, enumerate and resuspend at 2×10^8/ml.

Note: RBC lysing buffer contains 0.16 M Tris-HCl, 0.17M NH_4Cl, pH 7.0.

Prepare PKH26 (Sigma) stock solution (4 µM) fresh in Diluent C (Sigma)

4. Add 1 ml of cells to a 15 ml polypropylene tube containing 1 ml Diluent C. Add an equal volume of PKH26 (4 µM) and mix gently by inversion.

Note: Prepare PKH26 (Sigma) stock solution (4 µM) fresh in Diluent C (Sigma). Mix gently to prevent clumping of cells. DO NOT VORTEX CELLS. [Note: the PKH26-GL kit from Sigma contains the PKH26 dye and the Diluent C for general cell labelling applications].

5. Incubate for 1 min at r.t. with gentle inversion, add 2 ml FBS and mix by inversion. Stand for 2 min, add 10 ml PBS and centrifuge (250 x g, 5 min).

Note: The FBS stops the labelling reaction. Mix by gentle inversion of the tube. DO NOT VORTEX.

6. Discard supernatant, transfer cell pellet to a clean tube and wash X3 in 10 ml PBS.

7. Prepare control cells in an identical procedure in Diluent C without the PKH26 dye. Retain an aliquot of each for phenotypic analysis.

Note: CONTROL cells are incubated in Diluent C alone to show effects of Diluent only.

8. Resuspend the control and PKH26-labelled cells in PBS at 5×10^8 cells/ml and transfer 250 µl aliquots into 1 ml syringes with 26G needles.

Note: Transfer more than required for injection as there is residual ('dead') volume in the syringe.

Inject 50 - 200 µl into the tail vein

9. Inject 50 - 200 µl into the tail vein of a recipient mouse.

Note: Wipe tail with 70% alcohol to sterilize and to cause vasodilation making vein easier to find.

10. Resuspend untreated, control and PKH26-labelled cells for phenotypic analysis in staining buffer (PBS, 2% FCS, 0.1% azide) and aliquot 100 µl/well in v-bottomed, polypropylene plate.

Note: Stain donor cells prior to injection to determine phenotype of PKH26-labelled cells and to demonstrate that labelling does not selectively deplete specific subpopulations of cells.

Note: Stain untreated, control and PKH26-labelled cells: unstained, isotype, single-colour and multicolour samples.

11. Add appropriate antibodies (e.g. LYT2-FITC, THY1.2-biotin) or isotype controls, mix and incubate for 30 min at 4°C. Add 100 µl of staining buffer, mix and centrifuge 250 x g, 5 min.

12. Discard supernatant by rapid inversion of the plate, wash cells in 200 µl staining buffer and resuspend in streptavidin-APC. Incubate for 15 min at 4°C, wash X2 in staining buffer and resuspend in 500 µl in tubes for flow cytometric analysis.

Note: For single-laser cytometers, use FITC, PKH26 and PE-APC (or PE-CY5) for 3-colour analysis. For multilaser (488nm, UV, red) cytometers, use FITC, PKH26, APC (or CY5), ALLO-7 (APC-CY7) and/or Cascade Blue for multicolour analysis.

13. Sacrifice recipient mice by accepted method and remove pancreas, lymphoid tissues and peripheral blood (retro-orbital bleed).

Sacrifice recipient mice

14. Tease tissues apart and filter as previously described. Dice pancreas with scissors and digest with 5 ml collagenase (2 mg/ml in HBSS) 15 min at 37°C, filter through gauze and wash X3 in PBS. Enumerate cells and resuspend at 1×10^7/ml in staining buffer.

Note: Shaking vigorously for 10 seconds helps with the digestion of the pancreatic tissue.

15. Stain lymphoid and pancreatic cells using 96-well plate as described above. For peripheral blood, aliquot 100 µl whole blood into tubes, stain with antibodies and wash with 2 ml staining buffer. After final wash, lyse RBC with a commercial lysing reagent, wash in PBS and resuspend in 500 µl PBS for flow cytometric analysis.

Note: The commercial lysing reagent lyses RBC and fixes leukocytes in one step.

It is possible also to fix cells in 1 ml paraformaldehyde (1% in PBS) for analysis next day.

Flow
Cytometric
Analysis

16. Align flow cytometer as per standard protocol. Obtain FSC/SSC signals for unstained cells, gate lymphocyte population and adjust PMT voltages to set autofluorescence signals (log) on scale.

17. Obtain FSC/SSC signals for unstained cells from a normal mouse (no PKH26), gate lymphocyte population and adjust PMT voltages to set autofluorescence signals (log) on scale.

18. Check fluorescence for each single-colour control, set compensations. Check PKH26 fluorescence for PKH26-labelled donor cells (Fig. 1) and set compensations as required.

19. Establish a 2d-plot of SSC and PKH26 and gate the PKH26(+) cells (Fig. 2).

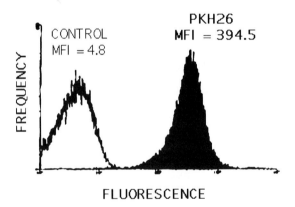

Fig. 1. Labelling of donor spleen cells with the fluorescent tracking dye PKH26. Overlay histogram of CONTROL (Diluent C treated) and PKH26-labelled donor spleen cells from a diabetic female NOD mouse. Cells were analysed for fluorescence on a standard FACS Vantage (Becton Dickinson Immunocytometry Systems, San Jose, CA) with excitation at 488 nm and fluorescence collected as the FL2 signal with a 575/26 bandpass filter. The mean fluorescence intensity (MFI) for control and PKH26-labelled cells is shown. "Reprinted from Journal of Immunological Methods, Volume 170, Andrew J. Beavis and Kenneth J. Pennline. Tracking of murine spleen cells in vivo: detection of PKH26-labeled cells in the pancreas of non-obese diabetic (NOD) mice. Pages 57-65, 1994 with kind permission of Elsevier Science-NL, Sara Burgerhartstraat 25, 1055 KV Amsterdam, The Netherlands"

SPLEEN

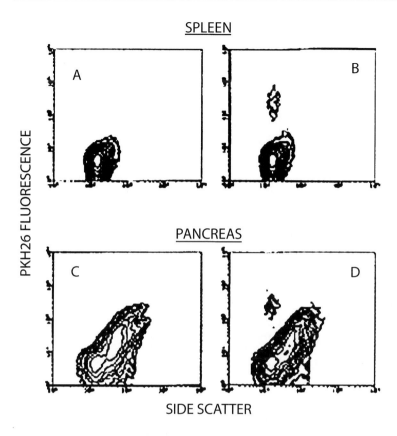

Fig. 2. Detection of PKH26(+) donor spleen cells in the lymphoid and pancreatic tissues of a recipient NOD mouse. Spleen and pancreas were harvested from a recipient NOD mouse 7 days following injection of control or PKH26-labelled donor spleen cells. Single cell suspensions were prepared from these tissues and analysed for fluorescence by flow cytometry using a standard FACS Vantage (Becton Dickinson Immunocytometry Systems, San Jose, CA). The PKH26(+) donor cells can be clearly resolved from the autofluorescence of recipient mouse cells by flow cytometric analysis using 2-d dot plots of side scatter (SSC) and PKH26 fluorescence for spleen (A,B) and pancreas (C,D) from recipient NOD mice injected with control (A,C) or PKH26-labelled (B,D) cells. "Reprinted from Journal of Immunological Methods, Volume 170, Andrew J. Beavis and Kenneth J. Pennline. Tracking of murine spleen cells in vivo: detection of PKH26-labeled cells in the pancreas of non-obese diabetic (NOD) mice. Pages 57-65, 1994 with kind permission of Elsevier Science-NL, Sara Burgerhartstraat 25, 1055 KV Amsterdam, The Netherlands"

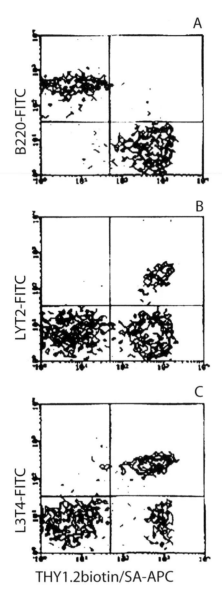

THY1.2biotin/SA-APC

Fig. 3. Two parameter contour plots showing the phenotype of PKH26(+) donor cells identified in the pancreas of a recipient NOD mouse. The pancreas was isolated from a recipient NOD mouse 7 days following injection of PKH26-labelled donor spleen cells. A single cell suspension was prepared, stained with monoclonal antibodies as described and analysed for fluorescence by flow cytometry using a standard FACS Vantage (Becton Dickinson Immunocytometry Systems, San Jose, CA). The PKH26(+) cells were gated from a 2-D plot of SSC and PKH26 (see Figure 2) and quadrant markers were set using isotypic control antibodies and SA-APC (not shown). (A) THY1.2-biotin/SA-APC vs B220-FITC; (B) THY1.2biotin/SA-APC vs LYT2-FITC; (C) THY1.2biotin/SA-APC vs L3T4-FITC. "Reprinted from Journal of Immunological Methods, Volume 170, Andrew J. Beavis and Kenneth J. Pennline. Tracking of murine spleen cells in vivo: detection of PKH26-labeled cells in the pancreas of non-obese diabetic (NOD) mice. Pages 57-65, 1994 with kind permission of Elsevier Science-NL, Sara Burgerhartstraat 25, 1055 KV Amsterdam, The Netherlands"

20. Acquire listmode data for unstained, isotypes, single- and multicolour samples at these instrument settings.

21. Quantitate the phenotype of donor cells prior to injection using a FSC/SSC gate. For cell tracking of donor cells in recipient mice, use a plot of SSC/PKH26 to identify cells of donor or recipient mouse origin and gate these onto dot-plots. Establish quadrants using single-colour controls and then analyse multicolour data for quantitative, phenotypic analysis (Fig. 3). **Data Analysis**

Acknowledgements

This work was performed in collaboration with Dr. Kenneth Pennline, Director of Biopharmaceutical Support Services, Cytometry Associates, Brentwood, TN.

References

Beavis AJ and Pennline KP. J. Immunol. Methods. 170: 57-65; 1994

In vivo Biotinylation for Cell Tracking and Survival / Lifespan Measurements

KENNETH A. AULT AND CATHY KNOWLES

Introduction

In vivo and *in vitro* biotinylation has been used in several research settings to conveniently label cells for tracking and survival / lifespan measurements. It has the advantage of not requiring radioisotopes, and seems to have very little effect on the normal functioning of cells. In brief, the cells are labeled by covalently binding biotin succinimide to amino groups on their surface. This biotinylation is rapid, stable, doesn't subsequently transfer from cell to cell, and does not affect cell function. The biotinylated cells can be detected using a labeled avidin molecule. In most applications, fluorescent avidin is used and the labeled cell can be detected using either flow cytometry or fluorescence microscopy. It has the additional unique capability that the labeled cells can actually be retrieved from the circulation using immobilized avidin columns, or by florescence activated cell sorting.

In vitro biotinylation has been done both in experimental animals and in man. In this procedure, the cells are removed from the body, washed, labeled with biotin, and then reinjected. This provides the advantages of biotinylation but also involves the disadvantage that the cells must be manipulated in vitro and washed, which may significantly alter their function.

The protocol described here (*in vivo* biotinylation) has been used only in experimental animals since it involves direct injection of reactive biotin succinimide into the animal. This results in the labeling of the entire cohort of circulating cells, as well as plasma proteins. Although this appears from the evidence thus far available to be quite safe, it has not been used in man. We and others have used it primarily to label platelets since these cells are highly reactive and are demonstrably altered by collection and washing. The use of *in vivo* biotinylation permits them to be labeled and tracked without ever removing them from the body or manipulating them in any way.

Materials

Equipment

Flow cytometer equipped with 3 color PMTs

Solutions

- Avetin anesthetic is 1.6 gm/mL 2,2,2-tribromoethanol (Aldrich Chemical Co., Milwaukee WI) in tert-amyl alcohol (Fischer Scientific, Pittsburgh PA), diluted 1:80 in saline
- Sulfo-NHS-LC-biotin, 3 mg in 300 µL of saline

Procedure

1. Avetin anesthetic is 1.6 gm/mL 2,2,2-tribromoethanol (Aldrich Chemical Co., Milwaukee WI) in tert-amyl alcohol (Fischer Scientific, Pittsburgh PA), diluted 1:80 in saline and injected i.p. at a dose of 0.2 ml per 10 gm body weight

   Prepare Avetin Anesthetic

2. Dissolve Sulfo-NHS-LC-biotin from Pierce Chemical Co.(Rockford IL), 3 mg in 300 ml of saline

   Prepare Water Soluble Biotin Succinimide

3. A blood sample should be drawn before biotinylation. 10 50 µl is adequate

   Draw Baseline Sample

4. Inject 150 µl slowly (over about 1 minute) in a tail vein.

   Inject Intravenously

   Note: Complete injection into the vein is necessary for good labeling.

5. Each sample should be 10-50 µl.

   Draw Post Biotinylation Sample

   Note: The timing of samples will be determined by the application. For platelet studies we draw the first sample at 3-4 hours and subsequent samples every 12 hours for 3-4 days

Labeling of Samples

Washing

6. Dilute 5 μl of blood to 1 ml with saline. Mix, then centrifuge at 1700g for 5 minutes. Remove the supernatant, and resuspend again in 1 ml saline. Repeat this washing step three times.

Note: Thorough washing of the cells is necessary to remove biotinylated plasma proteins which will compete for the avidin. Inadequate washing at this step is the most common cause of poor labeling.

Labeling with Fluorescent Avidin

7. After the last wash, resuspend the cells in the residual saline (about 50 μl) and add 20 μl of Red670 streptavidin (Gibco BRL, Gaithersburg MD).

Note: Red670 fluorochrome is compatible with both fluorescein and phycoerythrin in flow cytometric applications. Other fluorochromes may be used depending upon the application.

8. Incubate 15 minutes at room temperature, then wash once as above.

Other Labeling Procedure

9. Continue with any required subsequent labeling procedures that are needed to identify the cells of interest.

Flow Cytometric Analysis

10. After labeling is complete, resuspend the cells in 0.5 ml and analyze on the flow cytometer.

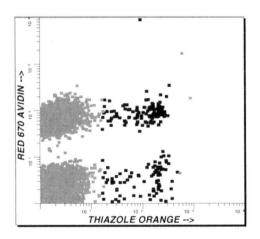

11. Gate on the cells of interest and display a fluorescence histogram for the avidin label.

Note: Use logarithmic amplification for the avidin fluorescence, and (usually) for other fluorescence markers as well.

12. Determine the boundary between biotin negative and biotin positive cells by comparing the pre-biotinylation and first post-biotinylation samples.

Data Analysis

13. To determine cell survival / lifespan, plot the percent biotin positive cells versus time and fit a regression line.

Note: It may be necessary to normalize the percent biotin positive cells if all of the cells were not completely biotin positive in the first post-biotinylation sample.

Assumptions and Cautions

The use of a labeling technique to measure the lifespan of circulating cells depends upon several assumptions. These are: 1) the labeling does not affect the lifespan of the cells of interest, 2) the label allows clear distinction between labeled and unlabeled cells for the duration of the measurement, i.e. in our

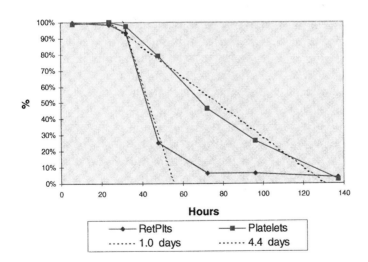

case there is no significant amount of biotin on the normal platelet, and the biotinylation results in a strong enough label to clearly resolve labeled platelets, 3) the entire cohort of circulating cells is labeled so that any unlabeled cells entering circulation can be accurately counted, 4) the labeled cells remain stably labeled for the duration of the measurement.

Biotinylation of platelets and erythrocytes has been used by other investigators as a technique to measure the lifespan of circulating cells.[1-6] These investigators have suggested that the biotinylation process does not appear to alter the function or the lifespan of red cells or platelets. Our results confirm this impression. We have not observed any adverse effects of *in vivo* biotinylation on the behavior or survival of mice. The biotinylated mice appeared to have normal platelet counts and normal levels of reticulated platelets. In addition we have not detected any change in hematocrit, reticulocyte counts, or leukocyte counts in biotinylated mice. Human platelets, biotinylated *in vitro* in whole blood, labeled at least as strongly as the murine platelets and yet showed no detectable defect in their function.[6] Finally, the estimate of platelet lifespan that we have obtained using *in vivo* biotinylation, is identical to that estimated by Corash et al using in vitro labeling with an entirely different fluorescent label.[7] Thus it seems very likely that the technique of *in vivo* biotinylation does not detectably alter the behavior of the platelets.

In vivo biotinylation strongly labels the entire cohort of circulating cells. The use of the technique to estimate the lifespan of circulating cells depends upon the assumption that the labeling is irreversible. Since the reaction of biotin succinimide with free amino groups results in a covalent alkyl-amide bond, the labeling should be stable in the absence of enzymatic activity that might cleave the bond. Studies by Cavill et al[5] have suggested that the labeling of red cells with sulfo-NHS-LC-biotin is not entirely irreversible. They reported that the label was stable in the short term, permitting measurement of red cell mass, but that it was "eluted" from the surface of the red cells over "several days" and thus could not be used to measure erythrocyte lifespan. Alternative possibilities are that biotinylated plasma proteins, or intrinsic platelet proteins, could be either shed, or internalized. If internalized, the biotin would not be accessible to the streptavidin label and thus there would be a decrease in labeling intensity. Our own data shows that the labeling decreases by about 50% in intensity over the first 24 hours of incubation, and then is stable *in vitro* for at least 3 days at 37°C. We could clearly distinguish labeled from unlabeled platelets even after the initial decrease in labeling intensity. Since our determination of platelet lifespan is complete within 3 days we feel it is unlikely that any possible elution of biotin from the surface of the platelet is significantly affecting our results. However this possibility must be considered in any application of this technique.

In some experiments, we have repeated biotinylated the same animals and have never observed an adverse effect. However, there have been reports in man of allergic hypersensitivity to biotin succinimide. Thus the possibility of biotin succinimide serving as a hapten and inducing an immune reaction should be kept in mind when using this technique.

Considerations for Different Applications

It is important to use saturating amounts of sulfo-NHS-LC-biotin in the biotinylation step to achieve the brightest possible labeling of the cells, and to use saturating amounts of labeled avidin in the labeling step to label all of the available biotin sites. The amounts given in the protocol have proven to be optimal for mice and for labeling platelets, but both should be titrated in preliminary experiments for your application.

We have observed that, in addition to labeling of circulating cells, some cells in tissues may also be labeled to a varying degree. For example, most bone marrow cells are labeled using this protocol. For this reason, we have seen that the new cells entering the circulation have significant levels of biotinylation for several hours after biotinylation. In our hands the first truly biotin negative cells were observed after 24-36 hours. This must be taken into account in interpreting the data, however it also suggests that this technique may be useful in tracking some cell populations in tissues as well as blood cells.

References

1. Heilmann E, Friese P, Anderson S, et al. Biotinylated Platelets: A new approach to the measurement of platelet lifespan. Br. J. Haematol 1993 Dec; 85(4):729-35.
2. Peng J, Freise P, Heilmann E et al. Aged platelets have an impaired response to thrombin as quantitated by P-Selectin expression. Blood 1994; 83:161-166.
3. Franco RS, Lee KN, Barker-Gear R et al. Non-isotopic measurement of the in vivo recovery and survival of two rabbit platelet populations using bi-level biotinylation. Blood 1993; 82:337a.
4. Suzuki T, and Dale GL. Biotinylated erythrocytes: in vivo survival and in vitro recovery. Blood 1987; 70:791-795.
5. Cavill I, Fisher TJ, and Hoy T. The measurement of the total volume of red cells in man: a non-radioactive approach using biotin. British J. Haematol. 1988; 70:491-493.
6. Ault, KA, Knowles C. In vivo biotinylation demonstrates that reticulated platelets are the youngest platelets in circulation. Exp. Hematol. 1995; 23:996-1001.
7. Corash L, Lin C, Reames A, et al. Post-transfusion in vivo platelet survival following photochemical treatment: A murine model. Blood 1993; 82:401a.

Enrichment and Selection for Reporter Gene Expression with MACSelect

MINDY WILKE-DOUGLAS AND UTE BEHRENS-JUNG

Introduction

Stable or Transient transfection of cells with genes of interest is widely used by many investigators to study regulation of gene expression, functionally analyze proteins, and perform site directed mutation analyses of expressed proteins. By utilizing reporter genes, as discussed in the previous chapters, transfected cells can be tracked, analyzed, and isolated by flow cytometry. One of the inherent problems faced by transient and stable transfection methodology is the low frequency of tranfectants.

Weeding out nontransfected cells from transfected cells has a major influence on the successful outcome of the transfection. In the past, vector-encoded protein conferring drug resistance has been used with transfection. In the case of transient transfection for analysis of transgenes, drug selection is many times inappropriate because of the short expression period of most transgenes. These are usually analyzed by bulk assays that dilute out the positive signal with the huge background of the untransfected cells.

For stably integrated transfectants drug selection is both expensive and time-consuming. In addition, the drugs may eliminate low expressers from the experimental range. A new and commercially available approach to enrich for either transient or stable transfectants has been developed by Miltenyi Biotec. The method utilizes Miltenyi's super-para-magnetic bead/antibody conjugate technology that selects a specific cell surface marker protein encoded by a co-transfection vector.[1] Transfected cells, formerly <2% of the starting population, can be highly enriched up to 90% by this drug-free technique. Further purification can easily be achieved by cell sorting from the homogeneous population (Figure 1). Two co-transfection vectors are currently available. The pMACS 4 vector encodes for the human CD4 surface protein with a truncated cytoplasmic domain that is expressed from the SV40 promoter/enhancer. Strong surface expression occurs within a few hours and the co-transfected cells may be separated 3-72 hours after transfection depending on the cell type and transfection method.

The pMACS K^k co-transfection vector encodes for the mouse MHC class I surface molecule H-2K^k with a truncated cytoplasmic domain that is expressed from the constitutive H-2K^k promoter. The pMACS K^k tranfected cells can be separated 8-72 hours after transfection depending on cell line and method.

Either vector can be used to co-transfect with other DNA or reporter genes such as green fluorescent protein with standard transfection methods, e.g. Electroporation, calcium phosphate precipitation, lipofection, and DEAE dextran transfection. Because both vectors have been cytoplasmically truncated, the signal transduction capacity of the proteins has been eliminated. This approach can be used with cell surface maker antibodies or other probes in a multicolor flow cytometric assay on living single cells to verify cell identity, cell cycle status, cellular function, or other parameter. The ability to transfect, select, and analyze on the same day makes measurements of early gene expression events, cell death, or growth inhibition at this stage possible that otherwise would have been missed.

Fig. 1. The enrichment of transfected cells was analzyed by flow cytometry. ▶ Debris and dead cells were excluded by the indicated gates. CHO cells were co-transfected with pEGFP™ and pMACS 4 in a molar ratio of 1:2 (A) or 1:4 (B) and enriched 24 hours after transfection using MACSelect 4 Microbeads.

Note that current GFP vectors are engineered for extremely high expression, which in most cases does not correlate to the expression levels of a typical gene of interest. This can be seen in panel A where 28.8% of the transfected cells are expressing detectable levels of GFP (A,1) and although these can be enriched 92.7% (A,III), many GFP expressing cells are not expressing CD4 at levels high enough to be isolated (A,II). In order to increase the levels of CD4 in cells cotrasfected with GFP an excess of pMACS 4 was used (B). CD4 can now be detected on the majority of GFP expressing cells but, this also leads to an increase in the number cells expressing CD4 only. Thus, more CD4 single positive cells are found in the positive fraction which reduces the purity of GFP cells in the positive fraction (III,B).

To mimic the expression of a normally expressed gene, CHO cells were cotransfected with pMACS 4 and pMACS K^κ in a molar ratio of 1:2. The transfected cells were enriched 24 hours after transfection from 2.6% to 50% of the population using MACSelect 4 MicroBeads (C)

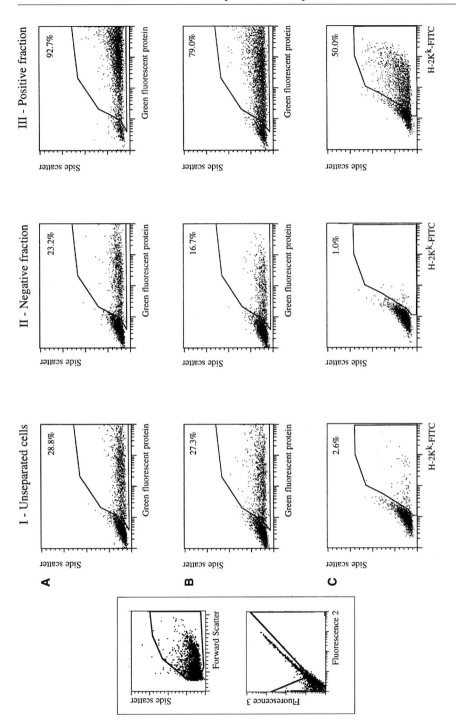

The following protocol is an example of the usefulness of MACSelect, with which enrichments of 10- to 20- fold are easily achieved. Low numbers of transfected cells, e.g., <2% can be enriched up to 40 to 50% so that background of untransfected cells is greatly reduced. Higher percentages of transfected cells, e.g., 10%, can be enriched to a homgeneous population of >95% transfected cells.

Green fluorescent protein (GFP) is often used as a reporter gene in transfection. In most currently available vectors a thoroughly engineered GFP gene is expressed under the control of the extremely strong CMV-promoter. Cotransfection leads to high expression of GFP in all transfected cells including cells carrying only a few copies of transfected GFP vector (Figure, 1A). This unusually high expression of GFP may not correlate with the expressioin rate of the cotransfected marker so that a careful optimization for cotransfection efficiency is necessary. We here propose a protocol in which an excess of pMACS vector is cotransfected with a vector encoding GFP so that a good cotransfection ratio of pMACS to GFP is achieved (Figure 1, B).

The GFP reporter system may not be ideal to mimic the expression rate of a typical "gene of interest". Therefore, we propose the use of a normally expressed reporter gene in a second protocol. the H-2K^{K}gene under the control of its own promoter may serve as an example for such a marker, expressed at a level more typical for a "gene of Interest". This example shows efficient enrichment of very low numbers of H-2K^{K}positive cells by MACSelect 4 MicroBeads after cotransfection with pMACS K^K and pMACS 4 (Figure1, C).

Subprotocol 1
MACSelect 4 and H-2K^k for Cotransfection and Analysis

Materials

Equipment

- Hemacytometer
- Electroporator
- Centrifuge

- VarioMACS, MiniMACS, MidiMACS, or SuperMACS magnet (Miltenyi Biotec)
- MS$^+$/RS$^+$ (up to 10^7 transfected cells) or LS$^+$/VS$^+$ (up to 10^8 transfected cells) separation columns (Miltenyi Biotec)
- Flow Cytometer

Solutions

- Growth medium specific to your cells
- Phosphate Buffered Saline (PBS), pH 7.2
- PBS with 5 mM EDTA
- **PBE** - PBS with 0.5% bovine serum albumin and 5 mM EDTA

Preparation

- pMACS 4 vector (#703-01 Miltenyi Biotec)
- pMACS K^k vector (#704-01 Miltenyi Biotec)
- MACSelect 4 Microbeads (#701-01 Miltenyi Biotec)
- pEGFP™ vector (Clontech)
- Anti human CD4-PE conjugate monoclonal antibody (Coulter/Immunotech)
- Anti mouse H-2K^k-FITC conjugate monoclonal antibody (#851-01 Miltenyi Biotec)

Procedure

1. Plate CHO cells in 9 cm petri dish.

2. Grow CHO cells to 80-90% confluence (~1.5-2.0 x 10^6 cells).

3. Remove medium from dish.

Grow CHO cells

Note: This procedure is based on one plate of cultured CHO cells. Plate as many dishes as you will need for your purposes.

4. Wash adherent cells with PBS without EDTA, removing PBS completely.

Wash cultured cells

5. Add 1 ml PBS supplemented with 5 mM EDTA.

6. Shake gently to disperse cells. Incubate for 5 minutes at room temperature, shaking gently twice during the incubation.

Treat with EDTA to remove cells from dish

Transfer cells for centrifugation

7. Add 10 ml PBS and transfer cell suspension to conical centrifuge tube.

8. Centrifuge cells for 10 minutes at 50 x g. Discard the supernatant.

Resuspend cells and transfer for electroporation

9. Resuspend cells in 500 µl of basal cell culture medium (without serum or supplements) and transfer to an electroporation cuvette (0.4 cm gap width).

Add DNA

10. Add 20 µg of mixed plasmid DNA - 5 µg of pMACS 4 and 15 µg of pMACS K^k - and incubate for 10 minutes at room temperature.

Electroporate

11. Electroporate the cells using 975 µF at 210 V.

Incubate for DNA transfer

12. Immediately after the pulse, place the electroporation cuvette in a 37°C incubator for 10 minutes.

Plate cells for culture

13. Transfer the electroporated cells into a 9 cm petri dish containing 10 ml prewarmed culture medium.

14. Aliquot 1.5 ml of electroporated cells in media into a separate 6 cm dish and add 3.5 ml of culture media as a control.

Culture for a day

15. Culture cells for 20-24 hours.

16. Remove medium from the 9 cm petri dish.

Wash cells

17. Wash cells with PBS without EDTA. Remove PBS completely.

18. Add 600 µl of **PBE**.

Incubate with MACSelect 4 MicroBeads

19. Add 80 µl MACSelect 4 MicroBeads to petri dish.

20. Shake gently to disperse liquid and incubate for 15 minutes at room temperature, shaking gently twice during incubation.

21. Tap or shake dish to detach cells completely.

Harvest labeled cells

22. Add 1.2 ml of degassed PBE (final volume 2 ml) and resuspend the cells carefully to prepare a single cell suspension.

Note: The quality of the single **cell suspension** is very important for the efficient selection of transfected cells.

23. Remove 100μl of the cell suspension as an unseparated control for subsequent analysis.

24. Place a MS⁺/RS⁺ column in the magnetic field of an appropriate MACS separation magnet.

Ready separation column on magnet

25. Prepare the column by washing with 500 μl of degassed PBE.

26. Apply the cell suspension to the column in 500μl aliquots.

Apply tagged cell suspension to separation column

27. Collect the negative cells which pass through the column (cells that do not have the MicroBeads attached).

Collect negative cell fraction

28. Wash the column 4 times with 500 μl of degassed PBE to remove the negative cells.

29. Remove the column from the magnet and place it over a suitable collection tube.

Remove column from magnet

30. Pipette 1 ml of degassed PBE onto the column and flush out the positive cells with the plunger supplied with the column.

Collect positive cell fraction

31. Stain cells with anti H-2K^k-FITC. (See Section 3 Chapter 2 for staining proctocols)

Stain cells with anti H-2K^k-FITC

32. Set up flow cytometer to measure fluorescein (excitation 488nm, emax 525-530nm).

Set up flow cytometer

33. Determine separation efficiency by analyzing the cell fractions and compare the H-2K^k staining of unseparated cells to the positively selected cell fraction.

Determine separation efficiency

34. Harvest the cells from the 6 cm dish of control cells as above with PBE (see points 3-6).

Analyze cotransfection efficiency

35. Stain cells with CD4-PE and H-2K^k-FITC antibodies.

36. Set up flow cytometer with filters for fluorescein and PE (excitation 488nm, emax 575nm). Compensate for green vs red spectral spillover. (see Section 2, Chapter 7)

37. Run H-2K^k-FITC and CD4-PE stained cells to analyze the cotransfection efficiency and transfection rate.

38. Collect data as a two dimensional plot of green (H-2K^k-FITC) vs. orange (CD4-PE) for each sample.

Subprotocol 2
MACSelect 4 and GFP for Cotransfection and Analysis

▣▣ Materials

Equipment

- Hemacytometer
- Electroporator
- Centrifuge
- VarioMACS, MiniMACS, MidiMACS, or SuperMACS magnet (Miltenyi Biotec)
- MS$^+$/RS$^+$ (up to 10^7 transfected cells) or LS$^+$/VS$^+$ (up to 10^8 transfected cells) separation columns (Miltenyi Biotec)
- Flow Cytometer

Solutions

- Growth medium specific to your cells
- Phosphate Buffered Saline (PBS), pH 7.2
- PBS with 5 mM EDTA
- **PBE** - PBS with 0.5% bovine serum albumin and 5 mM EDTA

Preparation

- pMACS 4 vector (#703-01 Miltenyi Biotec)
- pMACS K^k vector (#704-01 Miltenyi Biotec)
- MACSelect 4 Microbeads (#701-01 Miltenyi Biotec)
- pEGFP vector (Clontech)
- Anti human CD4-PE conjugate monoclonal antibody (Coulter/Immunotech)
- Anti mouse H-2K^k-FITC conjugate monoclonal antibody (#851-01 Miltenyi Biotec)

▧▧ Procedure

1. Plate CHO cells in 9 cm petri dish.

2. Grow CHO cells to 80-90% confluency (~1.5-2.0 x 10^6 cells).

Grow CHO cells

Note: This procedure is based on one plate of cultured CHO cells. Plate as many dishes as you will need for your purposes.

3. Remove medium from dish.

4. Wash adherent cells with PBS without EDTA, removing PBS completely.

Wash cultured cells

5. Add 1 ml PBS supplemented with 5 mM EDTA.

6. Shake gently and incubate for 5 minutes at room temperature, shaking gently twice during the incubation.

Treat with EDTA to remove cells from dish

7. Add 10 ml PBS and transfer cell suspension to conical centrifuge tube.

8. Centrifuge cells for 10 minutes at 50 x g. Discard the supernatant.

Transfer cells for centrifugation

9. Resuspend cells in 500μl of basal cell culture medium (without serum or supplements) and transfer to an electroporation cuvette (0.4 cm gap width).

Resuspend cells and transfer for electroporation

10. Add 20 μg of mixed plasmid DNA – 15 μg of pMACS 4 and 5 μg of pEGFP™ and incubate for 10 minutes at room temperature.

Add DNA

11. Electroporate the cells using 975 μF at 210 V.

Electroporate

12. Immediately after the pulse, place the electroporation cuvette in a 37°C incubator for 10 minutes.

Incubate for DNA transfer

13. Transfer the electroporated cells into a 9 cm petri dish containing 10 ml prewarmed culture medium.

Plate cells for culture

14. Aliquot 1.5 ml of electroporated cells in media into a separate 6 cm dish and add 3.5 ml of culture media as a control.

15. Culture cells for 20-24 hours.

16. Remove medium from the 9 cm petri dish.

Culture for a day

Wash cells	17. Wash cells with PBS without EDTA. Remove PBS completely.
	18. Add 600 μl of **PBE.**
Incubate with MACSelect 4 MicroBeads	19. Add 80 μl MACSelect 4 MicroBeads to petri dish.
	20. Shake gently to disperse liquid and incubate for 15 minutes at room temperature, shaking gently twice during incubation.
	21. Tap or shake dish to detach cells completely.
Harvest labeled cells	22. Add 1.2 ml of degassed PBE (final volume 2 ml) and resuspend the cells carefully to prepare a single cell suspension. The quality of the single **cell suspension** is very important for the efficient selection of transfected cells.
	23. Remove 100 μl of the cell suspension as an unseparated control for subsequent analysis.
Ready separation column on magnet	24. Place a MS$^+$/RS$^+$ column in the magnetic field of an appropriate MACS separation magnet.
	25. Prepare the column by washing with 500 μl of degassed PBE.
Apply tagged cell suspension to separation column	26. Apply the cell suspension to the column in 500 μl aliquots.
Collect negative cell fraction	27. Collect the negative cells which pass through the column (cells that do not have the MicroBeads attached).
	28. Wash the column 4 times with 500 μl of degassed PBE to remove the negative cells.
Remove column from magnet	29. Remove the column from the magnet and place it over a suitable collection tube.
Collect positive cell fraction	30. Pipette 1 ml of degassed PBE onto the column and flush out the positive cells with the plunger supplied with the column.

31. Set up flow cytometer to measure GFP. Use filter set up for fluorescein.

Set up flow cytometer

32. Determine separation efficiency by analyzing the cell fractions and compare the green fluorescent of unseparated cells to the positively selected cell fraction.

Determine separation efficiency

33. Harvest the cells from the 6 cm dish of control cells as above with PBE (see points 3-6).

Analyze cotransfection efficiency

34. Stain cells with CD4-PE antibody.

35. Set up flow cytometer with filters for GFP and PE. Compensate for green vs. red spectral spillover (see compensation chapter).

36. Run CD4-PE stained cells to analyze the cotransfection efficiency and transfection rate.

37. Collect data as a two dimensional plot of green (GFP) vs. red (CD4-PE) for each sample.

References

1. Siebenkotten G, Petry K, Behrens-Jung U et al. Employing surface markers for the selection of transfected cells. In: Cell Separation: Methods and Applications. Marcel Dekker, in press.

Use of PKH Membrane Intercalating Dyes to Monitor Cell Trafficking and Function

REBECCA Y. M. POON, BETSY M. OHLSSON-WILHELM,
C. BRUCE BAGWELL, AND KATHARINE A. MUIRHEAD

Introduction

Background: A Family of Lipophilic Membrane Dyes

Since their introduction as improved agents for *in vitro* and *in vivo* cell tracking,[34,56,79] the lipophilic membrane intercalating fluorochromes known as PKH dyes have been used to study a wide variety of cell types and biological processes. Originally developed by Horan and colleagues at Zynaxis Cell Science as part of a larger class of multifunctional drug delivery molecules known as Zyn-Linkers, this family of patented fluorescent dyes and cell labeling reagents became available in kit form for research use in 1989 and was purchased by Phanos Technologies in 1995. PKH dyes are therefore found in the literature under a variety of names, including Cell Linker dyes, Zyn-Linkers, Zyn-Linker® dyes, cell tracking dyes, and PKH dyes. A number of suppliers are also found in the literature, including Zynaxis Cell Science (which no longer sells these reagents), Sigma and Dainippon Pharmaceuticals Laboratory Products Division (both of which continue to distribute PKH kits for Phanos).

Despite the diverse nomenclature, all PKH dyes share certain common features, and in many ways resemble membrane lipids. They are small organic molecules with molecular weights of 300-1000 D which have polar head groups and, typically, two relatively long hydrocarbon tails. However, PKH dyes differ from membrane lipids in that their head groups are fluorescent, their tails contain only saturated hydrocarbon bonds, and the chemical linkages between tails and head group are more chemically and metabolically stable than those found in natural lipids.

Due to their similarities to membrane lipids, PKH dyes rapidly partition into the lipid bilayer of cell membranes, become incorporated in relatively large numbers without detrimental effect on cell function or immunogenicity, and remain stably associated with labeled cells for weeks to months, both *in vitro* and *in vivo* (Tables 1 and 2). Since they do not require specific cell surface receptors for binding but intercalate into the lipid regions of cell membranes, labeling with PKH dyes is applicable to any particle that contains lipid bilayers or other hydrophobic binding sites. These include mammalian, plant, and bacterial cells as well as viruses, liposomes, and emulsions (Table 1).

Because the PKH dyes partition into the lipid regions of cell membranes, leaving critical cell surface receptors unaltered, most cell types can be labeled to relatively high levels (10^6 - 10^8 molecules per cell) without deleterious effect on physiological function. However, labeling with PKH dyes is based on rapid partitioning of these lipophilic molecules into the membrane rather than a

Table 1. Examples of Cell Types Successfully Labeled With PKH Dyes

Cell Type	References
Hematopoietic cells (CD34+ stem cells, CFU-s, progenitor cells, culture initiating cells)	30, 33, 46, 48, 49, 52, 82, 90, 91, 94, 95, 96
Immune cells (lymphocytes, monocytes, macrophages, thymocytes, splenocytes, cytotoxic or autoimmune effector cells, dendritic cells)	5, 7, 18, 21, 29, 35, 40, 41, 43, 47, 50, 51, 55-57, 59, 69, 74, 75, 97, 104, 105, 108
Other blood and/or bone marrow cells (erythrocytes, neutrophils, platelets, stromal or neoplastic cells)	1, 2-4, 9, 53, 58, 85, 93, 98, 99
Cultured cells and/or cell lines (tumor cells and cell lines, smooth muscle cells, hybridomas, T-cell lines and clones)	2-4, 11, 12, 22, 26, 27, 37, 62, 63, 81, 89
Embryonic Cells	15, 36, 42, 60, 68
Endothelial cells	19, 25, 87, 102
Epithelial cells	17, 54
Neurons	10, 73, 92
Bacteria, parasites	16, 28, 71, 72, 83
Other "bioparticles" (viruses, protoplasts, phytoplankton, erythrocyte ghosts, fluorocarbon emulsions)	20, 77, 80, 100

saturation reaction of the type observed for ligand-receptor binding. Overlabeling can result in altered membrane integrity and/or cell function, and it is therefore important to verify that labeling conditions chosen do not affect cellular functions of interest.

A number of factors can influence the mechanism (membrane labeling *vs.* particle ingestion), intensity, and homogeneity or heterogeneity of labeling with PKH dyes. For general membrane labeling, specialized diluents have been formulated to optimize the solubility of these highly lipophilic dyes in aqueous solutions eliminating the need for addition of detergents or organic solvents. Use of such "GL" diluents minimizes dye aggregation and maximizes homogeneity and reproducibility of cell labeling. General membrane labeling occurs very rapidly (within 30 seconds to 5 minutes; see Subprotocol 1 for detailed methods). Final fluorescence intensity (mean fluorescence) is determined primarily by dye concentration and cell (membrane) concentration present during the staining process, but may also be influenced by cell type specific variables such as protein:lipid ratio and phospholipid composition.

Table 2. Examples of Cell Functions Monitored Using PKH Dyes

Function	References
In vitro	
Drug sensitivity/resistance	11, 22, 29
Differentiation	15, 36, 60, 68, 88
Proliferation	8, 12, 46, 48, 54, 82, 91, 94, 95, 103, 104, 106
Conjugate formation, adhesion, and/or fusion	4, 20, 26, 49, 69, 75, 81, 88, 96
Cytotoxicity, immunotherapy	21, 24, 32, 39, 41, 45, 62, 104
Phagocytosis	13, 21, 28, 63, 69, 71, 93, 98, 99
Cell-cell communication	2, 37, 89, 96
In vivo	
Migration/homing/engraftment	5, 7, 18, 31, 33, 43, 45, 47, 50, 51, 52, 55, 59, 61, 65, 74, 76, 90, 92, 108
Recirculation/cellular lifetime	55, 56, 77, 78, 86, 105
Proliferation/growth control	17, 42, 44
Differentiation, embryogenesis	17, 35, 42, 44, 84
Adhesion	25, 107
Blood flow assessment	9
Cytotoxicity, immunotherapy	6, 27, 53, 65, 97
Immunity/antigen presentation	51, 61, 85

Coefficient of variation (CV) of the resulting fluorescence distribution is determined by both physicochemical variables (e.g. rate and homogeneity of mixing between dye and cell suspensions) and biological variables (e.g. variability in size and/or membrane morphology among cell types present within the sample).

Alternatively, preferential labeling of phagocytic cells can be achieved by using a "PCL" diluent formulated to cause rapid aggregation of the PKH dyes into small particles which are inefficient at general membrane labeling but readily ingested by phagocytosis. Intensity and homogeneity of phagocytic cell labeling is a function of cell concentration, size and concentration of dye particles, state of phagocyte metabolism/activation, and length of time over which phagocytes are allowed to ingest dye particles (see Subprotocol 3 for detailed methods).

Overlabeling of cells can result in loss of membrane integrity and/or cell viability, adverse effects on cell function, and/or technical difficulties in adjusting color compensation to eliminate signal cross-over of the PKH dye into secondary detectors when multicolor analysis is being performed. Published studies (Tables 1 & 2) can be useful guides for selection of appropriate staining conditions, but it is important for each laboratory to verify that staining conditions, particularly dye concentration and cell concentration, are optimal for their cell type of interest. Typically, this involves preliminary evaluation to establish that staining conditions and cell:dye ratios used will reproducibly achieve homogenous labeling among all cells in a preparation and to maximize fluorescence intensity without affecting cell functions of interest, with the extent of verification being dependent on the complexity of the cell functions to be monitored (see for example references 25, 34, or 97).

Several of the PKH dyes are available in kit form with either GL diluent for general membrane labeling or PCL diluent for phagocytic cell labeling. All can be excited using the 488 nm argon ion laser of a standard clinical flow cytometer. PKH2 and PKH67 are efficiently excited at 488 nm and emit green fluorescence (Fig. 1; emission maximum~500 nm) which is optimally detected using filters similar to those used for fluorescein (FL-1 detector in many cytometers). These green PKH dyes are excellent choices for use in multiparameter studies with antibodies or viability probes which emit in the orange-to-red region (e.g. propidium iodide, 7-aminoactinomycin D, PKH26, Cychrome, ECD, Per-CP, Quantum Red, R-Phycoerythrin, Texas Red, Tricolor, CY5, *etc.*). PKH2, the first commercially available PKH dye, has been widely used for short-to-medium term *in vitro* and *in vivo* studies, *i.e.* studies lasting up to 10-20 days. The more recently introduced PKH67 has longer hydrocarbon tails, resulting in even more stable membrane labeling and minimal cell to cell transfer. PKH67 is the green dye of choice for use in studies of longer duration (*i.e.* >20 days), for studies utilizing dye dilution as an indicator of cell proliferation[11b] (see Section B below), and for studies with cell types which have been reported to exhibit low level dye transfer with PKH2 and/or in which a pre-incubation period to minimize dye transfer [22] is not practical[11b].

PKH26 has the longest tails, greatest membrane stability, and longest *in vivo* half life of the commercially available PKH dyes. It emits orange fluorescence (Fig. 1; emission maximum ~567 nm)

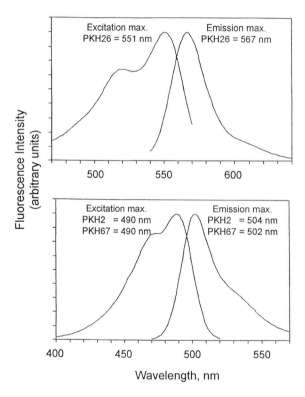

Fig. 1. Spectral Characteristics of Commercially Available PKH dyes. Excitation and emission spectra as determined for 10 μM solutions of PKH2, PKH26, and PKH67 in ethanol

which is optimally detected using filters similar to those used for rhodamine or phycoerythrin (FL-2 detector in many cytometers). It is most often excited at 488 nm but gives an even stronger signal on instruments where exciting lines of 514 nm or 560 nm are available, since these are closer to its absorbance maximum (551 nm). Because of its excellent membrane retention properties and *in vivo* half life, PKH26 has been widely used for *in vivo* cell tracking, assays in which dye dilution is used to monitor cell proliferation, or other long term studies (›60 days). PKH26 can be combined with PKH2 or PKH67 to simultaneously track two different cell types within a mixed population, or used in multiparameter studies with UV, green or longer red antibody labels or viability probes (e.g. Hoechst, SYTOX, fluorescein, Cy5 or allophycocyanine).

Fig.1 summarizes the spectral properties of the commercially available PKH dyes and can serve as starting point in selecting the PKH dye of choice for combination with other fluorescent probes. Subsequent sections of this chapter describe selected applications of the various PKH dyes in more detail.

Subprotocol 1
Cell Staining Using PKH Cell Tracking Dyes:
General Cell Labeling

Materials

Equipment

- Flow Cytometer
- Fluorescence Microscope
- Fluorescence Image Analysis System

Solutions

- PKH2-GL, PKH26-GL, PKH-67-GL Kits (kit content: Diluent A or C, plus PKH2 or PKH67 or PKH26 dye solution in ethanol at 10^{-3} M) and Cell Census Plus System™ Kit
- Phosphate Buffered Saline (PBS)

- Fetal Bovine Serum or Fetal Calf Serum (FCS)
- Cell Preparation Reagents: Cell Dissociation Reagents (Enzymatic or Non-Enzymatic)
- Erythrocyte Removal Reagents: Erythrocyte Lysing Buffer or Density Media for Cell Separation
- Trypan Blue
- Complete Culture Medium: *e.g.* RPMI with 10% FCS

Procedure

Preparation of Single Cell suspension

Note: Adherent cells in culture can be stained directly by flooding the culture vessel with a working dilution of the dye. However, this is not recommended when uniform staining of all cells in a population for quantitative fluorescence measurement is required.

For adherent cells in culture

1. Prewarm all reagents.

Note: Enzymatic (1 x Trypsin-EDTA, Sigma cat # T3924) and non-enzymatic dissociation reagents (Cell Dissociation Solution, Sigma cat # C5789 in HBSS, Sigma cat # C5914 in PBS) are available.

2. Remove all medium from the culture vessel by aspiration.

3. Rinse the adherent cell layer with a balanced salt solution without calcium or magnesium. Repeat once to remove all traces of serum.

4. Add enough cell dissociation reagent to flood the monolayer (about 5 ml for 75cm^2 flask). Rock vessel to ensure monolayer is covered.

5a. When using non-enzymatic reagent, incubate cells for 5-10 minutes.

5b. When using trypsin-EDTA, incubate cells at 37 deg. C. for approx. 2 minutes. Aspirate to remove trypsin-EDTA solution, continue incubation at 37 deg. until cells detach from the surface. Progress can be checked by examining on an inverted microscope.

Note: Time required to dislodge cells depends on the cell type, population density, the effective removal of all traces of serum in step 3 and the enzymatic activity of the trypsin-EDTA when used. Trypsin causes cell damage and exposure time should be kept to a minimum.

6. Tap the vessel sharply against the palm of your hand to dislodge the cells. Strongly adherent cells may require additional time to become dislodged.

7. Add complete growth media containing serum to the cells to inhibit further trypsin action if used. Pipette cells gently to further disperse cell clumps.

8. Pellet cells from dissociation reagent. Wash cells two more times before use.

For solid lymphoid organs such as lymph nodes or spleen

1. Dissociate mechanically.

2. Allow large pieces of debris or organ capsule to settle; remove cell suspension leaving behind debris.

3. Centrifuge to collect cells.

Note: Red blood cells (RBC) in lymphoid cells suspensions or in whole blood must be removed for studies on lymphocytes. The large number of RBC in these cell preparations would otherwise overwhelm the dye staining step.   **Removal of Erythrocytes**

Using Red Blood Cell Lysing Buffer (Sigma cat # R7757, 0.83% ammonium chloride in 0.01M Tris)

Note: The ammonium chloride lysing procedure described was optimized for mouse spleen. These steps can be adapted to whole blood.

1. Add 1 ml of buffer at room temperature to cell pellet from one mouse spleen.

2. Gently mix for 1 minute.

Note: If lysis is complete, the cell suspension should look clear.

3. Add 15-20 ml of medium or balanced salt solution.

4. Centrifuge at 500g for 7 minutes and discard supernatant.

Note: If RBC are found in the cell pellet, steps 1-4 may be repeated.

5. Cells should be washed two more times before proceeding to next section.

Using Accuspin System-Histopaque 1077 Cell Separation System (Sigma cat #A7054, A0561)

1. Warm tubes containing the cell separation medium to room temperature.

2. Examine tubes for presence of media above the polyethylene barrier (frit). Centrifuge the tubes briefly, 1 minute at 400 x g, to return all media to below the frit.

3. Add 3.0 ml whole blood or RBC containing cell suspension to tube.

Note: The polyethylene barrier (frit) in the Accuspin tubes makes addition of the blood or cell suspension a simple and quick operation rather than painstaking layering over a liquid surface.

4. Centrifuge at 800 x g for 15 minutes.

Note: After centrifugation, the mononuclear cells will form an opaque interface between the plasma on top and the Histopaque on the bottom. This interface will be formed above the frit.

5. Carefully aspirate off plasma to within 0.5 cm of the opaque interface layer containing the mononuclear cells.

6. Transfer the mononuclear cell layer to a conical centrifuge tube.

7. Fill tube with phosphate buffered saline (PBS). Pellet cells at 400 x g for 10 minutes.

8. Repeat washing cells two more times.

PKH Dye Loading   **Note:** Steps 1 - 3 (cell preparation for dye loading) and Steps 4 - 5 (dye dilution) need to be performed in parallel.

1. Wash cells in serum-free and protein-free medium for a minimum of three times, if this has not already been performed in cell preparation.

Note: All serum components, lipids or protein, must be removed for uniform and reproducible dye staining. Presence of lipids and salts from aqueous wash buffer causes dye molecules to be sequestered in micelles and become unavailable for cell staining. Aspiration of wash media or wash buffer is recommended over decantation.

2. Perform an accurate cell count.

3. Pellet cells from wash media and resuspend in **Diluent** to the concentration of 5×10^6 cells per ml.

Note: For general membrane labeling, PKH dyes are provided as a concentrated dye stock in kits with a patented non-ionic **Diluent**. The kits are available from the Sigma catalog under these names: PKH2-GL; PKH26-GL and PKH67-GL. The Cell Census Plus System^{TM} Kit (Sigma cat #CCPS-1) has additional components for flow cytometric proliferation assays. Cells should not be kept in **Diluent** for more than 1 to 2 minutes. Optimal conditioning time in **Diluent** is 30 seconds.

4. Calculate the quantity of dye required for the number of cells (steps 2 and 3) according to the following relationship: one volume of dye at working dilution (2 x final) per one volume of lymphocytes at 5×10^6 per ml (2 x final).

Note: The optimal concentration of dye is dependent on the cell type. A titration should be performed for each cell type or cell line. For PKH-26 and PKH 67 microbead standards are available for use a fluorescent intensity bench marks in the optimization for lymphocyte staining. For normal human peripheral blood lymphocytes (PBL), 1.5 to 2 µM final concentration is optimal for a final cell concentration at 2:5 x 10^6 per ml. Saturation staining is neither necessary nor desirable. Incorporate only enough dye to generate sufficient signal to noise ratio and homogeneous staining. Too much dye will affect membrane integrity, compromising cell yield and potentially causing adverse effect on cell function.

5. Prepare the necessary volume of working dye dilution in **Diluent**. PKH dye stocks are provided at 1 mM in ethanol.

Note: Dye stock should be kept tightly capped when not in use to prevent evaporation.

Note: Dye will adhere to plastics.

Note: Use only polypropylene tubes for the working dye dilution and the dye loading step. Prepare working dilution of dye during the last centrifugation of cells. In preparing the working dye dilution, transfer the calculated quantity of PKH dye in a single volume as loss will occur through binding to pipette tips.

Note: Dye dilution should be used within 15 minutes of preparation.

6. Mix cells and working dilution of PKH dye at one to one volume ratio by adding cells into dye. Mix rapidly but gently by pipetting up and down.

Note: Rapid mixing is critical for homogeneous staining.

7. Incubate cells in dye for three minutes.

8. Stop dye partitioning reaction by adding an equal volume of fetal calf serum (FCS).

9. Fill tube with medium containing 10% FCS (complete medium).

10. Pellet cells and aspirate to remove dye solution.

11. Resuspend cell pellet in complete media and transfer suspension to a fresh tube to complete washing steps.

12. Repeat wash steps at least two more times before using in culture or other assay applications. The cells are ready for examination on a flow cytometer. Figure 2 gives an example of flow histograms of stained cells.

Note: If stained cells are to be cultured for proliferation measurement, as in the Cell Census Plus™ System, they should be examined in the flow cytometer at this point to ascertain that the necessary staining intensity (between the third and fourth decade on a 4-decade log scale) and a cv of <20% have been obtained (Fig. 2). A cell count should be performed at the end of the staining steps to calculate the % cell yield. This can be a useful indicator of proper cell handling. High recovery (80-90%) can be expected for normal human PBL. An aliquot of the PKH labeled cells should be fixed in paraformaldehyde in PBS and stored at

4 degree C for a time zero control which is required for data analysis in ModFit. PKH-stained and fixed cells are stable for >2 weeks. PKH stained cells can be treated with 0.1 % saponin in PBS for membrane permeabilization and the detection of intracellular antigen(s) with antibodies. This treatment does not affect the membrane retention of the PKH molecules.

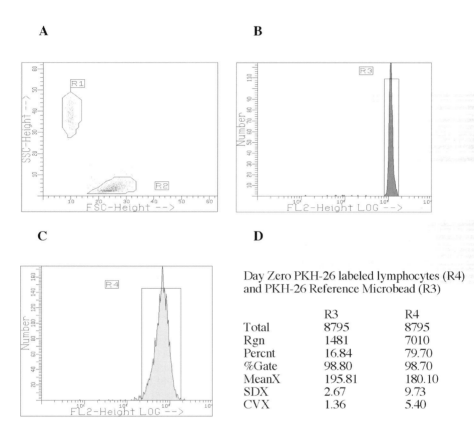

A

B

C

D

Day Zero PKH-26 labeled lymphocytes (R4)
and PKH-26 Reference Microbead (R3)

	R3	R4
Total	8795	8795
Rgn	1481	7010
Percnt	16.84	79.70
%Gate	98.80	98.70
MeanX	195.81	180.10
SDX	2.67	9.73
CVX	1.36	5.40

Fig. 2. Mixture of PKH-26 labeled human peripheral blood mononuclear cells on day zero and PKH-26 Reference Microbeads. Note the relative intensity of labeled cells to microbeads and small cv indicating homogeneity of dye loading. (A) Light scatter dotplot. (B) FL-2 fluorescence of Reference Microbeads gated by light scatter on singlet events (R1 gate). (C) FL-2 fluorescence of labeled PBMC gated on lymphocytes (R2 gate). (D) histogram statistics for B and C

Troubleshooting

Refer to the Trouble Shooting Guide (Table 3) for suggestions on possible causes and solutions to common staining problems.

Table 3. Trouble shooting guide for PKH2, PKH26 and PKH67 fluorescent cell linker dyes

Problem	Possible Cause	Solution
Cell Clumping	Dye concentration too high.	Lower labeling concentrations.
	Serum not added to stop reaction.	Serum, other protein or complete medium must be used to stop labeling reaction.
	Platelets present in whole blood sample.	Centrifuge sample at low speed to remove platelets before staining.
	Adherent cells not disaggregated.	Mechanical (aspirate through needle) or enzymatic (trypsin) dispersion prior to staining.
	Poor initial cell viability of sample.	Incubate with 0.002% DNAse for 30 minutes at 37°C before staining.
	Cells exposed to dye too long.	Expose cells to dye for 2-5 minutes.
Poor Staining Intensity	Filters incorrect for observing dye.	Check filter set-up. Check cell pellet after labeling; if pellet is pink or yellow, cells are stained.
	Working dye stock prepared too long before adding to cells - dye aggregating.	Prepare dye immediately prior to labeling.
	Salt content of labeling solution too high - dye aggregating.	Centrifuge cells and remove as much supernatant as possible to minimize physiological salts after suspension in Diluent.
	Serum present during labeling interfering with dye incorporation.	Wash cells in serum-free buffer 1-2 times prior to resuspension in Diluent for labeling.
	Staining intensity varies with cell type based on cell size/surface area, lipid to protein ratios of membranes and lipid component of membranes.	Adjust dye concentration for each cell type and experimental application. Analyze sub-populations independently.

Table 3. (Continued)

Problem	Possible Cause	Solution
	Dye concentration too low.	Increase dye concentration.
	Cell concentration too high.	Reduce cell concentration to a range of 10^6 to 10^9 cells/ml.
Patchy or Punctate Staining	Salt content of labeling solution too high – dye aggregating.	Centrifuge cells and remove as much supernatant as possible to minimize physiological salts after suspension in Diluent.
	Some localization of dye on certain cell membranes causes a patchy appearance.	May be cell type dependent, verify that cell function(s) of interest are not affected.

Subprotocol 2
Using PKH Microbeads as Internal Count Standards

Materials

Equipment

– Flow Cytometer

Solutions

– PKH26 Reference Microbeads (Sigma cat # P7458 and as component in Cell Census Plus System^{TM}, Sigma CAt # CCPS-1)
– PKH67 Reference Microbeads (Sigma # P2477)

Procedure

The use of PKH-26 microbeads as internal count standards is based on the same principle as admixing fluorescent cell standards with unknown cell samples. This method has been described (24, 92) and was given the name Standard Cell Dilution Assay (92). The microbeads present a stable alternative to preparing cell standards and can also serve as a uniform fluorescence intensity bench mark.

Sample Preparation

1. Shake the bottle of PKH26 microbeads vigorously to obtain a homogeneous suspension.

2. Add a volume of microbeads equal to that of the test cell sample.

Note: The test sample should be an unmanipulated aliquot of the culture. Caution: Wash steps introduce potential cell loss and volume changes. The concentration of microbeads and expected cell numbers should be similar (on the order of 2×10^5 per ml). Otherwise, adjust the volume ratio. The lot-specific microbead concentration is provided by the manufacturer on a Certificate of Analysis. The mixed sample may be fixed. The microbeads are stable in 2% paraformaldehyde for up to 2 weeks.

Data Acquisition

3. Set up the cytometer. Set a light scatter gate around the microbead singlet population (Figure 3). Set the cytometer counter to acquire a set number of microbead events, e.g. 5,000.

Note: Consult instrument manufacturer's manual for general instructions for proper instrument set up. See section B for discussion on proper selection of PMT voltage and color compensation adjustment in multiparameter analysis using PKH dyes and other fluorochromes.

Sample calculations for the data of Figure 4 are shown

Example for the calculation of absolute cell counts

Lot specific PKH-26 microbead concentration (B_0) = 200,000 per ml

Note: The value 200,000 per ml for B_0 is used as an example. The correct lot specific concentration may be found on the Certificate of Analysis provided by the manufacturer.

Number of microbeads counted bitmap R1 (B) = 1208

Note: The cytometer can be set to collect a fixed number of events, e.g. 5,000, in the microbead bitmap (R1) during data acquisition. The exact value is obtained from the cytometer histogram statistics.

Volume of cell culture admixed with microbeads (V_c) = 1 ml
Volume of microbeads admixed with cell culture (V_B) = 1 ml
Number of events counted in the cell bitmap R2 (C) = 6552

Note: Value obtained from cytometer histogram statistics.

Absolute cell number per ml in culture (C_0)
= C x B_0/B x V_B/V_C
= 6552 x 200,000/1208 x 1/1

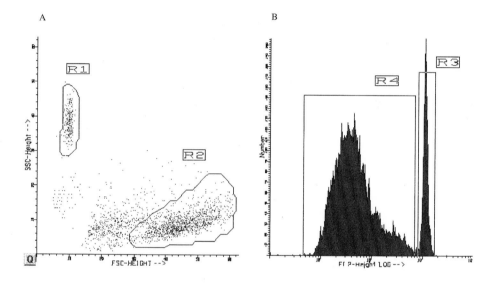

C

Human PBL, Day 4, 2.5 ug/ml PHA 01185b41		
	R3	R4
Total	10710	10710
Gated	7789	7789
Rgn	1208	6552
Percnt	11.28	61.18
%Gate	15.51	84.12
MeanX	195.48	105.94
SDX	2.94	24.74
CVX	1.50	23.35

Fig. 3. Light scatter dotplot of 1:1 mix of PKH-26 reference microbeads and human peripheral blood lymphocytes cultured with 2.5 µg/ml PHA for 4 days.

A) Light scatter dotplot. R1 is microbead singlet gate. R2 is proliferating live cells.

B) PKH-26 fluorescence of events in R1 and R2 gates corresponding to peaks marked as regions R3 and R4 respectively.

C) Histogram statistics used for sample calculation

Subprotocol 3
Cell Staining Using PKH Cell Tracking Dyes: Phagocytic Cell Labelling

▪▪ Materials

Equipment

- Flow Cytometer
- Fluorescence Microscope
- Fluorescence Image Analysis System

Solutions

- PKH dye kits PKH2-PCL; PKH26-PCL
- (PKH67 dye stock from the PKH67-GL kit may be used with Diluent B)
- 70% ethanol.

▪▪ Procedure

In Vivo Labeling of Resident Peritoneal Macrophages

Preparation of PKH dye aggregates

1. Remove 0.10 ml PKH dye (1.0 mM) from vial and mix thoroughly with 0.90 ml absolute ethanol. This will provide 100 µM working stock dye solution. Dye stock should be kept tightly capped when not in immediate use to prevent evaporation.

2. Remove the metal tab of one bottle of Diluent B. Because the diluent bottles have been overfilled to assure at least a 10 ml volume, withdraw the contents of the bottle using a 20cc syringe with needle, expel the overfill volume and reinject 10 ml back into the vial.

Note: Diluent B is a proprietary buffer formulation to induce dye aggregate formation. Phagocytic cell labeling occurs through the formation of dye micro-aggregates or particulates. The aggregate formation inhibits the partitioning of the dye into the cell membrane of non-phagocytic cells, such as lymphocytes, but facilitates dye uptake by phagocytic cells.

3. Remove 0.1 - 0.2 ml of the dye working stock (100 μM) with a 1cc syringe and a 24 gauge needle. Inject 0.05 ml of the dye into the diluent bottle. If the needle will not penetrate the rubber stopper, the stopper may be removed. However, it is essential that the contents of the diluent vial remain sterile as any contamination will induce inflammation into the peritoneal cavity.

4. Replace the stopper, if necessary, and shake the bottle vigorously to mix the dye and diluent. Allow the solution to stand for at least 15 minutes.

Note: PKH dye prepared in Diluent B may also be used in the *in vitro* labeling of other phagocytic cells, such as neutrophils and monocytes, following steps 6-12 in the General Cell Labeling protocol. The amount of dye and incubation time must be optimized by titration assay for each cell type. The amount of dye for *in vitro* labeling should be 10x-100x higher than for in vivo labeling.

In vivo labeling

Note: This procedure was optimized for the labeling of resident peritoneal macrophages (MΦ) in Balb/c mice, approximately 20 gm body weight. The peritoneum of naive Balb/c mice (housed under virus free conditions) contains 20-30% MΦ .[72] Labeling conditions, including dye concentration and volume to be injected should be optimized for other mouse strains which are larger or have more peritoneal MΦ .

1. Withdraw 0.5 ml of the diluted dye into a fresh sterile syringe for each animal to be injected. If possible, keep the rubber stopper on the diluent vial and flame the stopper before inserting each needle. Flame the needle again after withdrawing the dye, then flame the needle cap and place it over the needle.

Note: In the study of MΦ recruitment, inflammation induced by the labeling procedure must be minimized by observing stringent sterile procedure as described at each step. In addition, **Diluent B** has been tested for lipopolysaccharide (LPS) content and has been found to be less than 0.25 EU/ml.

2. Repeat step 1 for each syringe to be filled. Use a fresh syringe and needle for each animal. Shake the dye/diluent mixture before removing each sample for injection.

3. Inject the contents of each syringe intraperitoneally (i.p.). Before each injection, swab the abdominal fur with a gauze pad soaked in alcohol. This procedure will minimize the incidence of inflammation from the i.p. injection.

Harvest labeled macrophages from peritoneum

1. To harvest the peritoneal cells, sacrifice the animal (e.g. by cervical dislocation), and wet the abdominal fur with 70% ethanol.

Note: Labeled peritoneal MΦ may be harvested by peritoneal lavage at any time from 2 hours to 21 days after the i.p. injection of PKH dye.

2. Make a small incision in the inguinal area cutting through the fur, but not through the dermis. Grasp the abdominal fur and retract the fur towards the shoulders. This procedure will expose the intact abdominal cavity.

3. Fill a 5cc syringe with ice-cold PBS (larger volumes up to 10 ml may be preferred for larger mice), forcefully inject the PBS into the abdominal cavity, and withdraw the syringe. The injection site should be the inguinal fat pads. The abdominal fat will seal the injection site when the syringe is withdrawn.

4. Massage the abdominal cavity to loosen peritoneal cells adhering to the abdominal wall and viscera.

5. Insert the needle into the abdominal cavity, about midline, and gently withdraw the lavage fluid. With some experience, the user will be able to collect 90% of the injected fluid.

6. Place the contents of the syringe into a 15cc polypropylene tube. The tube should be kept on ice to minimize adherence and/or clumping of the MΦ.

7. Centrifuge the peritoneal cells for 5 minutes at 350 x g, at 4 deg. C. The cells can be resuspended and washed in saline, immunofluorescence wash buffer (for immunofluorescence staining) or media (for functional assays).

8. After step 7 and optional immuno-fluorescence staining or other treatment(s) for functional assays, cells are ready for the flow cytometer or other instruments for fluorescence detection.

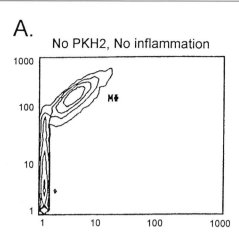

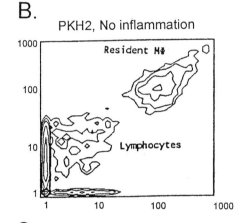

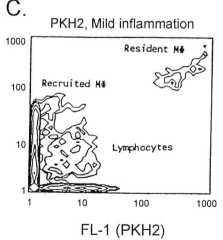

FL-1 (PKH2)

Fig. 4. In Vivo Labeled Peritoneal Macrophages. Balb/C peritoneal macrophages (MΦ) were labeled with PKH2 according to the In Vivo phagocytic cell labeling protocol. PE-MAC-1 was used to counterstain harvested cells to positively identify macrophages either resident at the time of in vivo labeling or recruited due to inflammation induced in the procedure.

A) **MAC-1 staining of cells from peritoneal lavage in untreated animals.** Resident peritoneal MΦ exhibit bright positive MAC-1 fluorescence and relatively high green autofluorescence, while lymphocytes exhibit low-moderate MAC-1 fluorescence and low green autofluorescence.

B) *In vivo* **PKH2 labeling and absence of inflammation.** I.p. injection of PKH2 in Diluent B causes preferential labeling of peritoneal MΦ. Resident MΦ exhibit bright positive staining for both PKH2 and MAC-1, while lymphocytes exhibit low levels of PKH2 staining and low-moderate MAC-1 fluorescence.

C) *In vivo* **PKH2 labeling and resultant mild inflammation.** Mild inflammation caused by failure to maintain sterile technique during i.p. injection causes an influx of recruited MΦ that are PKH2 negative and, as previously reported (57) exhibit reduced MAC-1 staining compared with resident MΦ. Resident MΦ are present at reduced levels, due to the influx of recruited cells, but remain PKH2 bright and MAC-1 bright

Note: The labeled resident MΦ (in the peritoneum at the time of injection) can be distinguished from subsequently recruited MΦ, which are not labeled by the fluorescent PKH dye. Under the microscope, labeled cells appear patchy or spotted because the dye is localized in phagocytic compartments of these cells. The dye appears to be resistant to metabolic attack and has been found to remain with the cells for more than 21 days *in vivo*.

Note: The example in Figure 4 illustrates the differential staining expected between lymphocytes and resident peritoneal MF and the importance of maintaining sterility to avoid inflammation.

Flow Cytometric Assays of Cell Function Using PKH Dyes

A. *In Vivo* Lifetime, Migration and Recirculation

Ideally, membrane dyes to be used for studies of *in vivo* cell trafficking and function should exhibit rapid incorporation into cell membranes, prolonged membrane retention, lack of toxicity, and lack of immunogenicity. PKH3, a red dye excited at 595 nm, was shown to have no effect on osmotic fragility of rabbit red blood cells (RBCs) and to be well retained when *ex vivo* labeled RBCs were reinjected intravenously (half life for loss of cell-associated fluorophore >150 days[78]). Labeled RBCs exhibited lifetimes in circulation comparable to those obtained using radioisotope tracers (40-60 days) and disappeared from circulation at a constant rate throughout the 60 day study, suggesting that the labeled cells did not elicit an immune response. Lifetime in circulation was unaltered for a second preparation of labeled cells administered 60 days after completion of the initial lifetime study, further substantiating lack of immunogenicity.

Investigators at Zynaxis Cell Science used rabbit red cells for *in vitro* and *in vivo* studies of structural features that correlated with membrane binding and retention of various PKH dyes having different spectral characteristics (B. Jensen, personal communication). Reducing the number of hydrocarbon tails, reducing tail length(s), and/or incorporation of branched or unsaturated tails resulted in reduced membrane retention. Addition of negative charges to the head group improved water solubility but reduced membrane retention and staining intensity. PKH2, the first commercially available green fluorescent dye, was found to have a half life of 10-11 days on circulating rabbit red cells (manufacturer's product insert) and of 8-10 days on circulating sheep lymphocytes.[86] PKH26, the orange fluorescent dye introduced in kit form in 1990, was found to have membrane retention characteristics similar to PKH3, with a half life of ~158 days on circulating rabbit red cells (manufacturer's insert; Fig. 5) and of ~153 days on circulating sheep red cells.[105] As with PKH3, PKH2 and PKH26 were found to have no effect on circulating lifetimes and other functional properties of red cells and lymphocytes. Although there are as yet no published *in vivo* studies with the more recently introduced PKH67, *in vitro* studies (RYMP and KAM, unpublished data; M. Waxdal, P. Wallace, personal communication ref. 11b) suggest that it also will have minimal effects on cell function and lifetime in circulation.

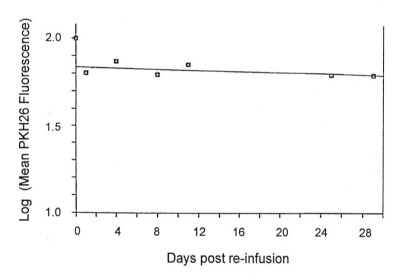

Fig. 5. **Prolonged retention of PKH26 in the membrane of circulating rabbit erythrocytes.** Rabbit erythrocytes were labeled *ex vivo* with PKH26 and reinfused using methods similar to those previously described [34, 78]. Blood samples were monitored by flow cytometry to determine percent PKH26+ cells remaining in circulation and mean intensity of PKH26+ cells as a function of time postreinfusion. Half life for elution of PKH26 from the membrane, determined from a first order plot of mean fluorescence *vs.* time, was 205 days for the example shown, with mean ± SD = 130 ± 55 days over 10 animals. PKH26 labeled erythrocytes also exhibited normal lifetimes in circulation (data not shown)

B. Assessment of Cell Proliferation by Dye Dilution

As summarized by O'Gorman et al.,[66] single cell proliferation analysis offers a number of advantages over conventional bulk assays for cell activation/proliferation such as deoxynucleotide (dNTP) precursor uptake (tritiated thymidine, TdR; bromodeoxyuridine, BrdU) or biochemical measurement of metabolic activity (MTT, resazurin, etc.). In a complex cell mixture where different functional subpopulations may respond to stimulation with different kinetics, with fewer or more divisions, by apoptosis, or not at all, average population measurements are inadequate to address relevant biological questions.

Flow cytometric methods that measure BrdU incorporation as an index of DNA synthesis are a significant improvement over bulk methods in analyzing the response of heterogeneous populations to biological stimuli. However, although dNTP uptake is a prerequisite for DNA synthesis and DNA synthesis is a prerequisite for cell division, dNTP uptake is at best an imperfect predictor of cell division. Using BrdU, Neckers et al.[64] showed that dNTP uptake can occur in the absence of cell division and that inhibitors of cell growth or division do not inhibit BrdU uptake to the same extent as cell growth. Also, BrdU incorporation has been detected in cells that are not in S-phase cells as identified by DNA content.

Early studies by a number of investigators suggested that decreases in PKH dye fluorescence intensity associated with cell division could be used to semi-quantitatively monitor proliferative responses within heterogeneous cell populations.[12, 34, 38] More recently, Yamamura et al. described a mathematical model[103] which enables this approach to be applied in a more quantitative fashion and in settings where it would be difficult and/or inappropriate to use cell counting or other traditional measures of cell growth and division.[14, 66, 101] In particular, the Cell Census Plus™ System (developed by Sigma, St. Louis, MO) combines reagents for general membrane labeling using PKH26 with PKH26 reference microbeads (developed by Flow Cytometry Standards Corp., San Juan, PR) and a specialized software model (developed by Verity Software House, Topsham, ME). This combination of reagents and analytical software enables quantitative assessment of subset specific responses without requiring complex isolation and purification procedures (see

Appendix A14 for detailed instructions in the use of the model and ModFit). The ability to quantify absolute recoveries of specific cell number is particularly useful in settings where cytolysis and/or apoptosis may be occurring as well as proliferation (see protocol 2 for methods).

Measurement of cell proliferation based on dye dilution makes the key assumption that if a cell bears N molecules of dye in its membrane prior to division, membrane and dye will be equally distributed between the daughter cells, each of which will then bear N/2 dye molecules. This in turn requires that the loss of dye from the membrane due to elution and/or metabolism should be slow compared with division-related decreases in number of dye molecules per cell. The dye dilution principle is readily illustrated by labeling a population of asynchronously dividing cells taken from a homogeneous cell line, placing the labeled cells back in culture, and then monitoring cell number and mean fluorescence intensity over time. As shown in Fig. 6, the fluorescence intensity (FI) of the population steadily decreases over time but peak width remains relatively constant, suggesting that division of dye among daughter cells is relatively equal.

While the natural temptation is to label cells as brightly as possible so as to increase the number of divisions that can be detected, some caution must be exercised, since excessive concentrations of dye in the membrane can lead to stacking of the fluorophores and quenching of fluorescence. This in turn can lead to non-linearity between number of dye molecules per cell and observed fluorescence intensity, *e.g.* an increase in fluorescence after the first division rather than a decrease, followed by a linear decrease in intensity for subsequent divisions (Fig. 7; P. Wallace, personal communication).

Note added in proof: Apoptotic or necrotic cells may in some cases exhibit altered fluorescence intensity[11b] and, where possible, should be gated out prior to using dye dilution to assess extent of cell proliferation.

If all cells in the population are dividing, if cells are not overlabeled, and if non-proliferation related dye losses are small, a plot of relative increase in cell number *vs.* relative decrease in fluorescence intensity (or increase in fluorescence index)[103] should give a slope of 1.0 and a correlation coefficient of >0.98 (Fig. 8). This type of correlation has been observed for many dif-

ferent cultured cell types, including K562 (human erythroleukemia), YAC-1 (murine lymphoma), U937 (human monocytic leukemia), HL60 (human promyelocytic leukemia) and A10 (rat
aortic smooth muscle cell line)[11,34,64,82] (R. Poon, Sigma, unpublished; B. Jensen, G. Kopia, & K. Muirhead, Zynaxis, unpublished).

In contrast to the relatively stable peak width seen over time
for homogeneous cell cultures (Fig. 6), PKH fluorescence distribution broadens dramatically over time for a complex cell po-

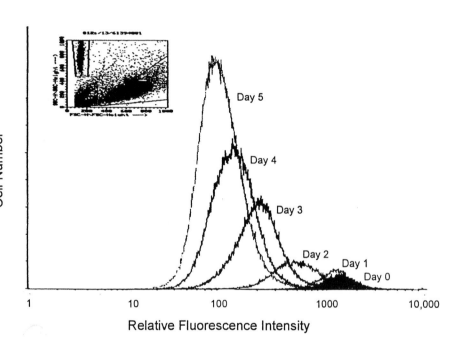

**Fig. 6. Use of PKH26 dye dilution to monitor cell proliferation in a human lym-
phoma cell line (reprinted from Ref. 103).** 8E5/LAV cells were maintained in
logarithmic growth phase by rapid subculturing in RPM1640+10% FCS. Exponentially growing cells were labeled at time 0 with 1.0 µM PKH26 using a general
membrane labeling protocol, cultured in RPM1640+10% FCS, and sampled at
approximately 24 hour intervals. Samples were pelleted and resuspended in
1.0 ml of 0.5% paraformaldehyde-PBS containing 7.5 µm diameter microbeads
at a concentration of approximately 2×10^5 beads/mL (upper left population in
insert), and analyzed by flow cytometry to determine cell number and PKH26
fluorescence intensity. As described in Ref. 103, an excellent correlation was
found between cell number (determined based on the internal bead standard)
and the cellular proliferation index calculated based on observed decrease in
fluorescence intensity

pulation in which subpopulations are responding differentially to a proliferative stimulus (Fig. 9). Quiescent lymphocytes, being more homogeneous in size than cultured ones, exhibit a tighter PKH26 distribution at time zero (compare unstimulated lymphocyte population in Fig. 9 with exponentially growing γE5/LAV cells in Fig. 6). However, by day 4 post-stimulation, the PKH26 distribution broadens dramatically (Fig. 9, right panel) as different subsets respond differentially, although it still includes a significant number of non-responding cells with fluorescence intensity similar to that observed at time zero.

Assuming that the initial fluorescence intensity distribution is symmetrical and that dye dilution is observed in each of the re-

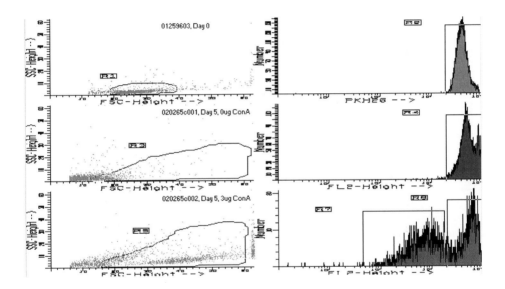

Fig. 7. Fluorescence histograms of Balb/c spleen cells labeled with PKH26 on Day 0 (top panel) and Day 5 after culturing in media without stimulus (unstimulated, baseline control for proliferation, center) or with 3 µg/ml ConA (proliferated sample, bottom panel). Note that fluorescence intensity distribution of unstimulated cultured-cells and ConA activated cells on Day 5 showed events in higher channel numbers than the Day 0 parent population. These data files were generated under standardized cytometer settings. The median fluorescence intensity in relative linear channels are: R2=4142, R4=7169, R7=698 and R8=5882. The light scatter patterns suggested absence of significant cell clumping, which was verified by visual inspection under the microscope. This phenomenon can, therefore, be interpreted as over staining and self-quenching of the sample on Day 0 and subsequent reversal of the quenching effect resulting in an underestimation of the proliferative response

sponding subpopulations just as it is in the more homogeneous cultured cells, a mathematical model can be used to predict the position of the PBL (peripheral blood lymphocytes) daughter generations from the known FI of the parent (G1) population. The observed histogram is modeled as the sum of multiple distinct Gaussians, each representing successive daughter genera-

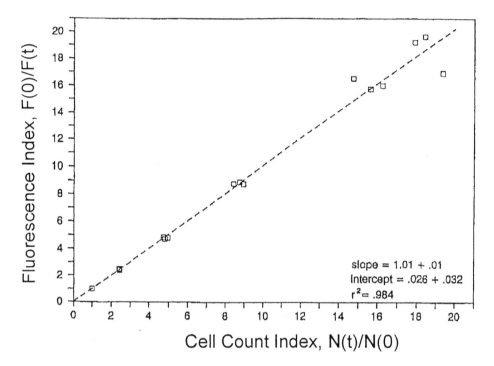

Fig. 8. Correlation of dye dilution with cell count in cultured YAC-1 cells. Asynchronous exponentially growing YAC-1 cultured lymphoma cells were labeled with PKH26 at time 0 using a general membrane labeling protocol (1×10^7/mL, 8 μM PKH26) and replaced in culture. Aliquots were analyzed at 18-24 hour intervals for cell growth and change in fluorescence intensity. Fluorescence intensity scales were standardized by running Coulter Standard Brite beads each day at the same instrument settings used for cell analysis and normalizing cellular fluorescence to a constant bead position. Relative decrease in fluorescence intensity (Fluorescence Index) was linearly correlated with relative increase in cell number (Cell Number Index) throughout the culture period, despite the fact that cell growth was non-exponential at the latest times analyzed (Cell Number Index ~20). Coefficients of variation (CV) of fluorescence remained approximately constant throughout the study (range: 20.1 - 22.5) suggesting that dye was approximately equally distributed between daughter cells at each cell division

tions with intensities of N/2, N/4, N/8, *etc.* (Fig.10). This mathematical model is described in a specific proliferation analysis subroutine (Cell Proliferation Model or Proliferation Wizard) for the ModFit data analysis software (ModFit v5.2 or ModFit LT v2.0; Verity Software House, Topsham, ME). Using this model and a standardized fluorescence intensity scale, the ModFit program derives from the PKH26 fluorescence intensity distribution several descriptors of response to stimulation. These include 1) extent of proliferation, as measured by number of daughter generations present in the culture, 2) relative abundance of successive daughter generations, 3) Proliferation Index (PI) for stimulated vs. control samples, and 4) non-proliferating fraction (NPF). The proliferation analysis software also estimates the goodness of fit of the deconvolution analysis using a reduced chi square statistic (Fig 10; see Appendix A14 for detailed description of data analysis).

Use of PKH dyes to monitor cell proliferation can be a particularly powerful tool when coupled with multiparameter flow cytometric analysis to dissect differential responses as a function of state of maturation and/or activation. Subpopulations of interest may be identified using cell surface antigens (e.g. CD4, CD8, various ligand/receptor pairs), intracellular cellular antigens (e.g. cytokines, signal-transducing molecules), and/or mar-

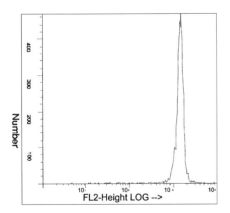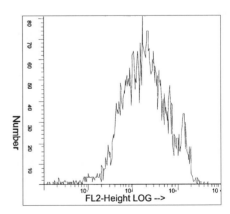

Fig. 9. Human peripheral blood monouclear cells were labeled with PKH-26 and cultured with ConA for four days. Histograms were obtained by light scatter gates on lymphocytes on Day 0 immediately after PKH26 dye loading (left panel) and Day 4 of exposure to ConA (right panel). Note dramatic peak broadening in functionally heterogeneous population after activation

kers of membrane integrity or asymmetry (propidium iodide, annexin V). Although the Cell Census Plus system was originally developed for use with PKH26, evidence to date suggests that PKH67 can be similarly utilized to monitor cell proliferation in many cases (11b; P. Wallace, R. Poon & K. Muirhead, unpublished data). These findings allow the choice of orange or green fluorescence to monitor dye dilution/cell proliferation.

Note that the assumption of Gaussian daughter cell distributions will be invalidated by starting with an asymmetric parent distribution, and that the ability to resolve and detect low frequency subpopulations undergoing multiple cell divisions will be a function of the symmetry and CV of the parent distribution. Whichever PKH dye is used, it is therefore important to identify staining conditions that give initial fluorescence intensity distributions which are as narrow and symmetrical as possible. Symmetrical distributions are readily achieved for homogeneous cell populations (e.g. quiescent or asynchronously dividing cell lines), with peak widths being characteristic of a given cell

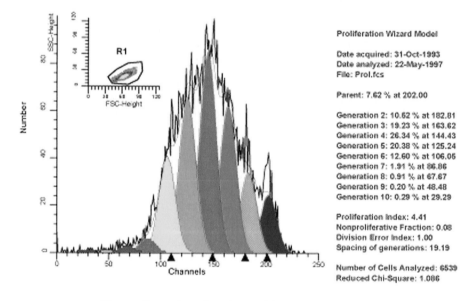

Proliferation Wizard Model

Date acquired: 31-Oct-1993
Date analyzed: 22-May-1997
File: Prol.fcs

Parent: 7.62 % at 202.00

Generation 2: 10.62 % at 182.81
Generation 3: 19.23 % at 163.62
Generation 4: 26.34 % at 144.43
Generation 5: 20.38 % at 125.24
Generation 6: 12.60 % at 106.05
Generation 7: 1.91 % at 86.86
Generation 8: 0.91 % at 67.67
Generation 9: 0.20 % at 48.48
Generation 10: 0.29 % at 29.29

Proliferation Index: 4.41
Nonproliferative Fraction: 0.08
Division Error Index: 1.00
Spacing of generations: 19.19

Number of Cells Analyzed: 6539
Reduced Chi-Square: 1.086

Fig. 10. Illustration of ModFit Proliferation Analysis printout. PKH26 labeled human peripheral blood mononuclear cells were stimulated in culture with ConA for 4 days. Analysis of the proliferation data was performed with light scatter gating for total lymphocytes excluding only dead cells and debris

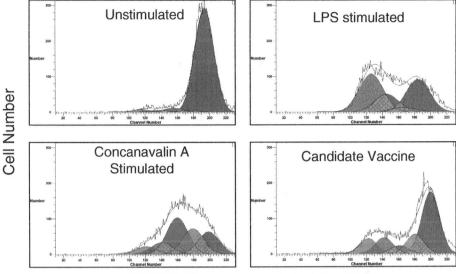

PKH26 Fluorescence Intensity (FL2)

Fig. 11. Assessment of Cell Proliferation by Dye Dilution: Effect of Various Stimuli on PKH26 Fluorescence Intensity Distribution of Cultured Splenocytes. Mouse splenocytes were harvested from animals previously immunized orally with a candidate vaccine. On day 0 post-harvest cells were labeled with PKH26 (final concentration 2 µM, 1×10^7 cells/mL) using a general membrane labeling protocol (Subprotocol 1), washed thoroughly, and placed in culture with medium alone (upper left), 2 µg/mL Concanavalin A (lower left), 50 µg/ mL LPS + 20 µg/mL dextran sulfate (upper right panel), or the same vaccine previously administered orally (lower right). On day 5 post-stimulation, cells were harvested, washed, and analyzed on a Becton Dickinson FACScan. Fluorescence amplification was adjusted to place the unstimulated PKH26 stained control sample (upper left) at a relative intensity of approximately 10^3 (HV=680). Note that different stimuli give rise to different degrees and patterns of broadening in the PKH26 distribution, indicative of differential proliferative responses. The Cell Proliferation Model (Sigma Chemicals, Verity Software House, Inc.) was used to quantitatively estimate extent of cell proliferation from the day 5 PKH26 fluorescence distributions. Gaussian distributions shown represent successive daughter generations used by the model to fit the observed PKH26 distribution, where each daughter generation is assumed to have a mean PKH26 fluorescence intensity half that of the preceding generation (see also Appendix A14).

type and determined primarily by variation in cell size. Use of reproducible staining conditions and technique typically give very reproducible staining intensity and CV for a given cell type (e.g., in the authors' experience, exponentially growing YAC-1 murine lymphoma cells typically give symmetrical fluorescence distributions with CV~15-20 whether they are stained with PKH26 or PKH67).

Lymphocytes responding to different stimuli give rise to distinct PKH dye intensity distributions (Figures 11 and 12) and therefore different PI values. The average response of the total population can be further dissected using fluorescent antibody markers to identify cell types involved in the subset specific response to a given stimulus (Figures 13, 14 and Figures 12 and 25 should be in Appendix A14 for an example in 3-color analysis).

Particular care must be taken to establish appropriate fluorescence amplification and color compensation settings when using PKH26 in combination with fluorescein labeled antibodies for estimation of proliferative responses within specific subpopulations of cells. Overamplification of the PKH26 signal in an attempt to maximize the detectable number of cell generations can result in inability to fully compensate for crossover emission of fluorescein in the FL2 detector typically used to monitor PKH26 (Fig. 13).

This in turn can lead to an artificially high "PKH26" intensity distribution for antibody positive cells, a distortion of the overall PKH26 distribution, and, potentially, to a significant underestimate of proliferative responses within the antibody positive population (Fig. 14).

Appropriate FL2 fluorescence amplification settings can be chosen by running a completely unstained sample and placing the autofluorescence distribution on scale but within approximately the first decade. If this is done, PKH26 stained cells will typically fall near or in the lower fourth decade prior to cell division, and compensation settings can be chosen which fully correct for any spillover of FITC emission into the FL2 channel. Just as overamplification/undercompensation will lead to an underestimation of proliferation for antibody positive cells, overcompensation must also be avoided since this will lead to an artificially low "PKH26" intensity distribution for antibody positive cells and an overestimation of proliferative responses within that population. Final color compensation set-

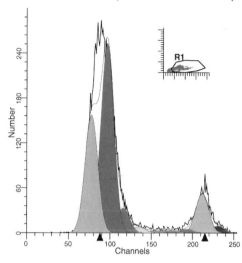

Proliferation Response of Total Balb/C Splenocytes to C57BL6

Proliferation Wizard Model

Date acquired: 2/ 2/96
Date analyzed: 2-Jul-1997
File: 020266rf.001

Parent: 1.85 % at 231.00

Generation 2: 10.01 % at 211.81
Generation 3: 0.96 % at 192.62
Generation 4: 0.89 % at 173.43
Generation 5: 0.59 % at 154.24
Generation 6: 1.58 % at 135.05
Generation 7: 6.31 % at 115.86
Generation 8: 48.00 % at 96.67
Generation 9: 29.81 % at 77.48
Generation 10: 0.00 % at 58.29

Proliferation Index: 12.69
Nonproliferative Fraction: 0.02
Division Error Index: 1.00
Spacing of generations: 19.19

Number of Cells Analyzed: 10709
Reduced Chi-Square: 2.664

S-Phase Assessment Not Active.

Fig. 12. Anti-Major Histocompatibility Antigen Response in Mixed Lymphocyte Culture. Balb/C splenocytes were labeled with PKH-26 and co-cultured with X-irradiated (1,200 rads) C57BL6 splenocytes for 6 days. PKH-26 fluorescence data were gathered using a light scatter gate to exclude dead cells and debris. The difference in proliferative pattern compared to mitogen stimulation in Figure 11 may be attributable to the difference between an oligoclonal and a polygonal response

tings should be established using the appropriate single color controls (cell labeled with PKH26 only or FITC-Ab only).

Where it is known that a specific subset will respond to stimulus used, appropriate fluorescence amplification and color compensation settings can readily be selected by comparing the PKH26 intensity distributions obtained for i) a sample which has been stained with PKH26 but not with that antibody and ii) all cells (antibody positive and negative) in a sample which has been stained with both PKH26 and antibody. These distributions should be identical (i.e. the presence or absence of antibody labeled cells should not affect the shape of the overall PKH26 distribution), and the correct compensation setting is the lowest one which achieves this result (Fig. 15 A & B).

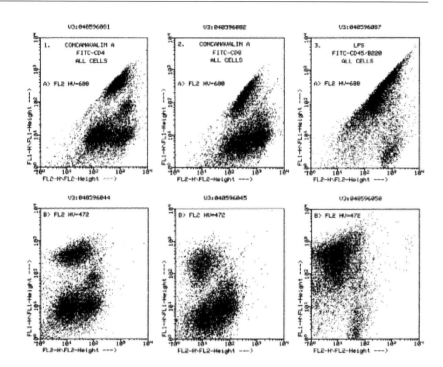

FITC-Mab Fluorescence Intensity (FL1)

PKH26 Fluorescence Intensity (FL2)

Fig. 13. Assessment of Cell Proliferation by Dye Dilution: Effect of Excessive Fluorescence Amplification. Mouse splenocytes were labeled with PKH26 as described in Figure 11 and placed in culture with either 2 µg/mL Concanavalin A (left and middle panels) or 50 µg/mL LPS + 20 µg/mL dextran sulfate (right panels). On day 5 post-stimulation, cells were harvested, stained with fluoresceinated monoclonal antibodies directed against CD4(left panels), CD8 (middle panels) or CD45R/B220 (right panels) cells, and analyzed on a Becton Dickinson FACScan.Upper Panels: PKH26 fluorescence amplification was adjusted to place the unstimulated PKH26 stained control sample (not shown) at a relative intensity of approximately 10^3 (HV=680). Compensation for crossover of fluorescein (FL1) into the PKH26 (FL2) channel was incomplete, even at settings of Fl2-90%FL1 (note strong diagonal distribution for antibody positive cells). This resulted in a falsely elevated FL2 signal and an underestimation of extent of proliferation in the Ab+ population (see Figure 14). Lower Panels: PKH26 fluorescence amplification was adjusted to place the autofluorescent unstained control sample (not shown) on scale but within the first decade of FL2 (HV=472). Compensation settings were adjusted to FL2-13.2%FL1 using samples labeled with PKH26 only or FITC-Ab only. Note that Ab+ distributions no longer appear diagonal and proliferative responses are evident in antibody positive as well as antibody negative populations

Where it is not known which cell subpopulation(s), if any, will respond, it is recommended that a positive control treated with a suitable mitogen be used to insure that appropriate amplification/compensation settings are being used. For this purpose any polyclonal mitogen which gives rise to a substantial proliferating fraction is appropriate. Additionally, when a mitogen is used as a positive control to demonstrate that the cell sample is capable of responding to some stimulus, it is preferable to match the kinetics of the mitogen with that of the experimental system. For instance, responses to Con A and PHA are typically detectable 4-6 days post-stimulation, whereas responses to PWM or microbial antigens may not be detectable until day 7-8 (R. Poon, unpublished data) and detection of low frequency responses to tetanus or PSA may require as long as 10 days[A].

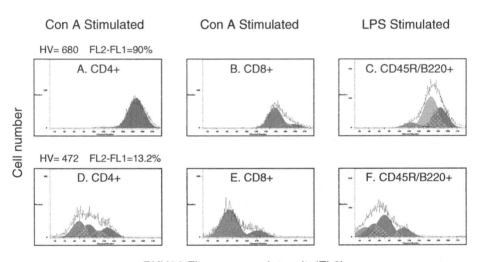

PKH26 Fluorescence Intensity(FL2)

Fig. 14. Effect of Inappropriate Fluorescence Amplification and/or Compensation on Estimation of Subset Proliferation Using PKH26 Dye Dilution. PKH26 fluorescence distributions collected as described in Figure 13 were analyzed using the Cell Proliferation Model. Overamplification of PKH26 fluorescence and failure to fully compensate for FITC crossover (upper panels) leads to underestimation of proliferative responses for antibody positive populations (upper panels; calculated Proliferation Indices of 1.0, 2.4 and 1.7 for stimulus/subset combinations, left to right). Use of appropriate amplification/compensation settings allows ready identification of the expected proliferative responses (lower panels; calculated Proliferation Indices of 2.8, 3.8 and 3.0 for stimulus/subset combinations, left to right)

A.

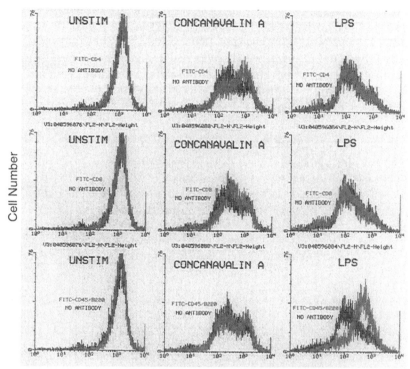

PKH26 Fluorescence Intensity (FL2)

Fig. 15A,B. Determination of Appropriate Fluorescence Amplification Settings for Two Color Analysis of Cell Proliferation using PKH26 Dye Dilution. Murine splenocytes were isolated, labeled with PKH26, stimulated in vitro, and counterstained with subset-specific antibodies as described in Figure 13, and analyzed by flow cytometry. Data overlays compare PKH26 fluorescence distributions collected for all cells (Ab+ and Ab-) from single color samples stained only with PKH26 (black) or dual color samples labeled with PKH26 and fluoresceinated antibody (turquoise). **Panel A.** Overamplification of PKH26 signals results in inability to compensate for crossover of fluorescein (FL1 antibody) into the PKH26 (FL2) channel, resulting in distortion of the PKH26 distribution in dual color vs. single color samples. Note that the distortion is worse for stimulus/Ab combinations in which a large fraction of the Ab+ population is proliferating (e.g. LPS and CD45R/B220, lower right)

B.

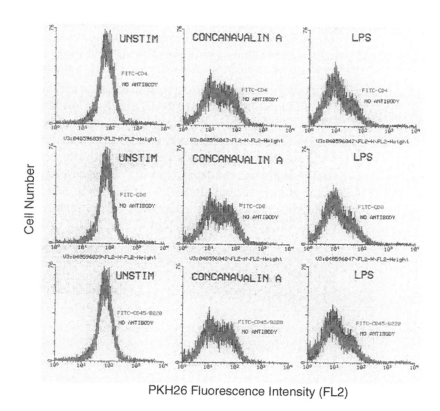

PKH26 Fluorescence Intensity (FL2)

Fig. 15. Panel B. When appropriate amplification of PKH26 signals is selected, the shape of the PKH26 distribution is identical, whether or not antibody is present

Note added in proof: Broader fluorescence distributions are sometimes obtained with PKH dyes than with probes which randomly label cellular proteins (*e.g.,* CFSE,[B] biotin-SE[C]). However, protein labeling can impair expression of critical receptors (*e.g.,* CD80[C] or CD4; AD Wells, pers. comm.) and abolish responses to some recall antigens (A. Givan, pers. comm.), whereas PKH dye labeling appears to leave receptor expression[A,] [D] and antigen-specific responses intact (Ref. D; P. Wallace, pers. comm.). Hence, each laboratory should verify that the labeling method chosen is appropriate for the type of cells and functional studies of interest.

C. Measurement of Cytotoxic Effector Function using PKH dyes

Immune cell mediated cytotoxicity can involve a wide range of cell types and mechanisms (antigen specific cytotoxic lymphocytes, CTL; natural killer or non-MHC restricted killer lymphocytes; antibody dependent monocyte or macrophage killing), but typically involves two discrete steps: conjugate formation and lytic killing. During conjugate formation, stable receptor-mediated associations are formed between effector and target cells. In some cases, conjugate formation can be monitored using cell size or fluorochrome-tagged antibodies specific to either the effector or target population to distinguish homologous from heterologous conjugates. In cases where size is not a good discriminator, where antibodies are not available and/or where use of antibody tags may interfere with functional receptors, the PKH dyes offer a more universal approach to conjugate enumeration. Using this strategy, conjugates are enumerated as double-labeled events. In addition to direct enumeration of the conjugates formed, it is also possible to analyze the phenotype of the effectors in the conjugates by using the third and fourth fluorescence channel in the flow cytometer, providing additional ability to resolve biologically relevant subsets within a complex effector population.

Release of ^{51}Cr from pre-labeled target cells has traditionally been used to monitor the second, lytic step in assays of cell-mediated cytotoxicity, since breach of target cell membrane integrity represents a common endpoint for many effector cell types and lytic mechanisms,. However, ^{51}Cr release is generally useful only for relatively short term assays (<10 hours), due to toxic effects of the radiolabel on target cells, increasing levels of spontaneous release over time, and the possibility of reutilization of released radioisotope by other cells within the effector: target mixture. Such limitations make ^{51}Cr release inherently unsuitable for longer term assays such as 18 hour NK killing or monocyte/macrophage mediated ADCC, which can require as long as 48-120 hours to complete. These limitations in combination with the safety/environmental issues associated with use of radioisotope have led to increasing use of alternative non-radioactive assays such as enzyme release (alkaline phosphatase, lactate dehydrogenase; [101] and colorimetric or fluorimetric assessment of reductive dye fixation by live cells (MTT, XTT, resazurin, etc.)[14]. However, like ^{51}Cr release, these are bulk assays for total lytic activity which cannot be used to distinguish different effector cell types and/or mechanisms within complex populations.

Single cell analysis by flow cytometry offers the capability to dissect cytotoxic events in considerably more detail. A number of investigators have used membrane dyes such as PKH2, PKH26, F-18 and D275 (DiO) to label effector and/or target cells. Uptake of fluorochromes such as propidium iodide (PI) or SYTOX, which are excluded by cells with intact membranes, has most frequently been used to enumerate target cells undergoing lysis. However, targets undergoing apoptosis can also be identified using reagents such as annexin V[11b] and antibodies to Fas (CD95, tranducer molecule for the apoptotic signal) (Yamamura personal communication). Different membrane dyes vary considerably in the times required to prepare labeled target or effector populations, ranging from 5-10 minutes for the PKH dyes [32, 79, 62] to 1-2 hours for F-18 [70] or overnight for D275/DiO [39]. As noted by Hatam et al. [32], the ability to rapidly prepare labeled target cells can be a significant advantage in settings where some spontaneity in test scheduling is required, such as the clinical laboratory[62].

Theoretically, either effector or target cells can be labeled with PKH dyes. However, the technical preference has generally been to label target cell populations, which are typically more homogeneous than effector cell populations and therefore give rise to a narrower fluorescence distributions and easier selection of analysis boundaries between targets and effectors. Addition of a viability dye (typically one which monitors membrane permeability) allows discrimination of live from "dead" cells within both effector and target cell populations (Fig. 16).

Percent specific cytotoxicity after 1 hour as measured by flow cytometry using PKH dye labeled targets was found to correlate well with that obtained at 4 hours in the Cr[51] release assay.[32,79] PKH dye labeling does not appear to alter either function or specificity of cytotoxic effector cells [79, 97] or susceptibility of target cells[32, 79] (Fig. 17). Note that in longer-term assays, target cell proliferation can be monitored by dye dilution and effector-mediated cytostasis can be detected even in the absence of cytotoxicity.

Single cell analysis by flow cytometry has several advantages over bulk analysis methods for standard short-term NK assays. Using propidium iodide as the viability dye enables more rapid readout of cell killing (1 hour vs. 4 hours for ^{51}Cr). Ability to quantify live (L) and dead effector cells as well as live *vs.* dead (D) target cells allows direct evaluation of effector:target (E:T) ratios and more accurate determination of cytotoxic potency. In this assay, E:T ratio is calculated as [(LE+DE)/LT+DT)]. Percent killing at a given E:T ratio is calculated as [DT/(DT+LT)$_{specific}$-DT/(DT+LT)$_{background}$]x100, where background values are those for control tubes lacking effectors.

PKH dyes have also proven useful in longer term cytotoxicity studies (e.g., 18 hour NK killing, monocyte/macrophage mediated tumor cytotoxicity, or "bystander" mediated killing). Flieger *et al.* [24] described the combined use of PKH2 labeled target cells, PKH26 labeled effector cells, and covalently labeled FITC cell standards to monitor absolute numbers of targets and effectors and determine the fate of tumor cell targets (lysis, phagocytosis) during a 5 day ADCC assay. Freeman *et al.*[27] found

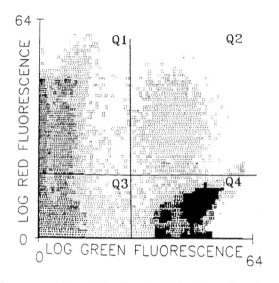

Fig. 16. **Flow cytometric analysis of murine NK/YAC samples using PKH1 and propidium iodide (reprinted from Ref. 79).** YAC tumor targets were labeled with PKH1 (green fluorescence) as described[79] , incubated with interferon-γ treated murine spleen cells (upper and lower panels, respectively) for 4 hours at 37°C. Viability (membrane permeability) was determined for both effector (PKH1 negative) and target (PKH1 positive) populations based on propidium iodide uptake (red fluorescence). Quadrant analysis allows quantification of live effectors (LE, unlabeled, quadrant 3), dead effectors (DE, PI+, quadrant 1), live targets (LT, PKH+, quadrant 4) and dead targets (DT, PKH+/PI+ quadrant 2). Similar studies have been carried out using PKH2 to study Fas-FasL interactions involved in cell mediated cytotoxicity and activation induced cell death[41]

that after ganciclovir treatment, uptake of apoptotic vesicles from HSV-TK transfected PKH26 labeled tumor cells led to by-stander killing of non-transfected tumor cells.

When interpreting two color data from assays in which effector cell populations may contain phagocytes (e.g. monocyte/macrophage or neutrophils), it is important to remember that such effector cells may also be able to phagocytose PKH+ target cells (see Section C below). Events that are in fact phagocytized target cells may be incorrectly analyzed as intact live targets unless size or phagocyte specific surface markers are used to distinguish between the two possibilities. Similarly, where there is apparent "leakage" of dye from stained target cells to unstained effectors, microscopic observation should be used to determine whether phagocytosis and/or uptake of apoptotic vesicles arising from labeled target cells are occurring. The latter process can also give rise to events positive for both PKH dye and membrane integrity indicating dye. This is particularly important with target cells such as K562 which can give rise to significant numbers of apoptotic vesicles when culture conditions are suboptimal (Fig. 18).

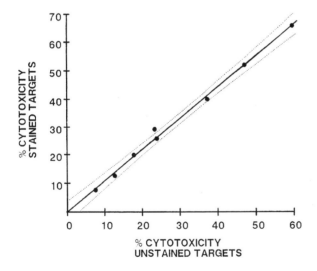

Fig. 17. Effect of PKH1 staining on susceptibility of YAC-1 tumor targets to NK-mediated cytolysis (reprinted with permission from Ref. 79). YAC-1 targets (unstained or stained with 5 μM PKH1) were incubated with murine NK preparations at varying E/T ratios [79] and percent cytolysis was determined by ^{51}Cr release. YAC-1 susceptibility to cytolysis was unaltered by PKH1 labeling (slope 1.10, intercept 0.47, correlation coefficient 0.9941; dashed lines indicated 95% confidence intervals for regression)

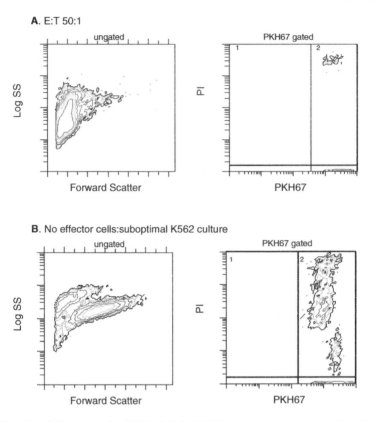

Fig. 18. **NK cytotoxicity assay using PKH67 labeled K562 targets (data courtesy of A. Palini and M. Waxdal, FAST Systems, Gaithersburg, MD).** K562 targets grown under optimal(early log phase, high surface: volume ratio) or suboptimal (late log phase, low surface:volume ratio) conditions were labeled with PKH67 using a general membrane labeling protocol (10 µM dye, 2 x 10⁷ cells/ mL) for use as target cells in an NK cytotoxicity assay. Samples were analyzed at constant instrument settings on a Coulter EPICS Elite flow cytometer using the LFL1 to collect PKH67 fluorescence and LFL3 channel to collect PI fluorescence.

A. K562 targets taken from optimal culture conditions were incubated for 4 hours with a monocyte depleted peripheral blood mononuclear effector cell preparation (E:T 50:1) and counterstained just prior to analysis with 5 µg/mL propidium iodide (PI) to identify dead (PI+) cells. K562 targets were not well resolved from effector cells by light scatter (left panel) but could be clearly resolved by gating on PKH67 fluorescence (right panel), allowing more accurate enumeration of % dead target cells.

B. K562 targets grown under suboptimal (high density, oxygen deprived) culture conditions exhibit not only bright PI+ dead cells ((right panel, quadrant 2, upper 4th decade of "PI" axis) but also subpopulations with elevated red autofluorescence (right panel, quadrant 2, 1st-2nd decade) and intermediate intensity PKH67 and PI fluorescence (right panel, quadrant 2, mid-3rd to mid-4th decade). The intermediate intensity population was determined by microscopic observation to be composed of small, membrane bounded, DNA containing vesicles, most likely apoptotic bodies, and was found to correlate with the light scatter population having reduced forward scatter and elevated right angle light scatter (left panel, upper left population)

D. Phagocytic Cell Tracking and Measurement of Phagocytosis by Flow Cytometry

PKH dyes have been used in two quite different ways to monitor phagocytic cell recruitment and/or function. Using salt based diluents which stimulate formation of PKH dye aggregates, several investigators have demonstrated selective labeling of phagocytic cells without effect on adhesion, migratory activity, cytokine release or triggering of activation enzymes.[1, 56, 107] Melnicoff et al. used PKH1 to monitor the fate of resident vs. recruited peritoneal macrophages in the presence of various inflammatory stimuli, [56, 57] and similar studies have been carried out using PKH2 (Fig. 19). Albertine and co-workers were able to selectively label circulating neutrophils with PKH26 *in vivo* using a similar approach, enabling measurement of migration to extravascular sites in response to chemoattractants.[1] Selective *in vivo* labeling is particularly advantageous for cell types such as neutrophils where isolation and *ex vivo* labeling may lead to undesired activation and altered responses.

Alternatively, general membrane labeling of bacteria, red cells and/or tumor cells has been used to prepare target cells for the study of mechanisms of phagocytic effector cell mediated killing. PKH2 has been used to label a variety of human and veterinary pathogens including *listeria, salmonella,* and *Haemophilus somnus* by investigators studying heterogeneity of phagocytic killing, adaptive bacterial responses, and bacterial effects on phagocytic function.[28, 71] Investigators have also used PKH2 or PKH26 labeled tumor cells or erythrocytes to study of Fc receptor mediated antibody dependent tumor cell phagocytosis by activated macrophages.[21, 63, 98, 99] As noted above, microscopy[21] or fluorescence quenching techniques[93] must be used to distinguish adherent from ingested particles.

E. Summary

The PKH membrane intercalating dyes enable rapid, high intensity labeling of a wide variety of cell types without significant effects on cell viability, function or immunogenicity (Table 1). They are applicable to monitoring a broad range of cell functions (Table 2), and can be combined with many other fluorescent probes for multiparameter cytometric studies (Fig. 1). Although

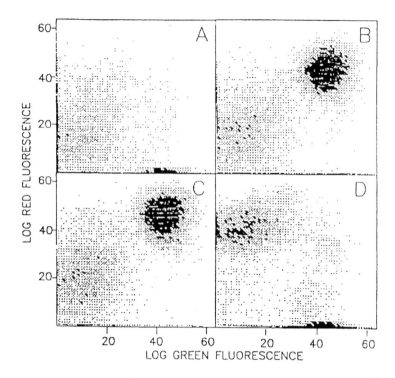

Fig. 19. Preferential in vivo labeling of resident peritoneal macrophages (reprinted from Reference 55 with kind permission of the Society for Leukocyte Biology and Alan R. Liss, Inc.).
Peritoneal cells were harvested 24 hours after i.p. injection of 0.2 μM PKH1 in sterile endotoxin-free buffered salt solution and analyzed by flow cytometry. Twenty - 40% of peritoneal cells became PKH1 bright, falling in the third decade, while approximately 20% were PKH1 dim, falling above channel 1 but within the first decade. PKH1 bright cells were identified as resident peritoneal macrophages and PKH1 dim cells as B cells by indirect immunofluorescence analysis using phycoerythrin-labeled goat anti-rat IgG (PE-GAR) as the second antibody.
Panel A. Less than 5% non-specific staining and minimal labeling of PKH1 bright cells was seen with PE-GAR alone (lowest dot level = 1 cell). Note: Green fluorescence amplification was set to display macrophage autofluorescence within the first decade of a 3-decade log scale; at these settings, lymphocyte autofluorescence fell in channels 0-3.
Panel B. 40.9% of peritoneal cells and essentially all PKH1 bright cells became double labeled after staining with monoclonal antibody F4/80 (macrophage specific). PKH1 dim cells did not stain with F4/80, suggesting that these cells were not resident macrophages.
Panel C. 41.7% of peritoneal cells and essentially all PKH1 bright cells became double labeled with MAC-1 (a macrophage and neutrophil marker), consistent with differential counts of <5% neutrophils in resident peritoneal cell populations. PKH1 dim cells did not stain with MAC-1, confirming that these were not macrophages or neutrophils.
Panel D. 39% of peritoneal cells were positively labeled by monoclonal antibody 14.8 (B lymphocyte specific). Essentially all PKH1 dim cells (20% of peritoneal cells) but only a minimal number of PKH1 bright cells stained with 14.8, indicating preferential uptake of dye by phagocytic cells when delivered in salt-containing diluent.

the focus of this manual is primarily on flow cytometric applications, the principles and staining protocols for PKH dyes described here are also applicable to the preparation of labeled cells for use with other fluorescence based cytometric techniques such as confocal microscopy[21] and fluorescence whole body imaging.[23] These reagents are therefore expected to continue to serve as powerful and versatile tools for biomedical researchers seeking to determine how changes observed at the genomic level lead to changes in cell trafficking and function at the tissue or whole body level.

Acknowledgements

The authors would particularly like to recognize the contributions of the following to the body of knowledge summarized herein: Dr. Bruce Jensen and Ms. Edie Gavin for their work on structural determinants of membrane binding and retention for the PKH dyes; Drs. Abraham Schwartz and Yasuhiro Yamamura for development of the first generation Cell Proliferation Model; Mssrs. Don Carpenter, H. Beal McIlvain, Robert Lorinc, and R. Clayton Flectcher for their expert assistance with the multiparameter splenocyte proliferation studies; and Dr. Myron Waxdal and Mr. Alessio Palini for the data shown in Fig. 19.

References

1. Albertine KH, Gee MH. In vivo labeling of neutrophils using a fluorescent cell linker. J Leukoc Biol. 1996;59:631-8.
2. Ashley DM, Bol SJ, Kannourakis G. Human bone marrow stromal cell contact and soluble factors have different effects on the survival and proliferation of paediatric B- lineage acute lymphoblastic leukaemic blasts. Leuk Res. 1994;18:337-46.
3. Ashley DM, Bol SJ, Kannourakis G. Measurement of the growth parameters of precursor B-acute lymphoblastic leukaemic cells in co-culture with bone marrow stromal cells; detection of two cd10 positive populations with different proliferative capacities and survival. Leuk Res. 1994;18:37-48.
4. Ashley DM, Bol SJ, Tucker DP, Waugh CM, Kannourakis G. Flow cytometric analysis of intercellular adhesion between B-cell precursor acute lymphoblastic leukemic cells and bone marrow stromal cells. Leukemia. 1995;9:58-67.

5. Audran R, Collet B, Moisan A, Toujas L. Fate of mouse macrophages radiolabelled with PKH-95 and injected intravenously. Nucl Med Biol. 1995;22:817-21.

6. Basse P, Herberman RB, Nannmark U, et al. Accumulation of adoptively transferred adherent, lymphokine-activated killer cells in murine metastases. J Exp Med. 1991;174:479-88.

7. Beavis AJ, Pennline KJ. Tracking of murine spleen cells in vivo: detection of PKH26-labeled cells in the pancreas of non-obese diabetic (NOD) mice. J Immunol Methods. 1994;170:57-65.

8. Bennett S, Por SB, Cooley MA, Breit SN. In vitro replication dynamics of human culture-derived macrophages in a long term serum-free system. J Immunol. 1993;150:2364-71

9. Ben-Nun J, Nemet P. Intraocular pressure and blood flow of the optic disk: a fluorescent blood cell angiography study. Surv Ophthalmol. 1995;39 Suppl 1:S33-9.

10. Betarbet R, Zigova T, Bakay RA, Luskin MB. Migration patterns of neonatal subventricular zone progenitor cells transplanted into the neonatal striatum. Cell Transplant. 1996; 5:165-78.

11a. Boutonnat J, Muirhead K, Barbier M, Mousseau M, Ronot X, Seigneurin D. The use of PKH26 to study proliferation of chemoresistant leukemic sublines. Anticancer Res. 1998; 18: 4243-4251.

11b. Boutonnat J, Barbier M, Murhead R, Mousseau M, Grunwald D, Ronot X, Seigneurin D. Response of chemosensitive and chemoresistant cell lines to drug therapy: simultaneous assessment of proliferation, apoptosis and necrosis. Cytometry 42: in press.

12. Boyd FT. Identification of growth inhibited cells by retention of a lipophilic fluorescent dye. Cell Growth Differ. 1993; 4:777-84.

13. Bratosin D, Mazurier J, Slomianny C, Aminoff D, Montreuil J: Molecular mechanisms of erythrophagocytosis: flow cytometric quantitation of in vitro erythrocyte phagocytosis by macrophages. Cytometry 1997; 30: 269-274.

14. Carmichael J, Degraff WG, Gazdar AF, Minna JD, Mitchell JB: Evaluation of a tetrazolium-based, semiautomated assay: assessment of chemosensitivity testing. Cancer Res 47: 936-942, 1987.

15. Chang IK, Yoshiki A, Kusakabe M, et al. Germ line chimera produced by transfer of cultured chick primordial germ cells. Cell Biol Int. 1995;19:569-76.

16. Chung JD, Conner S, Stephanopoulos G. Flow cytometric study of differentiating cultures of Bacillus subtilis. Cytometry. 1995;20:324-33.

17. Coleman WB, Wennerberg AE, Smith GJ, Grisham JW. Regulation of the differentiation of diploid and some aneuploid rat liver epithelial (stemlike) cells by the hepatic microenvironment. Am J Pathol. 1993;142:1373-82.

18. Constant SL, Wilson R. *In vivo* lymphocyte responses in the draining lymph nodes of mice exposed to *Schistosoma mansoni*: preferential proliferation of T cells is central to the induction of protective immunity. Cellular Immunity. 1992;139:1-17.

19. Dailey S. Rose D, Carabasi A et al. Origin of cells that line damaged native blood vessels following endothelial cell transplantation. Am. J. Surgery 1991; 162:107-110.

20. Dimitrov DS, Blumenthal R. Photoinactivation and kinetics of membrane fusion mediated by the human immunodeficiency virus type 1 envelope glycoprotein. J Virol. 1994;68:1956-61.
21. Ely P, Wallace PK, Givan A. *et al.* Bispecific-armed interferon-gamma primed macrophage mediated phagocytosis of malignant non-Hodgkins lymphoma. Blood 1996; 87:3813-3821.
22. Embleton MJ, Charleston A, Affleck K. Efficacy and selectivity of monoclonal-antibody-targeted drugs and free methotrexate in fluorescence-labelled mixed tumor-cell monolayer cultures and multicellular spheroids. Int. J. Cancer 1991; 49:566-572.
23. Farkas DL, Ballou BT, Fisher GW *et al.* Microscopic and mesoscopic spectral bio-imaging. Proceedings SPIE 1996; 2678:200-209.
24. Flieger D, Gruber R, Schlimok G, Reiter C, Pantel K, Riethmuller G. A novel non-radioactive cellular cytotoxicity test based on the differential assessment of living and killed target and effector cells. J Immunol Methods. 1995;180:1-13
25. Ford JW, Welling TH, Stanley JC, Messina LM. PKH26 and 125I-PKH95: characterization and efficacy as labels for in vitro and in vivo endothelial cell localization and tracking. J Surg Res. 1996;62:23-8.
26. Foty RA, Steinberg MS: Measurement of tumor cell cohesion and suppression of invasion by E- or P-cadherin. Cancer Res. 1997; 57: 5033-5036.
27. Freeman S. Abboud C, Whartenby K et al. The "bystander effect": tumor regression when a fraction of the tumor mass is genetically modified. Cancer Res. 1993;53:5274-5283.
28. Gomis SM, Godson DL, Beskorwayne T, Wobeser GA, Potter AA: Modulation of phagocytic function of bovine mononuclear phagocytes by Haemophilus somnus. Microb.Pathog. 1997; 22: 13-21.
29. Gonokami Y, Konno SI, Kurokawa M, Adachi M. Effect of methotrexate on asthmatic reaction in sensitized guinea pigs. Int Arch Allergy Immunol. 1995;106:410-5.
30. Gothot A, Pyatt R, McMahel J, Rice S, Srour EF: Functional heterogeneity of human CD34(+) cells isolated in subcompartments of the G0 / G1 phase of the cell cycle. Blood 1997; 90: 4384-4393.
31. Greenwood JD, Croy BA. A study on the engraftment and trafficking of bovine peripheral blood leukocytes in severe combined immunodeficient mice. Vet Immunol Immunopathol. 1993;38:21-44.
32. Hatam L, Schuval S, Bonagura VR. Flow cytometric analysis of natural killer cell function as a clinical assay. Cytometry. 1994;16:59-68.
33. Hendrikx PJ, Martens CM, Hagenbeek A, Keij JF, Visser JW. Homing of fluorescently labeled murine hematopoietic stem cells. Exp Hematol. 1996;24:129-40.
34. Horan PK, Melnicoff MJ, Jensen BD, Slezak SE. Fluorescent cell labeling for in vivo and in vitro cell tracking. Methods Cell Biol. 1990;33:469-90.
35. Hugo P, Kappler J, Godfrey D, Marrack P. A cell line that can induce thymocyte positive selection. Nature 1992; 360:679-681.
36. Ide H, Wada N, Uchiyama K. Sorting out of cells from different parts and stages of the chick limb bud. Dev Biol. 1994;162:71-6.
37. Imaizumi K, Hasegawa Y, Kawabe T, et al: Bystander tumoricidal effect and gap junctional communication in lung cancer cell lines. Am.J. Respir.Cell Mol.Biol. 1998; 18: 205-212.

38. Jensen B, Horan P, Poste G. Fluorescent analysis of cellular growth rate in adherent cell systems. Cytometry 1988; Suppl. 2:39.
39. Johann S. A versatile flow cytometry-based assay for the determination of short- and long-term natural killer cell activity. J. Immunol. Methods 1995; 185:209-216.
40. Johnsson C, Festin R, Tufveson G, Totterman TH: Ex vivo PKH26-labelling of lymphocytes for studies of cell migration in vivo. Scand.J.Immunol. 1997; 45: 511-514.
41. Kaneda R, Iwabuchi K, Onoe K. Dissociation of Fas-mediated cytotoxicity and FasL expresion in a cytotoxic T cell clone. Comparative analysis of Fas-mediated cytotoxicity between a T-hybridoma and a T-cell clone. Immunol. Letters 1997; 55:53-60.
42. Kanki JP, Ho RK: The development of the posterior body in zebrafish. Development 1997; 124: 881-893.
43. Khalaf AN, Wolff-Vorbeck G, Bross K, Kerp L, Petersen KG. In vivo labelling of the spleen with a red-fluorescent cell dye. J Immunol Methods. 1993;165:121-5.
44. Kraft DL, Weissman IL, Waller EK. Differentiation of CD3-4-8- human fetal thymocytes in vivo: characterization of a CD3-4+8- intermediate. J Exp Med. 1993;178:265-77.
45. Lacerda JF, Ladanyi M, Louie DC, Fernandez JM, Papadopoulos EB, O'Reilly RJ. Human Epstein-Barr virus (EBV)-specific cytotoxic T lymphocytes home preferentially to and induce selective regressions of autologous EBV- induced B cell lymphoproliferations in xenografted C.B-17 scid/scid mice [published erratum appears in J Exp Med 1996 Sep 1;184(3):1199]. J Exp Med. 1996;183:1215-28.
46. Ladd AC, Pyatt R, Gothot A, et al: Orderly process of sequential cytokine stimulation is required for activation and maximal proliferation of primitive human bone marrow CD34+ hematopoietic progenitor cells residing in G0. Blood 1997; 90: 658-668.
47. Ladel CH, Kaufmann SH, Bamberger U. Localisation of human peripheral blood leukocytes after transfer to C.B- 17 scid/scid mice. Immunol Lett. 1993;38:63-8.
48. Lansdorp PM, Dragowska W. Maintenance of hematopoiesis in serum-free bone marrow cultures involves sequential recruitment of quiescent progenitors. Exp Hematol. 1993;21:1321-7.
49. Leavesley DI, Oliver JM, Swart BW, Berndt MC, Haylock DN, Simmons PJ. Signals from platelet/endothelial cell adhesion molecule enhance the adhesive activity of the very late antigen-4 integrin of human CD34+ hemopoietic progenitor cells. J Immunol. 1994;153:4673-83.
50. Lehner T, Wang Y, Cranage M, et al. Protective mucosal immunity elicited by targeted iliac lymph node immunization with a subunit SIV envelope and core vaccine in macaques. Nature Medicine 1996; 2(2):767-775.
51. Lu Y, Bigger JE, Thomas CA, Atherton SS: Adoptive transfer of murine cytomegalovirus-immune lymph node cells prevents retinitis in T-cell-depleted mice. Invest.Ophthalmol.Vis.Sci. 1997; 38: 301-310.
52. Luens KM, Travis MA, Chen BP, Hill BL, Scollay R, Murray LJ: Thrombopoietin, kit ligand, and flk2/flt3 ligand together induce increased numbers of primitive hematopoietic progenitors from human

CD34+Thy-1+Lin- cells with preserved ability to engraft SCID-hu bone. Blood 1998; 91: 1206-1215.

53. Ma X, Weyrich A, Lefer D et al. Monoclonal antibody to L-selectin attenuates neutrophil accumulation and protects ischemic reperfused cat myocardium. Circulation 1993; 88:649-658.

54. Maines JZ, Sunnarborg A, Rogers LM, Mandavilli A, Spielmann R, Boyd FT. Positive selection of growth-inhibitory genes. Cell Growth Differ. 1995; 6:665-71.

55. Melnicoff MJ, Morahan PS, Jensen BD et al. In vivo labeling of resident peritoneal macrophages. J Leukoc Biol 1988;43:387-97.

56. Melnicoff MJ, Horan PK, Breslin EW, Morahan PS. Maintenance of peritoneal macrophages in the steady state. J Leukoc Biol. 1988;44:367-75.

57. Melnicoff MJ, Horan PK, Morahan PS. Kinetics of changes in peritoneal cell populations following acute inflammation. Cell Immunol. 1989; 118:178-91.

58. Michelson AD, Barnard MR, Hechtman HB, et al. In vivo tracking of platelets: circulating degranulated platelets rapidly lose surface P-selectin but continue to circulate and function. Proc Natl Acad Sci U S A. 1996;93:11877-82.

59. Mikecz K, Glant TT. Migration and homing of lymphocytes to lymphoid and synovial tissues in proteoglycan-induced murine arthritis. Arthritis R2heum. 1994;37:1395-403.

60. Moreno-Mendoza N, Herrera-Munoz J, Merchant-Larios H. Limb bud mesenchyme permits seminiferous cord formation in the mouse fetal testis but subsequent testosterone output is markedly affected by the sex of the donor stromal tissue. Dev Biol. 1995;169:51-6.

61. Morikawa Y, Tohya K, Ishida H, Matsuura N, Kakudo K. Different migration patterns of antigen-presenting cells correlate with Th1/Th2-type responses in mice. Immunology. 1995;85:575-81.

62. Muirhead, K,. Foxx K, Palini A, Stregevsky E, Ohlsson-Wilhelm B, Waxdal M. "A "Stat" Non-Isotopic NK Cytotoxicity Assay Using Cryopreserved Target Cells". Cytometry, Supplement 9, 101-102, 1998.

63. Munn D, McBride M , Cheung C. Role of low-affinity Fc receptors in antibody-dependent tumor cell phagocytosis by human monocyte-derived macrophages. Cancer Res. 1991;51:1117-1123.

64. Neckers LM, Funkhouser WK, Trepel JB et al. Significant non-S-phase DNA synthesis visualized by flow cytometry in activated and in malignant human lymphoid cells. Exp Cell Research 1995; 156:429-38.

65. Ogawa M, Tsutsui T, Zou JP, et al: Enhanced induction of very late antigen 4/lymphocyte function- associated antigen 1-dependent T-cell migration to tumor sites following administration of interleukin 12. Cancer Res. 1997;

66. O'Gorman MRG, Corrochano V, Poon RYM. Beyond tritiated thymidine: flow cytometric assays for the evalution of lymphocyte activation/proliferation. Clin. Immunol. Newsletter 1996; 16:164-172.

67. Peschhold K, Pohl T, and Kabelitz D. Rapid quantification of lymphocyte subsets in heterogeneous cell populations by flow cytometry. Cytometry 1994; 16:152-59.

68. Pin CL, Merrifield PA: Regionalized expression of myosin isoforms in heterotypic myotubes formed from embryonic and fetal rat myoblasts in vitro. Dev.Dyn. 1997; 208: 420-431.

69. Pricop L, Salmon JE, Edberg JC, Beavis AJ: Flow cytometric quantitation of attachment and phagocytosis in phenotypically-defined subpopulations of cells using PKH26-labeled Fc gamma R-specific probes. J.Immunol.Methods 1997; 205: 55-65.

70. Radosevic K, Garritsen H, VanGraft M et al. A simple and sensitive flow cytometric assay for the determination of the cytotoxic activity of human natural killer cells. J. Immunol. Methods 1990; 135:81-89.

71. Raybourne RB, Bunning VK. Bacterium-host cell interactions at the cellular level: fluorescent labeling of bacteria and analysis of short-term bacterium-phagocyte interaction by flow cytometry. Infect Immun. 1994;62:665-72.

72. Redman CA, Kusel JR. Distribution and biophysical properties of fluorescent lipids on the surface of adult Schistosoma mansoni. Parasitology. 1996;113:137-43.

73. Rivas RJ, Hatten ME. Motility and cytoskeletal organization of migrating cerebellar granule neurons. J. Neurosci 1995; 15:981-989.

74. Rosenblatt-Velin N, Arrighi JF, Dietrich PY, Schnuriger V, Masouye I, Hauser C: Transformed and nontransformed human T lymphocytes migrate to skin in a chimeric human skin/SCID mouse model. J.Invest. Dermatol. 1997; 109: 744-750.

75. Rosenman S, Ganji A, Tedder T and Gallatin W. Syn-capping of human T lymphocyte adhesion/activation molecules and their redistribution during interaction with endothelial cells. J. Leuk. Biol. 1993; 53:1-10.

76. Samlowski WE, Robertson BA, Draper BK, Prystas E, McGregor JR. Effects of supravital fluorochromes used to analyze the in vivo homing of murine lymphocytes on cellular function. J Immunol Methods. 1991;144:101-15.

77. Schlegel RA, Lumley-Sapanski K Williamson P. Single cell analysis of factors increasing the survivial of resealed erythrocytes in the circulation of mice. In The Use of Resealed Erythrocytes as Carriers and Bioreactors (Magnani M, DeLoach JR eds) Plenum Press, NY; 1992:133-138.

78. Slezak S, Horan P. Fluorescent In-Vivo tracking of hematopoietic cells. Part I. Technical considerations. Blood 1989; 74(6):2172-2177.

79. Slezak SE, Horan PK. Cell-mediated cytotoxicity. A highly sensitive and informative flow cytometric assay. J Immunol Methods. 1989;117:205-14.

80. Smith D, Kornbrust E, Lane T. Phagocytosis of a fluorescently labeled perflubron emultion by a human monocyte cell line. Art. Cells, Blood Subs. & Immob.Biotech. 1994;22(4):1215-1221.

81. Spotl L, Sarti A, Dierich MP, Most J. Cell membrane labeling with fluorescent dyes for the demonstration of cytokine-induced fusion between monocytes and tumor cells. Cytometry. 1995;21:160-9.

82. Srour EF, Bregni M, Traycoff CM, et al. Long-term hematopoietic culture-initiating cells are more abundant in mobilized peripheral blood grafts than in bone marrow but have a more limited ex vivo expansion potential. Blood Cells Mol Dis. 1996;22:68-81.

83. Taguchi H, Osaki T, Yamaguchi H, Kamiya S. Flow cytometric analysis using lipophilic dye PKH-2 for adhesion of Vibrio cholerae to Intestine 407 cells. Microbiol Immunol. 1995;39:891-4.
84. Takezawa R, Watanabe Y, Akaike T. Direct evidence of macrophage differentiation from bone marrow cells in the liver: a possible origin of Kupffer cells. J Biochem (Tokyo). 1995;118:1175-83.
85. Tamaki K, Saitoh A, Gaspari AA, Yasaka N, Furue M. Migration of Thy-1+ dendritic epidermal cells (Thy-1+DEC): Ly48 and TNF- alpha are responsible for the migration of Thy-1+DEC to the epidermis. J Invest Dermatol. 1994;103:290-4.
86. Teare GF, Horan PK, Slezak SE, Smith C, Hay JB. Long-term tracking of lymphocytes in vivo: the migration of PKH-labeled lymphocytes. Cell Immunol. 1991;134:157-70.
87. Thomas GA, Lelkes PI, Chick DM et al. Skeletal muscle ventricles seeded with autogenous endothelium. ASAIO J. 1995; 41:204-211.
88. Toda S, Yonemitsu N, Minami Y, Sugihara H. Plural cells organize thyroid follicles through aggregation and linkage in collagen gel culture of porcine follicle cells. Endocrinology. 1993;133:914-20
89. Tomasetto C, Neveu MJ, Daley J, Horan PK, Sager R. Specificity of gap junction communication among human mammary cells and connexin transfectants in culture. J Cell Biol. 1993;122:157-67.
90. Traycoff CM, Cornetta K, Yoder MC, Davidson A, Srour EF. Ex vivo expansion of murine hematopoietic progenitor cells generates classes of expanded cells possessing different levels of bone marrow repopulating potential. Exp Hematol. 1996;24:299-306.
91. Traycoff CM, Orazi A, Ladd AC, Rice S, McMahel J, Srour EF: Proliferation-induced decline of primitive hematopoietic progenitor cell activity is coupled with an increase in apoptosis of ex vivo expanded CD34+ cells. Exp.Hematol. 1998; 26: 53-62.
92. Trumble TE, Parvin D. Cell viability and migration in nerve isografts and allografts. J Reconstr Microsurg. 1994;10:27-34.
93. Van Amersfoort ES, Van Strijp JA. Evaluation of a flow cytometric fluorescence quenching assay of phagocytosis of sensitized sheep erythrocytes by polymorphonuclear leukocytes. Cytometry. 1994;17:294-301.
94. Veena P, Cornetta K, Davidson A, et al: Preferential sequestration in vitro of BCR/ABL negative hematopoietic progenitor cells among cytokine nonresponsive CML marrow CD34+ cells. Bone Marrow Transplant. 1997; 19: 1213-1221.
95. Verfaillie CM, Miller JS. A novel single-cell proliferation assay shows that long-term culture- initiating cell (LTC-IC) maintenance over time results from the extensive proliferation of a small fraction of LTC-IC. Blood. 1995;86:2137-45.
96. Verfaillie CM, Catanzaro P. Direct contact with stroma inhibits proliferation of human long-term culture initiating cells. Leukemia. 1996;10:498-504.
97. Wallace PK, Palmer LD, Perry-Lalley D, et al. Mechanisms of adoptive immunotherapy: improved methods for in vivo tracking of tumor-infiltrating lymphocytes and lymphokine-activated killer cells. Cancer Res. 1993;53:2358-67.

98. Wallace PK, Keler T, Coleman K, et al: Humanized mAb H22 binds the human high affinity Fc receptor for IgG (FcgammaRI), blocks phagocytosis, and modulates receptor expression. J.Leukoc.Biol. 1997; 62: 469-479.

99. Wallace PK, Keler T, Guyre PM, Fanger MW: Fc gamma RI blockade and modulation for immunotherapy. Cancer Immunol.Immunother. 1997; 45: 137-141.

100. Ward G, Miller L, Dvorak J. The origin of parasitophorous vacuole membrane lipids in malaria-infected erythrocytes. J. Cell Sci. 1993; 106:237-248.

101. Weidmann E, Brieger J, Jahn B, Hoelzer D, BergmannL, Mitrou PS: Lactate dehydrogenase-release assay: a reliable, nonradioactive technique for analysis of cytotoxic lymphocyte-mediated lytic activity against blasts form acute myelocytic leukemia. Ann Hematol 70: 153-158, 1995.

102. Williams SK, Kleinert LB, Rose D, McKenney S. Origin of endothelial cells that line expanded polytetrafluorethylene vascular grafts sodded with cells from microvascularized fat. J Vasc Surg. 1994;19:594-604.

103. Yamamura Y, Rodriquez N, Schwartz A et al. A new flow cytometric method for quantitative assessment of lymphocyte mitogenic potential. Cell. Molec. Biol. 1995;41(Suppl. 1):S121-S132.

104. Yamamura Y, Rodrigues N, Schwartz A et al. Anti-CD4 cytotoxic T lymphocyte (CTL) activity in HIV positive patients: flow cytometric analysis. Cell. And Mol. Biol. 1995; 41 (Suppl. 1):S133-S144.

105. Young AJ, Hay JB. Rapid turnover of the recirculating lymphocyte pool in vivo. Int Immunol. 1995;7:1607-15.

106. Young JC, Varma A, DiGiusto D, Backer MP. Retention of quiescent hematopoietic cells with high proliferative potential during ex vivo stem cell culture. Blood. 1996;87:545-556.

107. Yuan Y, Fleming B. A method for labeling rat neutrophils with a fluorescent dye for intravital microvascular studies. Microvascular Res.1990; 40:218-229.

108. Zeine R, Owens T. Direct demonstration of the infiltration of murine central nervous system by Pgp-1/CD44high CD45RB(low) CD4+ T cells that induce experimental allergic encephalomyelitis. J Neuroimmunol. 1992;40:57-69.

References added in proof:

A. Givan A, Fisher JL, Waugh M, Ernstoff M, Wallace PK. A flow cytometric method to estimate precursor frequencies of cells proliferating in response to specific antigens. J. Immunol. Meth.; in press.

B. Wells AD et al. Following the fate of individual T cells throughout activation and clonal expansion. J. Clin. Invest. 1997; 100: 3173:3183.

C. Darling D et al. *In vitro* immune modulation by antibodies coupled to tumor cells. Gene Therapy 1997; 4: 1350-1360.

D. Allsopp CE, Nicholls SJ, Langhorne J: A flow cytometric method to assess antigen-specific proliferative responses of different subpopulations of fresh and cryopreserved human peripheral blood mononuclear cells. J Immunol Methods 1998; 214: 175-186.

Section 5

Flow Cytometry of Nucleic Acids

ROCHELLE A. DIAMOND

One of the oldest cornerstones of flow cytometry is the measurement of nucleic acids. Next to phenotyping, cell cycle distribution and ploidy analysis have been the most extensively used flow cytometric methods in the clinical laboratory. Basic researchers have employed these nucleic acid measurements for years to look at the population dynamics of cell activation, various effects of drugs on the cell cycle, and differentiation/quiescence of developing cells. The ability to sort chromosomes has revolutionized genetic mapping.

Contributors Derek Davies and Ingrid Schmid cover some of the most commonly used fluorochromes and methods for analysis of DNA, cell cycle, and apoptosis. Howard Shapiro provides a method for RNA content. Mottley et al. look at how DNA is used and measured by botanists in the plant world. Similarly, Marie et al. utilize DNA to analyze marine picoplankton for marine biology studies. Wendy Schober and Jeff Bachant give us a consensus protocol for measuring yeast DNA, a staple in cell biology laboratories.

RNA Content Determination Using Pyronin Y

HOWARD M. SHAPIRO

Introduction

During the late 1970's, Darzynkiewicz and his colleagues described subcompartments of the cell cycle defined on the basis of two-parameter measurements of cellular DNA and RNA content using acridine orange (AO) under carefully controlled staining conditions.[1,2] In particular, they demonstrated that it was possible to distinguish between proliferating (G1) and quiescent (G0) cells with diploid DNA content, because the G1 cells have higher RNA content. When the technique was applied to peripheral blood lymphocytes stimulated with mitogens or alloantigens, it was noted that RNA content increased by about 20 hours after stimulation, well in advance of the onset of DNA synthesis. With the description of lymphocyte activation antigens in the early 1980's, there arose some motivation for making correlated measurements of DNA and RNA content and antigen display in activated lymphocytes. This was essentially impossible using AO, first, because the fluorescence of AO extended from the green to the far red spectral region overlapping the spectrum of all available antibody labels, and, second, because the acid denaturation needed to insure stoichiometric staining of DNA and RNA by AO would likely denature any surface antigenic determinants of interest. In a search for alternative dyes or dye combinations which might be usable for DNA and RNA staining and compatible with at least one antibody label, I examined the combination of Hoechst 33342 and pyronin Y.[3]

Hoechst 33342 had been shown by Arndt-Jovin and Jovin[4] to produce stoichiometric staining of DNA in live cells. Pyronin Y, a close molecular homolog of AO, had long been used as an absorption stain for RNA in combination with methyl green, which stains DNA and binds preferentially to A-T pairs, as does

Hoechst 33342. The Hoechst dye requires UV excitation and emits around 450 nm, while pyronin Y has an absorption maximum near 550 nm and an emission maximum around 575 nm; use of the combination requires a dual-beam flow cytometer with UV and green or blue-green excitation. In practice, pyronin Y proved to be weakly but adequately excitable at 488 nm, allowing it to be used in combination with fluorescein-labeled antibodies. Relatively high dye concentrations are necessary because both the Hoechst dye and pyronin Y may be pumped out of cells; this works relatively well with human and rat lymphoid cells, but not with mouse cells, which apparently have more active glycoprotein efflux pumps. In general, caution is required when using the Hoechst/pyronin technique on living cells, because, while the Hoechst dye may not kill them, pyronin Y can do, by forming complexes with RNA,[5] and possibly by poisoning mitochondria, in which it accumulates in response to the transmembrane potential gradient.

When cells are stained with fluorescent antibodies and subsequently fixed, Hoechst/pyronin Y staining is considerably easier. At present, I use pyronin Y from Polysciences (Warrington, PA); Sigma (St. Louis, MO) also offers the dye (product number P9172).

Unlike AO, which forms luminescent complexes with single-stranded RNA, pyronin Y is reported to form fluorescent complexes only with double-stranded RNA, such as ribosomal RNA.[6-8] However, the ribosomal RNA content of cells generally parallels their total RNA content, at least in the context of representing proliferative vs. quiescent states, and Darzynkiewicz et al have therefore given pyronin Y a clean bill of health as an RNA stain in fixed cells.

Pyronin Y can be used with fluorescein- and/or PE/Cy5 tandem- or PerCP- labeled antibodies, but substantial compensation is likely to be needed; emission filters used for phycoerythrin are suitable for measurement of pyronin Y. Also, while the Hoechst/pyronin technique has the disadvantage of requiring dual wavelength (UV and 488 nm) excitation, RNA content measurement alone in cells stained with both dyes may provide useful information and can be done in an instrument with a single 488 nm excitation beam.

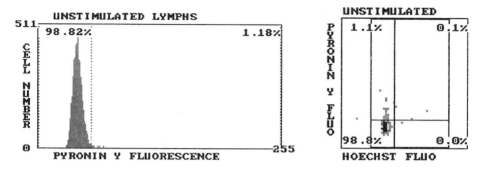

Fig. 1. T cells stained with Hoechst and Pyronin Y Fluo prior to stimulation

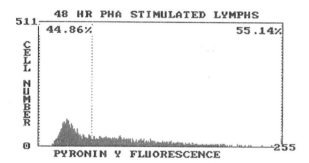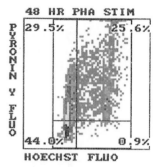

Fig. 2. T cells 48 hours after PHA stimulation

References

1. Traganos F, Darzynkiewicz Z, Sharpless T, Melamed MR. Simultaneous staining of ribonucleic and deoxyribonucleic acids in unfixed cells using acridine orange in a flow cytofluorometric system. J Histochem Cytochem 1977; 25:46.

2. Darzynkiewicz Z, Traganos F, Melamed MR. New cell cycle compartments identified by multiparameter flow cytometry. Cytometry 1980; 1:98.

3. Shapiro HM. Flow cytometric estimation of DNA and RNA content in intact cells stained with Hoechst 33342 and pyronin Y. Cytometry 1981; 2:143.

4. Arndt-Jovin DJ, Jovin TM. Analysis and sorting of living cells according to deoxyribonucleic acid content. J Histochem Cytochem 1977; 25:585.

5. Darzynkiewicz Z, Kapuscinski J, Carter SP, et al. Cytostatic and cytotoxic properties of pyronin Y: relation to mitochondrial localization of the dye and its interaction with RNA. Cancer Res 1986; 46:5760.

6. Darzynkiewicz Z, Kapuscinski J, Traganos F, Crissman HA. Applications of pyronin Y in cytochemistry of nucleic acids. Cytometry 1987; 8:138.

7. Kapuscinski J, Darzynkiewicz Z. Interactions of pyronin Y(G) with nucleic acids. Cytometry 1987; 8:129.

8. Traganos F, Crissman HA, Darzynkiewicz Z. Staining with pyronin Y detects changes in conformation of RNA during mitosis and hyperthermia of CHO cells. Exp Cell Res 1988; 179:535-44.

DNA Content Determination

DEREK DAVIES

Subprotocol 1
Propidium Iodide for DNA Content

The study of the cell cycle by flow cytometry is an extremely widespread application. By staining (fixed) cells with a stoichiometrically-binding DNA fluorochrome it is possible to assess the percentages of cells in G_1, S or G_2/M.

It is important to distinguish doublets of G_1 cells from true G_2/M cells, so some form of pulse shape analysis should be used. As a cell passes through the light source it creates a "pulse" of electrical energy. Generally we measure the height of this pulse which is equivalent to the maximum brightness of the cell. However we can also measure the width of the pulse i.e. the time that it takes to pass through the beam. Knowing these two parameters enables the area of the pulse also to be measured. By displaying pulse area versus pulse width or pulse height versus pulse width (depending on the cytometer), it is possible to separate single cells from clumps. The most obvious example is two G_1 cells that are stuck together. They will show a G_2 amount of DNA (the height of the pulse will be twice G_1). However, they will show an increased pulse width compared to a true G_2 cell.

There are several commercial programs available that will perform mathematical deconvolution of the DNA histogram eg Modfit (Verity) and MultiCycle (Phoenix Flow Systems) as well as the proprietary software available with your flow cytometer. In many non-clinical situations, the most valuable information can simply be reporting the percentage of cells in the proliferative phase (S and G_2/M) of the cycle.

▪▪ Materials

Equipment

Standard flow cytometer

Solutions

- 70 % ethanol is made up of 7 parts absolute alcohol and 3 parts distilled water.
- ribonuclease (100µg/ml DNase free, Sigma)
- propidium iodide (50µg/ml in PBS)

▪▪ Procedure

Harvest Cells   1. Harvest cells. Spin at 1200 rpm for 5 minutes.

Fix cells   2. Discard supernatant; add 1 ml 'fridge-cold 70% ethanol to 1×10^6 cells while vortexing the cell pellet gently.

3. Fix for at least 30 minutes at 4°C.

4. Spin at 2000 rpm for 5 minutes.

Note: 70% ethanol is made up of 7 parts absolute alcohol and 3 parts distilled water.
Once in 70% ethanol, samples may be kept for up to two weeks.

Remove fixative   5. Discard supernatant and resuspend cells in 1 ml PBS. Spin 2000 rpm for 5 minutes.

6. Repeat step 5 twice more.

Remove RNA   7. Add 100µl ribonuclease (100µg/ml, DNase free, Sigma). Leave at room temperature for 5 minutes.

Note: Ribonuclease removes RNA. This is necessary as propidium iodide binds to double stranded nucleic acids by intercalating in the double helix.

Stain cells with Propidium iodide   8. Add 400µl propidium iodide (50µg/ml in PBS).

Note: PI is a dye-binding dye that is potentially mutagenic so wear gloves.

Fig. 1

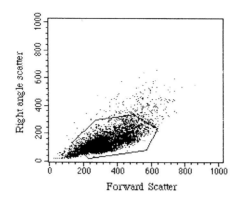

9. Acquire data on a flow cytometer with a 488nm light source, the cytometer should be triggered on the propidium iodide signal. Samples should firstly be gated by forward scatter against right angle light scatter (Fig. 1).

Acquire data by flow cytometer

10. A secondary gate should be placed around the single cell population on a pulse area versus pulse width dot plot (Fig. 2).

Gate

Note: To exclude falsely increased G2-M data, it is important to gate out the doublet cells which fall at the same intensity of PI as the G2-M cells.

Note: It is often better to run samples ungated and then gate in analysis mode in order to include all populations in data.

11. Propidium iodide fluorescence from cells that fall in the two gates should be acquired above 620nm (Fig. 3).

Analyze by DNA software

Fig. 2

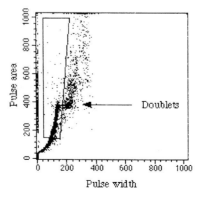

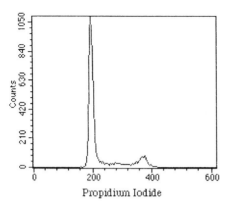

Fig. 3

Note: There are several commercial programs available that will perform mathematical deconvolution of the DNA histogram eg Modfit (Verity) and MultiCycle (Phoenix Flow Systems) as well as the proprietary software available with your flow cytometer. In many non-clinical situations, the most valuable information can simply be reporting the percentage of cells in the proliferative phase (S and G_2/M) of the cycle. Alternatively "markers" can be placed around the G_1, S and G_2/M peaks.

Subprotocol 2
Vital DNA staining with Hoechst 33342

There are occasions where it is important to know the cell cycle status of unfixed cells, for example if they are to be re-cultured or enzyme activity is to be assessed. The DNA-binding dye Hoechst 33342 is unique in that it easily enters live cells and it is therefore possible to use it for this purpose.

Hoechst 33342 enters viable cells in a time dependent and concentration dependent manner. The dye is cumulatively cytotoxic so the optimal period and optimal dye concentration for cell staining should be determined by a pilot experiment. For most mammalian cells between 5 and 15µg/ml for 20-60 minutes should be optimal. The dye concentration should not exceed 20µg/ml if the cells are to be re-cultured. If the cells are to be used, for example, to extract protein or RNA, the profile may be improved at the expense of cell viability.

Materials

Equipment

Flow cytometer with dual beam excitation. The primary laser is at 488nm to generate light scatter signals and excite propidium iodide, and the secondary at UV (351-363nm) to excite Hoechst 33342.

Solutions

Hoechst 33342 is kept as a 1mg/ml stock solution in distilled water.

Procedure

1. Incubate growing cells with 10µg/ml Hoechst 33342 for 30 minutes at 37°C.

   Stain DNA of live cells

   Note: Hoechst 33342 is kept as a 1mg/ml stock solution in distilled water.

2. Harvest cells in the appropriate manner, and wash once in PBS.

   Remove excess dye

3. Acquire data on a flow cytometer with dual beam excitation. The primary laser is at 488nm to generate light scatter signals and excite propidium iodide, and the secondary at UV (351-363nm) to excite Hoechst 33342. Red fluorescence from propidium iodide is measured above 620nm and blue fluorescence from Hoechst between 390 and 480nm. [Signals from both fluorochromes are measured in linear mode.] Hoechst fluorescence is linearly amplified, propidium iodide fluorescence logarithmically.

   Acquire data on a flow cytometer

   Note: Non-viable cells may also be excluded by the addition of propidium iodide (50µl of a 50µg/ml stock in PBS) Fig. 4.

   Viability Gate

4. Samples are also gated on a dot plot of forward scatter against right angle light scatter, (Fig. 5) and a dot plot of pulse width against pulse area of the Hoechst signal. (Fig. 6)

   Gate and Analyze

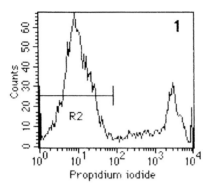

Fig. 4. Propidium Iodide negative cells indicates viability

Note: Pulse processing should be used to gate out cell clumps and doublets of G1 cells (see Propidium iodide DNA protocol).

Note: It is often better to run samples ungated and then gate in analysis mode in order to include all populations in data. The CV (coefficient of variation) of the G1 peak should be comparable to

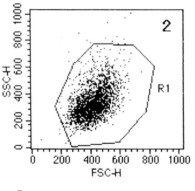

Fig. 5. Hoechst Staining of gated cells

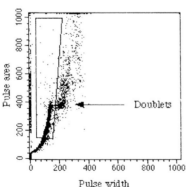

Fig. 6. Doublet Discriminations Plot

Fig. 7 Scatter Plot

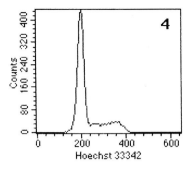

that obtained after ethanol fixation and propidium iodide staining (below 5% for cultured cells). (Fig. 7)

Subprotocol 3
Bromodeoxyuridine staining for cell cycle kinetics

The development of the BrdU technique for the detection of
synthesising cells revolutionised the study of cell cycle kinetics.
BrdU is an analog of thymidine and is incorporated into cells that
are actively synthesising new DNA. By altering the length of time
for which cells are exposed to BrdU different questions can be
addressed eg a short labelling time will allow rates of synthesis to
be compared or a cohort of cells can be labelled and their progression through the cell cycle monitored. It is important to distinguish doublets of G_1 cells from true G_2/M cells, so some form
of pulse shape analysis should be used. Analysis of the single cells
should show three populations - two populations of cells that are
BrdU-FITC negative (G_1 and G_2/M), and cells that are BrdU positive (S phase). The BrdU positive cells are those that have been
synthesising DNA during the time that BrdU has been present.
These data are usually reported as percentages of cells in the
three compartments - G_1, S and G_2/M. It is only correct to refer
to BrdU-positive cells as S phase cells if the incubation time with
BrdU has been short. Once the BrdU is present with the cells for a
longer period of time, some cells will have passed into G_2 or even
may have divided and be back in G_1, although they will still be
BrdU positive.

▨ ▨ Materials

Equipment

Standard flow cytometer

Solutions

- 70% ethanol.
- 2N Hydrochloric acid
- PBS-T is PBS plus 0.1% bovine serum albumin plus 0.2% Tween 20.

▨ ▨ Procedure

Label synthe-sising cells with bromo-deoxyuridine (BrdU)

1. Treat cells with 10μM bromodeoxyuridine for an appropriate length of time (30 minutes - 48 hours).

Note: BrdU is kept as 0.5ml aliquots of a 10mM stock (in RPMI or other suitable culture medium) at -20°C. Once thawed, the BrdU should be used immediately and any unused discarded.
BrdU is light-sensitive so incubation should be performed in the dark. Subseqently cells should not be exposed to strong light sources particularly UV light.

Harvest cells

2. Harvest cells. Spin at 1200 rpm for 5 minutes.

Fix cells

3. Discard supernatant; add 1 ml 'fridge-cold 70% ethanol to 1 x 10^6 cells while vortexing the cell pellet gently.

4. Fix for at least 30 minutes at 4°C.

5. Spin at 2000 rpm for 5 minutes.

Note: 70% ethanol is made up of 7 parts absolute alcohol and 3 parts distilled water.
Once in 70% ethanol, samples may be kept for up to two weeks.

Remove fixative

6. Discard supernatant and resuspend cells in 1 ml PBS. Spin 2000 rpm for 5 minutes.

7. Repeat step 5 twice more.

8. Add approx 1 ml of 2N hydrochloric acid for each 10^6 cells and leave at room temperature for 30 minutes with frequent agitation.

Expose incorporated BrdU

Note: 2N Hydrochloric acid is made up 70 parts distilled water to 30 parts concentrated acid. The acid serves to open the DNA helix to expose the incorporated BrdU.

9. Spin at 2000 rpm for 5 minutes.

Note: Spin the cells directly, without adding PBS.

10. Discard supernatant and resuspend cells in 1 ml PBS. Spin 2000 rpm for 5 minutes. Repeat.

11. Discard supernatant and resuspend cells in 1 ml PBS-T. Spin 2000 rpm for 5 minutes.

Note: PBS-T is PBS plus 0.1% w/v bovine serum albumin plus 0.2% v/v Tween 20. This serves to further permeabilise cells to facilitate entry of antibodies into the cell.

12. Discard supernatant and add 50µl anti-BrdU monoclonal antibody diluted in PBS-T. Leave for 20 minutes in the dark at room temperature.

Antibody staining of BrdU

Note: The appropriate dilution for the batch of antibody should be previously determined. Good staining is obtained using MAbs from Sera Lab (Clone ICR1) or Becton Dickinson (CloneB44.) As a positive control, growing HL60 or HeLa S3 cells treated with BrdU for 30 minutes can be used.

13. Add 1 ml PBS-T, spin at 2000 rpm for 5 minutes.

14. Discard supernatant and repeat step 13.

15. Discard supernatant and add 50µl FITC-conjugated rabbit anti-mouse immunoglobulins (polyclonal F(ab')$_2$ fragments) diluted 1:20 in PBS-T. Leave for 20 minutes in the dark at room temperature.

Fluorescent tagging of anti-BrdU antibody

16. Add 1 ml PBS, spin at 2000 rpm for 5 minutes.

17. Discard supernatant and repeat step 13.

18. Add 100µl ribonuclease (100µg/ml, DNase-free, Sigma). Leave at room temperature for 5 minutes.

Remove RNA

Note: Ribonuclease removes RNA. This is necessary as propidium iodide binds to double stranded nucleic acids by intercalating in the double helix.

Stain double-stranded DNA

19. Add 400μl propidium iodide (50μg/ml in PBS).

Note: Propidium iodide is a dye-binding dye that is potentially mutagenic so wear gloves.

Analysis

20. Analyse on a flow cytometer with a 488nm light source. The cytometer is triggered on the propidium iodide signal, and the samples are gated firstly on a dot plot of forward scatter against right angle light scatter, and secondly on pulse width against pulse area of the propidium iodide signal. (Fig. 8) Red fluorescence (Propidium iodide) is measured above 620nm. and green fluorescence (FITC) between 515 and 545nm. (Fig. 9)

Note: It is important to distinguish doublets of G_1 cells from true G_2/M cells, so some form of pulse shape analysis should be used.

Note: Analysis of the single cells should show three populations - two populations of cells that are BrdU-FITC negative (G_1 and G_2/M), and cells that are BrdU positive (S phase). The BrdU positive cells are those that have been synthesising DNA during the time that BrdU has been present. These data are usually reported as percentages of cells in the three compartments - G_1, S and G_2/M.

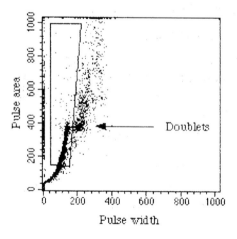

Fig. 8

Fig. 9

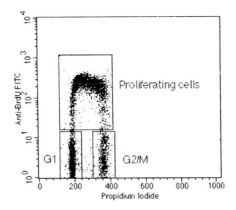

Note: It is only correct to refer to BrdU-positive cells as S phase cells if the incubation time with BrdU has been short. Once the BrdU is present with the cells for a longer period of time, some cells will have passed into G_2 or even may have divided and be back in G_1, although they will still be BrdU positive.

References

1. Arndt-Jovin DJ, Jovin TM. Analysis and sorting of living cells according to deoxyribonucleic acid content. J Histochem Cytochem 1977; 25:585-589.
2. Crissman HA, Steinkamp JA. A new method for rapid and sensitive detection of bromodeoxyuridine in DNA-replicating cells. Exp Cell Res 1987; 173: 256-261
3. Dean PN et al. Cell cycle analysis using a monoclonal antibody to BrdUrd. Cell Tissue Kinet 1984; 17: 427-436.
4. Gratzner H. Monoclonal antibody to 5-bromo and 5-iododeoxyuridine: a new reagent for detection of DNA replication. Science 1982; 218: 47-475.
5. Hamilton VT, et al. Flow microfluorometric analysis of cellular DNA: critical comparison of mithramycin and propidium iodide. J Histochem Cytochem 1980; 28:1125-1128.
6. Karawajewl, et al. Flow sorting of hybrid hybridomas using the DNA stain Hoechst 33342. J Immunol Methods 1990; 129:277-282.
7. Krishan A. Rapid flow cytofluorometric analysis of mammalian cell cycle by propidium iodide. J Cell Biol 1975; 66:188-195.
8. Lydon MJ, et al. Vital DNA staining and cell sorting by flow microfluorimetry. J Cell Physiol 1980; 102:175-181.

Dead Cell Discrimination

DEREK DAVIES AND CLARE HUGHES

Subprotocol 1
Propidium Iodide for dead cell discrimination

During phenotypic analysis, it is often desirable to exclude dead cells from the analysis as they tend to stain non- specifically and give misleading results. As cells lose their membrane integrity during cell death they become accessible to DNA binding dyes such as propidium iodide, 7AAD and TO-PRO-3. **Propidium iodide (PI)** binds to double stranded nucleic acids by intercalating in the double helix. The accessiblity is due to deterioration of the membrane during cell death. The dead cells are hence distinguished from live cells by bright staining with PI. Propidium iodide can be used in multicolour analysis in conjunction with fluorochromes such as FITC with a 488nm laser, with APC (red 635nm Laser) or AMCA (UV excitation). The Forward scatter signal is reduced during death in conjunction with more intense PI staining of the dead cells.

Materials

Equipment

Standard flow cytometer

Solutions

stock PI solution 50µg/ml, dissolved in PBS

▓▓ Procedure

1. Harvest cells. Spin at 1200 rpm for 5 minutes.   **Harvest Cells**

2. Discard supernatant; add 1 ml of PBS. Spin at 1200 rpm for   **Staining**
   5 minutes.

3. Add 50μl of propidium iodide (stock solution 50μg/ml, dissolved in PBS). Mix.

Note: **Propidium iodide** is a DNA-binding dye that is potentially mutagenic so wear gloves.

4. Leave for 2 minutes.

Note: The staining is almost immediate.

5. Analyse on a flow cytometer with a 488nm light source mea-   **Analysis**
   suring red fluorescence from propidium iodide above 620nm.

Note: This procedure can only be performed on unfixed cells.

Note: Propidium iodide can be used in multicolour analysis in conjunction with fluorochromes such as FITC with a 488nm laser, with APC (red 635nm Laser) or AMCA (UV excitation).

Note: The Forward scatter signal is reduced during death in conjunction with more intense PI staining of the dead cells.

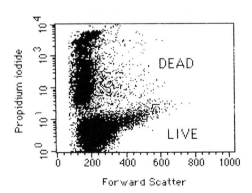

Subprotocol 2
Dead cell discrimination using To-Pro-3

TO-PRO-3 has similiar properties to propidium iodide in that it binds to double stranded nucleic acids by intercalating in the double helix. The accessiblity of the dye to the double helix of dead cells is due to deterioration of the membrane during cell death. The dead cells are hence distinguished from live cells by staining very brightly with TO-PRO-3. The difference between PI and TO-PRO-3 is their excitation and emission spectra.

TO-PRO-3, like PI, is a DNA-binding dye that is potentially mutagenic so wear gloves. It is available from Molecular Probes, as a powder, best storage at -70°C. TO-PRO-3 has an excitation maximum of 642nm and an emission maximum of 661nm. It can be used to exclude dead cells in multicolour analysis along with FITC, PE and tandem dyes, on a 2 laser benchtop machine (488nm argon ion laser / 635nm diode laser), which includes inter-beam compensation for tandem dye overlap with TO-PRO-3. The filter set up would be FITC: 530/30, PE: 585/42, tandem dye: 650 LP and TO-PRO-3: 670 LP. TO-PRO-3 can also be excited by a high powered 488nm laser, at 50mW or above. As the cells die the Forward Scatter signal is reduced and the dead cells stain very brightly with TO-PRO-3.

▪▪ Materials

Equipment

Analyse on a flow cytometer with a 633nm light source, measuring red fluorescence from TO-PRO-3 above 640nm.

Solutions

TO-PRO-3, final concentration 1µM, from a stock solution of 100µM, dissolved in dimethyl sulphoxide (DMSO).

▨ ▨ Procedure

1. Harvest cells. Spin at 1200 rpm for 5 minutes.   **Harvest Cells**

2. Discard supernatant; add 1 ml of PBS. Spin at 1200 rpm for   **Staining**
5 minutes.

3. Add TO-PRO-3, final concentration 1μM, from a stock solution of 100μM, dissolved in dimethyl sulphoxide.

Note: TO-PRO-3 is a DNA-binding dye that is potentially mutagenic so wear gloves.

Note: It is available from Molecular Probes, as a powder, best storage at -70°C.

Note: It is important to store stock solutions in polypropylene tubes.

4. Leave for 10 minutes.

5. Analyse on a flow cytometer with a 633nm light source, mea-   **Analysis**
suring red fluorescence from TO-PRO-3 above 640nm. Triggering on forward scatter.

Note: TO-PRO-3 can also be excited by a high powered 488nm laser, at 50mW or above.

Note: As the cells die the Forward Scatter signal is reduced and the dead cells stain very brightly with TO-PRO-3.

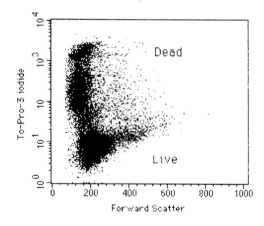

References

1. van Hooijdonk CAEM, Glade CP, van Erp PEJ. TO-PRO-3 iodide: A novel He-Ne Laser-excitable DNA stain as an alternative for propidium iodide in multiparameter flow cytometry. Cytometry 1994; 17: 185.
2. O' Brien MC, Bolton WE. Comparison of cell viability probes compatible with fixation and permeablisation for combined surface and intracellular staining in flow cytometry. Cytometry 1995; 19: 243.

Detecting Apoptosis

DEREK DAVIES AND INGRID SCHMID

Subprotocol 1
Hoechst / Propidium Iodide Staining for Apoptotic Cells

During the apoptotic process, the cell membrane changes its symmetry and its permeability to certain dyes. By late stage apoptosis, the cell membrane is permeable to most of the dyes used to assess viability (eg propidium iodide, 7 aminoacti-nomycin D). However in the earlier stages of apoptosis, the cell membrane exhibits a differential permeability. It is possible to exploit this differential permeability to two dyes, propidium iodide and Hoechst 33342, to distinguish mid stage from late stage apoptosis.

Differential staining of live and membrane-damaged cells

This step separates those cells that cannot pump **Hoechst** out efficiently (dead and apoptotic cells) from live cells which still have effective membrane pumps. The influx of Hoechst into cells is time-dependent. It is important that the time taken to run the sample is not too long after the addition of dye as eventually live cells will also take it up. The optimal time will depend on the cell type. For murine thymocytes and a human T cell line (J6) this is 5 minutes.

Differential staining of dead and apoptotic cells

Propidium iodide enters cells whose membranes have lost functionality. Dead cells will allow propidium iodide in, but apoptotic cells will still be able to exclude it. By displaying red (Propidium iodide) against blue (Hoechst) fluorescence, three populations should be seen. Double negatives (live cells), double positives (dead cells) and single blue positives (apoptotic cells).Quadrant analysis allows the percentage of cells found in the live, dead and apoptotic popualtions to be quantified.

▪ ▪ Materials

Equipment

Flow cytometry with dual beam excitation. The primary laser set at 488nm to excite propidium iodide, and the secondary set to UV (351-363nm) to excite Hoechst.

Solutions

A stock of 1mg/ml Hoechst in distilled water is used. From this a solution of 10μg/ml in PBS is made up immediately before the assay. 50μl propidium iodide (50μg/ml in PBS).

▪ ▪ Procedure

Prepare Cells

1. Take 1×10^6 cells and spin at 1200 rpm for 5 minutes.

2. Discard supernatant; resuspend pellet in 1 ml of culture medium.

Note: The culture medium can contain phenol Red but should not have more than 5% protein.

Differential staining of live and membrane-damaged cells

3. Add 100μl Hoechst 33342 (10μg/ml). Leave for 5 minutes at room temperature.

Note: The influx of Hoechst into cells is time-dependent. It is important that the time taken to run the sample is not too long after the addition of dye as eventually live cells will also take it up. The optimal time will depend on the cell type. For murine thymocytes and a human T cell line (J6) this is 5 minutes.

Note: A stock of 1mg/ml Hoechst in distilled water is used. From this a solution of 10µg/ml in PBS is made up immediately before the assay.

4. Add 50µl propidium iodide (50µg/ml in PBS).

Note: Propidium iodide enters cells whose membranes have lost functionality. Dead cells will allow propidium iodide in, but apoptotic cells will still be able to exclude it.

Note: Propidium iodide and Hoechst 33342 are DNA-binding dyes which are potentially mutagenic so wear gloves

5. Acquire data by flow cytometry with dual beam excitation. The primary laser set at 488nm to excite propidium iodide, and the secondary set to UV (351-363nm) to excite Hoechst. Collect fluorescence from propidium above 620nm and from Hoechst above 390nm. Samples should be run ungated, thresholding on Forward scatter so as not to exclude small cells.

6. Statistics can be obtained using a quadrant statistic analysis of a dot plot displaying Propidium Iodide and Hoechst signals Three populations should be seen. Double negatives (live cells), double positives (dead cells) and single blue positives (apoptotic cells). Quadrant analysis allows the percentage of cells found in the live, dead and apoptotic populations to be quantified. (Fig. 1)

Differential staining of dead and apoptotic cells

Acquire data by flow cyto-metry

Analyze

Fig. 1

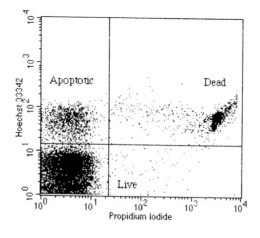

Subprotocol 2
7-Aminoactinomycin D (7-AAD) staining for detection of apoptotic cells

7-AAD is a non-vital DNA dye that is excluded by the cell membrane of live cells. Due to alterations in membrane integrity the dye can cross into early apoptotic cells and (late apoptotic or necrotic) dead cells resulting in dim DNA staining for early apoptotic cells and bright staining for dead cells. 7-AAD is suitable for detection of apoptotic cells in cell preparations which have been cell surface antigen stained with antibodies conjugated to fluorochromes with emission spectra that can be separated from the 7-AAD emission, e.g., FITC and PE.

▓▓ Materials

Equipment

Flow cytometer with following filter configuration (Fig. 2).

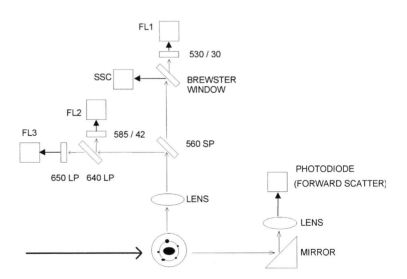

Fig. 2.

Solutions

- 7-AAD stock solution (1mg/ml) is prepared by first dissolving 1 mg of 7-AAD powder in 50 μl of absolute methanol, then adding 950 μl of 1 x PBS. 7-AAD is light sensitive and should be stored at 4°C in the dark.
- AD stock solution (1mg/ml) is prepared by dissolving 1 mg of AD powder first in 50 μl of ice cold absolute ethanol, then adding 950 μl of 1 x PBS. Sonicate the solution for 10 minutes in the cold; keep it overnight at 4°C before use.

▨▨ Procedure

1. Incubate PBS-washed cells with 20 μg/ml of 7-AAD for 20 minutes in the dark.

Label apoptotic cells with 7-aminoactinomycin D (7-AAD)

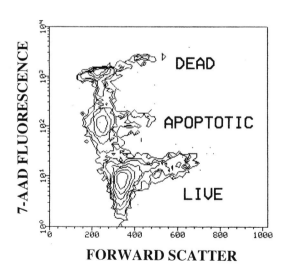

FORWARD SCATTER

Fig. 3. 7-Aminoactinomycin D staining for detection of apoptotic cells. Fresh, ex-vivo human thymocytes were cultured for 2h at 37°C in the presence of 10^{-5} M dexamethasone, stained with 7-AAD as described in the procedure, then analyzed. Live cells: 60%; early apoptotic cells: 26%; late apoptotic (dead) cells: 14%

Analyze cells 2. Analyze samples in the 7-AAD staining solution on a flow cytometer equipped with a 488 nm argon laser measuring red fluorescence above 650 nm.

Fix cells 3. Centrifuge cells at 250 x g for 5 minutes. Discard supernatant
(optional) completely; add 1 ml of 1% formaldehyde solution containing 20 μg/ml of actinomycin D (AD).

Note: Alternatively to analysis of unfixed cells, 7-AAD-stained samples can be formaldehyde fixed for biohazard considerations and to preserve them for later analysis.

Note: Samples can be stored for up to three days at 4°C in the dark and are analyzed on the flow cytometer in the formaldehyde/AD solution.

Subprotocol 3
Annexin V-FITC Staining for apoptotic cells

In mid to late apoptosis there are changes in the symmetry of the plasma membrane which involve externalisation of phosphatidylserine (PS) residues. Normally these are on the internal surface of the plasma membrane and as such are undetectable in cells whose membrane is undamaged. It is possible to detect the externalised PS residues by their binding to the protein Annexin V.

▓▓ Materials

Solutions

- **Ca-containing medium**
- Hepes Buffer
- 10mM Hepes/NaOH pH 7.4
- 140mM CaCl
- 2.5mM CaCl$_2$

▨ ▨ Procedure

1. Take 1×10^6 cells and spin at 1200 rpm for 5 minutes.

Prepare Cells

2. Discard supernatant; resuspend pellet in 1 ml of Hepes buffer. Spin at 1200 rpm for 5 minutes.

Put cells into Ca-containing medium

Note: Hepes Buffer:
10mM Hepes/NaOH pH 7.4
140mM CaCl
2.5mM CaCl^2

3. Resuspend in 1 ml Hepes buffer. Add 2.5µg Annexin-V FITC (Bender MedSystems, Vienna, Austria). Incubate in the dark at room temperature for 30 minutes.

Binding of Annexin to PS residues

4. Prior to running on the cytometer add 50µl propidium iodide (50µg/ml in PBS).

Stain with Propidium iodide

Note: PI is a DNA-binding dye which is potentially mutagenic so wear gloves.

5. Acquire data on a flow cytometer with a 488nm light source. Samples should be run ungated and the cytometer triggered on Forward scatter. Green fluorescence from Annexin-V FITC is collected at 515-545nm and red fluorescence from propidium iodide above 620nm. (fig. 4)

Acquire Data

Note: Generally data are displayed in a dot plot, enabling enumeration of live, dead and apoptotic populations.

Analyze

Fig. 4

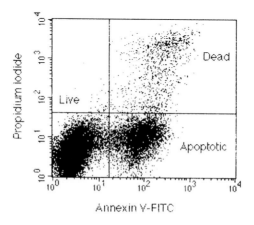

References

1. ORMEROD MG, et al. Increased membrane permeability of apoptotic thymocytes: a flow cytometric study. Cytometry 1993; 14:595-602.
2. ORMEROD MG, et al. Apoptosis in interleukin-3-dependent haemo-poietic cells. Quantification by two flow cytometric methods. J Immunol Methods 1992; 153:57-65.
3. Schmid I, Uittenbogaart CH, Keld B, et al. A rapid method for measuring apoptosis and dual-color mmunofluorescence by single laser flow cytometry. J Immunol. Meth. 1994; 170: 145-157.
4. Philpott NJ, Turner AJC, Scopes J et al. The use of 7-amino actinomycin D in identifying apoptosis: simplicity of use and broad spectrum of application compared with other techniques. Blood 1996; 87: 2244-2251.
5. VERMES I, et al. A novel assay for apoptosis. Flow cytometric detection of phosphatidyserine expression on early apoptotic cells using fluorescein-labelled Annexin V. J Immunological Methods 1995; 184:29-51.

Chromosome Preparation – Polyamine Method

DEREK DAVIES AND CLARE HUGHES

Introduction

Conventional karyotyping involves cytogenetic analysis of a metaphase spread; flow cytometry can also be used to produce a flow karoytype from a monodispersed suspension of chromosomes, stained with two DNA dyes with different base pair specification. Bivariate analysis on a dual laser cytometer will separate individual chromosomes according to their differences in DNA content (size) and their base pair ratio. The only chromosomes which remain grouped are chromosomes 9-12, their DNA content and base pair ratios being too similiar to be separated. Flow karyotypes have been produced for many types of human cells including fibroblasts, stimulated peripheral blood cells and lymphoblastoid cell lines. Other mammalian flow karyotypes have also been produced including rat, mouse and hamster.

Flow karyotyping, however, has limited clinical use due to the polymorphism that occurs generally across a normal population and hence the support of conventional cytogenetics is required. Aberrations from normal profiles can be detected if there are sufficient differences in DNA content of a chromosome e.g. trisomy 21, where there is a fifty percent increase in the chromosome 21 peak or significant changes in the base pair ratio due to translocations, for example the 22,11 translocation shown below is obvious from its derivative chromosomes. The 22,11 translocation in its unbalanced form has possible links with breast cancer. Other translocations deduced by flow include the 14,18 translocation found in ALL.

The real advantage of flow cytometry is that chromosomes can be sorted, and used for:
- The construction of cosmid and 'phage libraries. Large numbers are required, for example typically $2\text{-}5 \times 10^6$ chromo-

somes are required for a gene library. This has been an essential technique for the mapping of the human genome as part of the HUGO project.

- From small numbers of sorted chromosomes PCR can be performed either DOP-PCR or ALU-mediated PCR yielding amplified DNA fragments which can then be labelled with fluorochrome or biotin to produce chromosome painting probes. Chromosome paints can be used for:

- Whole chromosome painting

- Visualisation of translocations

- Identifications of human chromosomes in somatic cell hybrids

- Enumeration of chromosomes

Future directions of chromosome technology include FISH by flow cytometry. This involves preparing a metaphase suspension of chromosomes followed by hybridisation of a specific chromosome marker probe and detection of the chromosome profile and the labelled marker chromosome by flow cytometry. This will involve using a third fluorescent colour.

Flow cytometry provides an essential tool for the continued mapping of human chromosomes and is proving a valuable resource. The set up and operation of the flow cytometer for chromosome analysis is a time consuming and skilled operation where particular care and attention must be paid to machine and laser cleanliness and alignment. For these reasons it is best that a cytometer is dedicated to this type of analysis and sorting.

Materials

Equipment

Two multiline argon ion lasers can be used. The primary laser is tuned to UV wavelengths (351-363nm) and the secondary tuned to 457nm. 300mW of power is used for each laser.

The filter set up is FL1-390LP/480SP, FL3-520SP.

Solutions

- **Chromosome isolation buffer 1(CIB1)** - 20mM NaCl, 80mM KCl, 15mM Tris HCl, 0.5mM EGTA, 2mM EDTA, 0.15% W/V 2-mercaptoethanol, 0.2mM Spermine, 0.5mM Spermidine- pH 7.2 in autoclaved water.
- **CIB1 Solution** - 5 ml Digitonin (12mg/5 ml of distilled water dissolved by heating on a hotplate) 1 ml of CIB1 Isolation buffer 4 ml of distilled water.
- Spermine and spermidine, the polyamines, are to stablise the chromosomes. Digitonin is a detergent used to assist cell lysis.
- 50μg/ml PI and 0.1% Triton X-100 in PBS
- 0.05μg/ml colcemid
- chromosome suspension add 30μl **Hoechst 33258**, 40μl of 15mM $MgCl_2$ and 50μl **Chromomycin A3**
- 100μl sodium citrate (100mM) and 100μl sodium sulphite (250mM)

Preparation

- Prior to running on the cytometer the chromosome suspension could be put through a fine syringe needle (23-25G), to allow for monodispersion.
- The chromosome suspension can be stored at 4°C for several weeks with little deterioration of the flow karyotype
- Care should be taken when aspirating/vortexing the chromosome suspension, as chromosome damage could occur.

Procedure

1. Block semi-confluent healthy growing cells with 0.05μg/ml colcemid for 5-16 hrs, generally overnight is a good blocking time.

Block Cells

Note: Monolayer/suspension cell lines may be used. Blocking time depends on the growth rate of the cells.

2. The proportion of suspension cells in mitosis can be estimated by re-suspending the cell pellet in 50μg/ml PI and 0.1% Triton X-100 in PBS and measuring on a bench top flow cytometer, 40-60% of the cells should be in mitosis.

Harvest Cells

For monolayer cells the proportion in mitosis can be estimated using an inverted microscope.

Note: As mitotic cells are round, monolayer cells can usually be shaken off, if this is unsuccessful trypsin may be used.

3. Once the cells are in suspension, the cells are then centrifuged at 100g for 10 minutes in 50 ml tubes, discard the supernatant and then re-suspend the cells in fresh medium (appropriate for the cell type) before a further 10 minute centrifugation at 100g.

4. Discard the supernatant by inverting the tube. Remove any remaining moisture with a tissue.

Note: It is important to make sure the tube is free from moisture (as this will affect the salt concentration during the next stage).

5. Disaggregate the cell pellet by vortexing gently or by flicking the tube.

Swelling   6. Add 5 ml of hypotonic solution, mix gently and leave for 10-30 min at room temperature. At this stage, pool the contents.

Note: Lymphoblastoid cell lines require approximately 20 mins and fibroblastoid lines require 30 mins. Periodic monitoring under the light microscope is a good idea.

Note: Hypotonic solution: 75mM KCl

Disruption of the cell membrane   7. Centrifuge the tubes for 10 mins at 100g, carefully remove the supernatant with a pasteur pipette and agitate the tube gently to disaggregate the cells. Add 10 x the volume of the cell pellet in cold CIB1 solution, aspirate gently with a Pasteur pipette.

Note: Chromosome isolation buffer 1(CIB1):
20mM NaCl, 80mM KCl, 15mM Tris HCl, 0.5mM EGTA, 2mM EDTA, 0.15% W/V 2-mercaptoethanol, 0.2mM Spermine, 0.5mM Spermidine-pH 7.2 in autoclaved water.

Note: CIB1 Solution:
5 ml **Digitonin** (12mg/5 ml of distilled water dissolved by heating on a hotplate)
1ml of CIB1
4 ml of distilled water.

Spermine and spermidine, the polyamines, are to stabilise the chromosomes. **Digitonin** is a detergent used to assist cell lysis.

8. For 1 ml of chromosome suspension add 30μl **Hoechst 33258**, 40μl of 15mM $MgCl_2$ and 50μl **Chromomycin A3** mix and leave the sample at 4°C for at least 2hr, in the dark.

Chromosome staining

Note: Hoechst 33258 has an AT binding preference, **Chromomycin** has GC binding preference.

Note: Prior to running on the cytometer the chromosome suspension could be put through a fine syringe needle (23-25G), to allow for monodispersion. The chromosome suspension can be stored at 4°C for several weeks with little deterioration of the flow karyotype.

Note: Care should be taken when aspirating/ vortexing the chromosome suspension, as chromosome damage could occur.

9. 15 minutes prior to running on the cytometer add 100μl sodium citrate (100mM) and 100μl sodium sulphite (250mM) . Aggregates and intact nuclei form a pellet on centrifugation at 100g for 1 minute transfer the supernatant containing the chromosomes to a new tube.

Note: The addition of sodium citrate and sodium sulphite improve the profile.

10. Use dual beam excitation to excite the **Hoechst** and **chromomycin**. Collect **Hoechst** and **chromomycin** fluorescence in linear mode, triggering on the Hoechst fluorescence signal.

Note: Two multiline argon ion lasers can be used. The primary laser is tuned to UV wavelengths (351-363nm) and the secondary tuned to 457nm. 300mW of power is used for each laser.

Note: The filter set up is Primary Laser: 390LP/480SP, Secondary Laser: 520SP.

11. Fig. 1 is of a chromosome profile of a normal human lymphoblastoid cell line GM1416B. Below this is an abnormal cell line; GM6229, another lymphoblastoid line that has a 22,11 translocation (Fig. 2).

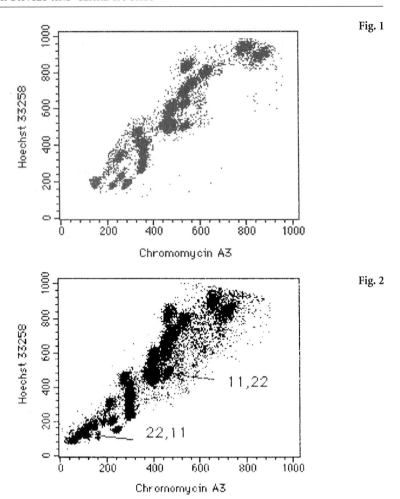

Fig. 1

Fig. 2

References

1. Green, DK. Analysing and sorting human chromosomes. Journal of Microscopy 1990; 159:237-245.
2. Young, BD. Chromosome analysis by flow cytometry: a review. Bas Appl Histochem 1984; 28: 9-19.

'D-Flowering' – The Flow Cytometry of Plant DNA

J. MOTTLEY, K. YOKOYA, AND A.V. ROBERTS

Introduction

The first flow cytometric study of DNA in higher plant cells was that described by Heller,[1] back in 1973. This study was carried out using broad bean (*Vicia faba*) and a Partec flow cytometer (Partec GmbH, Germany). Since that time, the use of flow cytometry on plants has blossomed, particularly for the estimation of DNA.

The potential applications of flow cytometry for determining DNA amounts in plants can be divided into those used in basic **research**, those used for quality control in **plant propagation** systems and those used in **plant improvement** programs.

Research

This information has uses for determining the relatedness of different genotypes,[2,3] and in genome mapping studies. Values for the 2C DNA amount in different species vary widely from 0.2 pg/2C for *Arabidopsis thaliana* to 127.4 pg/2C for *Fritillaria assyriaca*.[4] Up to the present time, Feulgen microdensitometry has been used primarily to obtain such values.

Determination of absolute nuclear DNA amounts

In both *Paspalum* and *Rosa* spp., the 2C DNA amount per unit set of chromosomes can vary significantly even within the same section, genera and/or species,[5,6], enabling the potential use of this property as a descriptor for different genotypes. In addition, absolute DNA amounts can be used for sexual identification in some dioecious species.[7]

Cell cycle analysis is one of the most important applications of flow cytometry in plants.[8] Such work can be extended further to

Analysis of the cell cycle

study cycle blocking agents, synchronization and chromosome isolation.[9] It should be noted, however, that mitotic nuclei cannot be analysed accurately by flow cytometry because the nuclear envelope breaks down and the chromosomes disperse during this phase.

Plant propagation

Detection of seeds with undesired ploidies
Flow cytometry allows the purity of seed lots to be assessed quickly and is of particular use for seeds produced from parents with different ploidies, such as sugar beet.[10]

Assessment of the genetic uniformity of micropropagated plants
The genetic uniformity of micropropagated somatic embryos and plantlets, which may be at risk due to *in vitro* culture, can be easily assessed using flow cytometry.[11,12,13,14,15] However, mutations involving small, but often significant, changes in DNA structure may not be detected using this technique.

Plant improvement

Determination of the ploidy of parents used in breeding programs
There is usually a strong correlation between nuclear DNA content and chromosome number within a given genus.[16] However, as found in *Paspalum* spp.,[6] the absolute DNA amount should be correlated with chromosome number in some standard specimens before changes in ploidy can be validated by flow cytometry.

The selection of plants with doubled chromosome numbers
Flow cytometry has been used to assess the increase in ploidy of plants following treatment with spindle inhibitors,[17] and through spontaneous change in *in vitro* culture.[18]

The selection of haploids
The high resolving power of flow cytometry makes it an ideal technique for discriminating haploid and diploid plants derived from anther culture.[19,20]

Sexual hybridization usually leads to offspring with predictable chromosome numbers which are usually the average of the two parents. Thus flow cytometry can be a very precise guide to this type of hybridity.[21] On the other hand, the genomes of somatic hybrids are typically very unstable and unpredictable, at least in *in vitro* culture. This instability results from chromosomal changes, either prior to or subsequent to protoplast fusion. Therefore, flow cytometry data on somatic hybrids should always be confirmed by other means, such as isoenzyme analysis,[22] or the use of molecular and genetic markers.[23]

The confirmation of hybridity of sexual and somatic hybrids

Subprotocol 1
DAPI Staining of Isolated Nuclei

▓▓ Materials

Equipment

In our case we use a Partec CA-III fitted with a 100W mercury high pressure HBO 100 lamp (Osram, Germany). The optimum excitation wavelength for DAPI staining is 359nm and it emits maximally at 461nm. The optical arrangement used for this stain includes KG 1, BG 38 and UG 1 excitation filters, a TK 420 dichroic mirror (to provide epi-illumination), a 40 x 0.8 quartz objective and GG 435, a long-pass barrier filter for blue fluorescence.

If you have a laser based cytometer then you must excite with a UV laser.

Solutions

- The sheath fluid is deionised water with 1 drop of Tween-20 added to inhibit algal growth.
- Partec solution A, or some other hypotonic medium containing non-ionic detergent, on ice. Polyvinylpyrrolidone (PVP-40) may be added to the isolation buffer to give a concentration of 1% (w/v).
- staining solution B.

Note: As with all DNA stains, care should be taken in the handling and disposal of these buffers.

Preparation

It is important to switch on the flow cytometer and lamp at least 2h before use to avoid drift in the measurements. This is not such a problem if an internal standard is used.

▓▓ Procedure

Prepare solutions

1. Ready Partec solution A, or some other hypotonic medium containing non-ionic detergent, on ice. Add PVP-40 to give a final 1%(w/v) prior to use.

Note: The "High Resolution DNA Kit, Type T (Partec GmbH, Münster, Germany) is used routinely. This kit consists of solution A for nuclei isolation and solution B for DAPI staining. Alternative solutions could be used ,[34] particularly where buffer composition, eg. concentration of DAPI, needs to be altered. Polyvinylpyrrolidone (PVP-40) may be added to the isolation buffer to give a concentration of 1% (w/v) prior to use. This is especially important when an external standard is used along with a test tissue high in content of interfering phenolics (Fig.1). This may not be required if an internal standard is used (Fig.2).

Note: As with all DNA stains, care should be taken in the handling and disposal of these buffers.

Set up flow cytometer

2. Set up flow cytometer.

Note: In our case we use a Partec CA-III fitted with a 100W mercury high pressure HBO 100 lamp (Osram, Germany). The optimum excitation wavelength for DAPI staining is 359nm and it emits maximally at 461nm. The optical arrangement used for this stain includes KG 1, BG 38 and UG 1 excitation filters, a TK 420 dichroic mirror (to provide epi-illumination), a 40 x 0.8 quartz objective and GG 435, a long-pass barrier filter for blue fluorescence.

The sheath fluid is deionised water with 1 drop of Tween-20 added to inhibit algal growth.

The same optical arrangement can be used for analyzing the DNA stain, Hoechst 33342.

If you have a laser based cytometer then you must excite with a UV laser.

It is important to switch on the flow cytometer and lamp at least 2h before use to avoid drift in the measurements, especially when external standards are used.

3. Cut tissue sample of approximately 10mg fresh weight. Place it together with an equal amount of the internal standard tissue - one over the other. Finely chop together with a sharp razor blade in 0.4 ml ice-cold solution A in a plastic Petri dish. **Prepare Tissue Sample**

Note: A 50mm^2 leaf disc suffices in most cases, although with parsley we use a small sprig of the leaf.

4. Incubate the mixture in the dark at room temperature for 15 mins.

5. Filter the homogenate through 50μm nylon mesh.

Note: Celltrics disposable filters are ideal for this purpose and can be purchased in various sizes.

6. Add 2 mls of staining solution B and incubate in the dark at room temperature for 5 mins. **Stain with DAPI**

Note: A sample of DAPI-stained nuclei is shown in Fig. 3.

7. Run samples. Set suitable flow cytometer parameters, especially 'amplification mode', 'speed', 'gain', 'lower limit' and 'upper limit' during/after running a test sample. **Run samples in flow cytometer**

Note: A linear (lin) **'amplification mode'** is often used so that all peaks are displayed in normal proportion. The alternative log mode for intensity values is more sensitive to small signals and, therefore, also displays more noise signal at lower intensity values.

A low **'flow speed'** (fl) is recommended (eg. 30 relative units) for maximum peak resolution and sharpness.

The voltage **'gain'** (g) can be set between 0 and 1000V and should be adjusted to give a good spread of the peaks across the whole intensity scale.

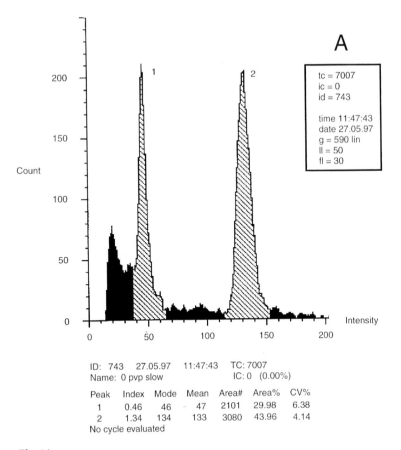

A

tc = 7007
ic = 0
id = 743

time 11:47:43
date 27.05.97
g = 590 lin
ll = 50
fl = 30

Count

Intensity

ID: 743   27.05.97   11:47:43   TC: 7007
Name: 0 pvp slow   IC: 0 (0.00%)

Peak	Index	Mode	Mean	Area#	Area%	CV%
1	0.46	46	- 47	2101	29.98	6.38
2	1.34	134	133	3080	43.96	4.14

No cycle evaluated

Fig. 1A.

Fig. 1. The neutralizing effect of PVP on the inhibition of DAPI staining by phenolics. (**A**) Shows the results of chopping parsley (*Petroselinum crispum*) leaf tissue (low phenolic content) with that from rose 'New Dawn' leaves (high phenolic content) in Partec Nuclei Isolation type T solution.
(**B**) Shows the results for nuclei prepared in a similar way to (**A**) except that 1%(w/v) PVP-40 was added to the Nuclei Isolation solution prior to use. Note the shift in peak mode values for both peaks in the two samples. The ratio of the mode values, however, remained constant. PVP had a similar effect on the mode values of rose leaf tissue when it was chopped alone but had no effect on those of parsley when chopped alone. This indicates that the PVP produced its effect by neutralising the quenching effects of phenolics, which are in high concentration in rose leaves but low in parsley leaves

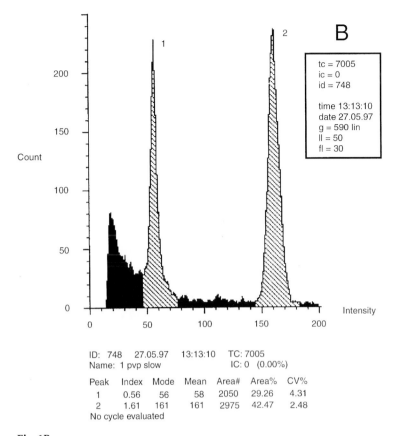

tc = 7005
ic = 0
id = 748

time 13:13:10
date 27.05.97
g = 590 lin
ll = 50
fl = 30

ID: 748   27.05.97   13:13:10   TC: 7005
Name: 1 pvp slow   IC: 0 (0.00%)

Peak	Index	Mode	Mean	Area#	Area%	CV%
1	0.56	56	58	2050	29.26	4.31
2	1.61	161	161	2975	42.47	2.48

No cycle evaluated

Fig. 1B.

One of the most important gated parameters is the **'lower limit'** (ll) discriminator for the channels used for displaying the data. If this is set too low, then low level noise signals from room light, sheath fluid and fluorescing cell debris will lead to an unacceptable noise-to-signal ratio and distinct peaks will not be detected. The **'upper limit'** is usually set at maximum unless peaks above a defined intensity are not required.

8. Process and plot the data. Record the data values: peak 'mode' values, 'area#' and 'CV%'.

Process and plot data

Note: We use Partec (DPAC version 2.1) software to process and plot data. The peaks to be analyzed are selected automatically by the software, or by the operator.

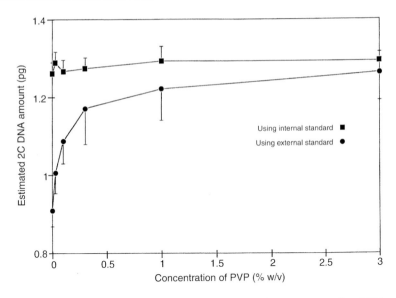

Fig. 2. Effect of different concentrations of PVP on the apparent nuclear DNA amount in leaf cells of 'New Dawn' rose. Mung bean (*Vigna radiata* (L.) Wilczek cv. Berken) was used as the external standard and parsley was used as the internal standard. Note that the calculated DNA amounts were influenced by PVP (ie. phenolics) only when an external standard was used

The most important parameter displayed is the peak 'mode', which is the channel number of a given peak with the highest count. This value is used to calculate DNA amounts, by simple ratio, relative to the peak mode values for the standard sample. Peaks are most accurately compared when the number of counts they represent, ie. their **'area#'** are similar. This value can also be used to compare the numbers of nuclei with a given DNA amount in the sample.

Peak sharpness is assessed quantitatively using the coefficient of variation (CV%)(on gaussian distributions this equals half the peak width at 67% of its height relative to the mean). This should be below 5% for the best estimates of DNA amounts.

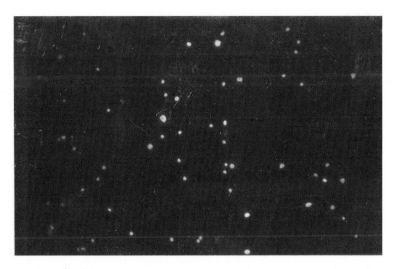

Fig. 3. Photomicrograph showing fluorescing nuclei of rose 'New Dawn' callus tissue. The nuclei were stained with DAPI and viewed at a magnification of 400 x in a Nikon Optiphot microscope equipped with a UV1A filter combination, providing an excitation wavelength of 365 μm

Subprotocol 2
Propidium Iodide (PI) or EB Staining of Isolated Nuclei

Materials

Equipment

In our case we use a Partec CA-III fitted with a 50W argon ion laser (Ion Laser Technology, USA). The optimum excitation wavelength for EB and PI are 526 and 536 nm and they likewise emit maximally at 604 and 620 nm. The best optical arrangement used for these stains consists of KG 1, BG 38 and BG 12 excitation filters, a TK 500 dichroic mirror, and a 40 x 0.8 quartz objective. In addition, a RG 570 barrier filter is used.

If you have a laser based cytometer then you must excite with a 488nm argon laser.

Solutions

- LB01 lysis buffer consists of:
  - 15 mM Tris 0.5mM spermine
  - 20 mM NaCl 0.1% Triton X-100
  - 80 mM KCl 2mM Na_2EDTA
  - 15 mM mercaptoethanol pH 7.5

The sheath fluid is deionised water with 1 drop of Tween-20 added to inhibit algal growth.
- 10 mg/ml DNase-free RNase A
- 10 mg/ml propidium iodide or ethidium bromide

Preparation

It is important to switch on the flow cytometer and lamp at least 2h before use to avoid drift in the measurements. This is not such a problem if an internal standard is used.

▓▓ Procedure

Prepare standard solutions

1. Prepare LB01 + PVP lysis buffer.

Set up flow cytometer

2. Set up flow cytometer.

Note: In our case we use a Partec CA-III fitted with a 50mW argon ion laser (Ion Laser Technology, USA). The optimum excitation wavelength for EB and PI are 526 and 536 nm and they likewise emit maximally at 604 and 620 nm. With our laser we use a 40 x 0.8 quartz objective lens for the excitation beam and a TK 560 dichroic mirror and a RG 610 barrier filter for the emitted light.
The sheath fluid is deionised water with 1 drop of Tween-20 added to inhibit algal growth.
If you have a laser based cytometer then you must excite with a 488nm argon laser.

3. Prepare 10mg/ml DNase-free RNase A in water.

4. Prepare 10mg/ml propidium iodide or ethidium bromide in water.

Note: The choice of EB or PI should be determined empirically, dependent on which of the two fluorochromes provides the better peak resolution and sharpness in histograms.

Note: As with all DNA stains, care should be taken in the handling and disposal of these buffers.

5. Cut tissue sample of approximately 10mg fresh weight. Place it together with an equal amount of the internal standard material - one over the other. Finely chop together with a sharp razor blade in 0.4 ml ice-cold LB01+PVP buffer in a plastic petri dish.

Prepare Tissue Sample

Note: A 50mm^2 leaf disc suffices in most cases, although with parsley we use a small sprig of the leaf.

6. Incubate mixture in the dark at room temperature for 15 minutes.

7. Filter the homogenate through a 20μm nylon mesh.

Note: Celltrics disposable filters are ideal for this purpose and can be purchased in various sizes.

8. To the filtrate add 2 mls of ice-cold LB01+PVP lysis buffer and 50 μg/ml (2.0μl of a 10 mg/ml stock in water) of each of DNase-free RNase and either ethidium bromide or propidium iodide. Incubate on ice in the dark for 1h. Filter through a 20 μm nylon mesh just prior to measurement.

Eliminate RNA and stain DNA

Note: Propidium iodide and ethidium bromide will bind both DNA and RNA. In order to specifically measure DNA, DNase-free RNase in saturating amounts must be added to eliminate the RNA.

9. Run samples. Set suitable flow cytometer parameters, especially 'amplification mode', 'gain', 'lower limit' and 'upper limit' during/after running a test sample.

Run samples in flow cytometer

Note: A linear (lin) **'amplification mode'** is often used so that all peaks are displayed in normal proportion. The alternative log mode for intensity values is more sensitive to small signals and, therefore, also displays more noise signals at lower intensity values.

A low 'flow speed' (fl) is recommended (eg. 0.4 µl/s) for maximum peak resolution and sharpness. The optimum flow speed is lower than with a UV light source because of the smaller area of focus.

The voltage 'gain' (g) can be set between 0 and 1000V and should be adjusted to give a good spread of the peaks across the whole intensity scale.

One of the most important gated parameters is the 'lower limit' (ll) discriminator for the channels used for displaying the data. If this is set too low, then low level noise signals from room light, sheath fluid and fluorescing cell debris will lead to an unacceptable noise to signal ratio and distinct peaks will not be detected. The 'upper limit' is usually set at maximum unless peaks above a defined intensity are not required.

Process and plot data   10. Process and plot the data. Record the data values: peak 'mode' and 'mean' values, 'area#' and 'CV%'.

Note: We use Partec (DPAC version 2.1) software to process and plot data. The peaks to be analyzed are selected automatically by the software, or by the operator.

The most important parameter displayed is the peak 'mode', which is the peak channel number with the highest count. This value is used to calculate DNA amounts, by simple ratio, relative to the peak mode values for the standard sample.

Peaks are most accurately compared when the number of counts they represent i.e. their 'area#' are similar. This value can also be used to compare the numbers of nuclei with a given DNA amount in the sample.

Peak sharpness is assessed quantitatively using the coefficient of variation (CV%)(on gaussian distributions this equals half the peak width at 67% of its height relative to the mean). This should be below 5% for the best estimates of DNA amounts.

A histogram of the flow cytometry of PI stained nuclei isolated from rose leaves is shown in Fig. 4.

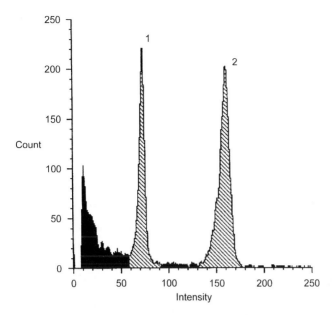

Fig. 4. Histogram showing the flow cytometry of propidium iodide stained nuclei isolated from the leaves of *Rosa fedtschenkoana* (*1*) and parsley (*Petroselinum crispum* 'Champion Moss Curled') (*2*)

Subprotocol 3
Isolation of protoplasts from rose callus and leaves

▨▨ Materials

Solutions

CPW13M
- 0.2mM KH_2PO_4 1.0mM KNO_3
- 10.0mM $CaCl_2 \times 2H_2O$ 1.0μM KI
- 1mM $MgSO_4 \times 7H_2O$ 71mM mannitol
- 0.1 μM $CuSO_4 \times 5H_2O$ pH 5.8

Enzymes
- 0.1 g/l Pectolyase Y23
- 2.0 g/l cellulase 'Onozuka' R10

CPW21S

– This has the same composition as CPW13M except that 60mM sucrose replaces mannitol.

▓▓ Procedure

Isolation of protoplasts from rose callus and leaves

1. Cut 1g of the tissue into pieces of approximately 1mm^3.

CPW13M

2. Add 4 mls of CPW13M solution and incubate at room temperature in the dark for 30 min to preplasmolyse the cells.

3. Allow the tissue pieces to sediment and pour off the supernatant.

Enzymes

4. Add 4 mls of cell-wall degrading enzyme mixture made up in CPW13M to the tissue pieces.

5. Incubate for 16h at a shaking speed of 30 rpm in the dark at 25°C.

Purification of protoplasts

6. Filter the digest through a 100μm nylon mesh and centrifuge the filtrate at 100 x g for 10 min.

CPW21S

7. Remove the filtrate, resuspend the protoplast pellet in 10 ml of CPW21S and centrifuge at 100 x g for 10 min.

Note: This has the same composition as CPW13M except that 60mM sucrose replaces mannitol.

8. Carefully, transfer the protoplasts in the uppermost layer using a pasteur pipette to a fresh tube.

9. Resuspend the protoplasts in 10 ml of CPW13M. Wash 3 x by repeated resuspension in CPW13M and centrifugation.

Protoplast lysis

10. Resuspend the final protoplast pellet for 15 mins in the dark in 0.4 mls of either Partec solution A (for DAPI staining) or LB01 + PVP lysis buffer (for EB/PI staining).

Staining of nuclei

11. Stain with either DAPI or EB/PI.

Note: For protocol, see sections on DAPI and EB/PI.
Less debris should show up when using protoplasts to produce nuclei suspensions than when the chopping method is used (Fig. 5).

12. Set up flow cytometer.

Note: For protocol, see sections on DAPI and EB/PI.

13. Run samples in flow cytometer.

Note: For protocol, see sections on DAPI and EB/PI.

14. Process and plot the data.

Note: For protocol, see sections on DAPI and EB/PI.

Set up flow cytometer

Run samples in flow cytometer
Process and plot data

COMPARISON OF FLOW CYTOMETRY WITH OTHER COMMON METHODS FOR DNA MEASUREMENTS IN PLANTS

Flow cytometry is based on the use of fluorochromes, or endogenous fluorophores, and the organelles or cells to be analysed must be suspended in a liquid medium. It is a rapid, precise and convenient method of analysing several thousand structures in only a few minutes and a sorting module can often be added to the instruments if required. Intercellular variation in DNA content is also determined and virtually any tissue can be analysed. However, the technique is technically sophisticated and requires equipment that is relatively expensive to purchase and maintain. Nevertheless, the use of flow cytometry for DNA measurements in plants has mushroomed since the mid 1980s.

Feulgen microdensitometry involves the use of mitotic nuclei stained *in situ* with Feulgen's reagent. It is one of the earliest and most widely used methods for quantifying DNA. However, it involves several time-consuming steps which are subject to experimental variation. Usually fewer replicate measurements are, therefore, performed with this method so it is statistically less accurate than flow cytometry. Most measurements of plant DNA amounts have relied on this method,[3] which is still the standard method by which others are compared. Good correlations for DNA amounts obtained with this method and flow cytometry

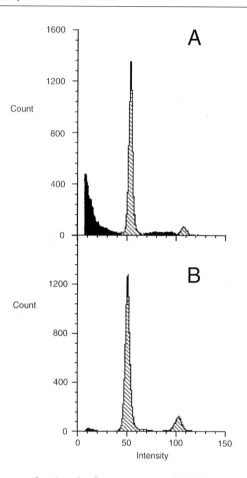

Fig. 5. Histograms showing the flow cytometry of DAPI-stained nuclei from rose 'Frensham' callus prepared by either (**A**) the chopping method, or (**B**) the isolation and subsequent lysis of protoplasts from the same callus. Note the higher amount of fluorescence, caused by debris (in black), in (**A**) compared to (**B**)

have been found using plants with a wide range of nuclear DNA levels.[3]

Chemical analysis involves the assay of DNA using spectrophotometric or fluorimetric techniques. The extinction of the DNA at specific wavelengths is determined either with or without the stoichiometric reaction of extracted DNA with specific reagents. Some of these methods involve the use of the same fluorochromes that are used in flow cytometry, such as 4',6-diamidino-2-phenylindole (DAPI) and ethidium bromide (EB).[24] With this method, the average amount of DNA per cell is determined after DNA is extracted from a known number of cells. However, losses do occur on extraction and purification of DNA from plant cells. Therefore, an internal standard such as radiolabelled DNA should be added to assess these losses. In addition, plant cells often contain interfering substances, the methods of analysis are relatively more time-consuming and they do not measure the intercellular variation in DNA content.

Chromosome counting is cell-based and involves the squashing and staining of meristematic tissues, such as root tips, so that metaphase chromosomes in dividing cells can be counted. This technique only reveals the chromosome number, which may not always be directly correlated to DNA amounts.[6] In addition, it is laborious and counts are subject to error due to overlapping chromosomes, especially if they are small and/or numerous. Intercellular variation in ploidy, eg. in chimeras, can be determined with this technique but not as easily as with flow cytometry.

Indirect methods for estimating the ploidy levels of plants have also been developed mainly for use by breeders. These methods only require the use of a microscope and a few replicate measurements are usually enough to confirm the ploidy status of a particular plant. The methods depend on the correlation of certain morphological characters, such as plant phenotype, pollen size and the size of guard cells and their chloroplast numbers with increased ploidy. These relationships have been found to hold true in several species, such as roses, sugar beet, *Arachis* spp. and Magnoliaceae.[25,26,27,28] These methods have a low resolving power when differences in amounts of DNA between two individuals are small. In addition, only the ploidy level of specific cell layers are determined by most of these methods so they are unreliable for characterizing chimeric tissues.

FLOW CYTOMETRY STANDARDS FOR PLANT MATERIAL

As with all other tissues, it is necessary in most investigations of plant tissues to compare the peaks obtained using flow cytometry with those of a suitable standard which has a known nuclear DNA amount. These standards can be either **external** or **internal** and consist of either **plant** or **animal** nuclei. The principles of running standards for plant material are similar to that of all tissues, but plant material can present us with special problems that need to be taken into account whenever the type of standard best to use with plant material is considered.

External standards are run separately before and/or after the test sample. This enables many test samples to be compared to a single or a few standard runs and this may save time. External standards may be less expensive if the standard material used is costly to purchase or prepare (eg. erythrocytes). There are two major disadvantages of using external standards for plant material, however. First, as with other types of material, fluctuations in instrument settings between runs, even though only slight, may affect the relationship between the test sample data and that for the standard. This can lead to unacceptable errors. Secondly, phenolic compounds released from damaged plant tissues, during chopping or grinding can have major effects on the peak positions obtained (Fig.1). This problem is particularly prevalent in woody tissues and can be alleviated by incorporating polyvinyl pyrrolidone (PVP-40, 10 g/l) into the chopping medium,[2] (Fig. 2).

Internal standards do not suffer from the problems associated with external standards because both test and internal standard nuclei should be equally affected by any pertubations which may occur during sample preparation or analysis (Fig. 1). Consequently, the relationship between peak positions of test and standard should remain constant. Internal standards should, therefore, be used whenever possible. It is desirable that the peak heights of the internal standard and test samples are similar and that the peaks do not overlap. It is important to make sure that the peaks of the standard species lie on the same linear amplification scale as the test sample. Use of logarithmic amplification to compare standard and test species with widely different nuclear DNA values on a single histogram is possible but provides less peak resolution.[29] Logarithmic amplification is, however, useful for detecting minor peaks and for displaying widely spaced peaks resulting from endoreduplication.[11]

Plant vs animal internal standards: many workers are of the opinion that a suitable plant standard can be used whenever a plant species is investigated using flow cytometry. This is partly due to the fact that internal standards should be treated in exactly the same way as the test material, regarding nuclei isolation and staining. Animal standards are usually cells added to samples just prior to staining, so are not subjected to the same isolation treatments as the plant cells themselves.

There is little evidence, however, that plant standards offer any real advantage over animal standards. In fact, plant standards can be subject to considerable variability both between different cultivars, between different plants of the same cultivar and between different tissues of a single plant. The nuclei of plant cells often undergo considerable endoreduplication during maturation of tissues.[30,31] This is evident in multiple peaks being produced (Fig. 6), some of which may overlap with those of the test samples. Consequently, young tissues of known developmental stage or age are preferred and they should be grown under controlled conditions. Plants may, however, be more convenient and cheaper to obtain than other standards. Several plants with a wide range of nuclear DNA values have been investigated as possible plant standards.[3] The 2C DNA amounts of these range from 1.01 pg in *Oryza sativa* cv. IR36 to 34.64 pg for *Triticum aestivum* cv. Chinese Spring.

Animal standards that have been used in plant work include erythrocytes of trout (5.05 pg/2C) and chicken (2.33 pg/2C),[11] human male leucocytes (7.0 pg/2C),[32] and human fibroblasts (7.94 pg/2C).[33] If animal internal standards are used then it is worth checking that metabolites, such as phenolics, released from the plant samples do not interfere differentially with the animal standards. However, DNA values found for various plant species using animal internal standards were found to be comparable to those obtained using Feulgen microdensitometry involving plant standards.[32,34,35]

THINGS TO CONSIDER

Instrument characteristics

Our laboratory uses the Partec CA-III flow cytometer equipped with a HB-100 high pressure mercury UV light source and a 50 mW argon ion laser source. The instrument is fitted with 2

Fig. 6. Histogram showing both diploid and tetraploid nuclei in the same tissue resulting from endoreduplication of mung bean root tips (**A**) and root bases (**B**). In each case, peak 1 corresponds to the G1 nuclei of diploid cells. Peak 2 results from a combination of G2 nuclei from diploid cells and predominantly G1/G0 nuclei from tetraploid cells. The G2/M nuclei from the tetraploids are found in peak 3. Note the increased proportion of tetraploid nuclei in the roots as they mature from tip to base

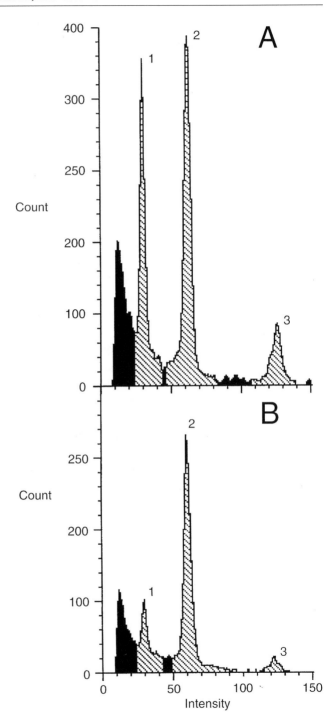

photomultipliers: a standard PMT 1 used for DAPI measurements, and a red sensitive (PMT red S20) for EB/PI measurements. The system makes use of 'Kohler' illumination in combination with a high numerical aperture microscopic objective, which results in minimum adjustment requirements, coupled with high sensitivity, accuracy and peak resolution. The optical arrangement also provides a large depth of focus, which is particularly important for use with large plant protoplasts.

Fluorochromes

The use of various fluorochromes for plant DNA measurements have been reviewed extensively,[36,37] so will only be summarised here. In general, most workers in this field choose between fluorochromes on the basis of the following characteristics:

- **Base specificity**
  Several fluorochromes, preferentially bind to specific base pairs. For example DAPI and Hoechst 33342 preferentially bind to AT pairs, whilst mithramycin prefers GC pairings. Both DAPI and Hoechst 33342 have been used extensively for the determination of plant DNA. One major difference between these two fluorochromes is that Hoechst 33342 penetrates intact membranes whereas DAPI (like EB and PI) does not. This only becomes important for DNA analysis, however, when non-permeabilized whole cells or protoplasts are stained.
  DAPI gives better peak resolution, with lower CV% values (ie. peak sharpness), compared to EB or PI. This is illustrated well in the sex determination of dioecious plants.[7] There is, however, some doubt as to the accuracy of absolute DNA values obtained using base-specific fluorochromes because different genotypes vary in their AT:GC ratios.[32] This is obviously important when internal standards of unknown AT:GC ratios are used to compare with test genotypes. It is irrelevant, however, when ploidy levels are being measured within the same genotype so it may be desirable to restrict the use of base-specific fluorochromes to work on ploidy determination.

- **Optical requirements**

  The fluorochrome of choice may be determined by the characteristics of the instrument that is available or can be added to existing facilities. The HB-100 light source used in most Partec instruments, in conjunction with suitable filters, can provide fluorescent light excitation from short UV up to red. This means that a number of commonly-used fluorochromes, such as DAPI, EB, PI and Hoechst 33342, can be used with this instrument. The standard Partec photomultiplier (PMT 1) detects emitted light between 350-700 nm. However, the PMT red S20 photomultiplier is more suited to use with red-emitting fluorochromes, such as EB and PI. Laser-based systems (mainly argon ion lasers) are more expensive and not all of them produce UV excitation wavelengths required for use with some fluorochromes. Some laser-based systems also require water cooling and daily alignment, which makes them less convenient to use. If a laser-based system is to be used then a UV laser providing an excitation wavelength of 360 nm is required for DAPI staining. EB/PI stained samples require a 488 nm argon ion laser (eg. 15 mW) and orange/red detection above 560 nm.

- **Fluorescence quenching**

  The red autofluorescence of chlorophyll in green tissues may interfere with the detection of EB/PI. Consequently, DAPI may be preferable for use in chloroplast-containing protoplasts or where chloroplast residues are a significant contaminant of isolated nuclei.

General problems associated with the use of plant material

Plant material can be particularly difficult to prepare for flow cytometry because of the inherent properties of the material, namely the presence of interfering **metabolites**, adhering **cell walls, plastids** and **vacuoles.** Protocols for the preparation of plant material for flow cytometry, therefore, always involve steps to eliminate or reduce such interfering factors as much as possible.

Starch can lead to glutinous preparations of nuclei, which are difficult to pass through the instrument tubing. Phenolic compounds bind tenaciously to chromosomal proteins and may interfere with chromatin stability or fluorochrome binding, or may absorb UV light directly. Cell walls need to be removed to produce suspensions of protoplasts or nuclei. They usually need to be broken physically by chopping or maceration, which can itself lead to nuclear damage and consequential decrease in yields and increase in CV% of the peaks obtained. Alternatively, they can be enzymatically removed by rather time-consuming methods to produce protoplasts.

As mentioned previously, the presence of pigments in plastids, such as chlorophyll and carotenoids, may give rise to autofluorescence and/or quenching problems which often interferes with the detection of added fluorochromes. However, they may be useful for characterizing and counting populations of intact protoplasts or organelles in the absence of such fluorochromes.

Vacuoles from cells of pigmented organs, such as petals, may also contain interfering anthocyanins. In addition, vacuoles are responsible for the large size of plant cells, which make them difficult to pass through flow cytometers. They also contribute to the fragility of protoplasts and the positional constraints of nuclei, the effects on DNA measurement of which are mentioned in later sections of this article.

Errors involved in the estimation of DNA amounts

Flow cytometry histograms of DNA amounts in nuclei, cells or protoplasts record a range of values around each peak. The differences expressed in these ranges may arise because of real differences or inaccuracies in the measurement of DNA. Flow cytometers are usually programmed to display parameters that include coefficient of variation (CV) of the sample. This expresses the range of measurements about the peak regardless of scale:

$$CV = SD \ / \ mean \ channel \ number \ x \ 100\%$$

where SD is the standard deviation. The SD can be used as a starting point in calculations for tests of significance such as the Student's-t test and analysis of variance. If the flow cytometer does not display it, the SD can be derived using the above formula.

The mean and SD can be more precisely estimated if samples are replicated. If an internal standard is used, the DNA measurements of the unknown can be converted directly to pg in each sample. The estimate of the mean and *SD* based on the replicated samples will then take into account errors in the calibration of channels from the assumed DNA amount of the standard. The *SD* can be calculated from the following formula or its derivatives:

$$SD = \sqrt{\frac{\sum (x - \bar{x})^2}{n - 1}}$$

x is the mean of a sample, $\bar{x}$ is the mean of the sample means and n is the number of samples.

If an external standard is used, a combined estimate of the *SD* is needed which allows for sampling errors of both the standard and the unknown. It can be estimated using the following formula:[2]

$$combined\ SD = \sqrt{\frac{SD_u^2}{\bar{x}_s^2} + \frac{SD_s^2 * \bar{x}_u^2}{x_s^4}} * A_{pg}$$

where: $\bar{x}_s$ and $\bar{x}_u$ are the channel numbers of the standard and unknown, respectively, SD_s^2 and SD_u^2 are the variances of the standard and unknown, respectively, and A is the DNA amount of the standard expressed in pg.

The number of samples that should be run will depend on the context of the investigation. If, for example, the objective is to detect variations in ploidy in a population of plants, the differences in DNA amount are likely to be sufficiently large to require no replication. If the objective is to study variations in DNA amount due to aneuploidy, replication is likely to be needed. The maximum number of samples that can be run will be determined by expediency but the minimum number can be assessed by the size of standard deviations obtained in test runs.

It must be remembered that the error expressions described above do not take account of potential sources of error such as assumed DNA amounts of the standards and inappropriate choice of fluorochrome.

Tissue chopping method

The preparation of suspensions of isolated nuclei by chopping,[34] grinding or slicing tissues in a hypotonic medium containing a non-ionic detergent is the most widely used method for preparing material from a wide range of plants and tissues for flow cytometry. The method is simple and rapid so allows sequential preparation and analysis of fresh material. It also does not suffer so much from lack of dye penetration, cytoplasmic DNA staining and chlorophyll autofluorescence that may be encountered with whole cells. Virtually any tissues can be used, although soft tissues containing a high proportion of healthy cells are preferred. Roots usually contain fewer interfering compounds but it should be checked that they are true roots and not rootstocks, to maintain consistency of genotype between root and shoot. Leaves often contain high levels of phenolics which may interfere with the action of fluorochromes. We routinely incorporate polyvinylpyrrolidone (PVP-40) at a concentration of 10g/l into the isolation buffers used with both DAPI and PI/EB. This significantly improves dye binding to the DNA and can improve the sharpness and resolution of the peaks, especially with tissues containing a high content of phenolics, such as rose leaves.

The tissue:buffer ratio is kept low when the material is chopped to dilute any interfering compounds released from the cut tissues. Each edge of the razor blade is used to chop only about 5 samples before being renewed, to keep the cuts clean. Squeezing the homogenate through a Pasteur pipette may help release the nuclei from broken cells and increase the yield of isolated nuclei.

The optimum fluorochrome concentration, especially of DAPI,[3] and time that the plant material should be incubated in isolation and staining solutions, should be determined individually for each tissue. This is because optimum conditions depend on many factors, such as the number of nuclei, the nuclear DNA amount, the degree of chromatin condensation and the concentration of any blocking compounds present in the tissues. However, in our experience, the protocols described here for staining with DAPI, ethidium bromide or propidium iodide give almost instantaneous staining and further penetration increases by only about 1-2% if left for several hours. Isolated nuclei can be fixed in 3:1 (w/v) ethanol:acetic acid,[38] or 100% ethanol,[6] and stored for future analysis if required.

Use of protoplasts

Protoplasts can be used whenever measurements on single cell populations are required or chopping of tissues is problematic. Their use often leads to purer preparations of nuclei with smaller background signals compared to the chopping method,[38] (Fig. 5). However, it has been found that a shift may occur towards lower ploidy values in the nuclei of some tissues analysed by this method.[12,39]

Protoplasts can be isolated from many kinds of tissues but callus, cell suspensions and leaves are usually favoured. It is not appropriate to review the methods for protoplast isolation here, but general protocols are available.[40] Some general points regarding the use of protoplasts for flow cytometry will be noted.

Plant protoplasts possess a large central vacuole which makes them fragile and highly sensitive to pressure changes and frictional forces. A chilled hypertonic medium can be used to reduce the size, and hence fragility, of protoplasts,[41] Protoplasts from leaves are more delicate than those from cell cultures, so extra care should be taken when preparing leaf protoplasts and running them in the flow cytometer. In any case, it is inevitable that whenever a protoplast suspension is prepared there will be a proportion of isolated nuclei and/or chloroplasts released from damaged protoplasts. This will affect the accuracy of the data if whole cell counts based on nuclear staining or chlorophyll autofluorescence are being made.

The preparation of protoplasts is time-consuming, taking from about 2 - 12 hours depending on the tissues used. This inevitably means that the physiological state of the protoplasts will be different from the starting tissue. This should always be taken into account but is especially important for cell cycle studies.

Plant cell nuclei are embedded in a thin layer of cytoplasm surrounding the vacuole and so are positioned at the periphery of cells. This means that when whole protoplasts are detected by flow cytometry using DNA-specific fluorochromes, then these positional effects result in greater CV% values for the peaks obtained.[42]

Poor dye penetration (e.g., of DAPI, EB and PI) through intact cell membranes, and chlorophyll autofluorescence in photosynthetic cells, means that work with protoplasts may not always be quantitative, unless the right choice of fluorochromes is made.

Both these problems, however, can be overcome by fixing the protoplasts in 3:1 ethanol:acetic acid,[38] or 100% ethanol,[6] and stored for future analysis if required.

Acknowledgements

The authors would like to thank Dr. V. Ost of Partec GmbH for invaluable advice on the use of the flow cytometer and sample preparation. Gratitude is also extended to Mr Keith Eley for preparing the flow histograms for publication and Dr. K. Kandasamy for photographic work.

References

1. Heller F O. DNA-bestimmungen an keimwurzelzellen von *Vicia faba* mit hilfe der impulszytophotometrie. Ber Deutsch Bot Ges 1973; (86): 437-441.
2. Dickson E E, Arumuganathan K, Kresovich S et al. Nuclear DNA content variation within the Rosaceae, American Journal of Botany 1992; (79): 1081-1086.
3. Bennett M D, Leitch I J. Nuclear DNA amounts in Angiosperms. Annals of Botany 1995; (76): 113-176.
4. Bennett M D, Smith J B, Heslop-Harrison J S. Nuclear DNA amounts in angiosperms, Proceedings of the Royal Society of London, 1982; B(216): 179-199.
5. Bennett M D, Cox A V, Leitch I J. Angiosperm DNA C-values database. 1997. http://www.rbgkew.org.uk/cval/database1.html.
6. Jarret R L, Ozias-Akins P, Photok S et al. DNA contents in *Paspalum* spp determined by flow cytometry. Genetic Resources and Crop Evolution 1995; (42): 237-242.
7. Dolezel J and Gohde W. Sex determination in dioecious plants *Melandrium album* and *M rubrum* using high-resolution flow cytometry. Cytometry 1995; (19): 103-106.
8. Galbraith D W. Flow cytometric analysis of the cell cycle. In *Cell Culture and Somatic Cell Genetics of Plants* (I K Vasil, ed), 1984;(1) 765-777. Academic Press Inc.
9. Dolezel J, Lucretti S, Macas J. Flow cytometric analysis and sorting of plant chromosomes. In P E Brandham and M D Bennett (eds.) Kew Chromosome Conference IV, pp.185-200. Royal Botanic Gardens, Kew. 1995.
10. De Laat, A.M.M., Gohde W. and Vogelzang M J D C. Determination of ploidy of single plants and plant populations by flow cytometry. Plant Breeding 1987; (99): 303-307.

11. Arumuganathan K. and Earle E D. Estimation of nuclear DNA content of plants by flow cytometry. Plant Mol Biol Reporter 1991; (9): 229-233.

12. Uijtewaal B A. Ploidy variability in greenhouse cultured and *in vitro* propagated potato (*Solanum tuberosum*) monohaploids (2n = x = 12) as determined by flow cytometry, Plant Cell Reports 1987; (6): 252-255.

13. Awoleye F., Van Duren M., Dolezel J, Novak, F J. Nuclear DNA content and *in vitro*-induced somatic polyploidization in cassava (*Manihot esculenta* Crantz) breeding. Euphytica 1994; (76): 193-202.

14. Kubalakova M., Dolezel J, Lebeda A. Ploidy instability of embryogenic cucumber (*Cucumus sativus* L) callus cultures. Biologia Plantarum 1996; (38): 475-480.

15. Sree Ramulu K, Dijkhuis P. Flow cytometric analysis of polysomaty and *in vitro* genetic instability in potato. Plant Cell Reports 1986; (3): 234-237.

16. Fahleson J, Dixelius, J, Sundberg E et al. Correlation between flow cytometric determination of nuclear DNA content and chromosome number in somatic hybrids within Brassicacaea. Plant Cell Reports 1988; (7): 74-77.

17. Leblanc O, Duenas M, Hernandez, M et al. Chromosome doubling in *Tripsacum*: the production of artificial, sexual tetraploid plants. Plant Breeding 1995; (114): 226-230.

18. Van Duren M, Morpurgo R., Dolezel J. et al. Induction and verification of autotetraploids in diploid banana (*Musa acuminata*) by *in vitro* techniques. Euphytica 1996; (88): 25-34.

19. Bhaskaran S, Smith RH, Finer J J. Ribulose biphosphate carboxylase activity in anther-derived plants of *Saintpaulia ionantha* cultivar Shag. Plant Physiol 1983; (73): 639-642.

20. Sharma D P, Firoozabady E, Ayres, N M et al. Improvement of anther culture in *Nicotiana*: Media, cultural conditions and flow cytometric determination of ploidy levels. Zeitschrift fur Pflanzenphysiologie 1983; (111): 441-451.

21. Sabharwal P S, Dolezel J. Interspecific hybridization in *Brassica*: application of flow cytometry for analysis of ploidy and genome composition in hybrid plants. Biologia Plantarum 1993; (35): 169-177.

22. Chaput M-H., Sihachakr D, Ducreux G. et al. Somatic hybrid plants produced by electrofusion between dihaploid potatoes: BF15 (H1), Aminca (H6) and Cardinal (H3). Plant Cell Reports 1990; (9), 411-414.

23. Squirrell J, Mottley J, Roberts A V. Genetic analysis of putative rosaceous somatic hybrids. NERC 17th Annual Meet. Tree Biotech. Liason Group 1997. Nottingham University, UK.

24. Van Lancker M, Gheyssens L C. A comparison of frequently used assays for quantitative determination of DNA. Analytical Letters 1986; (19): 615-623.

25. Mochizuki A. and Sueoka N. Genetic studies on the number of plastid in stomata. 1. Effects of autopolyploidy in sugar beets. Cytologia 1956; (20): 358-366.

26. Lloyd D. The induction, *in vitro*, of chromosomal variation in *Rosa*. PhD thesis, University of East London, London, UK, 1986.

27. Singsit C, Ozias-Akins P. Rapid estimation of ploidy levels in *in vitro*-regenerated interspecific *Arachis* hybrids and fertile triploids. Euphytica 1992; (64): 183-188.
28. Masterson J. Stomatal size in fossil plants: evidence for polyploidy in majority of angiosperms. Science 1994; (264): 421-424.
29. Bergounioux C, Brown S C. Plant cell cycle analysis. In *Methods in Cell Biology* 1990; 563-573. Academic Press.
30. DeRocher E J, Harkins KR, Galbraith D W et al. Developmentally-regulated systemic endopolyploidy in succulents with small genomes. Science 1990; (250): 99.
31. Galbraith D W, Harkins K R, Knapp S. Systematic endopolyploidy in *Arabidopsis thaliana*. Plant Physiology 1991; (96): 985.
32. Dolezel J, Sgorbati S, Lucretti S. Comparison of three DNA fluorochromes for flow cytometric estimation of nuclear DNA content in plants. Physiol Plantarum 1992; (85): 625-631.
33. Fasman G D. Deoxyribonucleic acid content per cell of various organisms. In Fasman G.D. (ed.) *Handbook of Biochemistry and Molecular Biology* (3rd edition) pp. 284-311. CRC Press, Cleveland, 1976.
34. Galbraith D W, Harkins K R, Maddox J M et al. Rapid flow cytometric analysis of the cell cycle in intact plant tissues. Science 1983; (220): 1049-1051.
35. Arumuganathan K, Earle E D. Nuclear DNA content of some important plant species. Plant Molecular Biology Reporter 1991; (9): 210-220.
36. Waggoner A S. Fluorescent probes for cytometry. In: Melamed, M.R., Lindmo, T. and Mendelsohn, M.I., eds. *Flow Cytometry and Sorting* (2nd edition) Wiley-Liss Inc., New York, pp. 209-225, 1990.
37. Haugland R P. Spectra of fluorescent dyes used in flow cytometry. In Z. Darzynkiewicz, ed, *Methods in Cell Biology*. Academic Press Inc 1994; (42): 641-663.
38. Ulrich I, Ulrich W. High-resolution flow cytometry of nuclear DNA in higher plants. Protoplasm 1991; (165): 212-215.
39. Dolezel J, Binarova P, Lucretti S. Analysis of nuclear DNA content in plant cells by flow cytometry. Biologia Plantarum 1989; (31):113-120.
40. Davey M, Power J B. Isolation, culture and regeneration of protoplasts. In (Dixon, R.A. and Gonzales, R.A., eds) Cell Culture: a practical approach, pp. 27-39. IRL Press, 2nd Ed, Oxford, New York, Tokyo, 1994.
41. Harkins KR, Galbraith D W. Flow sorting and culture of plant protoplasts. Physiologia Plantarum 1984; (60): 43-52.
42. Fox M H, Galbraith D W. Application of flow cytometry and sorting to higher plant systems. In Flow Cytometry and Sorting, pp 633-650. Wiley-Liss Inc, 1990.

Suppliers

PARTEC FLOW CYTOMETERS and CELLTRICS FILTERS
PARTEC GMBH, OTTO-HAHN-STR. 32, MUNSTER, D-48161, GER-
MANY [*phone* (GERMANY) 2534 5270; *fax* (GERMANY) 2534 8588]

CYTOMATION INC., 400, EAST HORSETOOTH ROAD, FORT
COLLINSCO, 80525, USA [*phone* (USA) 970 226 2200; *fax* (USA)
970 226 0107]

PLANT CELL WALL-DEGRADING ENZYMES, **Pectolyase Y23**
(EC 3.2.1.15 and EC 4.2.2.10)

SEISHIN PHARMACEUTICAL CO., LTD., 4-13, KOAMICHO, NIHON-
BASHI, CHUO-KU, TOKYO, JAPAN [*phone* (JAPAN) 03-669-2876]

PLANT CELL WALL-DEGRADING ENZYMES, **Cellulase
'Onozuka' R10** (EC 3.2.1.4)

YAKULT PHARMACEUTICAL IND. CO. LTD., 1-1-19, HIGASHI-SHI-
NABASHI, MINATO-KU, TOKYO, 105, JAPAN [*phone* (JAPAN)
03-574-6766 DNASE-FREE RNASE]

SIGMA CHEMICAL CO., , P.O. BOX 14508, ST. LOUIS, MISSOURI,
USA [*phone* ST. LUIS (USA) 63178-9916]

Flow Cytometry Analysis of Marine Picoplankton

D. MARIE, N. SIMON, L. GUILLOU, F. PARTENSKY, AND D. VAULOT

Introduction

In the last decade, the use of flow cytometry (FCM) has become more and more popular among limnologists and marine biologists, both for laboratory studies and field research. FCM allows the analysis of phytoplanktonic cells that are too dim to be discriminated by epifluorescence microscopy. Its major advantages are to provide rapid and accurate measurements of individual particles and to allow the discrimination between auto- and heterotrophic populations as well as between cells and detritus or suspended sediments. FCM is particularly well suited for the study of the smallest size class of the plankton (below 2 μm), called picoplankton. Picoplankton is composed by 4 major groups: heterotrophic prokaryotes, prochlorophytes (*Prochlorococcus*),[1] cyanobacteria (*Synechococcus*)[2] and eukaryotes. These small organisms dominate the biomass in the open ocean, reaching respective concentration ranges of 10^6 - 10^5, 10^5 - 10^3, 10^5 - 10^3 and 10^4 - 10^2 cells per ml. The geographical distribution of these organisms, their biological characteristics (carbon and pigment content), and their dynamics in relation to the biotic factors are of major interest for the oceanographers. Initially used to discriminate and enumerate the different populations of phytoplankton, the application of flow cytometry has been extended to physiological analyses (e.g. DNA analysis) and more recently to phylogenetic analyses with the help of fluorescent molecular probes.

We present here four methods that are useful for the analysis of marine picoplankton both for natural samples and cultures.

1. Photosynthetic picoplankton. Photosynthetic picoplankton possess naturally fluorescing pigments (chlorophyll, phycoerythrin). Therefore the straight analysis of marine samples allows one to obtain information on the abundance, cell size and pigment content of the major photosynthetic picoplankton groups (prochlorophytes, cyanobacteria and eukaryotes).[3] This type of analysis can be performed either on unfixed samples on board ships or on preserved samples that are brought back to shore.

2. Heterotrophic bacteria and cell cycle analysis of photosynthetic prokaryotes. Since bacteria do not contain naturally fluorescing pigments, they need to be stained prior to analysis. Nucleic acid stains are very useful in this respect. Initially the UV-excited dyes DAPI or Hoechst 33342 have been used for this purpose.[4,5] However, recently, 488 nm-excited dyes such as YOYO-1, PicoGreen or SYBR™ Green-I have been introduced [6,7,8] that make this type of analysis possible on small low-cost flow cytometers. Moreover, nucleic acid stains provide information on the cell cycle distribution of photosynthetic prokaryotes, which allows to estimate directly growth rates in the ocean. [9,10]

3. Taxonomy of eukaryotic picoplankton. The taxonomy of small eukaryotic picoplankton is still poorly known nowadays. Most species have very few morphological features and can hardly be discriminated, even at the class level, by classical methods such as optical microscopy. Fluorescent oligonucleotide probes targeted to 18S rRNA appear as very promising tools for this purpose[11,12] allowing the identification of specific groups within complex communities.

4. DNA content and G-C% of isolated eukaryotic nuclei. For many eukaryotic marine strains, taxonomy is also very uncertain at finer levels, typically the species level. Species differences can be resolved by assessing their DNA content and G-C%. This has proved very useful for example to resolve the taxonomy of the ubiquitous genus *Phaeocystis*.[13] Another application of DNA content determination in microalgae is to assess ploidy levels in order to resolve the sexual cycle.[14] Such determination can be achieved on cultures using the range of DNA stains which are currently available and which present different G-C% sensitivity.

Subprotocol 1
Abundance and cell characteristics of photosynthetic picoplankton

Flow cytometry is now routinely used for the analysis of marine oceanic samples. Ideally, samples should be first analyzed fresh on board ships. Alternatively, if fresh analysis is not possible, a simple method for the preservation of marine samples, that interfere minimally with the cellular properties of phytoplanktonic cells, was developed by Vaulot et al.[15] The combined analysis of the light-scattering parameters and of the fluorescence of natural photosynthetic pigments (chlorophyll, phycoerythrin) allows the identification of different groups that differ in terms of size and pigment contents. Several aspects are critical to successful analysis of picoplankton samples:

- careful sample preservation (if necessary)

- good discrimination of populations from noise

- accurate identification of populations

- careful determination of flow sample rate.

Figure 1-3: Example cytograms were obtained for samples collected at 2 different depths, 60 m (Fig. 1, 2) and 5 m (Fig. 3), in oligotrophic waters (Pacific Ocean) containing *Prochlorococcus* (*Proc*), *Synechococcus* (*Syn*) and picoeukaryotes (Euk). Data acquisition is triggered by red fluorescence to reduce interferences from non fluorescent particles. Subpopulations are interactively defined with gates and identified by the combination of all recorded parameters. The *Synechococcus* population is discriminated from other phytoplankters by its orange fluorescence (due to the presence of phycoerythrin) on the orange versus red fluorescences cytogram (Fig. 1). *Prochlorococcus* cells that are smaller and less fluorescent are distinguished from picoeukaryotes on the bivariate distribution of the SSC (a function of size) versus red fluorescence (Fig. 2). The low red chlorophyll fluorescence of *Prochlorococcus* in surface waters does not allow complete separation of them from noise (Fig. 3). Fluorescent microspheres (0.95 μm beads) are added as internal reference. All other particles are non photosynthetic detrital particles.

▪ ▪ Materials

Equipment

- FACSort flow cytometer
- Pipetmen and tips for 1 to 1000 µl
- Waterbath
- 1.8 ml Cryovials
- Vortex mixer
- Disposable 0.2 µm-filter units and plastic syringes

Reagents

- Paraformaldehyde powder (SIGMA P-6148)
- Glutaraldehyde: 25% aqueous solution (SIGMA G-6257)
- RNase A : SIGMA (R-4875)
- RNase B: SIGMA (R-7884)
- SYBR^{TM} Green-I: Molecular Probes (Ref S-7563)
- 0.95 µm Yellow-Green fluorescent beads: Polysciences (Ref 71825)

Solutions

Fixatives solution
- 10 ml of para-formaldehyde 10%
- 200 µl of glutaraldehyde 25%

SYBR^{TM} Green-I
- 1% of the commercial solution in distilled water (store frozen in aliquot of 1 ml)

RNase mixture 0.1 g/l
- 50% RNase A
- 50% RNase B

0.95 µm Beads solution
- 10^6 beads per ml in 0.2 µm filtered seawater

Sheath fluid
- 0.2 µm filtered seawater

Citrate stock solution
- 1 M potassium citrate in distilled water

Preparation

- paraformaldehyde (see previous section)
- RNase. The mixture of RNase A and B is boiled for 10 min at 90°C to degrade any contaminating DNase

▧▧ Procedure

1. Put 1 ml of sample in a pre-labeled cryovial.

2. Add 100 µl of a mixture of paraformaldehyde 10% and glutaraldehyde 0.5%.

Sample collection and preservation

Note: The preservation is not necessary if samples are run immediately. The mixture of fixatives may be aliquoted and preserved at -20°C.

3. Wait for 15 min at room temperature then freeze the samples in liquid nitrogen and transfer at -80°C until further analysis.

Fig. 1

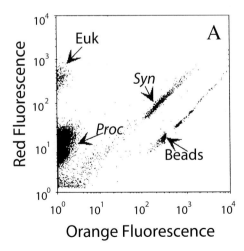

Note: The samples can be stored at -80°C for a period ranging from a few days to more than one year. They degrade very quickly if kept at -20°C.

Flow cytometric analysis

4. Preserved samples are quickly thawed in a water bath at 37°C.

Note: Dense culture samples can be diluted in 0.2 μm-filtered seawater in order to avoid coincidence on the flow cytometer.

5. Add 10 μl of the bead solution and 1 ml of the sample in a pre-labeled flow cytometric tube.

Note: The beads solution should be prepared daily. Beads are electrostatically charged and can stick to the wall of the tube. Moreover a progressive degradation of their fluorescence occurs when kept at room temperature.

Note: 0.2 μm-filtered seawater is preferred to distilled water as sheath fluid because the latter induces changes on both forward and side scatters as well as cell counts.

Instrument settings

6. Run a characteristic sample in order to adapt the configuration and the settings (see Comments).

Note: On the FACSort flow cytometer, for a natural sample collected at low depth, typical settings are FSC=E02, SSC=450, Green=650, Orange=650, Red=600. All parameters are collected on logarithmic scale. The discriminator is set on the red fluorescence and the threshold at 0.

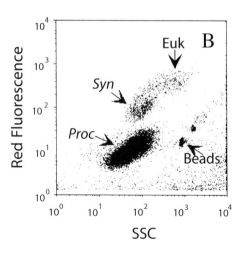

Fig. 2

7. Weigh a tube containing a sample. Start the acquisition. Start the chronometer as the injection of the sample begins.

Flow rate calibration

8. Simultaneously remove the sample tube and stop the chronometer. Weight the sampling tube.

Note: The chronometer indicates the time of injection allowing the evaluation of the flow rate. The difference between weights before and after the injection gives the delivered volume (see Comments).
A Pipetman-1000 or 100 can be used instead of a balance. (see Notes and Comments for more details)

Fig. 3

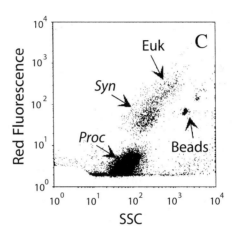

Comments

Instrument choice

Sensitivity is of critical importance when analyzing oceanic picoplankton. *Prochlorococcus* are very dim in surface oligotrophic waters and may be easily missed if the instrument sensitivity is poor or if optimal settings are not used. In our hand, the best commercially available instrument for picoplankton analysis is the FACSort from Becton Dickinson because of its excellent sensitivity and its compact design allowing easy use on board ships. Still it does not allow to completely resolve *Prochlorococcus* in most oceanic surface waters (Fig. 3). A custom modification of the laser focalization and of the forward scatter detection has been described[16] to allow *Prochlorococcus* detection even in extremely oligotrophic waters. Another critical aspect of picoplankton analysis is the necessity to obtain absolute (and not relative) abundance of the different cell populations. Unfortunately, most available instruments are not set to deliver well-defined sample volumes. Therefore in all cases it is necessary to precisely estimate the volume of sample analyzed. On the FAC-Sort, the most accurate method consist in determining the flow rate very precisely and then recording the time of analysis for each sample.

Preservation

Best results are obtained on fresh samples run immediately after collection. Fresh samples can be stored at 4°C for up to 12 hours with minimal effect. Fixation will always result in cell loss (about 10%), in change of scatters signals and in a sharp increase of orange fluorescence (up to two fold). The choice of the preservative may depend of the nature of the cells. Glutaraldehyde (0.1 to 1%) and paraformaldehyde (0.5 to 3%) are the most common chemicals used for preserving seawater samples. Glutaraldehyde at high concentrations (> 0.5%) generates crystallization, inducing noise that emits in the green region of the spectrum. The use of paraformaldehyde is an acceptable alternative to glutaraldehyde since it does not induce autofluorescence of the preserved cells. We found that a mixture of paraformaldehyde 1% and glutaraldehyde 0.05% (final concentrations) was an acceptable compromise.

Preparation of the fixative solution requires special care. Paraformaldehyde is a polymerized formaldehyde having poor crosslinking properties and cannot be used in that state as a fixative. By heating paraformaldehyde in water, the polymer dissociates into formaldehyde which is more water soluble and an efficient fixative. It is very difficult to completely dissolve the paraformaldehyde powder, it must be vigorously mixed in distilled water for 2 hours or more at 70°C. The solution obtained can be clarified, after cooling at room temperature, by addition of small amounts of sodium hydroxide 1N. The pH is then adjusted, but the solution must be always filtered on Whatman paper filters (Grade 113) and then through 0.2 m-filter before aliquoting and storage. We did not observe any significant difference between analysis of samples fixed with paraformaldehyde freshly prepared or preserved at -20°C for up to one year. However, after thawing, paraformaldehyde aliquots must be kept at +4°C and should not be used beyond one week.

Calibration of the flow rate

Most flow cytometers do not record the volume of the sample that has been analyzed; they only record the duration of the analysis. Therefore it is necessary, in general, to determine the sample flow rate very precisely. Some instruments can be equipped with automatic sampling devices that deliver known volumes with high reproducibility. Even in this case, the nominal volume must often be calibrated and corrected for a dead volume that must be determined. To calibrate such systems, a suspension of fluorescent beads can be used. After enumeration of the beads suspension by epifluorescence microscopy, 5 to 10 replicates are analyzed by FCM under fixed delivery conditions. The actual volume (V) delivered is given by:

$$V = A / S$$

where:
A = number of beads analyzed
S = number of beads per ml determined by epifluorescence microscopy in the initial suspension.

The use of fluorescent microspheres as internal standards at known concentration has sometimes been reported to calculate the analyzed volume. We do not recommend this method, however, because the electrostatic properties of beads are generally modified by seawater and they tend to stick to the plastic tubes or into the sample line, modifying their initial concentration.

A precise method for the calibration of the flow rate is described hereafter for a FACSort flow cytometer. After setting the rate (HI, MED, LO) to be calibrated, the outer sleeve of the injection needle is removed to inactivate the vacuum and to avoid aspiration of liquid. A tube containing two ml of 0.2 um-filtered seawater is set and the lower arm is immediately moved to the central position. Simultaneously a chronometer is started. After 10 min or more, the injection of the sample is stopped by moving the arm left or right, the sample tube is quickly removed and the chronometer stopped. The remaining volume is measured and the rate is deduced as follows:

$R = (Vi - Vf) / T$
where:
R = Rate (μl. min-1)
Vi = Initial volume (μl)
Vf = Final volume (μl)
T = Time (min)

Instead of using volume measurements, a balance can be used for weighing the tube containing filtered seawater before and after running and the following formula can be used:

$R = (Wi - Wf) / (T * d)$
where:
R = rate (μl. min-1)
Wi = initial weight (mg)
Wf = final weight (mg)
T = Time (min)
d = density of the solution (seawater; typically 1.036).

The method described here is designed for a FACSort flow cytometer but can be adapted to a majority of instruments. On the FACSort, the flow rate remains relatively constant over a large period. Nevertheless, it can be affected by environmental parameters such as room temperature and must be calculated daily at the beginning and at the end of the enumeration experiments. If it is suspected that the rate varies or drifts, it must be determined every 5 or 10 samples, since its determination is critical for abundance estimates. Instability may occur in the flow rate when aggregates or big cells are present in the sample which may clog the flow cell. A pre-filtration through a 10 μm nylon mesh is necessary in such cases.

Detection of phytoplanktonic cells

The detection and identification of phytoplanktonic groups is the main difficulty encountered by the operator. The intensity of the cellular parameters varies throughout the whole water column and the PMT voltages have to be adjusted frequently, depending on the size of organisms of interest, and for a given organism or community (e.g. picoplankton) depending on the depth sampled. The forward and the side scatters as well as the intensity of fluorescence from natural pigments usually increase with depth, due to photoacclimatation processes.[18]

The analysis of natural samples presents difficulties even for an experienced operator and one should be careful to collect all events of interest. The acquisition of small photosynthetic cells such as *Prochlorococcus* may be difficult particularly in surface samples of oligotrophic ocean waters (Fig. 3). The chlorophyll content is too low to be detected by the majority of the existing instruments.

If different depths have to be analyzed at a given site, it will be better to start with the deeper sample since the chlorophyll content per cell is then maximum. The threshold is set on the red detector and the red PMT voltage is increased until obtaining a clear separation between noise and the lowest population.

If only surface samples are available, the best approach is to start with a flow cytometric tube containing 0.2 μm-filtered seawater. The acquisition is started, the discriminator set on the red fluorescence (chlorophyll) and the threshold set at the minimum value. The red PMT voltage is increased until noise can be detected. The total number of events per second must be maintained below 100. Then a surface sample is run and the red PMT value is decreased if necessary in order to obtain a total number of events below 1000 per second.

In both cases, it is critical to adjust the PMT values in order to use all the dynamic range of the logarithmic scales (see Fig. 2).

Sheath fluid

Filtered seawater is preferred to distilled water, because this latter can induce modifications of the refractive indexes of the cells resulting in changes of the measured forward (FSC) and side (SSC) scatters.

Data acquisition

Samples are collected as listmode files and routinely 20,000 to 40,000 events are recorded typically during 1 to 4 min on a FAC-Sort flow cytometer using the high (HI) sample flow rate. List mode storage gives more flexibility in data analysis, while requiring large storage space (typically a cruise would require 1 Go of disk space). We usually transfer all data to another computer to keep the instrument available for more analyses. In marine samples, rare cells (e.g. nano- or microplankton) are difficult to study and require the analysis of large sample volumes.

Data processing

Absolute cell concentrations for each population in a given sample are computed as follows:

$C_{pop} = T * N_{pop} / R * (V_{total} / V_{sample})$
where:
C_{pop} = Concentration of population in cell $\mu l-1$
N_{pop} = Number of cells acquired
T = Acquisition time (min)
R = Sample flow rate ($\mu l. min-1$) as determined for the sample series.
V_{total} = Volume of sample plus additions (fixatives, beads, etc...) (μl) V_{sample} = Volume of sample (μl)
   Data were processed by the custom-designed software CYTO-WIN (Vaulot unpublished) running under Windows, available freely at http://www.sb.roscoff/Phyto/cyto.html. All parameters are reported relative to the beads added to the samples:

$X_{rel} = X_{pop} / X_{beads}$

where: X_{pop} is the average value of a cell parameter (scatter or fluorescence) for given population and X_{beads} the same parameter for the beads. Before ratioing, both X_{pop} and X_{beads} must be expressed as linear values (not channels) after conversion from the logarithmic recording scale. Since beads and cells have very different refractive index, such ratioing does not constitute a size characterization. For example in Fig. 2, although *Synechococcus* and beads have similar size (about 1 μm), the latter have a 10 times larger side scatter.

Subprotocol 2
Abundance of heterotrophic bacteria. Cell cycle analysis of photosynthetic prokaryotes

▓▓ Materials

The following methods are used both for the enumeration of prokaryotes in marine assemblages and for the cell cycle analysis of photosynthetic prokaryotes. Flow cytometric analysis of bacteria, that have generally a very low DNA content, requires the combination of highly fluorescent stains and sensitive instruments. Nucleic acid stains are used for this purpose. However since they stain both DNA and RNA and since bacteria may have a relatively high RNA content when grown under optimal conditions, RNA must be removed enzymatically. For years, only the UV-excited dyes Hoechst 33258 (HO258), Hoechst 33342 (HO342) and DAPI could be used for analysis of phytoplanktonic cells. Recently, new dyes from Molecular Probes Inc, (Eugene, Oreg.) TOTO-1, TO-PRO-1, YOYO-1, YO-PRO-1 and PicoGreen have been introduced for the detection of small amounts of nucleic acids on electrophoretic gels. TOTO-1 and YOYO-1 are cyanine dyes that are chemically different but possess similar optical properties. They can be excited by blue light, emit in the green region of the spectra, are cell impermeant and can be used on fixed cells. TOTO-1 seems to exhibit strong affinity for CTAG sequences, while YOYO-1 was found to have two binding modes: at low concentrations it appears to be intercalating and at high concentrations, external binding occurs. PicoGreen has a strong affinity for double-stranded DNA and was designed to quantify this molecule in solution. The fluorescence of these dyes is proportional to DNA concentration and does not depend on the G-C content. They are, however, very sensitive to the ionic strength of seawater and cannot be used directly on natural samples. Nevertheless the quality of DNA distributions obtained with YOYO-1 or PicoGreen on cultured samples after dilution in a low hypotonic buffer such as Tris-EDTA (10 mM Tris-HCl, 1 mM EDTA, pH 7.2), make them useful for culture studies.[7] Even more recently, Molecular Probes has released a new family of nucleic acid dyes (SYBR family) for gel staining purposes. SYBR[TM] Green-I (SYBR-I) has a strong affinity for double-stranded DNA, but also binds with single-stranded nucleic acids with lower affinity. It can be excited by UV-light but is optimally excited at 495 nm. It has the advantage of being less mutagenic than ethidium bromide or propidium iodide. Since it has a very high fluorescence yield and is not sensitive to ionic strength, it appears ideally suited for analysis of marine samples.[8] The method involves initial fixation by aldehyde fixatives (see above), and samples can be either analyzed immediately or after preservation in liquid nitrogen for delayed analysis.

We only present here the method with the SYBR-I stain. However, for cultures, SYBR-I can be replaced either by PicoGreen or YOYO-1.[7]

Fig. 4,5: Flow cytometric analysis of a natural seawater sample collected in the Pacific Ocean after staining with SYBR-I. Fig. 4 represents the DNA fluorescence versus chlorophyll content and shows the distribution of the cell cycle of *Prochlorococcus* and of the bacteria. Fig. 5 represents the distribution of the SSC (as a function of the size) versus DNA-fluorescence of the prokaryotic fraction. Two bacterial populations referred as BI-like and B-II-like groups can be discriminated. *Synechococcus* and picoeukaryotes were present at low concentrations but are not visible on these graphs due to the level selected and because *Prochlorococcus* and heterotrophic bacteria are largely outnumbering them.

Equipment

- FACSort flow cytometer
- Pipetmen and tips for 1 to 1000 µl
- Waterbath
- 1.8 ml Cryovials
- Vortex mixer
- Disposable 0.2 µm-filter units and plastic syringes

Reagents

- Paraformaldehyde powder (SIGMA P-6148)
- Glutaraldehyde: 25% aqueous solution (SIGMA G-6257)
- RNase A : SIGMA (R-4875)
- RNase B: SIGMA (R-7884)
- SYBR^{TM} Green-I: Molecular Probes (Ref S-7563)
- 0.95 µm Yellow-Green fluorescent beads: Polysciences (Ref 71825)

Solutions

Fixatives solution
- 10 ml of para-formaldehyde 10%
- 200 µl of glutaraldehyde 25%

SYBR^{TM} Green-I
- 1% of the commercial solution in distilled water (store frozen in aliquot of 1 ml)

RNase mixture 0.1 g/l
- 50% RNase A
- 50% RNase B

0.95 µm Beads solution
- 10^6 beads per ml in 0.2 µm filtered seawater

Sheath fluid
- 0.2 µm filtered seawater

Citrate stock solution
– 1 M potassium citrate in distilled water

Preparation

– paraformaldehyde (see previous section)
– RNase. The mixture of RNase A and B is boiled for 10 min at 90°C to degrade any contaminating DNase

▨ ▨ Procedure

Sample collection and preservation

1. Put 1 ml of sample in a pre-labeled cryovial.

2. Add 100 µl of a mixture of paraformaldehyde 10% and glutaraldehyde 0.5%.

Note: The mixture of fixatives may be aliquoted and preserved at -20°C.

3. Samples are fixed for 15 min at room temperature. If samples cannot be analyzed immediately, after fixation, freeze them in liquid nitrogen and transfer them at -80°C until further analysis.

Note: The samples can be stored at -80°C for a period ranging from a few days to more than one year. Otherwise they may be preserved at -20°C for a few weeks.

Fig. 4

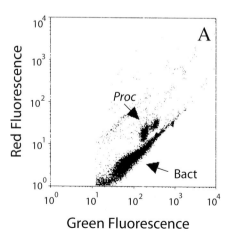

4. If samples have been preserved, they are thawed for 5 min at 37°C.

Sample preparation

Note: In order to avoid coincidence on the flow cytometer, culture samples can be diluted in 0.2 μm-filtered seawater or in a less hypotonic buffer for the cell cycle analysis.

5. In a pre-labeled FACSort tube, mix:
   - 250 μl of sample
   - 2.5 μl of RNase mixture
   - 7.5 μl of citrate
   - 5 μl of beads solution

Note: Incubation must be performed in the dark and analysis in a darkened room because most of the fluorescent dyes are light-sensitive

6. Incubate for 30 min at 37°C.

7. Add 2.5 μl of SYBR-I solution.

Staining

8. Incubate 15 min at room temperature in the dark.

9. Analyze the samples on the flow cytometer.

Flow Cytometric analysis

Note: Sample rate is set to medium speed (MED) in order to avoid coincidence. Typical settings are FSC=E01, SSC=450, Green=650, Orange=700, Red=630. All parameters are collected on logarithmic scale. The discriminator is set on the green fluorescence and the threshold at 200.

Fig. 5

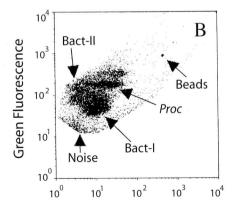

Comments

(see also Comments of the previous section)

Staining

Common problems are excessive noise and broad G1-like peaks. Noise can result from inadequate fixative solutions (see previous section). Make sure that all your solutions, including buffers, dyes or detergents are free of contaminating microorganisms. However, dye stock solutions must not be sterilized by filtration through 0.2 μm because some types of membranes can adsorb dyes. Broad G1-like peaks generally result from too high dye concentration, inducing nonspecific binding. Signal compensation may be necessary to better discriminate photosynthetic cells from heterotrophic bacteria.

In the first seconds of analysis, instability may occur or acquisition may look noisy (high numbers of events per second). A sample of stained sheath fluid can be used to initially equilibrate or to clean the sample line between two samples.

Data acquisition

Logarithmic amplification gives the multidecade dynamic range necessary for the analysis of natural populations of planktonic cells, that present a wide and complex distribution. Nevertheless for cell cycle analysis, acquisition needs to be performed both on logarithmic and linear scales. For the cell cycle analysis of photosynthetic cells, a T-connector is set on the output of the green photomultiplier to record simultaneously both signals for the green fluorescence of the SYBR-I. Then we disconnect the output from the orange photomultiplier and use the corresponding analog-to-digital converter (used previously for the orange fluorescence) to collect the linear signal for the green fluorescence.

Data processing

Data analysis is performed with cytowin as described in previous protocol. Care must be taken to include correction factors to account for sample dilution with the different solutions added.

Subprotocol 3
Taxonomy of eukaryotic picoplankton (In situ hybridization)

From knowledge of the ribosomal DNA sequences of microorganisms, nucleic acid probes specific for taxa can be designed.[19] They are complementary to a region of the ribosomal RNA molecule which is unique to the target group and can be used as "phylogenetic stains" once they have been labeled with a fluorochrome.[20] Oligonucleotide probes will hybridize to their homologous strand on the rRNA molecule within preserved cells. Labeled cells are detected by the probe-conferred fluorescence. This method is now commonly used for the identification of bacteria (see review from ref.[19]), but can also be used for the detection and identification of phytoplankton. Such probes are especially useful for the study of the smallest algae (picoeukaryotes) for which identification requires much time and expertise using traditional techniques because morphological characters are not readily available. Routinely used on cultured species,[11,12] whole-cell hybridization has yet to be applied to natural samples where the cell population of interest (a given species or genus for example) is part of a complex community. The latter application will require improvement in the sample preparation protocol to avoid cell loss and in the fluorescent reporters used to increase positive to negative signal ratios.

Cells are discriminated on SSC versus green fluorescence (FITC) cytograms. *Emiliania huxleyi* (Prymnesiophyceae) has been chosen as an example for *in situ* hybridization with a probe specific for the Prymnesiophyceae (FITC-PRYMN01, Fig. 6) and as a control, with a probe specific for the division Chlorophyta (FITC-CHLO01, Fig. 7) that should not label these cells. Fluorescent microspheres (0.95 μm beads) were added as internal reference.

Monoparametric distributions of the green fluorescence relative to FITC, are superimposed on Fig. 8. Data were collected independently for the autofluorescence, the negative (FITC-CHLO01), the positive (FITC-PRYMN01) and specificity (FITC-PRYMN01 / PRYMN01) controls.

Materials

Equipment

- FACSort flow cytometer
- Hybridization oven
- Centrifuge for 15 ml tubes
- Microcentrifuge
- Pipetmen and tips for 1 to 1000 µl
- 1,5 ml eppendorf tubes
- 15 ml tubes

Solutions

PBS
- Phosphate Buffered Saline (Sigma P 3688) pH 7.4

Permeabilising agents
- Paraformaldehyde 10% pH 7.2 (see above for preparation)
- 70% Ethanol in PBS, filtered through 0.2 µm

Hybridization buffer
- NaCl 0.9 M
- Tris HCl (pH 7.8) 20 mM
- SDS 0.01 %
- Formamide X % (concentration ranging from 0 to 50% has to be determined experimentally)

Probes

- Detailed protocol for the design and labeling of taxon specific oligonucleotide probes in Amman et al.[22]
- FITC labeled probes are kept at -80°C in 50 µl aliquots in distilled water. Working stocks are 50 ng/µl.
- Labeled probe used as examples:
  - FITC-CHLO01 specific for Chlorophyte algae
  - FITC-PRYMN01 specific for Prymnesyophyceae algae
- Unlabeled probe: Unlabeled-PRYMN01

▨▨ Procedure

1. For each species prepare 4 pre-labeled 15 ml tubes in order to measure: **Whole-cell hybridization**
   - Autofluorescence (no probe)
   - The fluorescence conferred by the specific probe (e.g. FITC-PRYMN01)
   - The fluorescence conferred by the non-specific probe (e.g. FITC-CHLO01)
   - The fluorescence conferred by a mixture of the specific probe labeled and unlabeled (e.g. FITC-PRYMN01 + PRYMN01)

Note: If the concentration of the target population is low (10^3, 10^4 cells/ml) in the original sample, concentration procedures (ultrafiltration or centrifugation) should be used prior to fixation.

Fig. 6

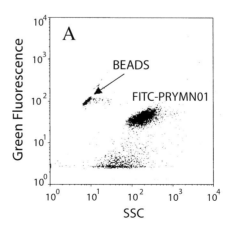

Fresh paraformaldehyde (less than one week old) has to be used.

2. In each tube add 5 ml of an exponental culture of picoeukaryotes at a concentration of about 10^5 cells/ml

3. Add 500 µl of paraformaldehyde 10% in each tube. Incubate for 1 h at 4°C.

4. Spin down the cells at 4000 x g for 3 min at 4°C

5. Remove the supernatant and immediately resuspend the cells in 500 µl of cold (-20°C) ethanol:PBS (70:30, vol:vol).

Note: This step modifies the natural fluorescence properties of photosynthetic cells. Photosynthetic pigments are damaged by alcohol treatments

6. Transfer each sample in Eppendorf tubes.

7. Spin down the cells at 4000 x g for 3 min and resuspend them in 20 µl of hybridization buffer.

Note: At that stage, samples can be stored at -80°C until needed for hybridization experiments and analyses

8. Add 1 µl of the chosen probe (as defined in step 1) to a 20 µl aliquot of the cells suspension.

Note: Alternative detergent can be tested in the hybridization buffer such as:
Triton (0.01 to 0.1%) (Sigma T 9284).

Fig. 7

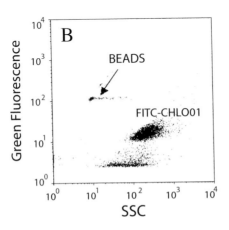

Tergitol (0.05 to 0.1%) (Sigma NP-40).
CHAPS (0.1%) (Sigma C 5070).

9. Incubate 3 h in the dark at 46°C.

10. Stop the hybridization by adding 500 µl of ice cold PBS pH 9.0. Keep the samples on ice.

Note: After hybridization, a washing step might be needed to remove nonspecific staining

11. Analyze the samples within 24 h.

Note: After 24 h an increase of the autofluorescence of the cells is observed as well as a decrease of the intensity of the fluorescence corresponding to specific labeling.

Analysis of samples

 Comments

Probe labeling

Fluorescein (FITC, Ex= 490nm, EM= 525nm) is currently used for probes labeling for *in situ* hybridization with algal cells. Purification of the oligonucleotides, is an essential step since remaining unlabeled oligonucleotides will compete with labeled probes and thus lower the fluorescence intensity of hybridized cells.

Fig. 8

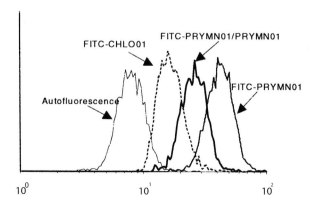

Signal intensity

It has been reported that the intensity of the labeling depends on the physiological state of the cells, although this idea is controversial and may not be true for all groups. We observed that the intensity of the specific signal is optimal when cells are harvested in exponential growth stage. The intensity of the signal is also size dependent. For example for the chlorophytes *Chlamydomonas concordia* (8 to 12 µm) and *Micromonas pusilla* (1 to 3 µm) the ratio between specific and non-specific probe-conferred signals are 10 and 3 respectively. A range of controls are necessary to distinguish a truly specific signal from autofluorescence and non-specific signals. Cells which do not grow under optimal conditions, may have fewer ribosomes and therefore show reduced fluorescence signals.[11] Differences in the cell permeability for probes among different species could be another reason for unequal labeling of cells. We use the following positive and negative controls:

- A sample without any probe is used to record the autofluorescence and all other fluorescence values will be ratioed to autofluorescence.
- A positive control constituted by a specific probe: the PRYMN01 and CHLO01 probes are used as positive controls for Prymnesiophyceae and chlorophyte cultures respectively.
- A negative control constituted by a non-specific probe: the CHLO01 and PRYMN01 probes are used as negative controls for Prymnesiophyceae and chlorophyte cultures respectively.
- A specificity control is constituted by a mixture of FITC-labeled and unlabeled specific probe in equal or varying proportions. If the probe is really specific, then 18S rRNA target sites should be blocked by the unlabeled probe and the intensity of the green fluorescence of FITC should decrease when the concentration of unlabelled probe increases.[12]

Until now, the use of this technique both on cultured and natural seawater samples has been limited by problems linked to low signal intensity. Several solutions have been proposed such as the use of multiple probes[21] or the use of indirect labeling (e.g. biotin labeled probes in conjunction with fluorescently labeled streptavidin). In all cases, very strict controls are needed before establishing that a specific probe labels a given strain of cultured algae.

Subprotocol 4
DNA content and G-C% on isolated eukaryotic nuclei for taxonomy and ploidy studies

An other potential use of nucleic acid-specific dyes, is the determination of the DNA content and the G-C% of phytoplankton cells. This is performed on unialgal cultures. DNA quantification has to be done on isolated nuclei, without fixation, in order to obtain stochiometric binding of the dyes. Ploidy level can be analyzed either on isolated nuclei or on fixed cells, when nuclei cannot be isolated. The isolation of nuclei, by osmotic pressure modification or by the action of detergents, presents two major advantages. First it facilitates the access of DNA to large stain molecules that normally cannot cross easily cellular membranes. Second, it removes the cytoplasm and avoids non-specific binding.

Addition of an internal reference is necessary to obtain a precise measurement of the DNA content of unknown species. Nuclei of reference and unknown cells must be mixed together prior to staining. The reference nuclei must be carefully chosen such that their DNA content are close to that of the planktonic cells. Chicken red blood cells (CRBC, 2.33 pg, 42.7% G-C) can be used for large algal cells. Algae with known DNA content and G-C% can be used when CRBC fall outside the range (e.g. *Phaeocystis* strain PCC 64 = 0.21 pg for the 1C, 54% G-C[13]). The DNA-related fluorescence should be acquired on both linear and logarithmic scales. The linear scale is optimal to accurately estimate the DNA content. The logarithmic scale is useful when the unknown species and the internal reference display a large difference in terms of dye-fluorescence intensity (Fig. 9). The cytogram of log Side Scatter (SSC) versus log DNA-fluorescence, is used for the identification of isolated nuclei and to gate out the debris. The adequacy of the preparation and staining procedure is evaluated by the coefficient of variation (CV) of the G1-like peak which must be as low as possible. Ideally, each sample should be run in duplicate, allowing for quality control of staining variability. Five to ten replicates should be performed for each species and averaged for reliable statistical results. The ratio of the peak positions of the sample to the standard is used for the calculation of the ploidy level. Routinely 5 000 to 20 000 events excluding noise are collected.

Three dyes are useful for such analysis. Propidium Iodide (PI) displays no base specificity and therefore is useful for assessing DNA content. Hoechst (or DAPI) dyes are A-T-specific while Chromomycin A3 (CA3) is G-C specific. Parallel staining of samples with these 3 dyes allows to estimate G-C% .[23] Propidium Iodide (PI, Ex=493 nm, Em=639 nm) binds to double-stranded nucleic acids by intercalation and has broad excitation bands both in UV and in blue-green regions that make it usable on the majority of the flow cytometers. PI is sensitive to the ionic strength and may be used in low hypotonic buffers. We have used it on whole cells, when analysis cannot be performed with HO342 or CA3. Staining is optimum at low concentrations ranging from 1 to 5 µg/ml.

The Hoechst bis-benzidimide dyes (HO342 and HO258) or DAPI are low cost non-intercalating dyes that have a high specificity for double-helical DNA, and bind preferentially to sequences with A-T bases. They emit blue fluorescence when excited by ultraviolet (UV) light at ~350 nm. HO342 can be used with a large variety of marine species. It cannot be used to estimate absolute DNA content or to compare cells that differ in their proportion of A-T bases. Nevertheless, it can be used on entire cells for ploidy measurements, by comparing the peak positions of an internal standard and of the species of interest, or of a mixture of species. HO342 fluorescence emission can be enhanced by addition of citrate (10 to 50 mM) or sodium sulfite (1 to 5 mM). It may be used at very low concentrations, up to 1 µg/ml, on fixed phytoplanktonic cells and is preferred to its homologue HO258 or to DAPI, that generally present unspecific binding. We observed that RNase treatment improves both the intensity of the signal and the coefficient of variation of the G1-like peak of HO342-stained cells, especially for the prokaryotic fraction of picoplankton.

Chromomycin A3, like Mithramycin, is a non-intercalating fluorescent antibiotic that presents a strong affinity for the 2-amino group of guanine in DNA when complexed with magnesium. It is optimally excited at 457 nm and very poorly at 488 nm. Therefore it is not suitable for single wavelength lasers that equip most small flow cytometers. Stock solutions of CA3 containing $MgCl_2$ (100 mM) show no deterioration for months when stored at -20°C. CA3 can be used on aldehyde-fixed cells without interfering with chlorophyll emission. It also gives good results with diatoms, after fixation by alcohols. It can be used over a wide pH range, from 6 to 8.5, with concentrations of 5 to 100 µg/ml, both in low and high hypotonic buffers, in presence of 30 mM magnesium. CA3 may be a potential health hazard and should be used carefully.

Flow cytometric analyses can be conducted on an EPICS 541 flow cytometer (Coulter, Hialeah, FL) equipped with a Biosense flow chamber and an Argon laser (Coherent). For each dye, the optical configuration of the EPICS 541 flow cytometer is illustrated in Fig. 10.

Fig. 9: *Phaeocystis* strain Rosko A was analyzed in presence of CRBC as internal standard after staining by the A-T-specific dye HO342. DNA fluorescence was collected as linear and logarithmic signals (Fig. 9). The linear fluorescence is necessary for the calculation of the ratio of modal DNA fluorescence between the species and the internal standard. The logarithmic signal allows to obtain a large dynamic range necessary to visualize a mixture of nuclei with very different DNA contents. Figure 11 shows the superimposition of histograms obtained independently for HO342, CA3 and PI. The amplification of the signal was adjusted for each dye to set the modal position of the internal reference (CRBC) to a fluorescence value of 70.

Materials

Equipment

- To quantify DNA content and GC%, analysis has to be performed with three different dyes, that have different excitation spectra. This requires the use of a tunable laser, that can be set in UV (353-357 nm), violet (457 nm) and blue (488 nm) lines that are available on argon laser
- Pipetmen and tips for 1 to 1000 µl
- 1.5 ml eppendorf tubes
- 10 µm nylon mesh

Reagents

- Hoechst 33342: stock solution 1 mg/ml in distilled water (Molecular Probes, H-1399)
- Chromomycin A3: stock solution 1 mg/ml in 100 mM magnesium chloride (Sigma, C 2659)
- Propidium Iodide: stock solution 1 mg/ml in distilled water (Sigma, P 4170)

Solutions

Nuclei isolation buffer
- Sorbitol 125 mM
- Potassium Citrate 20 mM
- Magnesium Chloride 30 mM
- Hepes: 55 mM
- EDTA: 5 mM
- Triton X-100: 0.1%

RNase mixture 0.1 g/l
- 50% RNase A
- 50% RNase B

Sodium bisulfite
- Stock solution 1 M in distilled water prepared daily and stored at 4°C

Preparation

RNase
- The mixture of RNase A and B is boiled for 10 min at 90°C to degrade any contaminating DNase

Alsever solution
- 72 mM NaCl
- 27 mM Sodium Citrate
- 114 mM Glucose
- 1% Triton X-100
- Adjust pH to 6.1 with Citric Acid 10%

CRBC (Chicken Red Blood Cells)
- Collect 1 ml of blood
- Add 3 ml of Alsever solution
- Mix for 1 min
- Centrifuge 1000 x g for 3 min
- Rince with 3 ml of Alsever solution
- Centrifuge 1000 x g for 3 min
- Resuspend in 3 ml of Alsever solution
- Store in small aliquots at -20°C.

 Procedure

1. Add 50 µl of sodium meta-bisulfite to 10 ml of isolation buffer.   **Isolation of nuclei**

Note: Addition of a protectant such as sodium bisulfite or β-mer-capto-ethanol is recommended to avoid rapid degradation of

Fig. 9

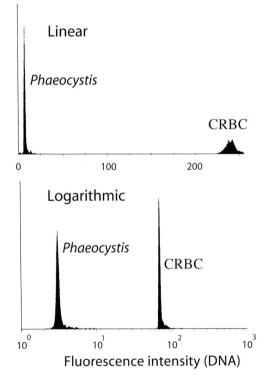

isolated nuclei. Sodium bisulfite is used in place of β-mercapto-ethanol because of the high toxicity of the latter.

Solutions containing sodium bisulfite are unstable and must be prepared just before the experiment.

2. Prepare 3 separate Eppendorf tubes, one for each dye.

3. Add 2 to 100 µl of the cell suspension to the necessary volume of isolation buffer, such that the final volume is 1 ml.

Add internal reference (CRBCs)

4. Add 5 µl of the internal reference. If CRBC are used thaw an aliquot and dilute it 50-fold with Alsever solution.

Note: The internal reference and the sample may be mixed before addition of the dye. The choice of the internal reference may depend on the nature of the sample (see Comments).

Staining

5. Add the stains, e.g. respectively 5, 40 and 30 µl/ml of stock solutions of Hoechst 33342, Chromomycin A3 or Propidium Iodide.

Note: RNase can be used with intercalary stains such as Propidium Iodide. Thus add 5 µl/ml of the RNase stock solution.

6. Keep the sample on ice and wait for 5 min for Hoechst, 15 min for Chromomycin or Propidium Iodide.

7. Filter through 10 µm nylon.

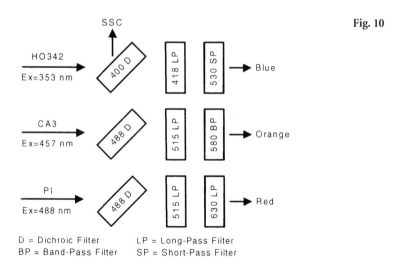

Fig. 10

8. Analyze on the flow cytometer at room temperature.

Note: Distilled water is used as sheath fluid.

Flow Cytometric analysis

Comments

Internal standard

The absence of convenient biological standard for the estimation of the DNA content of marine species constitute a major problem. The main difficulty of the procedure is therefore to choose an internal standard. CRBC standards generally give coefficients of variation below 5%. Nevertheless, in some cases, we observed a rapid degradation of the internal standard, due to the high hypotonicity of the isolation buffer or to some chemical compounds contained in the initial sample

DNA quantification

In the example given in Fig. 9, the position of the 1C peak of *Phaeocystis* relative to CRBC was 0.043 with HO, 0.19 with CA3 and 0.133 with PI. These ratios differ significantly, indicating that the G-C% of *Phaeocystis* is quite different from that of CRBC. As the intercalating dye PI binds proportionally to the base-pairs content, the ratio between the peak position of the sample to the standard is used to estimate the genome size. HO342 and CA3 bind to sequences of respectively 5 consecutive

Fig. 11

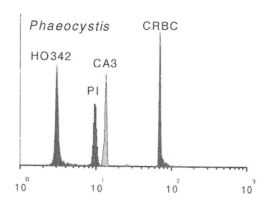

A-T bases and 3 consecutive G-C bases.[23] described a non-linear relationship between changes in fluorescence intensity and base composition. As A-T% (or G-C%) increases, the HO342 (or CA3) signal increases in a non-linear manner. Godelle developed a curvilinear model that allowed the determination of the G-C%. Two simplified estimates of the base-composition of a unknown species can be obtained from a known reference as follows:

$$\text{A-T\%}_{species} = \text{A-T\%}_{reference} * (R_{HO342} / R_{PI})1/5$$

$$\text{G-C\%}_{species} = \text{G-C\%}_{reference} * (R_{CA3} / R_{PI})1/3$$

where:
R_{PI} = Intensity species / intensity reference obtained with PI
R_{HO342} = Intensity species / intensity reference obtained with HO342
R_{CA3} = Intensity species / intensity reference obtained with CA3

Isolation of nuclei

With phytoplanktonic cells, it is often difficult to isolate nuclei without damaging them (e.g. for diatoms). The method described above has been developed in order to obtain optimal results on a wide range of photosynthetic cells. However, slight modifications in the composition of the isolation buffer may be required for some species. For example higher detergent concentrations may help for some difficult samples. Observations with epifluorescence microscopy will help the operator to optimize of the protocol. With HO or DAPI, variations in light emission can be observed when dye concentration is too high (nuclei appear white), when nuclei are partially degraded (variable color) or when the pH is too acid (yellow fluorescence). Aberrant morphology of nuclei or entire cells are usually an indicator of inadequate buffer composition.

This study was funded in part by contract MAS 3 - CT95 - 0016 (MEDEA) from the European Community and by the French programs Biodiversité Marine and Réseau Biodiversité ACC-SV7. The EPICS 541 and FACSort flow cytometers were funded in part by CNRS-INSU and the Région Bretagne.

References

1. Chisholm SW, Olson RJ, Zettler ER, Waterbury J, et al. A novel free-living prochlorophyte occurs at high cell concentrations in the oceanic euphotic zone. Nature 1988; Lond. 334: 340-343.
2. Waterbury JB, Watson SW, Guillard RRL. et al. Widespread occurrence of a unicellular, marine planktonic, cyanobacterium. Nature, 1979; Lond. 277: 293-294.
3. Olson RJ, Vaulot D, Chisholm SW. Marine phytoplankton distributions measured using shipboard flow cytometry. Deep Sea Res. 1985; 32: 1273-1280.
4. Button DK, Robertson BR. Kinetics of bacterial processes in natural aquatic systems based on biomass as determined by high-resolution flow cytometry. Cytometry 1989; 10: 558-563.
5. Monger BC, Landry MR. Flow cytometric analysis of marine bacteria with Hoechst 33342. Appl. Environ. Microbiol. 1993; 59: 905-911.
6. Li WKW, Jellett JF, Dickie PM. DNA distribution in planktonic bacteria stained with TOTO or TO-PRO. Limnol. Oceanogr. 1995; 40(8): 1485-1495.
7. Marie D, Vaulot D, Partensky F. Application of the novel nucleic acid dyes YOYO-1, YO-PRO-1 and PicoGreen for flow cytometric analysis of marine prokaryotes. Appl. Environ. Microbiol. 1996; 62: 1649-1655.
8. Marie D, Partensky F, Jacquet S, Vaulot D. Enumeration and cell cycle analysis of natural populations of marine picoplankton by flow cytometry using the nucleic acid stain SYBR Green I. Appl. Environ. Microbiol. 1997; 63: 186-193.
9. Vaulot D, Partensky F. Cell cycle distributions of prochlorophytes in the North Western Mediterranean Sea. Deep Sea Res. 1992; 39: 727-742.
10. Vaulot D, Marie D, Olson RJ. et al. Growth of *Prochlorococcus*, a photosynthetic prokaryote, in the equatorial Pacific Ocean. Science 1995; 268: 1480-1482.
11. Simon N, Lebot N, Marie D, et al. Fluorescent *in situ* hybridization with rRNA-targeted oligonucleotide probes to identify small phytoplankton by flow cytometry. Appl. Environ. Microbiol. 1995; 61: 2506-2513.
12. Lange M, Guillou L, Vaulot D, et al. Identification of the class Prymnesiophyceae and the genus *Phaeocystis* with ribosomal RNA-targeted nucleic acid probes detected by flow cytometry. J. Phycol. 1996; 32: 858-868.
13. Vaulot D, Birrien J-L, Marie D, et al. *Phaeocystis spp.*: morphology, ploidy, pigment composition and genome size of cultured strains. J. Phycol. 1994; 30: 1022-1035.
14. Edvardsen B, Vaulot D. Ploidy analysis of the two motile froms of *Chrysochromulina polylepis* (Prymnesiophyceae). J. Phycol. 1996; 32: 94-102.
15. Vaulot D, Courties C, Partensky F. A simple method to preserve oceanic phytoplankton for flow cytometric analyses. Cytometry 1989; 10: 629-635.

16. Dusenberry JA, Frankel SL. Increasing the sensitivity of a FACScan flow cytometer to study oceanic picoplankton. Limnol. Oceanogr. 1994; 39: 206-210.

17. Campbell L, Vaulot D. Photosynthetic picoplankton community structure in the subtropical North Pacific Ocean near Hawaii (station ALO-HA). Deep Sea Res. 1993; 40: 2043-2060.

18. Partensky F, Blanchot J, Lantoine F. et al. Vertical structure of picophytoplankton at different trophic sites of the tropical northeastern Atlantic Ocean. Deep Sea Research I, 1996; 43: 1191-1213.

19. Amann, RI, Ludwig W, Schleifer K-H. Phylogenetic identification and *in situ* detection of individual microbial cells without cultivation. Microbiol. 1995; Rev. 59: 143-169.

20. DeLong EF, Wickhan GS, Pace NR. Phylogenetic stains: ribosomal RNA-based probes for the identification of single cells. Science 1989; 243: 1360-1363.

21. Lee S, Kemp PF. Single-cell RNA content of natural marine planktonic bacteria measured by hybridization with multiple 16S rRNA-targeted fluorescent probes. Limnol. Oceanogr. 1994; 39: 869-879.

22. Amann RI. *In situ* identification of microorganisms by whole-cell hybridization with rRNA-targeted nucleic acid probes. Molecular Microbial Ecology Manual. Kluwer Academic Publishers 1995; 3.3.6: 1-15.

23. Godelle B, Cartier D, Marie D, et al. Heterochromatin study demonstrating the non-linearity of fluorometry useful for calculating genomic base composition. Cytometry 1993; 14: 618-626.

Yeast DNA Flow Cytometry

WENDY SCHOBER-DITMORE

Introduction

Yeast cells are abundant and relatively simple to manipulate therefore they have often been used to explore the process of cell growth and division. In the presence of mutations drugs or temperature conditions, DNA synthesis in these yeast cells, is slowed or inhibited. One of the tools that will reveal these changes uses DNA-binding fluorochromes, which are then examined by flow cytometry. To monitor the change in cell cycle of the yeast, the investigators synchronize the yeast culture, treat it via temperature or drugs, release it from synchrony, fix an aliquot from various time points as the cells come out of arrest, stain with DNA binding dyes and compare the staining patterns to control cells. The staining patterns reveal changes in the cell cycle progression.

It is essential that an asynchronous cell sample control which is representative of the cells to be tested is stained with the protocol described. This sample will confirm propidium iodide (PI) staining and RNase efficiency as well as confirm the intensity relationship between G0/G1 and G2/M. Typically, yeast cell cycle peaks have much wider coefficient of variation (CV) than those of mammalian cells, but as is the case with mammalian cells, the mean fluorescent intensity (MFI) of the G2/M population should be approximately 2 times that of the G0/G1 population. This asynchronous population and the corresponding G0/G1 and G2/M MFI values are a general measure of staining quality and a comparison for the experimentals.

Materials

Equipment

Flow Cytometer equipped with a 488 nm laser line which excites the propidium iodide (PI) and the emission is collected $\geq$ 590 nm after passing thru the appropriate laser blocking filters.

Solutions

- 10% EtOH
- phosphate buffered saline (PBS).
- RNase A buffer is 0.2 M Tris-HCl, 20mM EDTA with RNAse A added to a final concentration of 0.1%. (RNase is heat treated and DNase free)
- propidium iodide (PI) at 50 µg/ml final concentration in PBS (stock solution made at 500 µg/ml in H_2O)

Procedure

Fixation 1. Centrifuge 1×10^6 yeast cells and aspirate supernatant. Using a pipette, carefully resuspend cells in 5 ml of 70% EtOH. Fix 30 min to overnight on benchtop.

Note: For comparing samples, it is important that the cell number / ml be kept fairly constant between samples. If the cell concentration in the samples differs too widely, fluorescent intensities from sample to sample can drift.

Rehydration 2. Spin down fixed cells and rehydrate in 1 ml of phosphate buffered saline (PBS).

Note: Rehydration can take approximately an hour if the cells have been in EtOH for some time. The cells are stable at this point at 4 degrees for up to two weeks; if the cells are to sit it's a good idea to add NaN3 to a final concentration of 5 mM.

RNAse treatment 3. Wash rehydrated cells into 100 µl RNase A buffer for RNAse treatment. Incubate the samples at 37° for 4 hours.

Note: The RNAse treatment can be extended overnight which may produce narrower CV's (coefficient of variation) but in general is not necessary.

4. Wash the samples into 100 µl of PBS with propidium iodide (PI) at 50 µg/ml final. Staining occurs rapidly (within an hour). **Propidium iodide staining**

Note: This is often a convenient place to leave the samples overnight at 4 degrees in the dark.

5. Add 900 µl PBS to the samples for a dilution of 1:10. **Dilution**

6. Sonicate briefly to break up aggregates of cells. Use a fairly low setting and give each sample a 10-15 second pulse. **Sonicatication**

Note: If sonicatication is too high the cells can become completely disrupted, but are fairly forgiving. The cells can be sonicated, allowed to sit overnight, and still return good staining patterns the next day.

7. A 488 nm laser line excites the propidium iodide (PI) and the emission is collected $\geq$ 590 nm after passing thru the appropriate laser blocking filters. The instrument is set to initiate (trigger) all events from the linear PI signal, which helps eliminate inconsistencies due to the relatively small particle size or orientation in the beam. **Flow Cytometry**

8. Doublet discrimination is performed by gating on only the single cells, which can be discriminated from doublets and triplets using a linear PI signal vs. a peak PI signal in a dual parameter histogram. See figure 1. Typically, the single cells form a diagonal in this histogram when the PMT voltages and gains are set correctly to reveal the fluorescent events. The debris will be in the bottom left corner of this histogram and the double/triplets will be off the diagonal to the right, higher on the linear axis. This pattern and the parameters used to produce it, varies depending on the instrument manufacturer. A narrow gate is set around this singlet population and these singlet events are sent to a single parameter linear histogram.

Note: With proper sonication, doublets are rarely a problem in yeast sample.

Reference

Hutter KJ, Eipel HE. Microbial Determinations by Flow Cytometry. J Gen Microbial 1979 Aug; 113(2):369-75

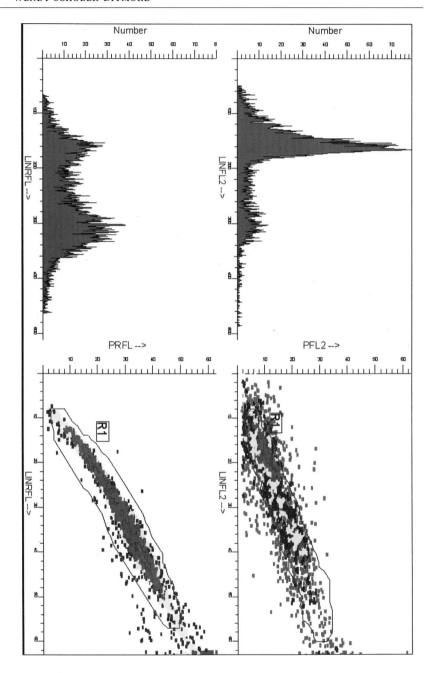

Fig. 1. Left: Linear single parameter (512 resolution) PI histograms resulting from the doublet discrimination gates (R1) set on the dual parameter linear PI vs. peak PI histograms on the right.

Section 6

Intracellular Physiological and Antibody Probes

ROCHELLE A. DIAMOND

Intracellular fluorescence measurements on flow cytometers can generate a variety of *in situ* data for physiological assays, functional assays, enzyme kinetic assays, as well as intracellular antigen detection. Many of these assays can be performed on live cells under normal physiological conditions. The stimulants or agonists can then be added in real time and the cell's responses recorded to determine population dynamics. Geoffrey Osborne *et al* present an intracellular calcium flux measurement which has become a standard for many immunologists and cell biologists looking at cellular response to ligand binding or other stimuli. Mike Fox describes two different methods to measure intracellular pH. Norberg *et al* use the intracellular pH to measure apoptosis. Rybak and Murphy use a ligand acidification kinetic assay on several kinds of cells.

Ingrid Schmid and Joe Dynlacht peer into fixed and permeabilized cells to measure intracellular antigens such as cytokines and nuclear matrix proteins. Mary Yui provides a protocol for measuring an intracellular cytokine. Sue DeMaggio describes a new line of commercial cytoenzymology products that open up off-the-shelf reagents to probe cell physiology.

Calcium Flux Protocol: INDO-1 Stained Thymocytes

GEOFFREY W. OSBORNE AND PAUL WARING

Introduction

Calcium kinetics in cells can be measured by flow cytometry in a convenient and sensitive fashion using the UV excitable dye INDO-1. The technique relies on the emission characteristics of the fluorochrome which shifts from a blue/green higher wavelength when unbound to calcium to a purple lower wavelength when bound to calcium. This enables a ratio of lower to higher wavelength fluorescence emission to be generated which is indicative of the calcium flux state of the cells, and most importantly is independant of cell size and total dye loading.

Much of the early work in this area was carried out by Grynkiewicz and colleagues,[1] with significant contributions also being made by Carl June and Peter Rabinovitch, whose work[2] is recommended for those interested in a detailed, theoretical insight into flow cytometric calcium analysis. The protocol presented here is a modification of their methods and has been used extensively on cells from experimental animals.

This protocol tests thymocytes with a range of concentrations of thapsigargan, however it may be substituted by other compounds of research interest. Thapsigargan is a plant derived sesquiterpine which irreversibly inhibits the endoplasmic reticulum calcium ATPase pump[3] but does not inhibit the plasma membrane calcium ATPase. This results in rapid cytosolic rises in calcium.[4]

Materials

Equipment

- Flow cytometer equipped with He-Cad or water cooled Argon Ion Laser with UV excitation wavelenghts 355 and 361
- Filters to collect fluorescence from
  - violet emission: / 405 nm ± 15 nm
  - Blue emission / 490 nm ± 15 nm
  - separated with a dichroic / 420-460 nm.
- Also a UV blocking filter UVLB(351-365) or a 380 long pass is used to block laser light in fluorescence path.

Solutions

- FCS free F15 medium
- 10 nM Thapsigargan

Preparation

Add 100 µl DMSO to the prepackaged Sigma 50 µg vial of INDO-1. This gives a stock solution of 0.5 µg/ml. Dilute to a final concentration 10 µg/ml.

Procedure

Prepare cells

1. Thymocytes at 2×10^7/ml at 37 °C in 1% FCS / F15 Media.

2. Pipette 1.9 ml of FCS free F15 medium in the required number of tubes and place in heating block or waterbath at 37 °C.

3. Add 100 µl DMSO to the prepackaged Sigma 50 µg vial of INDO-1. This gives a stock solution of 0.5 µg/ml. Dilute to a final concentration 10 µg/ml.

Note: Procedure to test a range of thapsigargan concentrations on thymocytes.
Keep prepared **INDO-1** at room temperature during experiment, but store at -20 °C.

4. Dilute from frozen stock solution to desired concentration. Try 10 nM Thapsigargan as a starting concentration. It should give a good flux.

Prepare Agonist (Thapsigargin)

Note: Keep **thapsigargan** on ice during experiments.

Note: *Caution:* **Thapsigargan** sticks to plastics very strongly so don't leave it in plastic containers and expect to get reproducable results!

5. Take 100 μl of cells, add 1.2 μl INDO-1 and incubate for 15 to 20 minutes at 37 °C.

Incubate cells

Note: For P815 or Jurkat cell lines, cells are resuspended at 1×10^7/ml and incubate in the presence of INDO-1 for 1 to 1.5. hours. Final cell concentration 1×10^6/ml.

6. Transfer cells into the 1.9 ml of FCS free F15.

Set up flow cytometer

7. Set up flow cytometer to store scattered light, fluorescence ratio and time parameters. Set total count to 400,000 cells. Set the cytometers temperature control unit to 37 °C.

Note: Collect fluorescence from violet emission using 405 nm +/- 15 nm, blue emission using 490 nm +/- 15 nm separated with a dichroic from 420-460 nm. Also a UV blocking filter UVLB(351-365) or a 380 long pass is used to block laser light in fluorescence path.

8. Put sample on flow cytometer, and based on the choosen testing time, calculate sample flow rate to use. For 40 minute time course (2400 secs) and 400,000 cells, adjust sample flow rate to give a flow rate of approximately 166 cells/second. Analyze first with respect to fwd and 90 light scatter from the primary lasers excitation beam. Then adjust the detectors which are collecting the INDO-1 signal which is generated by the secondary UV light emitting laser, to obtain a discrete "negative" population.

Note: Calibrate the fluorescence ratio to a known channel number on the flow cytometer. Do this for each sample by adjusting the ratioed parameter voltages such that the mean fluorescence ratio falls in the lower third of the linear ratio scale. The fine detail of this proceedure is instrument dependent and it is advisable to be thoroughly familiar with the ratioing of input signals before proceeding further.

Each sample needs to be well mixed initially, and ideally, continually during the run.

9. Establish a baseline for the sample by measuring the ratio of low/high fluorescence. Allow at least 5 minutes to make sure that variations are minimal (initial horizontal mean line in Fig. 1)

Test varying concentrations of agonist

10. Remove sample from the instrument and quickly add agonist, in this case 100 μl of 10nM thapsigargin. Replace sample on instrument and continue to save data until either the total count or time limit have been reached.

Note: Thapsigargin should give a flux as seen in Figure 1. If you don't see this type of response consider increasing the incubation time with INDO-1. If this fails, add a squirt of 70% ethanol to the sample and continue to run. If there is still no flux, you must assume the INDO-1 has not loaded.

Note the lack of change that the addition of an agonist has to the forward scattered light from the cells over the duration of the time course, Fig. 2.

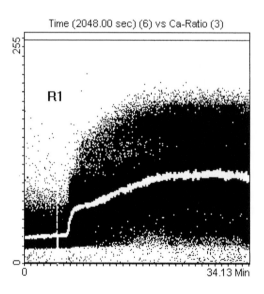

Fig. 1

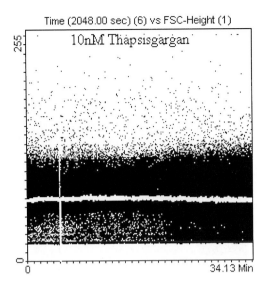

Fig. 2

11. 15 minutes prior to the end of the sample time of each sample, start incubating the next sample with INDO-1. Repeat the steps from 9 to 11 for as many concentrations of agonist as you are interested in.

12. When all the test samples have been run, do step 9 again then add 20 μl ionomycin to the remainder of the 2 ml's of sample. A "large" flux as seen in Fig. 3 should be occur.   **Establish maximum response**

Note: Note the dramatic affect the ionomycin has on the forward scattered light from the cells, Fig. 4. This can cause a decrease in the number of events triggering the instrument, if this scattered light parameter is used for trigger threshold, and may cause you to run out of "time" before reaching the desired number of events for collection.

13. Calculate the mean ratio fluorescence values using software such as the free program WinMDI (which created the data shown in these figures), and export these values and their corresponding linear channel values into a spreadsheet package. Then apply the formula given in the text section of this protocol to each mean fluorescence value to obtain calcium concentration.   **Data processing**

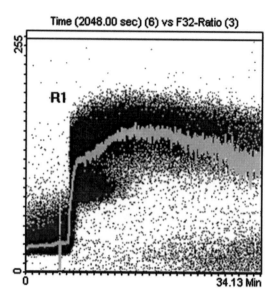

Fig. 3

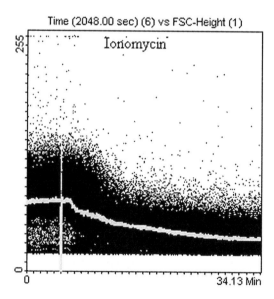

Fig. 4

Note: It cannot be over emphasised that a bit of trial and error will be needed when using this protocol with other cell types or other test compounds. For example, a range of concentrations of ionomycin will elicit a response, or a range of laser powers can be used as you're dealing with a ratio of fluorescence emission. The "take home" message here is really to be flexible in your approach to the problems you may encounter. Different test compounds may act more quickly that the one used in this protocol, allowing you to decrease the time course required to record a maximal flux, or provide you with the opportunity to try adding other compounds in the hope of reversing the flux. Consider all the variables and it should quickly become evident that this protocol can readily adapted to a wide range of investigations.

Instrument and Technical Considerations

Cells are typically incubated with the acetoxymethyl ester of INDO-1 which upon internalization is cleaved by esterases.[5] Depending on the type of cells and whether the cells have come directly from an animal/donor, or are cell lines, there can be wide variations in the levels of esterase activity. This needs to be taken into consideration when choosing a relative incubation time, and if there is little or no background knowledge regarding the esterase activity of the chosen cell type, then a range of incubation times must be trialed and the optimum established using an agent known to elicit a calcium flux.

Ideally the flow cytometer to be used can generate the ratio of the INDO-1 emitted light as a parameter, which reflects the ratio of free to bound calcium. This ratio can, if necessary, be calculated post data acquisition if the signals for low and high INDO-1 emissions are collected as raw data. However this is not the preferred method as it increases the data analysis time, increases the file size by storing two parameters instead of one, and most significantly, increases the difficulty in calibrating the INDO-1 ratio.

It is very important to calibrate the instrument so that a known ratio of INDO-1 is at a known channel on the ratio scale, thus providing a reference point for all fluxes which occur subsequently. For cell types with an unknown ability to respond to calcium agonists, the ratio in this laboratory is always set "low" on the ratio scale, i.e. 1:1 ratio calibrated to channel 100 on a 1024 channel linear scale. If the characteristic response to an agent is low, the ratio may be adjusted such that even very small fluxes can be detected. For example, the 1:1 ratio previously mentioned may be calibrated to channel 400, and thus a "small" flux of 2:1 will result in a shift to channel 800 on the 1024 channel scale. The one caveat on calibrating the position of the known ratio is that the ratio of maximal response of the cells, when treated with agonist, should fall on scale.

The usual agonist for obtaining a maximal reponse is ionomycin and it is worth remembering that ionomycin is particular good at sticking to sample lines, so they should be thoroughly cleaned by running DMSO or a strong detergent of choice, through the instrument for 10 minutes after ionomycin has been used in a sample. It is also often advisable to use ionomycin as the last sample of the run to avoid contamination of subsequent samples should your cleaning proceedure being less that perfect. The addition of a squirt of 70% ethanol to a sample can provide an alternative rapid method of checking whether the cells have loaded with INDO-1, without requiring the rigorous cleaning needed following ionomycin addition. The 70% ethanol effectively damages the integrity of the cell membrane allowing the INDO-1 to bind to calcium.

Another point that should be considered, is that certain experimental compounds may give a larger flux than ionomycin and if this is the case then this should then become your standard for maximal response (Rmax). See Data Analysis below.

Another desirable feature of the flow cytometer is to be able to record "time" as a parameter, thus enabling time versus calcium flux plots to be generated "on the fly". As with ratio, this parameter can be generated post data acquisition, based on the begin and end times stored within the data file. This is less than ideal as it in no way accounts of variations in sample flow rate, which can be significant over a long time course if the sample is not continually mixed.

Data Analysis

For some experiments it is enough to simply plot time versus fluorescence ratio, however if one wishes to calculate the size of the calcium movements in the cell, the following formula should be used

$$[Ca^{2+}]_i = K_d^* \frac{(R - R_{min})}{(R_{max} - R)} * \frac{S_{f2}}{S_{b2}}$$

where Kd is the effective dissociation constant, and R, Rmin and Rmax are the fluorescence intensity ratios at resting, zero and maximal intracellular calcium levels respectively and S_{f2}/S_{b2} is the ratio of blue fluorescence of the INDO-1 when calcium free or bound, respectively.[1]

Calculate Rmin, and S_{f2}/S_{b2} using a spectrofluorimeter, then the intracellular calcium concentration can be calculated from the mean fluorescence ratio recorded for each time point. Practically, it's best to import the mean ratio values into a spreadsheet program and let the computer calculate the $[Ca^{++}]_i$ and generate a plot of time versus $[Ca^{++}]_i$ Fig.8.1 and Fig. 8.3. Rather than leaving the analysis at this stage, it can be really informative to consider the scattered light parameters which should always be stored in your data file, as they can provide some valuable information about the way your calcium agonist is acting. As an example, in this thymocyte protocol, if one compares the forward scatter level for both thapsigargan and ionomycin, versus time, then the forward scatter tends to decrease as the time period increases when treated with ionomycin. Taking this approach further, three dimensional plots of scatter versus ratio versus time provide an elegant indication of how calcium fluxes within cells responding to agonists affects the cell morphology.

References

1. Grynkiewicz G, Poenie M, Tsien R.Y. A New Generation of Ca2+ Indicators with Greatly Improved Fluorescence Properties. The Journal of Biological Chemistry. 1985; 260 (6): 3400-3450.
2. June CH, Rabinovitch PS. Intracellular Ionized Calcium. In: Methods in Cell Biology. Vol 41. 2nd Edition, Part A. San Diego: Academic Press, 1994: 149-174.
3. Thastrup O, Cullen PJ, Drobak B et al. Academic Press, 1990. PNAS 87. 2466-2470.
4. Waring P, Beaver J. Cyclosporin A rescues thymocytes from apoptosis induced by very low concentrations of thapsigargin: effects on mitochondrial function. Exp. Cell Res 1996; 227: 264-276.
5. Haugland R.P. Handbook of Fluorescent Probes. 6th Edition. Molecular Probes: 549.

Intracellular pH Measured by ADB

MICHAEL FOX

Introduction

This protocol is used to measure the intracellular pH (pH_i) of cells using the dye 2,3-dicyano-1,4-hydroquinone diacetate (ADB). This dye readily passes through the plasma membrane and is cleaved by cellular esterases to form free dicyanohydroquinone (DCH) in the cell. The DCH shifts its fluorescence spectrum dependent upon the intracellular pH.[1] The ratio of fluorescence in two different bands is proportional to pH_i. DCH is excited by the argon UV laser lines, which limits its usefulness to users with UV lasers. It does give high resolution pH_i measurements with coefficients of variation (CVs) of 3-4% and can measure differences in pH_i of <0.05 pH units.[2] A disadvantage is that it readily leaks out of cells, so the timing of measurements must be carefully controlled.

It is important to understand that absolute values of pH_i are difficult to obtain. Relative values of pH_i are highly accurate with this method, however. One of the main reasons that absolute values are difficult is that the calibration depends upon knowing the intracellular potassium concentration. If that is in error, the pH_i values will be in error. However, relative values of pH_i are independent of the potassium concentration in the **HPC** buffer. Thus, it is preferable to quote pH_i values in terms of ΔpH_i compared to some standard cells.

▨ Materials

Equipment

- Flow cytometer with FL1 and FL2 detectors and UV laser capability to measure wavelengths between 420-440 nm and 470-485 nm. This can be done with bandpass filters or combinations of long pass and short pass filters. Setup used here is:
  - 418 nm longpass to block the scattered laser
  - 50% beamsplitter (from Corion)
  - 440 nm shortpass in front of FL1
  - 477 nm band pass (15nm wide) in front of FL2
- Tabletop centrifuge
- Pipettor and tips

Solutions

Normal Saline (NS) Buffer Preparation
- 5 mM Kcl
- 145 mM NaCl
- 0.5 mM $MgSO_4$
- 1 mM $CaCl_2$
- 1 mM Na_2HPO_4
- 5 mM glucose
- 10 mM HEPES
- 10 mM PIPES
- adjust pH with 1 M NaOH to pH 7.3 while stirring.

High Potassium Calibration (HPC) Buffers
- 120 mM KCl
- 30 mM NaCl
- 0.5 mM $MgSO_4$
- 1 mM $CaCl_2$
- 1 mM Na_2HPO_4
- 5 mM glucose
- 10 mM HEPES
- 10 mM PIPES
- adjust pH with 1 mM NaOH to pH 6.6, 7.0, 7.5 and 8.0 (individual buffers)

Stock ABD (2,3,-dicyano-1,4-hydroquinone diacetate) available from Sigma
- dissolve in dimethyl formamide (DFM) for a final concentration of 2 mg/ml
- store in the freezer (-20° C) for several weeks

Stock Nigericin
- Dissolve in 100% ethanol to a final concentration of 500 µg/ml. Store in the freezer (-20°C).

Procedure

Note: Grow cells in culture dishes for calibration and test samples.

Grow cells

1. For attached cells, plate out appropriate numbers of cells 12-24 hr in advance of the experiment to have 0,5 - 1 x 10^6 cells for each calibration and test sample.

2. Five samples are used for the calibration procedure; use 1 T-25 flask of cells for each sample.

3. For suspension cultures, the appropriate number of cells can simply be removed from the flask at the time of the experiment.

Prepare fresh buffers for calibration and measurement

Note: Buffers may be prepared in excess and used for several weeks. The pH should be adjusted at room temperature prior to each use, however. Buffers should be used at room temperature but can be stored at 4° C.

4. Prepare 100 ml **Normal Saline (NS) Buffer** (see **Solutions** for recipe).

5. Prepare 100 ml **High Potassium Calibration (HPC) Buffer** and split into 4 samples.

Note: The high potassium buffer should be adjusted to the intracellular potassium concentration in the cells of interest. This is usually around 120-140 mM. Adjust the NaCl concentration to maintain a total NaCl plus KCl of 150 mM.

6. Adjust pH of each **HPC Calibration Buffer** to 6.6, 7.0, 7.5, and 8.0 respectively using 1 M NaOH while vortexing. This range should cover most experimental results.

7. Adjust laser to UV lines (351-364 nm) with 50-100 mW power.   **Set up Flow Cytometer**

8. Insert filters for FL1 and FL2 detectors to measure wavelengths between 420-440 nm and 470-485 nm. This can be done with bandpass filters or combinations of longpass and shortpass filters. I use a 418 nm longpass to block the scattered laser, a 50% beamsplitter (From Corion), a 440 nm shortpass in front of FL1 and a 477 nm bandpass (15 nm wide) in front of FL2 (see Fig. 1).

9. Set up histograms to measure FL1, FL2, forward angle light scatter (FALS), and the ratio FL1/Fl2 using your software. All signals are processed with linear amplification. (See Fig. 4 for examples of FL1, FL2 and RATIO histograms.)

Note: The ratio of fluorescence intensity in wavelength band FL1 to that in wavelength band FL2 is proportional to pH_i. Your instrument must be able to calculate a ratio signal from two photomultiplier tubes, either in hardware or in software. The RATIO signal must be calculated on each cell as it passes through the laser beam.

10. Set gates from FALS to FL1 to FL2 to RATIO (Fig. 2). Gates on FL1 and FL2 should be set to eliminate any cells at the very top or bottom ends of the histogram, *e.g.* channels 7 and 250 for a 256 channel histogram. Appropriate gates on sample histograms are shown in Fig. 4.

Fig. 1. Example of a filter setup for measuring pH_i using DCH. This is only one possible filter configuration that would work. Combinations of dichroic and bandpass filters would also work

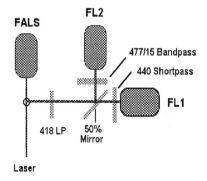

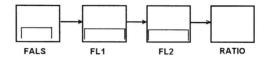

Fig. 2. Gating diagram for measuring pH$_i$

Note: Gating of histograms is necessary to exclude any cells that would go off-scale and pile up in the first few or last few channels. If included, they would skew the RATIO histogram. The intensity of the histograms can vary considerably at different pH values, which makes this necessary.

Calibrate ratio signal **Note:** The ratio measurement of your flow cytometer should be calibrated to assure that it gives a linear response. This need be done only once, not with every experiment.

11. Using the built-in pulser on your instrument, or high-resolution beads, adjust the FL1 high voltage to get a histogram in channel 50 (out of 256). Then adjust the FL2 HV to get a ratio histogram near the upper range (channel 225 or so), and collect and save FL1, FL2 and RATIO histograms. Adjust FL2 HV to get a new ratio histogram about 25 channels lower and save. Repeat this procedure to collect 6-10 sets of histograms (Fig. 3). Then calculate the actual ratio of FL1/FL2 and do a linear regression of the calculated ratio to the RATIO channel numbers from the RATIO histograms. This will show the linearity of the response and the actual ratio value for a given RATIO channel number.

Stain control and calibration samples **Note:** A control sample must be stained and run in order to set the voltages and gains on FL1 and FL2. After they are set with this sample, they cannot be changed during the course of the experiment.

12. Trypsinize cells from a flask. Pour medium from the flask into a 15 ml centrifuge tube, rinse flask once with ~2 ml trypsin, pour into tube, then add ~2 ml trypsin and incubate at 37° C for 3-5 min. After cells are detached, add to tube, then centrifuge in a tabletop centrifuge for 3-5 min at ~400 g (1500 rpm).

Fig. 3. Calibration of RATIO signal using a pulser. Each RATIO histogram corresponds to the ratio of the FL1 histogram and one FL2 histogram. The RATIO channels correspond to the calculated ration of FL1/FL2

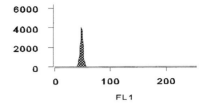

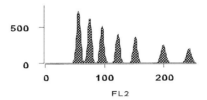

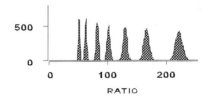

13. Decant or aspirate supernatant, then add 1 ml of **NS buffer** and resuspend cells by gentle vortexing or pipetting.

14. Add 5 μl of **Stock ADB** (10 μg/ml or 41 μM final concentration) to the cell suspension and incubate at room temperature for 10 min.

Note: The staining time is fairly critical since the intensity of staining varies considerably with time. This is due to the fact that ADB is hydrolyzed to DCH over time and the DCH leaks out of the cells fairly rapidly.[2]

15. Filter the sample through 40 μm nylon mesh. Run sample immediately on flow cytometer to set up high voltages. Adjust FL1 HV to get a peak in ~channel 60. Adjust FL2 HV to get a RATIO signal in channel 140-150. Do not make any adjustments after this is set.

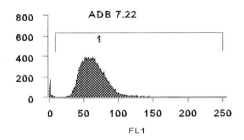

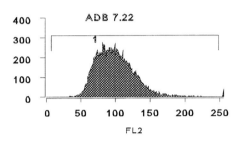

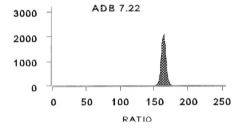

Fig. 4. Typical example of histograms from a calibration sample at pH 7.22. Upper) FL1 histogram; Middle) FL2 histogram; Bottom) RATIO histogram of FL1/FL2

Stain calibration samples and analyze

16. Trypsinize and centrifuge as in step 12 and decant or aspirate supernatant. Stagger each of the 4 calibration samples by 5 min to allow for time to run each sample on the flow cytometer after staining.

   Note: These samples determine the calibration curve that relates RATIO channel number to pH_i.

17. Add 1 ml **HPC buffer** (*e.g.* pH 6.6) to pellet and resuspend by gentle pipetting or vortexing.

18. Add 5 µl of **Stock ADB** and 10 µl **Stock Nigericin** to the cell suspension and incubate at room temperature for 10 min.

Note: Nigericin is added to adjust the pH_i to be the same as the external pH. Nigericin equilibrates the intracellular pH to the extracellular pH, provided that the potassium concentration in the buffer is the same as the potassium concentration in the cells.

19. Filter the sample through 40 μm nylon mesh and run 20,000 cells through the flow cytometer. CVs of the ratio histogram should be 2-4%.

20. Repeat steps 17-19 for each calibration buffer.

21. Perform steps 12-14 on test samples. Be sure to stagger samples so that the staining time is fairly constant. **Stain test samples and analyze**

22. Run the test samples through the flow cytometer.

23. Do a linear regression analysis of the mean RATIO values *vs* pH of the calibration buffers. The data should fall on a straight line with a high degree of accuracy ($R^2 = .99$) (See Fig. 5). **Data Analysis**

24. Calculate the test sample pH values from the linear regression analysis.

Note: Calculate the calibration curve from the calibration samples and then determine the pH_i of test samples by comparing the RATIO channel to the calibration curve.

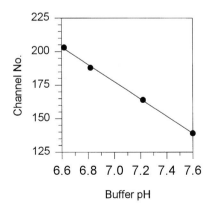

Fig. 5. Typical calibration curve of RATIO channel number vs buffer pH using ADB

References

1. Valet G, Raffael A, Moroder L, et al. Fast intracellular pH determination in single cells by flow-cytometry. Naturwissenschaften 1981; 68: 265-266.
2. Cook JA, Fox MH. Intracellular pH measurements using flow cytometry with 1,4-diacetoxy-2,3-dicyanobenzene. Cytometry 1988; 9: 441-447.

Measuring Intracellular pH Using SNARF-1

MICHAEL FOX

Introduction

This protocol is used to measure the intracellular pH (pH_i) of cells using the dye carboxy-seminaphthorhodafluor (SNARF-1). The acetoxy methyl ester form of this dye readily passes through the plasma membrane and is cleaved by cellular esterases to form free SNARF-1 in the cell. SNARF-1 shifts its fluorescence spectrum dependent upon the intracellular pH. The ratio of fluorescence in two different bands is proportional to pH_i. The pK_a is about 7.5, so SNARF-1 is better for measuring pH values between about 7 and 8. In contrast, DCH(see Protocol) is better for more acidic pH values. SNARF-1 is excited by the argon laser lines at 488 or 514 nm, though 514 nm excitation gives a stronger signal. It gives high resolution pH_i measurements with coefficients of variation (CVs) of 2-4% and can measure differences in pH_i of <0.05 pH units.[1] In contrast to DCH, it does not leak rapidly out of cells, so cells can be preloaded with SNARF-1, then treated with various agents and the pH_i measured immediately after the treatment. This makes it more flexible for measuring pH_i. We have found, however, that it consistently gives a higher value of pH_i than that obtained by DCH[1], and DCH is closer to the absolute value.

Intracellular pH is important in cell cycle regulation and plays an important role in cancer therapy by hyperthermia and by many drugs whose activities are pH-dependent. The accurate measurement of pH_i is very important in studying these agents. Heating cells causes significant changes in pH_i which can be measured with this technique.

It is important to understand that absolute values of pH_i are difficult to obtain. Relative values of pH_i are highly accurate with this method, however. One of the main reasons that absolute values are difficult is that the calibration depends upon knowing the intracellular potassium concentration. If that is in error, the pH_i values will be in error. However, relative values of pH_i are independent of the potassium concentration in the **HPC** buffer. Thus, I prefer to quote pH_i values in terms of ΔpH_i compared to some standard cells. Furthermore, the absolute values of pH_i determined by SNARF-1 are higher than those obtained by DCH (ADB Protocol) by about 0.4 pH units[1]. The reason for this is not clear, but it is a good reason to use relative pH_i values instead of absolute pH_i values.

Materials

Equipment

- Flow cytometer
- Tabletop centrifuge
- Pipettor and tips

Solutions

Normal Saline (NS) Buffer Preparation
- 5 mM KCl
- 145 mM NaCl
- 0.5 mM $MgSO_4$
- 1 mM $CaCl_2$
- 1 mM Na_2HPO_4
- 5 mM glucose
- 10 mM HEPES
- 10 mM PIPES
- adjust pH with 1 M NaOH to pH 7.3 while stirring.

High Potassium Calibration (HPC) Buffers
- 120 mM KCl
- 30 mM NaCl
- 0.5 mM $MgSO_4$
- 1 mM $CaCl_2$
- 1 mM Na_2HPO_4
- 5 mM glucose

- 10 mM HEPES
- 10 mM PIPES
- adjust pH with 1 mM NaOH to pH 6.6, 7.0, 7.5 and 8.0 (individual buffers)

Stock SNARF-1 AM (carboxy-seminaphthorhodafluor AM ester) available from Molecular Probes in 50 µg vials
- add 50 µl dimethyl sulfoxide (DMSO) for a final concentration of 1 mg/ml
- make up just prior to use

Stock Nigericin
- Dissolve in 100% ethanol to a final concentration of 500 µg/ml. Store in the freezer (-20°C).

Procedure

Note: Grow cells in culture dishes for calibration and test samples.

1. For attached cells, plate out appropriate numbers of cells 12-24 hr in advance of the experiment to have 2 - 5 x 10^5 cells for each calibration and test sample.

   Grow cells

Note: The fraction of cells stained adequately goes down if the cell concentration goes above 5 x 10^5 cells at the stain concentration used here.

2. Five samples are used for the calibration procedure; use 1 T-25 flask of cells for each sample.

3. For suspension cultures, the appropriate number of cells can simply be removed from the flask at the time of the experiment.

   Prepare fresh buffers for calibration and measurement

Note: Prepare fresh buffers for calibration and measurement. Buffers may be prepared in excess and used for several weeks. The pH should be adjusted at room temperature prior to each use, however. Buffers should be used at room temperature but can be stored at 4° C.

4. Prepare 100 ml **Normal Saline (NS) Buffer** (see **Solutions** for recipe).

5. Prepare 100 ml **High Potassium Calibration (HPC) Buffer** and split into 4 samples.

Note: The high potassium buffer should be adjusted to the intracellular potassium concentration in the cells of interest. This is usually around 120-140 mM. Adjust the NaCl concentration to maintain a total NaCl plus KCl of 150 mM.

6. Adjust pH of each **HPC Calibration Buffer** to 6.6, 7.0, 7.5, and 8.0 respectively using 1 M NaOH while vortexing. This range should cover most experimental results.

Set up Flow Cytometer

7. Adjust laser to 514 nm with 500 mW power.

8. Insert filters for FL1 and FL2 detectors to measure wavelengths between 550-590 nm and >610 nm. This can be done with bandpass filters or combinations of longpass and shortpass filters. I use a 530 nm longpass to block the scattered laser, a 590 dichroic, a 550 nm longpass in front of FL1 and a 610 nm longpass in front of FL2 (Fig. 1).

Note: Set up the laser, optical filters, and gating for the histograms to be collected.

9. Set up histograms to measure FL1, FL2, forward angle light scatter (FALS), and the ratio FL1/FL2 using your software. All signals are processed with linear amplification. (See Fig. 4 for examples of FL1, FL2 and RATIO histograms.).

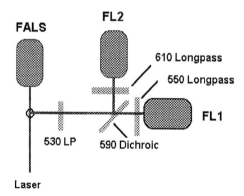

Fig. 1. Example of a filter setup for measuring pH$_i$ using SNARF-1

Note: The ratio of fluorescence intensity in wavelength band FL1 to that in wavelength band FL2 is proportional to pH_i. Your instrument must be able to calculate a ratio signal from two photomultiplier tubes, either in hardware or in software. The RATIO signal must be calculated on each cell as it passes through the laser beam.

10. Set gates from FALS to FL1 to FL2 to RATIO (Fig. 2). Gates on FL1 and FL2 should be set to eliminate any cells at the very top or bottom ends of the histogram, *e.g.* channels 7 and 250. Appropriate gates on sample histograms are shown in Fig. 4.

Note: Gating of histograms is necessary to exclude any cells that would go off-scale and pile up in the first few or last few channels. If included, they would skew the RATIO histogram. The intensity of the histograms can vary considerably at different pH values, which makes this necessary.

Note: The ratio measurement of your flow cytometer should be calibrated to assure that it gives a linear response. This need be done only once, not with every experiment.

Calibrate ratio signal

11. Using the built-in pulser on your instrument, or high-resolution beads, adjust the FL1 high voltage to get a histogram in channel 50 (out of 256). Then adjust the FL2 HV to get a ratio histogram near the upper range (channel 225 or so), and collect and save FL1, FL2 and RATIO histograms. Adjust FL2 HV to get a new ratio histogram about 25 channels lower and save. Repeat this procedure to collect 6-10 sets of histograms (Fig. 3). Then calculate the actual ratio of FL1/FL2 and do a linear regression of the calculated ratio to the RATIO channel numbers from the RATIO histograms. This will show the linearity of the response and the actual ratio value for a given RATIO channel number.

12. Trypsinize cells from a flask. Pour medium from the flask into a 15 ml centrifuge tube, rinse flask once with ~2 ml tryp-

Stain control and calibration samples

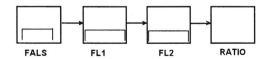

FALS   **FL1**   **FL2**   **RATIO**

Fig. 2. Gating diagram for measuring pH_i

sin, pour into tube, then add ~2 ml trypsin and incubate at 37°C for 3-5 min. After cells are detached, add to tube, then centrifuge in a tabletop centrifuge for 3-5 min at ~400 g (1500 rpm).

Note: A control sample must be stained and run in order to set the voltages and gains on FL1 and FL2. After they are set on this sample, they cannot be changed during the course of the experiment.

13. Decant or aspirate supernatant, then add 1 ml of **NS buffer** and resuspend cells by gentle vortexing or pipetting.

14. Add 5 µl of **Stock SNARF-1** (5 µg/ml or 11 µM final concentration) to the cell suspension and incubate at room temperature for 10 min.

Note: The staining time is not critical. Times up to an hour can be used and still obtain excellent values of pH_i, since the dye leaks out of the cells slowly.[1]

Fig. 3. Calibration of RATIO signal using a pulser. Each RATIO histogram corresponds to the ratio of the FL1 histogram and one FL2 histogram. The RATIO channels correspond to the calculated ration of FL1/FL2

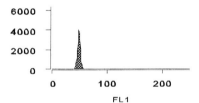

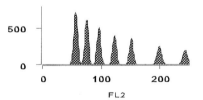

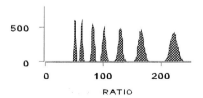

15. Filter the sample through 40 μm nylon mesh. Run sample immediately on flow cytometer to set up high voltages. Adjust FL1 HV to get a peak in ~channel 60. Adjust FL2 HV to get a RATIO signal in channel 140-150. Do not make any adjustments after this is set (Fig. 4).

Note: Run the control sample through the flow cytometer so that the high voltages can be set properly. Once the settings for FL1 and FL2 are made, they cannot be adjusted.

Note: Stain calibration samples and analyze them. These samples determine the calibration curve that relates RATIO channel number to pH_i

Stain calibration samples and analyze

16. Trypsinize and centrifuge as in step 12 and decant or aspirate supernatant.

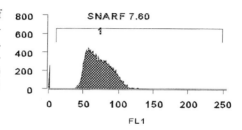
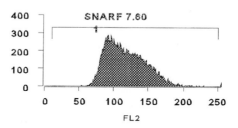
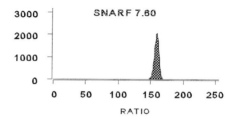

Fig. 4. Typical example of histograms from a calibration sample at pH 7.60. Upper) FL histogram; Middle) FL2 histogram; Bottom) Ratio histogram of FL1/FL2

17. Add 1 ml **HPC buffer** (*e.g.* pH 6.6) to pellet and resuspend by gentle pipetting or vortexing.

18. Add 5 µl of Stock **SNARF-1** and 10 µl of **Stock Nigericin** to the cell suspension and incubate at room temperature for 10 min.

Note: Nigericin is added to adjust the pH_i to be the same as the external pH. Nigericin equilibrates the intracellular pH to the extracellular pH, provided that the potassium concentration in the buffer is the same as the potassium concentration in the cells

19. Filter the sample through 40 µm nylon mesh and run 20,000 cells through the flow cytometer. CVs of the ratio histogram should be 2-3%.

20. Repeat steps 17-19 for each calibration buffer.

Stain test   21. Perform steps 12-14 on test samples.

samples and
analyze   22. Run the test samples through the flow cytometer.

Data Analysis   Fit a second order polynomial to the mean RATIO values *vs* pH of the calibration buffers. Actual values of pH_i for a given RATIO channel number are determined by the polynomial equation. Alternatively, simply graph the results (Fig. 5). The data should curve slightly, rather than lie on a straight line. The value of pH_i for a given RATIO channel number can be read directly from the graph.

Note: Calculate the calibration curve from the calibration samples and then determine the pH_i of test samples by comparing the RATIO channel to the calibration curve.

Note: It is important to understand that absolute values of pH_i are difficult to obtain. Relative values of pH_i are highly accurate with this method, however. One of the main reasons that absolute values are difficult is that the calibration depends upon knowing the intracellular potassium concentration. If that is in error, the pH_i values will be in error. However, relative values of pH_i are independent of the potassium concentration in the HPC buffer. Thus, I prefer to quote pH_i values in terms of ΔpH_i compared to some standard cells. Furthermore, the absolute values of pH_i determined by SNARF-1 are higher than those

Fig. 5. Typical calibration curve of RATIO channel numer vs buffer pH using SNARF-1

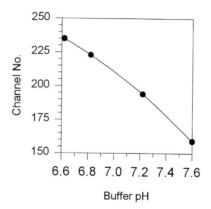

obtained by DCH (ADB Protocol) by about 0.4 pH units[1]. The reason for this is not clear, but it is a good reason to use relative pH_i values instead of absolute pH_i values.

24. Calculate the test sample pH values from the second order polynomial analysis or obtain graphically.

References

1. Wieder ED, Hang H, Fox MH. Measurement of intracellular pH using flow cytometry with carboxy-SNARF-1. Cytometry 1993; 14: 916-921.

FACS Analysis of pH and Apoptosis

GRANT W. MEISENHOLDER, JUDY NORDBERG,
AND ROBERTA A. GOTTLIEB

Introduction

Recent findings have generated interest in measuring intracellular pH (pH_i) in cells undergoing apoptosis. Studies have shown acidification to be an early feature of apoptosis in a number of systems[1] and prevention of acidification appears to be protective in some[2,3] but not all[4] systems. It is possible to obtain information about pH_i and apoptosis by three channel flow cytometry.

Measurement of pH_i is performed on live cells using carboxy-seminaphthorhodafluor-1 (c-SNARF-1), which is available as the cell-permeant acetoxy methyl (AM) ester. Because the AM ester can be cleaved in the presence of serum, cells must be loaded in serum-free media, using a concentration of 1-20 µM for 10-30 minutes. Retention varies among cell types, but in the Jurkat T-lymphoblast line, the dye is retained for over two hours.

C-SNARF-1 is analyzed using ratiometric technique with excitation at 488 nm and analysis at 575 (representing emission predominantly from the protonated form) and 620 nm (representing emission from the basic form).[5] Alternatively the second wavelength can be 610 nm, which is the isosbestic point (pH-insensitive). Ratiometric analysis allows for corrections in the extent of dye loading, and conversion of ratio to pH value is done by comparison to a set of pH standards which must be included with each experiment.[6] Normal cells are loaded with c-SNARF-1, then resuspended in a high-potassium buffer of defined pH (see Protocol) and the H^+/K^+ ionophore nigericin is added a few minutes before analysis. It is important to remember that the extracellular buffer and temperature will affect pH_i. We treat cells and load them in cell culture media in a CO_2 environment, and then transfer them to a bicarbonate-free buffer (transport buffer, see recipe below) for flow cytometry. The bicarbonate present in tissue culture media will be converted to CO_2 and H_2O and will alkalinize the cell, a process which is accelerated under the slight negative pressure of a FACS. Temperature also affects the pH value, and so it is important to use warmed media and to maintain near-physiologic temperatures of the samples.

For detection of apoptosis, cells are incubated with Annexin V-FITC in a calcium-containing buffer. This method relies on the observation that during apoptosis, phosphatidylserine is restricted to the inner leaflet of the plasma membrane lipid bilayer. However, during apoptosis, the bilayer becomes disordered, and phosphatidylserine becomes externalized.[7] Annexin V binds to

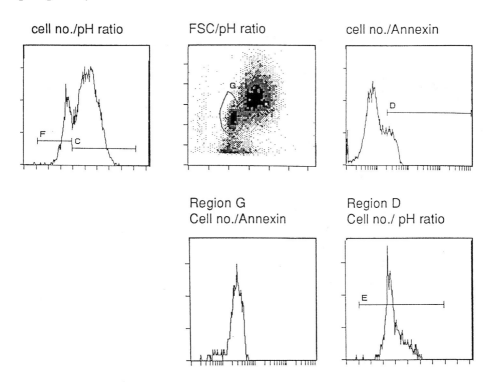

Fig. 1. Cells were induced to undergo apoptosis with anti-Fas antibody for three hours, then were labeled with c-SNARF-1-AM and Annexin V-FITC. The pH histogram (*top row, left*) shows an acidic subpopulation, seen also to have a lower forward scatter (*top row, center*). The distinction between regions F and C corresponds to pH 6.8. Analysis of the acidic subpopulation outlined in Region G shows strong labeling with Annexin V-FITC, indicating that acidified cells have externalized phosphatidylserine. The Annexin V-FITC bright cells (region D) are analyzed for pH (*below, right*), which shows that the annexin-positive cells are acidic

phosphatidylserine in the presence of calcium. One caveat is that cells with ruptured plasma membranes will also bind Annexin V; so most studies have required the simultaneous exclusion of propidium iodide. Retention of c-SNARF-1 also requires an intact plasmalemma and can be used with Annexin V-FITC. We have found a very tight correlation between cells which have undergone acidification and those which have externalized phosphatidylserine[8] (see Fig. 1).

C-SNARF-1 has been used in conjunction with Hoechst 33342 (DNA-binding) dye,[4] which is taken up more strongly by apoptotic cells. However, this has such a strong signal that it requires use of electronic compensation, and affects the pH calibration and therefore brings into question the reliability of the pH values obtained. However, it may be useful in cases where Annexin V is

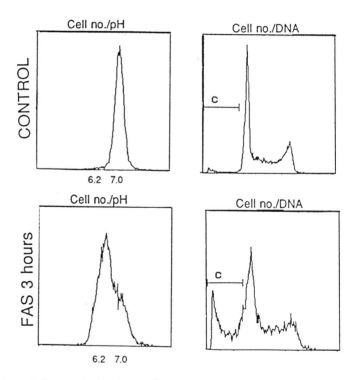

Fig. 2. Cells were induced to undergo apoptosis with anti-Fas antibody and analyzed for pH followed by fixation in ethanol and cell cycle analysis with propidium iodide. The majority of Fas-ligated cells have acidified (*bottom left*). Subsequent cell cycle analysis demonstrates an increase in subdiploid cells (region C, *bottom right*)

not suitable, for instance, when calcium-containing buffers cannot be used.

If additional evidence of apoptosis is desired, one can fix the cells in ethanol immediately after pH analysis, and then measure DNA content after propidium iodide staining. The c-SNARF-1 dye is washed out during fixation and will not interfere with PI staining. One drawback of this approach is that it is no longer possible to correlate pH and apoptosis on a cell-by-cell basis. However, we have found a reasonably good correlation between the percentage of acidified cells (using pH 6.8 as a cutoff) and the percentage of subdiploid cells[2] (see Fig. 2).

▦ Materials

Solutions

Serum Free Medium
- 103 mM NaCl, 5.4 mM KCl, 1.8 mM calcium gluconate, 0.4 mM MgSO$_4$, 5.6 mM Na$_2$HPO$_4$, 11 mM D-glucose, 24 mM NaHCO$_3$, pH 7.4.

Transport buffer
- 103 mM NaCl, 5.4 mM KCl, 1.8 mM calcium gluconate, 0.4 mM MgSO$_4$, 5.6 mM Na$_2$HPO$_4$, 11 mM D-glucose, 24 mM Hepes, pH 7.4.

Buffer for pH standards
- 125 mM KCl, 20 mM K$^+$ phosphate (adjusted to defined pH using mono- and di-basic phosphate salts See chart below).

Buffer for pH standards

pH	K$_2$HPO$_4$	KH$_2$PO$_4$	KCl	D-glucose	DI-H$_2$O
	1 M stock µl added	1 M stock µl added	1 M stock µl added	0.56 M stock µl added	
6.2	192	808	6250	500	q.s. to 50 ml
6.6	381	619	6250	500	q.s. to 50 ml
6.8	497	503	6250	500	q.s. to 50 ml
7.0	615	385	6250	500	q.s. to 50 ml
7.4	802	198	6250	500	q.s. to 50 ml
7.8	908	92	6250	500	q.s. to 50ml

▓ Procedure

Prepare cells 1. Wash cells twice by pelleting at 400 x g for 10 minutes, remove supernatant and resuspend in 2 mls of serum-free media. Repeat once and resuspend in same media at a concentration of 10^6 cells / ml in 12 x 75 mm Falcon tubes (#2058).

Prepare Snarf reagent 2. Dissolve carboxy-SNARF-1-AM (Molecular Probes, Eugene, OR) to a concentration of 1 mM in DMSO, add to cells to final concentration of 10μM. (Range 1-20 μM).

Incubate 3. Incubate 30 minutes (range 10-30 minutes) at 37°C and 5% CO_2.

4. Wash cells twice in transport buffer (see recipe on flow chart) and resuspend in transport buffer, all at 37°C.

5. Resuspend cells to be used for pH standards in K-phos buffers (pH range 5.8, 6.2, 6.6, 6.8, 7.0, 7.4, 7.8), and add nigericin (Sigma, St. Louis, MO) to standards (10 μM, from 10 mM stock prepared in ethanol).

Stain with Annexin V 6. For Annexin labeling, add Annexin V-FITC (Nexus Research, Maastricht, Netherlands) to cells (1 μg/mL) 5 min before flow cytometry.

Note: Annexin V needs at least 1.8 mM Ca^{++} for adequate binding. To reduce the reagent cost, samples can be analyzed in volumes of 200 μL in smaller tubes.

Determine pH 7. Determine pH by ratiometric analysis using excitation of 488nm and emission at 575 and 620nm (linear scales).

Note: Annexin V-FITC is read at 525nm (log scale) with 488nm excitation.

Fix for cell cycle analysis 8. For cell cycle analysis, add an equal volume of ice cold 100% EtOH dropwise while vortexing, incubate 10 min.

Stain with propidium iodide 9. Wash cells 1 x in PBS, resuspend in PBS with 10mg/ml RNAse A and 50μg/ml Propidium Iodide.

Analyse 10. Analyse cells with 488nm excitation and emission at >600nm.

References

1. Gottlieb RA. Cell acidification in apoptosis (Review). Apoptosis 1996; 1: 40-48.
2. Gottlieb RA, Nordberg JA, Skowronski E, Babior BM. Apoptosis induced in Jurkat cells by several agents is preceded by intracellular acidification. Proc. Natl. Acad. Sci. USA 1996; 93: 654-658.
3. Gottlieb RA, Dosanjh A. Mutant cystic fibrosis transmembrane conductance regulator inhibits acidification and apoptosis in C127 cells: Possible relevance to cystic fibrosis. Proc. Natl. Acad. Sci. U. S. A. 1996; 93: 3587-3591.
4. Li J, Eastman A. Apoptosis in an interleukin-2-dependent cytotoxic T lymphocyte cell line is associated with intracellular acidification. J Biol Chem 1995; 270: 3203-3211.
5. Basnett L, Reinisch L, Beebe DC. Intracellular pH measurement using single excitation-dual emission fluorescence ratios. Am J Physio 1990; 258: C171-C178.
6. Musgrove IA, Hedley DW. Measurement of intracellular pH. Methods in Cell Biology 1990; 33: 59-69.
7. Koopman G, Reutelingsperger CP, Kuijten GA, et al. Annexin V for flow cytometric detection of phosphatidylserine expression on B cells undergoing apoptosis. Blood 1994; 84: 1415-1420.
8. Meisenholder GW, Martin SJ, Green DR, et al. Events in apoptosis: Acidification is downstream of protease activation and Bcl-2 protection. J Biol Chem 1996; 271: 16260-16262.

Measurement of Ligand Acidification Kinetics for Adherent and Non-Adherent Cells

SHEREE LYNN RYBAK AND ROBERT F. MURPHY

Introduction

The use of flow cytometry for measuring the pH of endocytic vesicles is the focus of this chapter. The protocols described below are based on the dual fluorescence method first described by Murphy et al.[1] and are adaptations from previous work in the laboratory[2-5] and previous reviews.[6,7] Methods for measuring the pH of endocytic vesicles are described for both adherent and non-adherent cells. In addition, the use of two different endocytic markers, dextran and transferrin (Tf), are described. Transferrin is an iron-binding protein which binds to the Tf receptor on the cell surface. Upon reaching the acidic interior of the endosome, iron is released, while Tf remains bound to its receptor and recycles to the cell surface where it may bind more iron. Therefore, Tf is a convenient marker of early endosomes (2-5 min labeling) and recycling endosomes (10-12 min labeling). Dextran is a non-specific marker which enters the cell via fluid-phase endocytosis. Pulse-labeling cells for a period of time (10-60 min) followed by chasing for a long time allows dextran to be used as a probe for measuring the pH of late endocytic compartments such as late endosomes and lysosomes. The pH of earlier compartments may be measured with dextran if shorter pulse and chase times are used. However, it is sometimes difficult to obtain enough signal during such a short pulse.

There are several advantages of using flow cytometry to measure vesicular pH over other methods. First, flow cytometry allows for elimination of dead cells and debris from the final analysis. Second, the analysis of several thousand individual cells for any condition is quickly obtained. This allows for greater confidence in the final result. Third, due to the use of lasers, the signal:noise ratio is enhanced over that obtained with non-laser based instruments. Fourth, photobleaching is not a concern since each cell is only analyzed once.

Basic Approach

Measuring the pH of endocytic vesicles requires the following steps. First, cells are labeled with the marker of choice. The marker used depends on the endocytic vesicle of interest (see Introduction). Both experimental samples and samples whose endocytic vesicles are equilibrated (clamped) to neutral pH are prepared. Second, cells are analyzed by flow cytometry and the FITC/Cy5 ratio for each sample is determined. Third, the experimental FITC/Cy5 ratios are normalized to the pH clamped FITC/Cy5 ratio. Finally, the normalized FITC/Cy5 ratio is extrapolated to a previously-generated standard pH curve.

A full experiment for one cell type using dextran consists of performing either Protocol 1 or 2, Protocol 7, and Protocol 9. A full experiment for one cell type using Tf consists of performing either Protocol 3 or 4, either Protocol 5 or 6, Protocol 8, either Protocol 9 or 10, and Protocol 11.

Instrumentation

We have used an Epics Elite (Coulter Corporation, Miami, FL) dual-laser flow cytometer for the sample experiments described here. Sample flow rates, sheath pressure, and data rate were optimized for each cell type. Samples were kept on ice during analysis. A forward scatter threshold was used to select events to be recorded in list mode. In all experiments, FITC fluorescence was measured using 488-nm excitation (15 mW) and a 525-nm band-pass filter (25-nm bandwidth) while Cy5 fluorescence was measured using 633-nm excitation (10 mW) and a 675-nm band-pass filter (25-nm bandwidth). Spillover between the two fluorescences was negligible. Between 5000 and 10,000 events per condition are collected, with duplicate samples for each condition.

Before performing acidification measurements, a pH standard curve should be generated for each marker (see **Protocols 5 and 6**). It is essential to ensure that the cells are analyzed at the desired pH. On some flow cytometers, such as those using jet-in-air interogation, little mixing of sample and sheath fluids occurs during analysis. However, in other systems, mixing is unavoidable. It is important to note that when mixing does occur, the pH of the calibration sample can be altered. In the Elite using a cuvette with a 100 µm nozzle, this was a considerable problem which resulted in poor standard curves. To correct this problem, sheath buffers matched in pH to the pH of the sample to be analyzed were utilized. This sheath buffer was allowed to equilibrate throughout the fluidics of the flow cytometer. To ensure complete equilibration, the fluid exiting the nozzle was monitored until it was determined that the pH of the sheath fluid exiting the nozzle matched the pH of the sample, at which time the samples were analyzed. Although this is a long and tedious process, since the sheath buffer is changed for each pH value on the standard pH curve, the curve only needs to be generated once.

Fluorescent Dextran and Tf Conjugates

To measure the pH of endocytic vesicles, the ligand/marker of interest must be separately conjugated to a pH sensitive dye, usually fluorescein isothiocyanate (FITC), and a pH insensitive dye such as Cy5. Methods for conjugation of fluorescent dyes to dextran, ADDIN ENRef and Tf ADDIN ENRef have been previously described. FITC-dextran (70,000 M.W.) may be purchased from Sigma (St. Louis, MO) or Molecular Probes (Eugene, OR). However, whether the conjugates are prepared in the laboratory or purchased from an outside vendor, it is important to thoroughly dialyze conjugated dextrans and perform thin layer chromatography to ensure they do not contain free dye. [10]

The final concentration of each fluorescently-conjugated dextran typically ranges from 0.25 mg/ml to 4 mg/ml. It is important to test a range of concentrations to ensure that dextran-uptake is linear at the specific concentration used for subsequent experiments. Stock solutions of dextran are prepared in the growth medium of choice, without serum. For adherent cells, a 1 x stock is prepared, while for non-adherent cells a 2 x stock is prepared.

Stock solutions of fluorescently-conjugated Tf are usually around 1 mg/ml. The specificity of binding of the fluorescently-conjugated Tf must be determined by competition with unlabeled Tf. From this, an optimum labeling concentration is chosen, usually 2-10 µg/ml. Stock solutions of Tf (usually 5-10 µg/ml total labeled Tf) are prepared in the growth medium of choice, without serum. For adherent cells, a 1 x stock is prepared, while for non-adherent cells a 2 x stock is prepared.

Other Reagents

Iron-saturated Tf was purchased from ICN (Aurora, OH). The FITC-conjugated anti-Tf antibody was purchased from The Binding Site (Birmingham, England). Desferrioxamine mesylate (DFOM) was a gift from CIBA-Geigy (Suffern, NY). Sodium azide was purchased from Fisher Scientific (Pittsburgh, PA). All other reagents were purchased from Sigma (St. Louis, MO).

Miscellaneous Solutions

- PBS (8 mM Na_2HPO_4, 140 mM NaCl, 2.7 mM KCl, 1.5 mM KH_2PO_4, 0.9 mM $CaCl_2$, 0.5 mM $MgCl_2$, pH 7.4) is used for washing cells.

- RPMI, DMEM, Media 199 (or other growth medium; without serum) is used for cell labeling.

- Azide/2DG, 40 mM NaN_3 and 200 mM 2-deoxyglucose in PBS is used to deplete cellular ATP for pH calibration of dextran-labeled samples.

- Buffers for Dextran pH Calibration: 100 mM NH_4 Acetate, 100 mM HEPES, 20 mM NaN_3; pH 8.0, 7.4, 6.5, 6.0, 5.0, 4.5

- Buffers for Tf pH Calibration: 100 mM HEPES, 100 mM MES, 20 mM NaN_3; pH 8.0, 7.4, 6.5, 6.0, 5.0, 4.5.

Cell Culture

Adherent cells are grown in 35 mm or 60 mm dishes, depending on the number of cells that fit on a dish. All manipulations of adherent cells are done directly in the dish, after which time they are gently scraped into the desired buffer. At least 3 x

10^5 cells are needed per condition. Non-adherent cells may be grown in 100 mm or 150 mm dishes. Approximately $1\text{-}2 \times 10^6$ cells are needed per condition. In the experiments outlined below it is noted that some incubations are performed at 37°C. This temperature should be adjusted to the appropriate growth temperature required for the cells under investigation.

Subprotocol 1
Acidification of Dextran by Adherent Cells

▪▪ Materials

Equipment

Standard Flow Cytometer

Solutions

1 x FITC-dextran/Cy5 dextran stock PBS

Preparation

- 2 mg/ml FITC-dextran
- 0.25 mg/ml Cy5-dextran

▪▪ Procedure

Wash Cells   **1.** Remove medium and wash once with 37°C PBS.

Label Cells   **2.** To each 35-60 mm dish add 500 µl-1000 µl of the 1 x FITC-dextran/Cy5 dextran stock (Do not begin labeling all cells at once; see below).

3. Immediately place cells in a 37°C incubator for 2-60 min.

Note: The incubation (pulse) time is determined by the vesicle of interest. For example, 2 min would be appropriate for early endosomes, 5 min for late endosomes, and 60 min for lysosomes. Several time points may be collected in a single experiment to monitor the change in pH as the dextran traffics through the endocytic pathway.

4. Prepare parallel samples which do not get labeled. Incubate these samples in medium only. These samples are used to determine the background cellular fluorescence.

5. After the desired incubation time, remove dishes and quickly remove excess dextran. Wash cells eight times with 37°C PBS (~ 2 ml/dish/wash) to remove dextran stuck to the plasma membrane and surfaces of the dish. Because dextran becomes viscous at lower temperatures, is important to perform several warm washes. If fewer washes or colder PBS is used, the fluorescent dextran will stick to the surface of the cell. Do this step as quickly as possible. Stagger the incubation of the cells so that it only requires 1.5 - 2 min/batch of dishes to complete these washes.
Remove non-endocytosed dextran

6. Add 1 ml of 37°C medium without serum and return cells to the 37°C incubator for 0-120 min.
Chase

Note: The incubation (chase) time is also determined by the particular vesicle whose pH you are trying to determine. For example, 0-2 min would be appropriate for earlier vesicles such as

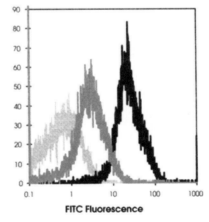

Fig. 1. Example ungated histograms of FITC fluorescence for A549 cells incubated with (*dark gray, black*) or without (*light gray*) 2 mg/ml FITC-dextran and 0.25 mg/ml Cy5-dextran for 30 min, washed, chased for 1 min, and clamped to pH 7.24 (*black*) or pH 4.5 (*dark gray*) as in **Protocol 7**. Note the high ratio of fluorescence to autofluorescence and the quenching of the FITC signal at low pH

FITC Fluorescence

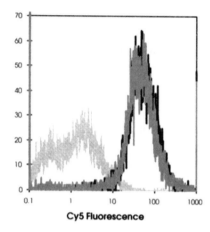

Fig. 2. Example ungated histograms of Cy5 fluorescence for A549 cells for the same samples as Fig.1. Note the high ratio of fluorescence to autofluorescence and the lack of quenching of the Cy5 signal at low pH

Cy5 Fluorescence

early endosomes, 5-10 min for late endosomes, and 60-120 min for lysosomes.

Final Wash   7. Place dish on ice and quickly wash twice with ice-cold PBS.

Note: Chill cells quickly to stop further membrane traffic.

Scrape Cells   8. Gently scrape cells into 400 µl-800 µl of PBS. The amount of PBS is determined by the final cell concentration desired (usually 10^6/ml) which ensures an optimal data/event rate. Keep samples on ice until ready to analyze.

Analyze   9. Analyze samples by flow cytometry, recording forward and side scatter (for gating live, single cells), and FITC and Cy5 fluorescences.

Subprotocol 2
Acidification of Dextran by Non-Adherent Cells

▓▓ Materials

Equipment

Standard Flow Cytometer

Solutions

1 x FITC-dextran/Cy5 dextran stock PBS.

Preparation

- 2 mg/ml FITC-dextran
- 0.25 mg/ml Cy5-dextran

▓▓ Procedure

1. Collect cells and centrifuge at 800 g for 1-2 min at room temperature.   **Collect, wash and count cells**

2. Remove medium and wash once with 37°C PBS.

3. Count cells. Dilute to 1×10^7/cells/ml with appropriate growth medium.

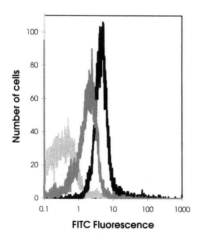

Fig. 3. Example ungated histograms of FITC fluorescence for K562 cells incubated with (*dark gray, black*) or without (*light gray*) 2 mg/ml FITC-dextran and 0.25 mg/ml Cy5-dextran for 30 min, washed, chased for 1 min, and clamped to pH 7.24 (*black*) or pH 4.5 (*dark gray*) as in **Protocol 7**. Note the lower fluorescence to autofluorescence ratio for K562 compared with A549 (see **Protocol 1**) and the similar quenching of FITC signal at low pH

Label cells   4. Add an equal volume of a 2 x FITC-dextran/Cy5 dextran stock.

5. Immediately place cells in a 37°C incubator for 2-60 min.

Note: The incubation (pulse) time is determined by the vesicle of interest, as described above for adherent cells.

6. Prepare parallel samples which do not get labeled. Incubate these samples in medium only. These samples are used to determine the background cellular fluorescence.

Remove non-endocytosed dextran   7. After the desired incubation time, remove cells from the incubator and aliquot cells (approximately 1-2 x 10⁶cells/aliquot) into 1.5 ml eppendorf tubes containing 1 ml 37°C PBS. Cells are then centrifuged in an Eppendorf microfuge at the greatest speed for the least amount of time that it takes the cells to be pelleted, without being damaged (K562 cells may be spun for 30 sec at 4000 rpm). Wash cells eight times with 37°C PBS to remove dextran stuck to the plasma membrane. Because these washes can take several minutes it may not be possible to determine early endosomal pH values using this method, unless the cells can be centrifuged for less time, and fewer samples are treated at once.

Chase   8. Resuspend final pellet of cells in 500 µl growth medium without serum and return cells to the 37°C incubator for 0 – 120 min.

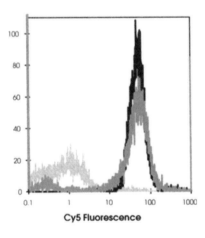

Cy5 Fluorescence

Fig. 4. Example ungated histograms of Cy5 fluorescence for K562 cells for the same samples as Fig. 3. Note the lack of quenching of the Cy5 signal at low pH

Note: The incubation (chase) time is also determined by the vesicle of interest, as described above for adherent cells.

9. Add 500 μl ice-cold PBS to cells. Centrifuge cells as described   **Final Wash**
   above and wash once more with ice-cold PBS.

Note: Chill cells quickly to stop further membrane traffic.

10. Resuspend cells into 400 μl-800 μl of PBS. The amount of PBS   **Resuspend**
    is determined by the final cell concentration desired (usually   **Cells**
    $1\text{-}2 \times 10^6$/ml) that ensures an optimal data/event rate. Keep
    samples on ice until ready to analyze.

11. Analyze samples by flow cytometry, recording forward and   **Analyze**
    side scatter (for gating live, single cells), and FITC and Cy5
    fluorescences.

Subprotocol 3
Acidification of Transferrin by Adherent Cells

▪▪ Materials

Solutions

- PBS
- serum-free medium
- 1 x FITC-Tf/Cy5 Tf stock
  When using a mixture of FITC- and Cy5-Tf, use a 9:1 ratio
  (9 times more FITC-Tf than Cy5-Tf).
- 4 μg/ml FITC-Tf and 2 μg/ml Cy5-Tf

▪▪ Procedure

1. Remove medium and replace with 1 ml serum-free medium.   **Wash cells**

2. Incubate 20 min at 37°C to remove endogenous Tf.

3. Remove medium and place cells on ice.

4. Wash twice with 4°C PBS to quickly chill cells.

5. To each 35-60 mm dish add 500 μl-1000 μl of the 1 x FITC-Tf/   **Label surface**
   Cy5 Tf stock.   **Tf-receptors**

Note: Choose a ratio of FITC to Cy5 to ensure that the signal: noise (the ratio of fluorescent signal from fully labeled cells to the autofluorescence of unlabeled cells) for both dyes is about equal.

Note: When using a mixture of FITC- and Cy5-Tf, use a 9:1 ratio (9 times more FITC-Tf than Cy5-Tf).

6. Prepare parallel samples which receive: (1) no labeled Tf (to determine the cellular autofluorescence) or (2) the same amount of labeled Tf in step 5, in addition to 1 mg/ml unlabeled Tf (blocked controls).

Note: Coincubation with an excess of unlabeled Tf inhibits specific binding of FITC-Tf and Cy-Tf, allowing measurement of the amount of non-specific binding.

7. Incubate on ice for 30 min.

Note: Incubation on ice allows only surface receptors to bind labeled Tf.

Remove unbound Tf

8. Wash three times with ice-cold PBS (~2 ml/dish/wash).

Note: Keep cells very cold to ensure that Tf does not internalize.

Allow Internalization

9. Add 1 ml of 37°C medium/dish. Place the labeled cells in 37°C incubator for 0-20 min Typical time points are: 0, 1, 2, 4, 6, 8, 10, 12, 14, and 20 min.

Note: Warming allows synchronized endocytosis of surface-bound Tf. A number of time points are normally collected for

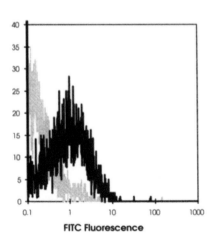

Fig. 5. Example ungated histograms of FITC fluorescence for A549 cells incubated with (*black*) or without (*light gray*) 4 µg/ml FITC-Tf and 2 µg/ml Cy5-Tf for 30 min on ice, washed, and scraped into PBS. Note the much lower ratio of fluorescence to autofluorescence for Tf as compared to dextran (see **Protocol 1**)

each experiment to monitor the pH of early and recycling endo-somes.

10. Remove cells from incubator at each desired time point and place on ice. Add 2 ml of ice-cold PBS to quickly chill cells.

Stop Internaliza-tion

11. If stripping surface Tf, follow **Protocol 6.**

Strip surface Tf, if desired

Note: The cells contain two populations of Tf: extracellular and intracellular. We would like to estimate the pH of intracellular Tf only. There are two methods to accomplish this. The first is to determine the percent of Tf on the surface of the cell at each time of incubation at 37°C and correct for it during calculation of pH values. This is done using an antibody against Tf (see **Protocol 5**). The second method is to strip away surface Tf by decreasing the pH of the medium (see **Protocol 6**).

12. Gently scrape cells into 400 µl-800 µl of PBS. The amount of PBS is determined by the final cell concentration desired (usually 10^6/ml) which ensures an optimal data/event rate. Keep samples on ice until ready to analyze.

Scrape cells

13. Analyze samples by flow cytometry, recording forward and side scatter (for gating live, single cells), and FITC and Cy5 fluorescences.

Analyze

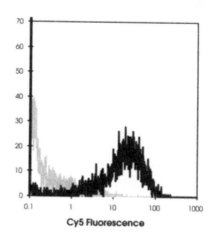

Fig. 6. Example ungated histo-grams of Cy5 fluorescence for A549 cells for the same samples as Fig. 5. Note the high ratio of fluorescence to autofluorescence

Subprotocol 4
Acidification of Transferrin by Non-Adherent Cells

▓▓ Materials

Solutions

- serum-free medium
- 4°C PBS
- 2 x FITC-Tf/Cy5 Tf stock

Preparation

- When using a mixture of FITC- and Cy5-Tf, use a 9:1 ratio (9 times more FITC-Tf than Cy5-Tf).
- 4 µg/ml FITC-Tf and 2 µg/ml Cy5-Tf.

▓▓ Procedure

Wash cells

1. Collect cells and centrifuge at 800 g for 1-2 min at RT.

2. Remove medium and replace with 1 ml serum-free medium.

3. Incubate 20 min at 37°C to remove endogenous Tf.

4. Centrifuge at 800 g for 1-2 min at 4°C.

5. Remove medium and place cells on ice.

6. Wash twice with 4°C PBS to chill cells, resuspending and centrifuging each time.

7. Count cells. Dilute to 1 x 10^7 cells/ml in serum-free medium.

Label surface Tf-receptors

8. Add an equal volume of a 2 x FITC-Tf/Cy5 Tf stock.

Note: Choose a ratio of FITC to Cy5 to ensure that the signal: noise (the ratio of fluorescent signal from fully labeled cells to the autofluorescence of unlabeled cells) for both dyes is about equal.

Note: When using a mixture of FITC- and Cy5-Tf, use a 9:1 ratio (9 times more FITC-Tf than Cy5-Tf).

Fig. 7. Example ungated histograms of FITC fluorescence for K562 cells incubated with (*black*) or without (*light gray*) 4 µg/ml FITC-Tf and 2 µg/ml Cy5-Tf for 30 min on ice, washed, and scraped into PBS

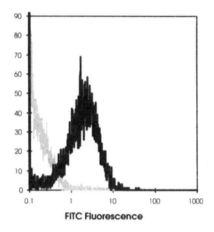

FITC Fluorescence

9. Prepare parallel samples which receive: (1) no labeled Tf (to determine the cellular autofluorescence) or (2) the same amount of labeled Tf in step 8, in addition to 1 mg/ml unlabeled Tf (blocked controls).

Note: Coincubation with an excess of unlabeled Tf inhibits specific binding of FITC-Tf and Cy-Tf, allowing measurement of the amount of non-specific binding

10. Incubate on ice for 30 min.

Note: Incubation on ice allows only surface receptors to bind labeled Tf.

11. Aliquot cells (approximately 1-2 x 10^6 cells/aliquot) into 5 ml sample tubes containing 4 ml PBS on ice marked with the desired incubation time. Centrifuge cells at 800 *g* for 1–2 min at 4°C. Remove PBS, resuspend pellet in 4 ml 4°C PBS. Repeat twice for a total of three washes.   **Remove unbound Tf**

Note: Keep cells ice-cold to ensure that Tf does not internalize.

12. Add 500 µl of 37°C medium per tube. Place the labeled cells in 37°C incubator for 0-20 min as marked on the tube. Typical time points are: 0, 1, 2, 4, 6, 8, 10, 12, 14, and 20 min.   **Allow Internalization**

Note: Warming allows synchronized endocytosis of surface-bound Tf. A number of time points are normally collected for each experiment to monitor the pH of early and recycling endosomes.

**Stop Interna-
lization**

13. Remove cells from incubator at each desired time point and place on ice. Add 3.5 ml of ice-cold PBS to quickly chill cells.

14. Centrifuge cells at 800 g for 1-2 min at 4°C, remove and discard supernatant.

**Strip surface
Tf, if desired**

15. If stripping surface Tf, follow **Protocol 6** below.

Note: The cells contain two populations of Tf: extracellular and intracellular. We would like to estimate the pH of intracellular Tf only. There are two methods to accomplish this. The first is to determine the percent of Tf on the surface of the cell at each time of incubation at 37°C and correct for it during calculation of pH values. This is done using an antibody against Tf (see **Protocol 5**). The second method is to strip away surface Tf by decreasing the pH of the medium (see **Protocol 6**).

**Resuspend
cells**

16. Resuspend cells in 400 µl-800 µl ice-cold PBS. Keep samples on ice until ready to analyze.

Analyze

17. Analyze samples by flow cytometry, recording forward and side scatter (for gating live, single cells), and FITC and Cy5 fluorescences.

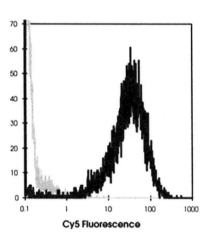

Fig. 8. Example ungated histograms of Cy5 fluorescence for the same samples as Fig. 7

Subprotocol 5
Determination of amount of surface transferrin

▨▨ Materials

See Protocols 3 or 4.

▨▨ Procedure

1. Prepare cells as described in **Protocol 3, Steps 1-4** or **Protocol 4, Steps 1-7.**

Prepare and Label cells

Note: Cells are labeled with just Cy5-Tf so that a FITC-conjugated antibody can later be used to detect surface Tf.

2. Label as in **Protocol 3, Step 5** or **Protocol 4, Step 8** except use ONLY Cy5-Tf, at a concentration equal to the sum of Cy5-Tf and FITC-Tf used above.

3. Prepare parallel samples that receive no Cy5-Tf ($2°$ Ab control).

4. Incubate on ice for 30 min.

5. Wash three times with ice-cold PBS.

6. Incubate at $37°C$ and collect the same time points as used in Protocol 3 or 4.

Allow Internalization

7. Wash twice with $4°C$ PBS to chill cells.

8. Add FITC anti-human Tf antibody to all samples and incubate on ice for 30 min.

Detect surface Tf

Note: Incubation on ice with anti-Tf antibody allows detection of surface Tf only (while not allowing any further Tf internalization).

9. Wash twice with ice-cold PBS.

10. Scrape or resuspend cells into 400 µl-800 µl PBS. Keep samples on ice until ready to analyze.

Resuspend cells

11. Analyze samples by flow cytometry, recording forward and side scatter (for gating live, single cells), and FITC and Cy5 fluorescences.

Analyze

Subprotocol 6
Removal of surface transferrin

▪▪ Materials

Solutions

- 0.15 M NaCl
- 0.05 M NaOAc
- 100 μM desferrioxamine mesylate (DFOM).
- PBS containing 100 μM DFOM.

▪▪ Procedure

Prepare and Label cells

1. Prepare and label cells as described in **Protocol 3, Steps 1-10** or **Protocol 4, Steps 1-14.**

2. Add 2 ml 0.15 M NaCl, 0.05 M NaOAc, 100 μM desferrioxamine mesylate (DFOM); pH 4.5 and incubate on ice 40 sec.

Note: Lowering the pH dissociates iron from Tf but the resulting apo-Tf has a high affinity for its receptor.

3. Wash three times with PBS containing 100 μM DFOM.

Note: Raising the pH allows apo-Tf to dissociate from surface receptors, leaving behind only intracellular Tf.

4. Keep a surface-labeled sample (not warmed to 37°C) that does not get stripped of surface Tf (to determine the total amount of bound Tf prior to warmup).

5. Continue with appropriate protocol , either **Protocol 3, Step 12** or **Protocol 4, Step 16.**

Subprotocol 7
Calibration of fluorescence ratios measured using dextran

▨▨ Materials

Solutions

See protocols 1 or 2.
- 200-400 µl of Azide/2DG
- Buffer for Dextran pH Calibration: see Miscellaneous Solutions in the chapter Introduction.

▨▨ Procedure

1. Label cells with FITC- and Cy5-dextrans and wash as described in **Protocol 1, Steps 1-5** or **Protocol 2, Steps 1-7.** Only a single labeling time is required; a time that generates high signal:noise, such as 30 min, should be chosen. Duplicate samples should be prepared for each pH value to be used.

Prepare and Label cells

Note: Initially, a standard pH curve should be generated using buffers with at least six pH values between 4.5 and 8. For subsequent experiments, only duplicate samples clamped to pH 7.4 are necessary (to adjust the ratios obtained in that experiment to the standard curve).

2. Incubate the washed cells in 200-400 µl of Azide/2DG for 10 min at 37°C to deplete cellular ATP.

Deplete ATP

Note: The goal is to equilibrate (clamp) internal dextran-containing compartments to the extracellular pH. The vacuolar H^+-ATPase will oppose this equilibration, so depletion of cellular ATP is desired. Azide inhibits oxidative phosphorylation, while 2DG inhibits glycolysis. The combination prevents ATP synthesis.

3. Add an equal volume of the appropriate Buffer for Dextran pH Calibration to each sample (see **Miscellaneous Solutions**) and add methylamine to 100 mM. Incubate for 10 min at 37°C.

Equilibrate samples to desired pH

Note: Methylamine is a weak base that can cross membranes in its unprotonated form. It becomes protonated in acidic compartments, raising their pH to the same value as the external pH. It is easier to use than hydrophobic ionophores, such as nigericin, that are difficult to keep in solution.

    4. Chill cells on ice until ready to analyze. DO NOT wash the cells.

    5. Analyze samples by flow cytometry.

Note: See **Protocols 1** and **2** for example histograms.

Calculate mean fluorescences

    6. Calculate the mean fluorescence per cell for all samples.

Note: Programs such as MFI or FCSTAB that generate tables of mean fluorescence for each sample are convenient for this step. Alternatively, mean fluorescence values can be read from displays for each sample and entered by hand into a spreadsheet. Microsoft Excel 5.0 spreadsheets that carry out the calculations described here are available at http://www.stc.cmu.edu/murphy-lab/protocols/flow/acidification.html

Subtract autofluorescence

    7. Calculate the average (between replicates) of the autofluorescence for each dye. Subtract this value from all other experimental values.

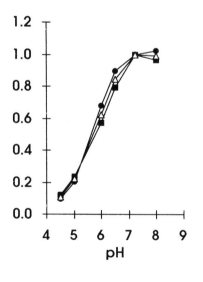

Fig. 9. Example calibration curves for A549 (l) and K562 (n) cells. Values shown are averages of six samples for each point. The average curve for both cell types is also shown (Δ)

Note: The mean fluorescence per cell is already an average (of all cells within a sample), but the average being calculated here (and in the next step) is the average of that value *between* replicate samples.

8. Average the mean fluorescence values for FITC [referred to as F(pH)] and Cy5 [referred to as C(pH)] for each pH value.

Average replicate samples

9. Calculate the ratio of F(pH) to C(pH) for each pH value.

Calculate fluorescence ratios

10. Divide the ratio for each pH value by the ratio for pH 7.24.

Normalize ratios to known pH

Note: Normalization sets the ratio equal to one for pH 7.4 so that values can be compared between experiments.

Subprotocol 8
Calibration of fluorescence ratios measured using transferrin

▓▓ Materials

Solutions

See protocols 3 or 4.
– ice-cold PBS
– 400 µl-800 µl of the appropriate Buffer for Tf pH Calibration.
– see Miscellaneous Solutions in the Chapter Introduction.

▓▓ Procedure

1. Label cells with FITC-Tf and Cy5-Tf and wash as described in **Protocol 3, Steps 1-5** or **Protocol 4, Steps 1-10**. Keep samples ice-cold to ensure that Tf does not internalize. Duplicate samples should be prepared for each pH value to be used. If the fluorescence-to-autofluorescence ratio is low for either fluorochrome (e.g., less than 10), it is useful to prepare unlabeled samples for each pH since autofluorescence can change with pH.

Prepare and Label cells

Note: Initially, a standard pH curve should be generated using buffers with at least six pH values between 4.5 and 8. It is not necessary to construct calibration curves for subsequent experiments, since the samples not warmed to 37°C (i.e., the 0 min samples) will provide a reference for comparison to the standard curve.

Wash 2. Wash cells twice with ice-cold PBS. Completely remove supernatant after final wash.

Note: It is important to only wash about six samples at a time to minimize the time between washing and analysis since Tf will eventually dissociate from its receptor. Complete removal of the PBS minimizes dilution (and resulting pH changes) of the pH buffer in the next step.

Equilibrate samples to desired pH 3. Add 400 µl-800 µl of the appropriate Buffer for Tf pH Calibration to each sample (see **Miscellaneous Solutions**).

Note: Since only surface Tf receptors are occupied with FITC-Tf and Cy5-Tf, adjusting external pH immediately alters the pH seen by the labeled Tf. No equilibration of internal compartments (as done in **Protocol 7**) is necessary.

Analyze 4. Immediately analyze samples by flow cytometry.

Note: See **Protocols 3** and **4** for example histograms.

Calculate mean fluorescences 5. Calculate the mean fluorescence per cell for all samples.

Note: Programs such as MFI or FCSTAB that generate tables of mean fluorescence for each sample are convenient for this step. Alternatively, mean fluorescence values can be read from displays for each sample and entered by hand into a spreadsheet. Microsoft Excel 5.0 spreadsheets that carry out the calculations described here are available at http ://www.stc.cmu.edu/murphylab/protocols/flow/acidification.html

Subtract autofluorescence 6. Calculate the average (between replicates) of the autofluorescence for each dye. Subtract this value from all other experimental values.

Note: The mean fluorescence per cell is already an average (of all cells within a sample), but the average being calculated here (and in the next step) is the average of that value *between* replicate samples.

Fig. 10. Example calibration curves for A549 (l) and K562 (n) cells. Values shown are averages of six samples for each point. The average curve for both cell types is also shown (Δ)

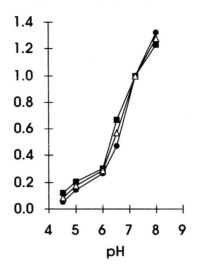

7. Average the mean fluorescence values for FITC [referred to as F(pH)] and Cy5 [referred to as C(pH)] for each pH value. — Average replicate samples

8. Calculate the ratio of F(pH) to C(pH) for each pH value. — Calculate fluorescence ratios

9. Divide the ratio for each pH value by the ratio for pH 7.24. — Normalize ratios to known pH

Note: Normalization sets the ratio equal to one for pH 7.4 so that values can be compared between experiments.

Subprotocol 9
Conversion of ratios to pH values for dextran or Tf if surface Tf was removed

▓▓ Materials

Microsoft Excel 5.0 spreadsheet available at:
http://www.stc.cmu.edu/murphylab/protocols/flow/acidification.html

▨▨ Procedure

Calculate mean fluorescences

1. Calculate the mean fluorescence per cell for all samples.

Note: A Microsoft Excel 5.0 spreadsheet that carries out all of the calculations described here is available at http://www.stc.cmu.edu/murphylab/protocols/flow/acidification.html

Subtract autofluorescence

2. Calculate the average (between replicates) of the autofluorescence for each dye. Subtract this value from all other experimental values.

Average replicate samples

3. If there is more than one sample per time point, average the mean fluorescence values for FITC [referred to as F(t)] and Cy5 [referred to as C(t)].

Calculate fluorescence ratios

4. Calculate the ratio of F(t) to C(t) for each time point and for the pH-clamped sample (dextran).

Normalize ratios to known pH

5. Divide the ratio for each time point by the ratio of the pH-clamped sample (dextran) or the *unstripped* surface-labeled sample (Tf).

Note: Normalization sets the ratio equal to one for pH 7.4 so that values can be compared between experiments.

Convert to pH

6. Convert the normalized ratios to pH values by interpolation onto the standard curve generated using **Protocol 7 or 8**.

Subprotocol 10
Conversion of ratios to pH values for Tf (if surface Tf was not removed)

▨▨ Materials

Microsoft Excel 5.0 spreadsheet available at:
http://www.stc.cmu.edu/murphylab/protocols/flow/acidification.html

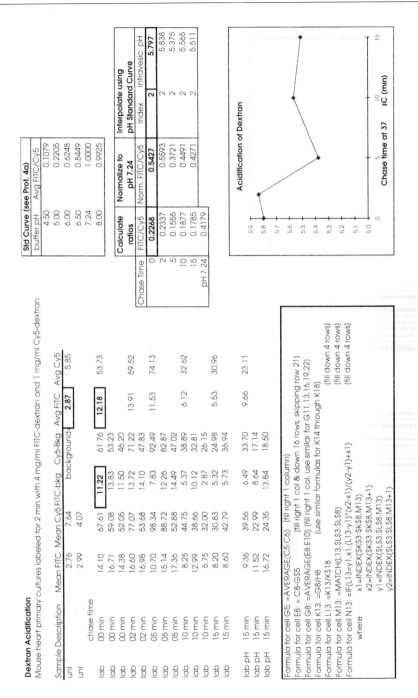

Fig. 11

▪▪ Procedure

Calculate mean fluorescences

1. Calculate the mean fluorescence per cell for all samples.

Note: A Microsoft Excel 5.0 spreadsheet that carries out all of the calculations described here is available at http://www.stc.cmu.edu/murphylab/protocols/flow/acidification.html.

Subtract autofluorescence

2. Calculate the average (between replicates) of the autofluorescence for each dye. Subtract this value from all other experimental values.

Note: The mean fluorescence per cell is already an average (of all cells within a sample), but the average being calculated here (and in the next step) is the average of that value *between* replicate samples.

Average replicate samples for acidification experiment

3. For the samples labeled with FITC-Tf and Cy5-Tf, average the mean fluorescence values for FITC [referred to as F(t)] and Cy5 [referred to as C(t)].

Average replicate samples for internalization experiment

4. For the samples labeled with Cy5-Tf and FITC-anti-Tf, average the mean fluorescence values for FITC [referred to as E(t)].

Normalize to time 0

5. Normalize all three sets of values to 100% at time 0.

Subtract fluorescence due to surface Tf

6. Subtract E(t) point-by-point from both C(t) and F(t) to obtain internal Cy5-Tf fluorescence versus time [$C_i(t)$] and internal FITC-Tf fluorescence versus time[$F_i(t)$], respectively.

Calculate ratio of internal fluorescence

7. For each time point, calculate the ratio of $F_i(t)$ to $C_i(t)$.

Convert to pH

8. Convert the internal ratios to pH values by interpolation onto the standard curve generated using **Protocol 8.**

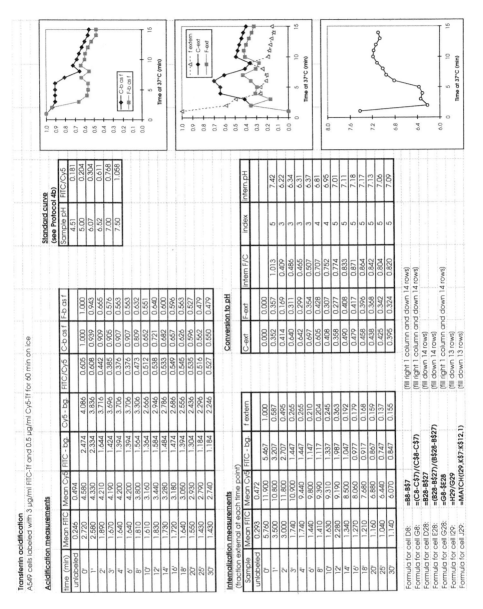

Fig. 12

Subprotocol 11
Calculation of specificity of Cy5-Tf and FITC-Tf

▨ ▨ Procedure

Calculate mean fluorescences

1. Calculate the mean fluorescence per cell for unlabeled samples, samples labeled with Cy5-Tf and FITC-Tf but not warmed to 37°C (unblocked samples), and samples incubated with Cy5-Tf, FITC-Tf and excess unlabeled Tf (blocked samples).

Average replicate samples

2. Calculate the average (between replicates) of the values from step 1.

Subtract autofluorescence

3. Subtract the average autofluorescence from the average fluorescence values for both fluorochromes for both the unblocked and blocked samples.

Calculate specificity

4. For each fluorochrome, divide the average fluorescence from the blocked samples by the average for the unblocked samples and subtract the result from 1. This is the fraction of the Tf conjugated to that dye that is binding specifically to surface receptors.

Validate experiment

5. Values for conjugates prepared in our laboratory are typically greater than 0.95. If the value is less than 0.95, pH values resulting from that experiment may be unreliable.

Note: Any non-specifically bound Tf may be processed differently by cells than Tf bound to its receptor. For example, it may be delivered to late endosomes and lysosomes, which may have a different pH. Signal from this mislocalized Tf will be averaged with signal from properly-localized Tf, resulting in inaccurate pH estimates.

▨ References

1. Murphy RF, Powers S, Cantor CR. Endosomal pH measured in single cells by dual fluorescence flow cytometry: rapid acidification of insulin to pH 6. J Cell Biol 1984;98:1757-1762.
2. Sipe DM, Jesurum A, Murphy RF. Absence of Na+,K+-ATPase regulation of endosomal acidification in K562 erythroleukemia cells. J Biol Chem 1991;266:3469-3474.

3. Sipe DM, Murphy RF. High resolution kinetics of transferrin acidification in BALB/c 3T3 cells: Exposure to pH 6 followed by temperature-sensitive alkalinization during recycling. Proc Natl Acad Sci USA 1987;84:7119-7123.

4. Cain CC, Murphy RF. Growth inhibition of 3T3 fibroblasts by lysosomotropic amines: correlation with effects on intravesicular pH but not vacuolation. J Cell Physiol 1986;129:65-70.

5. Roederer M, Bowser R, Murphy RF. Kinetics and temperature dependence of exposure of endocytosed material to proteolytic enzymes and low pH: evidence for a maturation model for the formation of lysosomes. J Cell Physiol 1987;131:200-209.

6. Wilson RB, Murphy RF. Flow-cytometric analysis of endocytic compartments. Methods Cell Biol 1989;31:293-317.

7. Murphy RF, Roederer M, Sipe DM, et al. Determination of the biochemical characteristics of endocytic compartments by flow cytometry and fluorometric analysis of cells and organelles. In: Yen A, ed. Flow Cytometry: Advanced Research and Clinical Applications. Boca Raton, FL: CRC Press, Inc., 1989:221-254.

8. Rybak, SL, Murphy, RF Primary cell cultures from murine kidney and heart differ in endosomal pH. J. Cell Physiol. 1998;176:216-222.

9. De Belder AN, Granath K. Preparation and properties of fluorescein-labelled dextrans. Carbohydr Res 1973;30: 375-378.

10. Preston RA, Murphy RF, Jones EW. Apparent endocytosis of fluorescein isothiocyanate-conjugated dextran by Saccharomyces cerevisiae reflects uptake of low molecular weight impurities, not dextran. J Cell Biol 1987;105:1981-1987.

Intracellular Antigen Detection by Flow Cytometry

INGRID SCHMID

Introduction

Flow cytometry has become a widely used technology; thus, interest in its application to the detection of intracellular antigens has increased; particularly, because flow cytometry permits the rapid and quantitative measurement of intracellular protein expression in single cells and can also provide simultaneous identification of cellular subpopulations that either differ in expression of surface antigens or DNA content. However, while cell surface immunophenotyping by flow cytometry has turned into a routine methodology, intracellular staining experiments are more complex and remain difficult. Staining of intracellular antigens for flow cytometry depends on methods that permeabilize the membranes of cells in solution to permit access of antibodies to intracellular components.

Most of the techniques developed for preparing cells for intracellular staining either use formaldehyde fixation (various concentrations and incubation times) followed by detergent treatment (e.g., Triton X-100, Tween 20, saponin) or use alcohol fixation (various concentrations of ethanol or methanol); some methods use detergents without prior cell fixation.[1,2] For a review of many published methods see ref. 3. Formaldehyde fixation in combination with permeabilization by detergent has been shown to be a superior cell preparation method[4,5] compared to alcohol fixation; mainly, because cell fixation with alcohol leads to cell clumping and subsequent cell losses, dramatically changes cellular light scatter characteristics, and often results in loss of cell surface antigen staining. However, as the interactions of cellular membranes and antigens with fixatives and detergents are still not completely understood, the optimal preparation method for staining of a given antigen in a certain cell type has to be determined experimentally. Furthermore, for staining of intracellular cytokines, cells have to be pre-treated with substances that interfere with the secretory pathway of the Golgi apparatus, e.g., monensin or brefeldin A, in order to accumulate sufficient cytokine within the cells for flow cytometric detection.[6]

Nonviable (dead) cells can interfere with accurate flow cytometric analysis, because they can bind antibodies nonspecifically. Often, dead cells can be distinguished on the flow cytometer from live cells by alterations in light scatter. This discrimination is frequently lost in permeabilized cell preparations. Whenever considerable numbers of dead cells are present they should be therefore removed prior to staining by Ficoll-Hypaque cell density separation. However, this step is time consuming and can lead to selective cell losses. Alternatively, and preferably, dead cells can be distinguished by up-take of DNA dyes due to loss of membrane integrity. Protocols have been described for use of this strategy in fixed and/or permeabilized cell preparations.[7,8]

Antibodies obtained from different sources can differ dramatically in their reactivity with a given intracellular antigen. It is advisable to use only reagents developed and tested for flow cytometry. For instance, antibodies used for detection of proteins on immunoblots may not show any reactivity in flow cytometric assays, because they may not be directed against the conformation-dependent epitopes present in intact cells. Variations in staining temperature and incubation time can improve reactivity of an antibody with the intracellular antigen. For cells that were permeabilized by saponin detergent the presence of saponin in the staining solution is required for antibody penetration.[2,8]

Nonspecific binding of antibodies is a considerable problem in staining of intracellular antigens, because the antibody can interact with the many components present inside a cell. Strategies to reduce nonspecific binding are: use of monoclonal antibodies which often result in clearer staining patterns compared to polyclonal antisera; use of directly-labeled antibodies to avoid the nonspecific binding of the second antibody; use of the biotin-streptavidin system in case indirect staining has to be done; careful titration of the staining reagents to find the optimal concentration; blocking steps, e.g., with heat-inactivated human serum or normal mouse serum; multiple, large volume washes with buffer solutions which contain low concentrations of detergents.

Antibodies utilized for intracellular staining experiments should be conjugated to fluorochromes with low molecular weights in order not to hinder antibody mobility. Thus, fluorescein isothiocyanate (FITC) is most commonly used; new low molecular weight dyes have recently become available, e.g., 7-amino-methylcumarin-3-acetic acid (AMCA) (Jackson Immunoresearch Laboratories, West Grove, PA), Cascade blue (Molecular Probes, Eugene, OR), and AlphaRed (Exalpha, Boston, MA). Phycobiliproteins such as phycoerythrin (PE) have much higher molecular weights but are able to cross into permeabilized cells. They are probably best used for staining experiments where the low expression level of the antigen requires staining with a fluorochrome with a bright fluorescence signal and where accurate quantification of the intracellular antigen is less critical. Fluorochromes with emission spectra which can be separated from each other must be used when cell surface antigen staining is combined with intracellular staining. Generally, staining for detection of expression of cell surface antigens is performed prior to cell permeabilization. Usually, there will be some loss of fluorescence intensity of cell surface antigen staining during cell permeabilization; its extent will vary depending on the cell type and surface antigen, the antibody used for staining, the fluorochrome attached to the antibody, and the permeabilization protocol. For instance, with FITC and PE cell surface staining, even staining of antigens that are expressed at low levels is generally preserved sufficiently to permit discrimination from background. Peridinin chlorophyll protein (PerCP^{TM}, Becton Dickinson Immunocytometry Systems (BDIS), San Jose, CA) shows a dramatic decrease in staining intensity with some permeabilization protocols and thus has to be used with caution as a fluorochrome for combining cell surface with intracellular staining.

Because of the high probability of nonspecific antibody binding, controls are essential for a successful intracellular staining experiment. Isotypic control antibodies must be used at the same protein concentration as the relevant antibody. Experimental controls include cells known to express the antigen in question and cells that do not express it. Furthermore, for characterization of a novel intracellular antigen it is advisable to verify the localization of the staining by fluorescent microscopy, because flow cytometry cannot provide this information.

In recent years, many commercial reagent systems for intracellular staining have become available that can also be used to stain whole blood. These include, but are not limited to ORTHO PermeaFix[TM] [9](Ortho Diagnostics System, Raritan, NJ), FIX & PERM[TM] (Caltag, South San Francisco, CA), FACS Permeabilization Solution[TM] (BDIS, San Jose, CA), and IntraPrep[TM] Permeabilization Reagent (Beckman Coulter, Miami, FL). Generally, these protocols are convenient and work well for many applications, e.g., FIX & PERM[TM][10] and FACS Permeabilization Solution[TM][11] have been shown to be useful for staining of intracellular cytokines, and can save time for reagent preparation. However, the reagent systems can be costly and owing to proprietary rights their components are not listed. Thus, optimization of the intracellular staining protocol for a given cell type and/or antigen is not possible. Furthermore, none of the above mentioned commercially available systems is recommended for combining intracellular antigen staining with DNA staining due to high coefficients of variation on DNA distributions.

Materials

Equipment

Standard Flow Cytometer with filter setup appropriate for the fluorochrome needed in the staining protocol as in Fig. 2.

Solutions

- buffered formaldehyde solution
- 0.2% Tween 20
- Staining buffer containing 50% heat-inactivated human serum or 20% mouse serum.

Preparation

buffered formaldehyde solution prepared by dissolving solid paraformaldehyde in 1 x PBS by heating it to 70°C in a fume hood.

Procedure

Fix cells 1. Add 0.875 ml of cold 1 x PBS to a pellet of PBS-washed cells in 12 x 75 mm polystyrene culture tubes and vortex. Add 0.125 ml of cold 2% formaldehyde solution. Vortex again.

Note: Suitable cell preparations include single cell suspensions from lymphoid tissues, mononuclear cells derived from blood, and cells grown in suspension cultures.
The preferred fixative is a buffered formaldehyde solution. Alternatively, 16% EM grade formalin is commercially available (Polysciences, Warrington, PA).

2. Incubate for 30 minutes at 4°C.

Note: For optimal preservation of light scatter differences between cell sub-populations, cells should be fixed for a minimum of 30 minutes; otherwise, e.g., for cell lines, shorter incubation times may be used. For optimized staining of certain intracellular antigens, it may be necessary to increase the concentration of the fixative, use an alternate detergent, e.g., saponin, or use an alternate method, e.g., ethanol fixation. The fixation and permeabilization procedure described here is suitable for combining intracellular staining with cell surface antigen staining.

Permeabilize cells 3. Centrifuge for 5 minutes at 250 x g. Discard supernatant. Add 1 ml of 0.2% Tween 20, pre-warmed to 37°C, and incubate for 15 minutes at 37°C.

Note: Tween 20 is a mild detergent that is needed for solubilization of the cell membrane after fixation.

Stain intracellular antigen 4. Centrifuge for 5 minutes at 250 x g. Discard supernatant. Add the appropriate dilution of fluorochrome-labeled antibody in staining buffer to the cells

Note: The staining buffer should contain serum, e.g., 50% heat-inactivated human serum or 20% mouse serum, to reduce non-specific antibody staining. The optimal antibody concentration as well as the best staining temperature and time has to be determined experimentally and may differ for various antibodies and intracellular antigens. For staining with an unlabeled antibody, repeat step 4 and 5 for the second antibody.

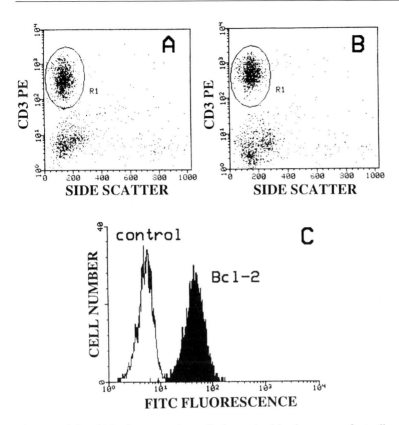

Fig. 1. Peripheral blood mononuclear cells from a healthy donor were first cell surface stained with CD3 phycoerythrin (PE)-conjugated antibody (Becton Dickinson Immunocytometry Systems (BDIS), San Jose, CA), then fixed and permeabilized as described in the protocol. Then, cells were stained with either mouse IGg1-fluorescein-isothiocyanate (FITC) control (BDIS) or with anti-Bcl-2 FITC monoclonal antibody (Dako, Carpinteria, CA). (A) Side scatter vs. CD3 PE dot plot showing the gate that was used for gating the control histogram in C. (B) Side scatter vs. CD3 PE dot plot showing the gate that was used for gating the Bcl-2 histogram in C. Note: setting a lymphocyte gate on forward vs. side scatter first, then selecting the CD3+ T cells would have been possible; however, for improved discrimination of CD3+ lymphocytes from other cell subsets, a gate was set using a combination of side scatter vs. CD3 fluorescence. This was done on the sample stained with the isotype control antibody and the sample stained with the relevant antibody, respectively, to permit the direct comparison of background staining with Bcl-2 staining on CD3+ T cells. (C) Overlay of the FITC fluorescence distribution of after staining with either control antibody or anti-Bcl-2. 100% of CD3+ T lymphocytes express Bcl

5. Wash two times with 1 x PBS containing 0.1% Tween 20 by centrifugation at 250 x g for 5 minutes.

Note: The presence of Tween in the washing buffer can reduce nonspecific antibody staining.

Analyze cells 6. Analyze samples on a flow cytometer equipped with a laser and filter set up appropriate for the fluorochrome that was used for antibody labeling.

Stain cells for DNA content (optional) 7. After the last washing step, resuspend the cells in buffer containing either 10 µg/ml of propidium iodide (PI) and 10 Kunitz units of ribonuclease A, or 20 µg/ml of 7-aminoactinomycin D (7-AAD), depending on the fluorochromes that were used for cell surface and/or intracellular staining.

8. Incubate for 20 minutes in the dark.

Note: The procedure described here is particularly suited for combining intracellular staining with DNA content analysis, because fixation with a low concentration of formaldehyde results in good CVs on G_0G_1 peaks. Use PI for DNA staining in combination with FITC-labeled antibodies, and 7-AAD in combination with PE-labeled antibodies.

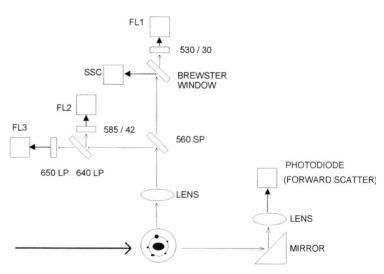

Fig. 2

9. Analyze cells in their respective staining solutions on a flow   **Analyze cells**
   cytometer equipped with a 488 nm argon laser measuring
   fluorescence above 620 nm for PI and above 650 nm for
   7-AAD.

References

1. Schroff RW, Bucana CD, Klein RA et al.. Detection of intracytoplasmic
   antigens by flow cytometry. J. Immunol. Methods 1984; 70:167-177.
2. Jacob MC, Favre M, and Bensa JC. Membrane permeabilization with
   saponin and multiparameter analysis by flow cytometry. Cytometry
   1991;12:550-558.
3. Clevenger CV and Shankey, TV. Cytochemistry II: Immunofluorescence measurements of intracellular antigens. In: Bauer KD, Duque
   RE, Shankey TV, eds., Clinical Flow Cytometry, Baltimore, Williams
   and Wilkins, 1993:157-175.
4. Clevenger CV, Bauer KD, and Epstein AI. A method of simultaneous
   nuclear immunofluorescence and DNA content quantitation using
   monoclonal antibodies and flow cytometry. Cytometry 1985; 6:208-214.
5. Schmid I, Uittenbogaart CH, and Giorgi JV. A gentle fixation and permeabilization method for combined cell surface and intracellular staining with improved precision in DNA quantification. Cytometry 1991;
   12:279-285.
6. Sander B, Andersson J, and Andersson U. Assessment of cytokines by
   immunofluorescence and the paraformaldehyde-saponin procedure.
   Immunol. Rev. 1991;119:65-93.
7. Riedy MC, Muirhead KA, Jensen CP et al.. Use of a photolabeling technique to identify nonviable cells in fixed homologous or heterologous
   cell populations. Cytometry 1991; 12:133-139.
8. Schmid I and Giorgi JV. Preparation of cells and reagents for flow cytometry, Discrimination of nonviable cells. In: Coligan JE, Kruisbeek
   AM, Margulies DH, Shevach EM, Strober W, eds., Current Protocols
   in Immunology, Vol 1, Unit 5.3, New York, John Wiley & Sons, 1996.
9. Francis C and Connelly MC. Rapid single-step method for flow cytometric detection of surface and intracellular antigens using whole
   blood. Cytometry 1996; 25:58-70.
10. Ferrick DA, Schrenzel MD, Mulvania T et al.. Differential production of
    Interferon-γ and Interleukin-4 in response to Th1 and Th2-stimulating
    pathogens by $\gamma\delta$ T cells in vivo. Nature 1995; 373:255-257.
11. Maino CV, Ruitenberg J, and Suni MA. Flow cytometric method for
    analysis of cytokine expression in clinical samples. Clinical Immunol-
    ogy Newsletter 1996; 16(6):95-98.

Flow Cytometric Measurement of Nuclear Matrix Proteins

JOSEPH R. DYNLACHT

Introduction

The nuclear matrix (NM) is important in defining nuclear structure and is the site of DNA and RNA synthesis, and regulation of gene expression.[1-4] All NMs possess three common elements: a lamina, residual nucleoli, and a dense, fibrous structure extending throughout the nuclear interior and believed to be contiguous with nuclear pore complexes.[1]

The NM is composed of intermediate filament-related proteins.[2] The NM contains many low abundance proteins, heterogeneous nuclear RNAs, and ubiquitous proteins, some of which are cell cycle specific[5] or proliferation-dependent.[6] The identity and function of many nuclear matrix proteins (NMPs) still remain to be elucidated.

The NM is normally highly insoluble and resistant to high salt buffers, non-ionic detergents, and nucleases. However, recent reports indicate that specific NMPs may be released from dying cells in a soluble form,[7,8] or may be degraded preferentially by proteases during apoptotic death.[9-17] Indeed, we[17] and others[7-9,13] have shown that solubilization and/or degradation of NMPs is among the earliest detectable events during apoptosis induced by drugs, radiation, and hyperthermia.

The following technique represents an alternative and less labor-intensive approach to measuring relative changes in NMP content than SDS-PAGE and Western blotting, and offers the opportunity for correlating changes in NMP content with cell cycle position or DNA fragmentation in single apoptotic cells. Potential applications of this technique may include the study of changes in NM composition during proliferation or differentiation and during cell death. Its usefulness for measuring relative or absolute changes in the content of different NMPs should, of course, be confirmed using SDS-PAGE, Western blotting and a scanning densitometer. Such testing for a particular NMP should be considered for each new application and cell type, in order to determine whether the measured changes in NMP immunofluorescence reflect changes in the ability of the antibody to bind to the epitope or a true change in NMP content. Furthermore, when the protocol is used to measure changes in NM composition during cell death, the investigator may find it useful or necessary to use Western blotting to determine whether the decrease in NMP content measured by flow cytometry represents solubilization of the NMP or degradation of the NMP by caspases.

To date, we have used flow cytometry to study relative changes in proteins comprising three NM substructures (the nuclear lamina, the internal nuclear matrix, and the nuclear pore complex) using monoclonal antibodies against major components of each substructure [lamin B, NuMA (nuclear mitotic apparatus protein)[18,19] and the ~270 kD nucleoporin Tpr,[20,21] respectively]. We have used the technique to document heat-induced changes in the NM[17,22], and for determining the kinetics of NMP degradation or solubilization during apoptosis or necrosis, respectively (Dynlacht et al.[23]). While very useful, it should be noted that measurements of decreases in NMP fluorescence during cell death studies should be interpreted with caution. For example, it is possible that cleavage and degradation of specific NMPs may occur prior to a measured loss of fluorescence of that NMP if the fragmented or degraded NMP contains an intact antigenic site which can still be recognized by the antibody.

OVERVIEW

Many labs have sought to use a one- or two dimensional SDS-PAGE and Western blotting approach for characterizing changes in NMP composition. However, these approaches are cumbersome and difficult for many labs to perform on a rapid and routine basis. These approaches also suffer because the analysis yields results from bulk preparations of cells, and no information is obtained at the single cell level.

This flow cytometric approach has been used to measure relative changes in NMPs in several human tumor cell lines, and may be valuable for studying changes in NM composition during cell differentiation and mitogenesis. It has proven useful for detecting degradation of specific NMPs in single cells undergoing apoptosis,[17] and for correlating NMP degradation with DNA fragmentation and cell cycle position. This is accomplished by dual-staining for NMPs and DNA.

Fixation and staining protocols are relatively straightforward, and require a minimum of manipulation. Cell samples should be fixed overnight and then stained and analyzed the next day. However, fixed samples have been stored up to three days prior to staining, with no reduction in fluorescence intensity noted.

Materials

Equipment

- 15 ml polystyrene centrifuge tubes
- pipetmen and tips
- refrigerated benchtop centrifuge
- timer
- electronic cell counter or hemocytometer
- pipets
- FITC and PI filter sets

Solutions

- **100% ethanol** and phosphate buffered saline (PBS, pH 7.4) for fixation
- **Tris-buffered saline solution (TBS)** : 20 mM Tris, pH 7.4, 150 mM NaCl
- **Goat serum**
- **Triton X-100**
- **Monoclonal antibodies against nuclear matrix proteins**
- **FITC-conjugated secondary IgG**
- 50 µg/ml stock solution of PI (in buffer containing 1% sodium citrate and 0.1% Triton X-100)

Preparation

- Place ethanol, PBS, and TBS in refrigerator several hours prior to fixation/staining. Thaw goat serum and prepare TgT buffer prior to spinning cells out of fixative.

- Prepare ice cold "TgT buffer" consisting of 0.01% Triton X-100 and Tris-buffered saline (TBS) (TBS; 20 mM Tris, pH 7.4, 150 mM NaCl, 20% goat serum)
- Each sample will require approximately 16 ml of TgT buffer for washing and labeling of cells, and 11 ml of TBS per sample in addition to that required to make up the TgT buffer.

Note: Each laboratory should determine optimal concentrations of primary and secondary antibodies for labeling cells.

If degradation of nuclear matrix proteins is being studied, be sure to prepare and fix untreated samples to be stained later with FITC-conjugated IgG only. These samples can be used as controls and are useful for comparisons with cell populations that are degrading or have degraded the nuclear matrix proteins being studied.

Procedure

1. Trypsinize, spin 10^6 cells out of medium and transfer cells to 15 ml tube.   **Count cells**

Note: Be sure to prepare a sample to be used as a secondary-only (FITC-IgG) labeled control.

2. Centrifuge 7 min at 1500 RPM (250XG) at 4° C; Aspirate medium.

3. Break up pellet and add 3 ml ice cold PBS.   **Fix cells in**

4. Add 4 ml ice cold 100% ethanol dropwise while vortexing   **57% ethanol**

5. Store samples overnight at 4° C. All subsequent procedures should be done at 4° C.

6. Centrifuge cells out of fixative and wash one time in 5 ml TgT buffer.

Note: Prepare ice cold "TgT buffer" consisting of 0.01% Triton X-100 and Tris-buffered saline (TBS) (TBS; 20 mM Tris, pH 7.4, 150 mM NaCl, 20% goat serum).

7. Resuspend cells in 150 µl of TgT buffer for 10 min.

8. Add 10 µl of primary antibody (vs. nuclear matrix protein) and incubate for 1 hr.   **Antibody Stain**

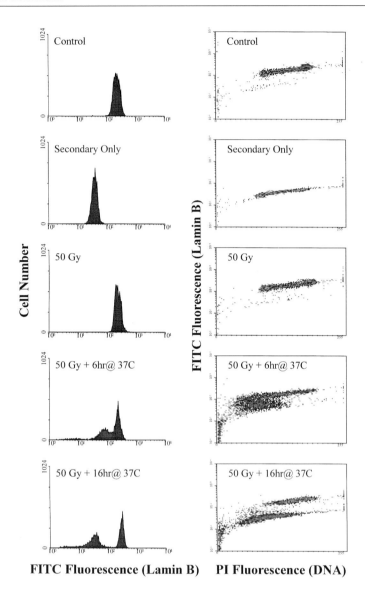

FITC Fluorescence (Lamin B)   PI Fluorescence (DNA)

Fig. 1. Flow cytometric detection of nuclear matrix protein (lamin B) degradation during apoptosis. Cells were irradiated with 50 Gy of X-rays and then incubated at 37° C for various times prior to fixation and staining for lamin B and DNA (see *Procedures Section*). Single-parameter histograms for cells stained with a monoclonal antibody against lamin B appear in the left column; dual-parameter histograms of the same samples stained with propidium iodide (to measure DNA content and fragmentation) and a monoclonal antibody against lamin B appear in the right column. Single- and dual-parameter histograms of cells stained with an FITC-conjugated secondary antibody only are also shown for reference

9. Wash cells twice in 5 ml TgT buffer; resuspend in 450 µl TgT and add 10 µl of secondary FITC-conjugated IgG; Incubate 1 hr.

Note: Incubation of cells with primary and secondary antibodies; Assuming a cell pellet=~40 µl, dilution of antibodies are 1:20 and 1:50 respectively, but optimal dilutions should be determined for each cell type. Agitate cells at least once every 15 min to prevent cells from settling.

10. Add 5 ml TBS and centrifuge; Resuspend in 5 ml TBS again and centrifuge; finally, resuspend in 500 µl TBS

11. Add 20 µl of a 50 µg/ml stock solution of PI (in buffer conatining 1% sodium citrate and 0.1% Triton X-100) to 180 µl of labeled cells.

To simultaneously measure cell cycle position or DNA fragmentation

Note: (DNA fragmentation is indicated by decreased stainability of a DNA-specific fluorochrome) (PI).

12. Analyze immediately with the 488 nm laser line set at 100 mW.

Analyze changes in nuclear matrix protein content using flow cytometry

Acknowledgements

I am grateful to Marvin Earles and Chad O'Nan for their excellent technical support. Flow cytometry was performed with the expert assistance of Jim Henthorn at the Flow Cytometry and Cell Sorting Core Facility, Warren Medical Research Institute, University of Oklahoma Health Sciences Center.

References

1. Berezney, R., Mortillaro, M. J., Ma, H., Wei, X., and Samarabandu, J. (1995) The nuclear matrix: A structural milieu for genomic function. In: International Review of Cytology. R. Berezney and K. Jeon, eds. Vol. 162. Academic Press, San Diego, CA, pp. 1-65.
2. de Jong L, van Driel R, Stuurman N. et al. Principles of nuclear organization. Cell Biol. Int. Reports 1990; 14(12): 1051-1074.
3. Fey EG, Bangs P, Sparks C, et al. Critical Reviews in eukaryotic gene expression, 1990; 1(2): 127-143.
4. van Wijnen AJ, Bidwell JP, Fey EG, et al. Nuclear matrix association of multiple sequence-specific DNA binding activities related to SP-1, ATF, CCAAT, C/EBP, OCT-1, and AP-1. Biochemistry1993; 32: 8397-8402.

5. Chaly N, Bladon T, Setterfield G et al. Changes in distribution of nuclear matrix antigens during the mitotic cell cycle. J. Cell Biol 1984; 99: 661-671.

6. Bidwell JP, Fey EG, van Wijnen AJ, et al. Nuclear matrix proteins distinguish normal diploid osteoblasts from osteosarcoma cells. Cancer Res 1994; 54: 28-32.

7. Miller TE, Beausang LA, Winchell L, et al. Detection of nuclear matrix proteins in serum from cancer patients. Cancer Res 1992; 52: 422-427.

8. Miller T, Beusang LA, Meneghini M, et al. Death-induced changes to the nuclear matrix: The use of anti-nuclear matrix antibodies to study agents of apoptosis. Biotechniques 1993; 15: 1042-1047.

9. Kaufmann SH. Induction of endonucleolytic DNA cleavage in human acute myelogenous leukemia cells by etoposide, camptothecin, and other cytotoxic anticancer drugs: A cautionary note. Cancer Res 1989; 49: 5870-5878.

10. Takahashi A, Alnemri ES, Lazebnik YA, et al. Cleavage of lamin A by Mch 2a but not CPP32: Multiple interleukin 1b-converting enzyme-related proteases with distinct substrate recognition properties are active in apoptosis. Proc Natl Acad Sci 1996; 93, 8395-8400.

11. Neamati N, Fernandez A, Wright S, et al. Degradation of lamin B1 precedes oligonucleosomal DNA fragmentation in apoptotic thymocytes and isolated thymocyte nuclei. J. Immunology 1995; 154: 3788-3795.

12. Weaver VM, Carson CE, Walker PR, et al. Degradation of nuclear matrix and DNA cleavage in apoptotic thymocytes. J. Cell Science 1996; 109: 45-56.

13. Voelkel-Johnson C, Entingh AJ, Wold WSM, et al. Activation of intracellular proteases is an early event in TNF-induced apoptosis. J. Immunology 1995; 154: 1707-1716.

14. Lazebnik YA, Takahashi A, Moir RD, et al. Studies of the lamin proteinase reveal multiple parallel biochemical pathways during apoptotic execution. Proc Natl Acad Sci 1995; 92: 9042-9046.

15. Martin SJ, Green DR. Protease activation during apoptosis: Death by a thousand cuts? Cell 1995; 62: 349-352.

16. Ucker DS, Obermiller PS, Eckhart W, et al. Genome digestion is a dispensable consequence of physiological cell death mediated by cytotoxic T lymphocytes. Mol Cell Biol 1992; 12: 3060-3069.

17. Dynlacht JR, Henthorn J, O'Nan C, et al. Flow cytometric analysis of nuclear matrix proteins: Method and potential applications. Cytometry 1996; 24: 348-359.

18. Yang CH, Lambie EJ, Snyder M. NuMA: an unusually long coiled-coil related protein in the mammalian nucleus. J Cell Biol 1992; 116(6): 1303-1317.

19. Compton D, Szilak I, Cleveland D. Primary structure of NuMA, an intranuclear protein that defines a novel pathway for segregation of proteins at mitosis. J Cell Biol 1992; 116: 1395-1408.

20. Pante N, Aebi U. Exploring nuclear pore complex structure and function in molecular detail. J Cell Science 1995; Supp 19: 1-11.

21. Cordes, V. C., Reidenbach, S., Rackwitz, H. R., Franke, W. W., Identification of protein p270/Tpr as a constitutive component of the nuclear pore complex-attached intranuclear filaments. J. Cell Biol. 1997; 136: 515-529.

22. Dynlacht JR, Story M D, Zhu W-G, and Danner J, Lamin B is a prompt heat shock protein. J. Cell Physiol. 1999; 178:28-34.

23. Dynlacht JR, Earles M, Henthorn J, Roberts ZV, Howard EW, Sparling D, Seno JD, and Story M D, Degradation of the nuclear matrix is a common element during both radiation-induced apoptosis and necrosis. Radiation Research (in press)

CellProbe Flow Cytoenzymology

SUSAN DEMAGGIO

Introduction

The CellProbe Reagent product line from Coulter Corporation has recently been introduced and offers a variety of fluorogenic enzyme substrates for intracellular measurement of enzyme activity in live cells by flow cytometry. This new technique will be invaluable in functional studies of cells and opens a new door in research possibilities. Since enzymes are involved in almost every cell function and process and are present in all cell types, the variety of applications for the use of flow cytoenzymology is vast. The ability of the technique to detect small changes in enzyme activity and concentration makes these substrates a powerful adjunct to current flow cytometry applications. In combination with surface markers, this technique can be used to measure the enzyme activity of particular subsets of cells. Other applications include cell lineage determination; monitoring of cellular processes such as apoptosis, signal transduction, maturation, and activation; cell to cell interaction including cell cytotoxicity, activation, and antigen presentation; host defense mechanisms such as oxidative burst, phagocytosis, and inflammation; cell migration in metastasis or inflammation; drug effects in disease; disease related cellular research in conditions such as HIV, leukemia / lymphoma, transplantation, and autoimmunity; and of course, the basic cellular biology research applications such as cell function, cell responses and cytokine production.

The reagent line offers a selection of synthetic substrates designed specifically for flow cytometry. Few researchers have been able to overcome the difficult technical challenges of this type of protocol, such as cell permeabilization, enzyme cross-reactivity, the need to maintain cell viability, and the sensitive nature of a kinetic assay.[1-3] Now the potential for cytoenzymology is available to the rest of us, through the availability of these substrates. They are "flow-ready" reagents, designed for rapid preparation and analysis. Analysis can be done on homogeneous suspensions or density gradient separated samples, tissues, and even whole blood, by taking advantage of the Coulter Q-Prep workstation and reagent system. The incubation steps are short allowing the analysis to be available is less than 45 minutes, compared to traditional enzyme analysis methods which can take several hours to days.

PREPARATION

Presented here is a basic procedure to get you started in cytoenzymology. A comprehensive monograph is available from Coulter on the CellProbe system, from which most of this protocol is taken, and each reagent includes an insert with the particular requirements of that enzyme substrate reagent. The list of Cell-Probe reagents is also included here but will change, so your representative should be consulted for product availability. This table includes the incubation times and PMT settings for the various enzymes.

CellProbe Reagent Incubation Times and PMT Settings

Product name/ Enzyme	Incubation Times in Minutes	Data Acquisition PMT Settings
BU.Butyryl Esterase	1	Low
Cl Ac.Cl Acetate Esterase	1	Low
L.Aminopeptidase	1	Low
Palmitate.Alkyl Esterase	1	Low
A.Aminopeptidase M	5	Low
D.Aminopeptidase A	5	Medium
G.Aminopeptidase	5	Medium
K.Aminopeptidase B	5	Medium
P.Pro Aminopeptidase	5	Medium
R.Aminopeptidase B	5	Medium
DCFH.Peroxides	5	High

CellProbe Reagent Incubation Times and PMT Settings (Continued)

Product name/ Enzyme	Incubation Times in Minutes	Data Acquisition PMT Settings
DCFH,PMA.Oxidative Burst	5	High
FDA.Esterase	5	High
FDA,NaF.Esterase	5	High
E.Coli .Phagocytosis	10	See Package Insert
AAPL.Elastase	10	High
AAPV.Elastase	10	High
AG.Cathepsin	10	High
FR.Kallikrein	10	High
Gal.Galactosidase	10	High
GFGA.Collagenase	10	High
GGL.Subtilisin	10	High
GL.Cathepsin D	10	High
Glu.Glucosidase	10	High
Glucuronide.Glucuronidase	10	High
GP.DPP IV	10	High
GPLGP.Collagenase	10	High
KA.DPP II	10	High
LL.DPP I	10	High
LY.Calpain	10	High
LY.DPP I	10	High
PO4.Acid Phosphatase	10	High
QS.Cathepsin D	10	High
RGES.Elastase	10	High
TP.Cathepsin	10	High
VK.Cathepsin	10	High
VS.Cathepsin	10	High

The standardization of fluorescence intensity is critical to consistent measurement of enzyme activity. The PMT high voltage must be standardized daily using a fluorosphere standard. Each day determine the high voltage required to establish the same target mean for each specific CellProbe reagent. Three different PMT settings, low, medium or high, have been defined to optimize fluorescence measurement of the different enzyme activity ranges which normally occur with cellular enzymes. The first protocol in this chapter will accomplish the standardization of the instrument.

Preparation of the sample is necessary to achieve a single-cell suspension, whether you are working with a whole blood sample, a density gradient separated homogeneous sample, a cellular body fluid, or a tissue sample. Gentle handling of samples will maintain their near-native state. It is necessary to remove extracellular enzymes prior to processing. Extraneous enzymes can hydrolyze the enzyme substrate, leading to falsely increased or decreased results. It is also necessary to ensure that the whole blood, cell suspensions, and body fluid samples are fresh, analyzed within 6 hours of collection, washed (prepared) within 4 hours and analyzed within the next 2 hours. Tissue samples must be analyzed immediately. Fixation of samples is possible but provides a different enzyme activity than live populations. Samples should be analyzed at a concentration of 3×10^6 cells per milliliter, so adjustments in cell concentration are necessary to ensure accurate results.

PROCESSING

Unlike surface markers, enzyme-substrate binding is a continuous, kinetic rate reaction where the measurement of fluorescence intensity is important. Therefore, consistency in technique and analysis provides the best results. To maximize your results use gentle techniques to maintain the cells in a native state. Live cells provide the best results. Practice consistent, standard techniques from day to day, as variations in technique cause erratic results. Remember the reaction is kinetic so you need to keep timing accurate. Times in the individual protocols have been optimized to give the best staining results. The results measured are mean fluorescent intensities, not percent positive; enzyme activity is found on virtually all cells. The measurement of cellular activity is found in the intensity of the staining. It is extremely helpful to know the cell biology and cellular interaction of the cells you are analyzing. Activity is dependent upon the particular cells of interest. Shifts in activity, both increased and decreased, can be seen when comparing to a normal cell population.

If you are planning to analyze a number of enzymes on a certain sample or a number of different cell samples for enzyme activity, it is helpful to organize the analysis into a panel of enzymes. Several different strategies for planning panels are suggested, depending upon the particular enzymes under study. It is helpful to group them according to incubation times. Set up all the equipment and label all the tubes prior to beginning, and use the time between sample washes in the preparation stage, to reconstitute your reagents. Keep your batch sizes small (10-20 tubes at the most); remember that it takes time to handle the tubes. If the enzyme incubation time is 1 minute, limit the tube number to 5. Add substrate to the tubes at intervals of 10-15 seconds or more to allow yourself more time for handling and processing. When preparing multiple enzymes on multiple specimens or large batches, make three blanks per specimen. For analysis, group samples by PMT voltages rather than by specimens. Use one blank to set the forward and side scatter for each PMT voltage. For small batches, analyze all prepared tubes on the same specimen before preparing and analyzing tubes on a different specimen. If you are using a Q-Prep, place it close to the flow cytometer so you can process one tube while analyzing another. The Multi-Q-Prep and TQ-Prep workstations are not recommended because of the critical timing, temperature and mixing requirements of kinetic assays.

A representative protocol for sample processing follows the sample preparation protocol. For your particular enzyme, follow the package insert carefully to provide accurate timing and temperature conditions for the enzyme and substrate of interest.

ANALYSIS

Data analysis on the flow cytometer is based on the intensity of staining, as mentioned previously, since all the cells should contain enzyme activity. Enzyme-substrate binding is a continuous kinetic reaction. Without consistency in the setup and analysis it is impossible to acquire comparable results. Several steps are necessary to assure this consistency. First, the standardization of the PMTs with the fluorescence standard is required to set high voltages which will give consistent readings from one day to the next. It is also necessary to establish baseline enzyme substrate characteristics or normal values by running normal cellular samples. A positive control cell should be analyzed to establish that the reagents are working properly, and to set the high values expected from the assay. It will monitor reagent stability, regulate technique, and standardize mean fluorescence measurement. A blank tube for each specimen is recommended next. This tube contains a processed specimen without a CellProbe reagent, and provides a means to set up the instrument scattergram, to see negative cell populations on the low end of the fluorescence histogram and to determine nonspecific fluorescence on each cell sample.

Of course, it is best to know the sample populations of interest and understand the cellular process involved when planning research studies using these enzyme substrate reagents. You can identify cell populations of lymphocytes, monocytes, macrophages, granulocytes, red cells, platelets, tumor cells or particular tissue cells. Certain cell populations may be involved in the cellular process under study, others may not. Do not limit the analysis to one cell population, but explore to see if altered enzyme levels occur in related cell populations. Data should be saved in list mode so reanalysis on different populations is possible at a later date.

Analysis can also show a bimodality or multimodality, broad distribution or narrow, increased or decreased enzyme activity when compared to a normal cell line. When analyzing, measure each peak of a bimodal histogram separately and record its mean peak fluorescence. It may be necessary to run more control cells or to add simultaneous surface marker analysis to determine why there are two or more peaks. It is possible that two cell populations of differing maturity exist in the same specimen, such as blast cells and normal granulocytes for example, or that mixed cell lineages are appearing in the same gate.

Analysis of cellular enzyme activity is new and promises to offer much insight into the intracellular activities of cells, but along with that new insight, many more questions will be raised about how cells function. It will be interesting to watch the development of flow cytometric analysis to enhance our understanding of the biology of life.

Materials

Equipment

Standard Flow Cytometer

Solutions

Cell Probe reagents

Preparation

Washed specimens

Subprotocol 1
CellProbe Reagent Standardization

Procedure

1. Create 3 instrument protocols for daily standardization to be used with a standardized fluorosphere such as Flow-Set fluorospheres (PN 6607007). These protocols will be referred to as Std. Low, Std. Medium and Std. High.

Instrument preparation

Note: IMPORTANT: DO NOT change the analysis regions from day to day. This will alter mean values and impact precision.

2. Create 3 instrument protocols for analysis of processed samples: substrates low, substrates medium, and substrates high.

3. Verify fluidic integrity of the flow system with fluorospheres.

Instrument standardization

4. Refer to Table 1 to determine whether the reagents being analyzed are low, medium, or high intensity. Select the appropriate standardization protocol: Std. Low, Std. Med. or Std High.

5. Insert the standard fluorospheres. Adjust the forward scatter (FSC) and side scatter (SSC) voltages and/or gain to yield a familiar light scatter histogram for your cell type.

6. Adjust the PMT setting of the fluorescence signal to place the fluorospheres in the target mean appropriate for the instrument protocol and CellProbe substrates being analyzed. Either peak intensity, mean intensity or mean fluorescence may be used. Adjust the target values accordingly. Use a consistent unit of measure from day to day for standardization and sample analysis.

Note: NOTE: These fluorosphere target mean values were determined for use on whole blood prepared with CellProbe reagents, lysed on a Q-Prep. High voltages will vary from the recorded values listed here based on PMT, filters, and instruments.

Sample of target values and High Voltage settings – yours MAY be different.

Target Mean for Flow-Set Fluorospheres

Established Target Mean	HV recorded	LOW
2.46+0.10	717	Medium
5.82+0.20	808	High
22.7+0.30	977	

7. Repeat this procedure for each of the standardization instrument protocols and respective target mean values: low, medium, and high. Note the PMT high voltage setting for each level.

8. Transfer the high voltage settings to the corresponding data acquisition protocols labeled Substrates Low, Substrates Medium, and Substrates High to be used for analysis. Use these data acquisition protocols to analyze your samples.

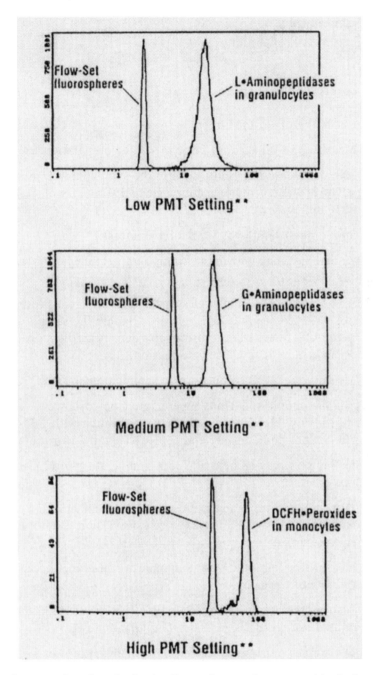

Fig. 1. Overlay of standardization fluorospheres and enzyme activity in three levels

Subprotocol 2
Specimen Preparation

▓ ▓ Procedure

1. Check the pH of the buffer.

Whole Blood 2. Measure the WBC count prior to sample preparation.

3. Mix the blood and gently pipette the sample (1 mlL into a polypropylene centrifuge tube and add 9 ml PBS or HBSS (1:10 dilution).

4. Spin diluted sample at 200 x for 10 mintues.

Note: Samples may be held at Room Temperature for 2 hours before processing.

Homogeneous Suspension or Density Gradient Separation Specimens 2. If a homogeneous, single cell suspension is available, proceed to step 4. To separate cells, gravity separate the sample for 30 minutes or separate according to a density gradient protocol to achieve a single cell suspension.

3. Measure the WBC count prior to sample preparation.

4. Mix the sample and gently pipette into a polypropylene centrifuge tube. Resuspend cells to 10 ml volume with PBS or HBSS. Spin diulted sample at 500 + 200 g for 5-10 minutes.

Note: Samples may be held at Room Temperature for 2 hours before processing.

Tissue Specimens 2. Tease tissue sample apart. Press tissue through a 30-150 μm nylon mesh to arrive at a single-cell suspension. Do not use enzyme digestion.

3. Measure the cell count prior to sample preparation. Check the viability of the sample.

4. Suspend the cells in a volume of 10 ml with PBS or HBSS. Spin diluted sample at 500 ± 200 g for 5-10 minutes.

Note: Samples must be processed immediately after washing.

5. Carefully remove supernatant with a transfer pipette and discard.

6. Suspend the sample pellet in 10 ml of PBS or HBSS. Mix **GENTLY**, using a transfer pipette.

7. Spin diluted sample at $500 \pm 200\,g$ for 5-10 minutes.

8. Repeat steps 5-7 for a total of three washes.

9. With a pipette, remove supernatant and resuspend the pellet in enough buffer to provide a cell concentration of $3.0 \pm 0.5 \times 10^6$ cells / mL. Check the cell count. Dilute if necessary to desired cell concentration.

Subprotocol 3
Sample Processing

Procedure

1. Prepare the washed specimen (from the last procedure) in the proper concentration of $3.0 + 0.5 \times 10^6$ cells/ mL. Check the cell viability before beginning the assay.   **Before Starting**

Note: Remember this is a kinetic assay where the measurement of intensity is important, therefore consistency in technique and analysis provides the best results. Follow the general guidelines in the introduction for best results.

2. Add 50 µL of the washed sample or CellZyme control cells to a labeled test tube, **GENTLY** pipetting samples to the bottom of the tube to minimize cell loss. Include 1 blank test tube per specimen.

3. Prewarm the sample at 37° C for 5-10 minutes. A waterbath is preferred to ensure homogeneous temperatures. If plastic tubes are used, increase the pre-incubation time to 7-10 minutes.

4. Add 25 µL Cell Probe reagent to the test tube in the bath. For the blank tube add 25 µL of buffer.

Note: Take caution to pipet the reagent straight into the tube with the pipet tip positioned as close to the bottom as possible without touching the cells to minimize splashing onto the sides of the tube.Use a stop watch to time the incubation – start it NOW.

Note: Allow 10-15 seconds between tubes for handling time.

Do Not Vortex   5. Mix by hand and return immediately to the water bath.

6. Incubate for the appropriate time for the specific substrate as indicate in the package insert, 1, 5, or 10 minutes at 37°C.

7. Place on crushed ice for at least 3 mintues but no longer than 20 minutes (5 minutes for plastic tubes).

8. Lyse whole blood sample in the Q-Prep on the 35 second cycle, or resuspend other samples in 1 ml cold buffer.

9. Hold prepared samples on ice. Analyze within 30 mintues of 37° C incubation (step 6).

Subprotocol 4
Data Analysis

▨▨ Procedure

Set up the Instrument   1. Set up the instrument with a fluorescence standard fluorosphere using the standardization protocols in Procedure 1. Verify that you have updated the data acquisition protocols to the daily high voltage recorded on the fluorescence standard for the appropriate substrates (low, medium or high).

2. Select the data acquisition protocol appropriate to the CellProbe reagent: low, medium or high.

Run Blank   3. Insert the CellProbe blank, the processed sample that does not contain a CellProbe reagent onto the instrument.

Note: Samples may be grouped according to specimen or by acquisition voltage.

4. Adjust the forward scatter voltage, side scatter voltage and / or gain to yield a familiar light scatter histogram appropriate for the sample. Once adjusted, these settings should not require adjustment for the remaining processed samples on the same specimen.

5. Draw gates on the light scatter populations of interest. Gate the fluorescence histograms based on these gates.

6. The background intensity is sample dependent. In general, background fluorescence for lymphocytes appears in the first decade. Unlike monoclonal antibodies, you should not set a 2% cut-off value: the blank is used only as a guideline to ensure the positive fluorescence is higher than the background.

Note: IMPORTANT: Before analyzing any population of interest, verify that the light scatter histogram is gated appropriately for the sample and you have selected the correct high voltage for the CellProbe reagent.

7. Insert the processed sample. Collect a minimum of 2500 gated events or for 60 seconds. Measure the mean fluorescence, mean intensity or peak intensity but use the same unit of measurement consistently to ensure instrument standardization, sample analysis, and comparative data analysis are consistent.

Run Samples

Note: Gated populations with multiple peaks may be significant; Measure the mean fluorescence of each peak. Bimodal peaks may occur in abnormal samples. Determine percent positives when pertinent to your study.

8. Change the data acquisition protocol according to the enzyme activity: low, medium or high before analyzing a new substrate.

References

1. Dolbeare FA, Smith RE. Flow cytometric measurement of peptidases with use of 5-nitrosalicyladehyde and 4-methoxy-b-naphthylamine derivatives. Clin Chem 1977;23:1485-1491.
2. Main-Berdel J, Valet G. Flow cytometric determination of esterase and phosphatase activities and kinetics in hematopoietic cells with fluorgenic substrates. Cytometry 1980; 1(3):222-228.
3. Watson JV. Enzyme kinetic studies in cell populations using fluorgenic substrates and flow cytometric techniques. Cytometry 1980;1(2):143-151.

Intracellular Cytokine Detection

MARY A. YUI

Introduction

The predominance of different T cell subsets in an immune response can have a profound influence on the outcome of that response. These subsets are defined in part by the pattern of cytokines produced. T cells can have very different functions, depending upon this pattern of cytokine secretion, with Th1 cells producing predominantly IL-2 and interferon-γ, while Th2 cells produce IL-4, IL-5, IL-6, IL-10 and IL-13.[1,2] As a result, the functional heterogeneity of these T cell populations require single cell detection of intracellular cytokines for many studies. Several single cell cytokine detection methods, including limiting dilution analysis, ELISPOT, immunohistochemistry, *in situ* hybridization, and flow cytometry have been developed.[3] Flow cytometry has advantages over the other techniques by virtue of its rapidity and the ability to perform simultaneous quantitative multiparameter analysis of surface markers with one or more intracellular cytokines in a large population of cells. Until recently, these flow cytometric methods have not been routinely utilized by immunologists due to a number of technical difficulties. However, several published modifications, and the availability of commercial kits, in conjunction with directly conjugated monoclonal antibodies, have greatly improved the reliability and ease of this method. It now makes flow cytometry a powerful tool for studying the responses of individual cells in an immune response.

Considerations

A number of factors need to be considered for this method to be successful. These include the appropriate stimulation conditions, the kinetics of cytokine expression, inclusion of a protein transport inhibitor, the fluorochrome conjugate and monoclonal antibody clone to be used, and, of course, appropriate controls.

Cell Stimulation

Intracellular cytokines are rarely detectable in freshly isolated cells, so a method of *in vitro* cell stimulation must be selected. One typical method for chemically stimulating T lymphocyte cytokine expression includes an *in vitro* incubation of freshly purified lymphocytes with phorbol myristate acetate (PMA) and ionomycin or calcium ionophore (CaI). Direct stimulation of the T cell receptor can be accomplished with plate-bound anti-CD3 and anti-CD28 or specific antigen-bearing antigen presenting cells, providing a more biologically relevant stimulus.[7] The kinetics of expression must also be determined as it will differ with stimulation protocol, species, cell populations and the cytokines of interest.[4-6]

Monensin or brefeldin A have been used to prevent the secretion of synthesized cytokines by lymphocytes. They are agents that block the exit of newly synthesized cytokines from the Golgi.[8] The efficacy of monensin vs. brefeldin A to inhibit cytokine secretion should also be considered for each system. They should be added for 2-16 hours during *in vitro* stimulation, although not longer than 16 hours due to toxicity. The action of monensin and brefeldin A is reversible, so staining in the presence of these inhibitors may be helpful.[3]

Fixation, permeabilization, and staining

Once the stimulated cells are harvested, they can be directly processed for FACS or they can be fixed and frozen in 10% DMSO for later staining, a useful method for clinical samples.[7] Cells should first be incubated with an antibody that will block Fc receptors on B cells, macrophages and immature thymocytes, to prevent non-specific binding of antibodies, then stained for surface antigens (e.g., CD4, CD8, CD3, Tcrβ, Tcrγδ, etc.). It should be noted that the mode of stimulation and the inhibitors used may effect expression of surface markers. CD4 and CD8 can be down-modulated from the surface of T cells with PMA and CaI stimulation[9], as can CD3 (personal observation). In addition, stimulation can cause cell death and changes in forward and side scatter of the live stimulated, blasting cells. This must be taken into account when gating for live lymphocyte populations.

The fixation of cells, without loss of antigenicity, and the subsequent permeabilization of membranes are crucial steps in this procedure. Typically, paraformaldehyde is used for fixation and saponin is used to permeabilize membranes to allow entry and exit of anti-cytokine antibodies.[5,8] Saponin must be present throughout staining and washing steps because of its reversibility.

Monoclonal antibodies specific for a wide variety of human, rat and mouse cytokines have become available in recent years making the use of polyclonal antisera with high background problems obsolete. In addition, directly fluorochome-conjugated monoclonal antibodies are now also available obviating the need for second step reagents, resulting in lower background staining in many cases.[7] Biotin-avidin systems are not as good due to decreased efficiency of entry into permeabilized cell membranes. Antibodies must be titered with either stimulated cell lines known to produce the target cytokine or with primary cells. Pharmingen (San Diego, CA) offers activated and fixed positive control murine cells.

Reagent kits, detailed protocols, and directly fluorochrome-conjugated monoclonal antibodies are now available commercially through several companies including Pharmingen, Biosource International (Camarillo, CA), R & D Systems, Inc. (Minneapolis, MN), Becton Dickinson (Palo Alto, CA), and Caltag Laboratories, Inc. (Burlingame, CA). Pharmingen in particular has a helpful technical protocol sheet and references in the catalog. Reagents can also be made in the lab from published protocols (see references).

Controls and analysis

Because intracellular cytokine staining often is seen as a shift in fluorescence rather than as a discrete population, the use of controls in distinguishing positive from negative cells is crucial for statistical analysis. Use of directly conjugated antibodies avoids the often problematic need for secondary reagent controls. The most useful controls are: (1) staining with an isotype-matched control antibody to an irrelevant antigen, (2) preincubation of conjugated antibody with recombinant cytokine, or (3) pre-incubation of cells with unconjugated antibody of the same clone, followed by staining with the conjugated antibody.

It may also be valuable to determine whether staining observed is in part due to exogenously secreted surface-bound cytokine. Using 2-color analysis of extracellular and intracellular IL-2 pools, Prussin and Metcalfe[7] showed that for human PBMC, all cells staining for extracellular cytokine also stained intracellularly, indicating that they were not detecting passively acquired IL-2.

Materials

Equipment

Standard Flow Cytometer

Solutions

- PharMingen Cytostain™ kit - containing, Cytofix/Cytoperm, Perm/Wash
- PharMingen Fc Block (anti-Fc Receptor antibody)
- Cell surface staining antibodies of your choice
- Directly conjugated anti-cytokine antibody of your choice
- Staining Buffer: Hank's balanced salt solution (no Phenol Red) with 0.25% BSA and 0.1% Na azide, pH 7.4, filtered, stored at 4° C

Preparation

- Ice bucket
- 96 well plates or staining tubes

Procedure

Intracellular Staining for Cytokines Using PharMingen's CytoStain™Kit

In Vitro lymphocyte stimulation and inhibition of cytokine secretion

1. Stimulate freshly isolated cells from blood or spleen.

Note: Careful optimization of stimulus, inhibitor (monensin vs. brefeldin A), and timing must be done for the best detection sensitivity.

Note: Stimulation and inhibition can be done simultaneously or the inhibitor can be added some time after stimulation. The inhibitor should not be in the cultures for more than 16 hours due to toxicity.

2. Add the appropriate concentration of inhibitor (brefeldin A or monensin).

3. Incubate 4-16 hours.

Harvest and transfer cells to 96-well staining plates or staining tubes

4. Transfer cells, 0.5-1 X 10^6 cells/well to 96-well staining plate (Corning # 25802) or to staining tubes of choice.

5. Centrifuge 4 min 250 x g in centrifuge with plate holder.

6. Remove supernatant by rapid inversion or with light suction.

Block Fc receptors

7. Resuspend cells in 1 µg Fc Block™ 10^6 cells in 100 µl Staining Buffer.

Note: For mouse cells use anti-FcγII/III clone 2.4G2 supernatant or purified antibody (e.g., Pharmingen Fc Block™. For human cells use an isotype matched control antibody.

8. Incubate 4°C 15 min.

9. Pellet cells by centrifugation 4 min at 250 x g and remove supernatant.

Cell surface staining

10. Add pre-titered conjugated antibody in Staining Buffer.

Note: Optimum antibody concentrations should be previously determined.

Note: Surface staining should be done before fixation due to potential loss of epitopes. Staining Buffer: Hank's balanced salt solution (no phenol red) with 0.25 % BSA and 0.1 % Na azide, pH 7.4, filtered, stored at 4°C

11. Incubate 4°C 30 min, dark.

Washes

12. Pellet cells by centrifugation at 250 x g and remove supernatant.

13. Resuspend cells in 250 µl Staining Buffer, pellet, discard supernatant. Repeat.

Fixation and permeablization

14. Resuspend cells in the remaining staining buffer and add 100 µ Cytofix/Cytoperm®, mixing quickly and thoroughly.

Note: Cells must be resuspended prior to the addition of fixative to avoid cell clumping. Add a small volume (~20 µl) of Staining Buffer if necessary.

15. Incubate 4°C 20 min, dark.

16. Wash 2 times using Perm/Wash®.

Note: The Perm/Wash® solution comes as a 10 x and must be diluted before use.

Intracellular cytokine staining

17. Resuspend cells in the appropriate dilution of conjugated anti-cytokine in 1X Perm/Wash®.

Note: Use the saponin-containing buffer for staining to allow entry of the antibody into the cell.

Note: Antibodies must be previously titered.

Note: For a specificity control, preincubate the antibody with an appropriate concentration of recombinant cytokine at 4°C for 20 min. If blocking with the same unconjugated antibody, do that step first then wash and continue staining with conjugated antibody.

18. Incubate 4°C 30 min, dark.

19. Wash 2 times with 1 x Perm/Wash® (>200 µl)/wash.

Washes

Note: It is important to use the saponin-containing buffer for all washes to allow unbound antibody to exit the cells.

Analysis   **20.** Resuspend cells in 0.3-0.4 ml staining buffer for analysis on the flow cytometer.

Note: Cytokine positive cells often appear as a shift in the population not as a separate population (Fig. 1).

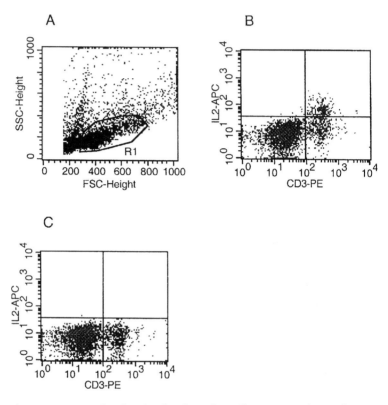

Fig. 1. IL-2 expression in stimulated CD3⁺ T cells. Murine spleen cells were isolated and stimulated with 12.5 ng/ml PMA and 190mM calcium ionophore, A23187, for 10 hours. Using reagents from Pharmingen's Cytofix/Cytoperm Plus® kit (Cat. #2300KK), GolgiPlug™ (brefeldin A) was then added and the cells incubated another 5 hours. Cells were harvested, blocked with 2.4G2 hybridoma supernatant, stained with 0.4μg anti-CD3-PE (Pharmingen, Cat. # 01085B), fixed, permeabilized, and stained with 0.2μg APC-conjugated anti-IL-2 (Pharmingen, Cat. #18179A) in 50μl Perm/Wash® solution. A. Forward scatter (FSC) vs. side scatter (SSC) with gating for live lymphocytes (R1). B. Some CD3⁺ T cells, but not CD3⁻ cells (predominantly B cells), express intracellular IL2 after stimulation. C. Preincubation of anti-IL2-APC with 2.5μg/ml recombinant mouse IL2 (Pharmingen, Cat. # 19211T) for 20 min, 4°C, blocks anti-IL2-APC staining and can be used as a control to set markers for statistical analysis

References

1. Mosmann TR, Sad S. The expanding universe of T-cell subsets: Th1, Th2, and more. Immunol. Today 1996; 17:138-146.
2. Carter L, Dutton R. Type 1 and Type 2: a fundamental dichotomy for all T-cell subsets. Curr. Opin. Immunol. 1996; 8:336-342.
3. Carter LL, Swain SL. Single cell analyses of cytokine production. Curr. Opinion Immunol. 1997; 9:177-182.
4. Coligan JE, Kruisbeck AM, Margulies DH, Shevach EM, Strober W. *Current Protocols in Immunology*. Greene Publishing and Wiley-Interscience, NY. 1994.
5. Sander B, Andersson J, Andersson U. Assessment of cytokines by immunofluorescence and the paraformaldehyde-saponin procedure. Immunol. Rev. 1991; 119:65-93.
6. Sander B, Höidén I, Andersson U, Möller E, Abrams JS. Similar frequencies and kinetics of cytokine producing cells in murine peripheral blood and spleen. J. Immunol. Methods 1993; 166:201-214.
7. Prussin C, Metcalfe DD. Detection of intracytoplasmic cytokine using flow cytometry and directly conjugated anti-cytokine antibodies. J. Immunol. Methods 1995; 188:117-128.
8. Jung T, Schauer U, Heusser C, Neumann C, Rieger C. Detection of intracellular cytokines by flow cytometry. J. Immunol. Methods 1993; 159:197-207.
9. Anderson ST, Coleclough C. Regulation of CD4 and CD8 expression on mouse T cells. J. Immunol. 1993; 151:5123-5134.

Section 7

Electronic Cell Sorting

ROCHELLE A. DIAMOND

The 1990's have ushered in an era of commercially available high speed sorters. Long strides have been made in sorting rare cell populations, low frequency transfectants, and single cell deposition robotics along with all kinds of other methodologies take cells well beyond the sorter.

Marty Bigos and Steve Merlin provide tips on standard and reproducible sterile sorts. Peter Lopez reviews the commercially available Mo-Flo high speed sorting instrument which is now offering the possibility of 4 way sorting. Sean Morrison provides a protocol for sorting rare cell populations. Bob Lief and Karen Chew present equipment for sorting onto slides and methodologies for what can be done with sorted cells.

Creating Standard and Reproducible Sorting Conditions

M. BIGOS, R.T. STOVEL AND D.R. PARKS

Introduction

This tutorial describes how to establish and reproduce sorting conditions for jet-in-air sorters (referred to as sorter). The criteria to be met by the standard sort conditions are stable sorting without signal measurement degradation. An operator familiar with her sorter should be able to use this protocol to determine standard conditions within one to two hours. Once standard conditions are determined, subsequent sort setups using these conditions should be on the order of 5 minutes.

1. Identify an optimal set of sorting conditions for each nozzle size used.

2. Record the set of jet measurements and instrument settings that give the conditions in Part I.

3. Setup for sorting by reproducing the set of measurements and instrument settings.

We have found this approach to be more reliable than allowing variable conditions and trying to compute matching drop delay settings. It is also more convenient and faster than evaluating experimental sort delay profile every time you sort. This tutorial is brief. References [1,2] provide more detailed descriptions of aspects of the material covered here.

Subprotocol 1
Identifying Optimal Sorting Conditions

We assume that you are familiar with the operation of the sorter. In particular you know how to:

- View the stream and drops with a strobe.

- Make measurements of the drop position.

- Use test signals or test particles to get deflected side streams.

- Know how to set the sheath pressure, the drop drive frequency, the drop charge phase, the sort mode, the number of deflected drops, and the drop delay. This information should be contained in the manuals provided by the manufacturer of your sorter.

There are five main variables to standardize: stream velocity, drop frequency, droplet breakoff position, sort delay time, and charge phase. These results are recorded in the protocol following. In addition, some sorters allow for the adjustment of the charge pulse shape this should be optimized for uniformity of the deflected streams, but will not be discussed here.

▓▓ Materials

Equipment

Standard Flow cytometer equipped to sort particles

▓▓ Procedure

Before Starting
1. Set up the sorter and align it properly for making measurements on test particles and typical biological material to be sorted (referred to as cells).

Note: It is also important to remove all air bubbles from the nozzle; bubbles absorb energy, leading to instability of the jet breakoff and requiring higher drop drive amplitudes. Flushing the nozzle with ethanol will usually accomplish this.

2. Set the pressure initially to the manufacturer's recommended value. This will result in a stream velocity on the order of 10 m/sec. for most current commercial instruments.

Set Initial Stream Velocity

Note: Stream velocity is roughly related to the square root of the sheath fluid pressure. Instruments designed for high speed sorting or sorting very large objects ("low speed sorting") will have jet velocities significantly different from this.

3. The drop drive frequency is proportional to the stream velocity divided by the stream diameter: $f = k(V/d)$. An optimal frequency minimizes noise caused by scattered light from the stream while maintaining stable drop formation. These conditions are met when $1/k$ is between 3.5 and 4.5. Choose a frequency within this range.

Set Drop Drive Frequency

4. Measure Drop Breakoff. With the drop drive on, measure the average spacing between drops by measuring the distance of over at least 5 drops. Set the drop drive amplitude so that the distance between the primary laser intercept and the drop breakoff, as shown in Standard View 1, (fig. 1) is within the range of drop lengths (also referred to as drop delay) allowed by the manufacturer.

Set Droplet Breakoff

Note: The shorter the breakoff distance, the more stable sorting will be. However too short a length will require a high drop drive amplitude which will degrade the measurement signal. Drop drive or clock triggering, if available, is useful for this. At normal operating conditions there should be no additional triggering of the instrument with the drop drive on.

5. Turn up the gains in both the forward and side scatter measurement channels so noise created by having the drop drive on is visible in the sorter display system (channel pulse displays are good for this). These gain values should be higher than normal operating conditions. Adjust the obscuration bars for both measurement channels to minimize the noise signals.

Adjust Laser Obscuration Bars

6. Preliminary Test of Conditions: Run cells at normal measurement conditions on forward and side scatter positions with the drop drive off and on at the amplitude set in step 4. There should be negligible difference between the two measurements in terms of "spreading", or increased coefficient

Test of Preliminary Conditions

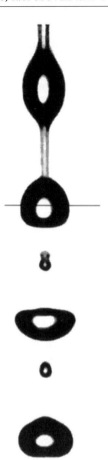

Fig. 1. Standard View 1. The first drop is just about to break off from the stream. The horizontal line is the measurement crosshair in the middle of the drop. This is the standard breakoff length

of variation. If not, then it is necessary to change the conditions to be able to reduce drop drive amplitude. Keeping the conditions constant, vary the frequency over the computed range in step 3 to find the shortest break off distance. There may be more than one. This may allow for some reduction in drop drive amplitude. If the measurements are still degraded, repeat step 4 with a longer breakoff.

7. Measure the position of drop N, where N is approximately half the breakoff length (in drops) determined in step 4. For example, if the breakoff length is about 16 drops, then N should be 8.

8. Using the sorter's test method of producing deflected side streams, adjust the charge phase so these streams are uniform. Note the changed view of the drop break-off area as in standard View 2 (Fig. 2).

Set Charge Phase Setting

9. Using particles that can be easily seen under a microscope, perform sorts onto a slide while varying the electronic sort delay, sorting the same number, between 20 and 50, for each delay. Move the slide to obtain little "puddles" corresponding to the different delays. This must be done in single drop deflection mode using the cloning or counter setting of the sorter for best accuracy. Vary the delay around the measured drop delay length in step 4, first coarsely, and then finely

Set Drop Delay

Fig. 2. Standard View 2. This is an example of what you might obtain as standard view 2. The crosshair is in the middle of the first detached drop. This position is duplicated during the setup by adjusting the drop charge phase

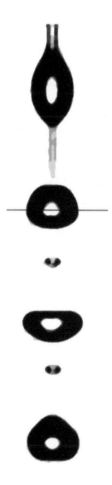

(most sorters allow delay settings that vary by drops and fractions of a drop). Examine the slide to find the delay setting which produces the best recovery.

Final Test of conditions

10. Reset the deflection mode to its normal setting and set the deflected drops to at least 2. Set the sorter drop delay as determined in step 9. Using cells, verify that sorting them does not significantly degrade the uniformity of the side streams. If it does, then a shorter breakoff, with more drop drive amplitude should be used. Repeat the procedure starting at step 4. If step 6 and the conditions here cannot be met, then either a nozzle of a different size is needed, or there is a more fundamental problem to solve.

Note: You have now determined a standard drop drive frequency, a standard breakoff length at Standard View 1, a standard position of drop N, a standard drop delay setting, and a standard charge phase which produces Standard View 2. These conditions should work well when sorting or cloning with 2 or more deflected drops selected.

Subprotocol 2
Setting Up to Sort Using the Standard Conditions

▦▦ Materials

Equipment

Standard Flow cytometer equipped to sort particles

▦▦ Procedure

After appropriate warm-up, (possible sterilization), and alignment of the sorter:

Set the standard drop drive frequency and drop delay

1. Set the standard drop drive frequency and drop delay.

2. Set up the deflection plates and side streams.

Set up the deflection plates and side streams

3. Set the drop breakoff point at the standard distance below the laser intercept, obtaining Standard View 1.

Set the drop breakoff point

4. Count down to drop N. If this is not the standard length, raise or lower the sheath pressure accordingly. To insure fast convergence to the standard position of drop N, overshoot by as much as the original deviation. For example, if drop N is further than the standard length, lower the sheath pressure until it is the same distance less than the standard length. Repeat steps 3 & 4.

5. Set the phase to produce your version of Standard View 2. You are now ready to sort.

Set the phase

SORT

References

1. Lindmo T, Peters DC, and Sweet RG. Flow sorters for biological cells. In: Melamed M, Lindmo T, and Mendelsohn, M, eds. Flow Cytometey and Sorting. 2nd Edition. NY: Wiley-Liss, 1990: 145-170.
2. Parks DR. Flow cytometry instrumentation and measurement. In: Herzenberg LA, Herzenberg LA, Blackwell C, and Weir DM, eds. Weir's Handbook of Experimental Immunology, Vol II. 5th Edition. MA. Blackwell Science, 1996: 47.1-47.12.

Sterilization for Sorting

STEVEN MERLIN

Introduction

In order to aseptically sort and collect cells for tissue culture work, it is imperative that thorough cleaning and sterilization procedures be implemented before and after the use of Fluorescent Activated Cell Sorter instruments. The intent of sterilization is to use an agent that effectively destroys all microorganisms (e.g., viruses, bacteria, fungi and their spores) on inanimate surfaces. The following protocol addresses the steps and reagents necessary to achieve sterilization. Adequate and routine cleaning is a prerequisite for any sterilization procedure. The exposure time of the microorganism and spores to the sterilant agent will result in greater likelihood of elimination. Generally, 10 minutes is sufficient time to destroy most organisms. However, in the case of cell sorter instruments, thin tubing, narrow channels and other areas that are generally difficult to access may harbor organic materials in addition to microorganisms. It may therefore be desirable to lengthen the exposure times to 20 or 30 minutes.

Hydrogen peroxide solution in a concentration of 6% to 10% has very high activity in its efficacy against vegetative bacteria, tubercle bacillus, bacterial spores, fungi, lipid and medium-sized viruses as well as non lipid and small viruses. Chlorine compounds such as household bleach (Sodium hypochlorite) are effective intermediate disinfectants suitable for fluidic systems. Bleach is highly corrosive when coming into contact with aluminum and other metal parts. Avoid using chlorine bleach products on metal surfaces of the cell sorter. Ethyl alcohol (70% ethanol) may have limited virucidal activity and is best used for cleaning non critical areas and metal parts because of its quick evaporation time.

Materials

Solutions

- Hydrogen Peroxide (H_2O_2), 6-10% stabilized
- Ethanol (Ethyl alcohol/ETOH) 70%
- Sodium Hypochlorite (Household Bleach) 5.25% Stock Solution 50,000 mg/L chlorine
- Sterile H_2O (filtered with 0.2 micron filter) Millipore MILLI-GS 0.22 micron Filter Unit, Catalog number IVGS0103F

Preparation

- Before mixing and handling disinfectant/sterilization agents, protect eyes, skin and clothing with appropriate splash goggles, latex gloves and lab coat respectively.
- Prepare fresh solution of hydrogen peroxide containing between 6-10% H_2O_2. Dilute ETOH in sterile water to give a final concentration of 70%.
- Household Bleach (Chlorox®) is commercially available as a 5.25% solution of sodium hypochlorite (50,000 mg/L of free available chlorine). Make a 1:10 dilution using sterile water to yield a 0.5% NaOCl solution containing approximately 5,000 mg/L of free available chlorine.Note: All dilutions of hydrogen peroxide and sodium hypochlorite should be made up immediately prior to use. This will prevent loss of germicidal action that can occur during storage.

Procedure

1. Remove the nozzle tip and soak in H_2O_2 for 10 minutes. A short sonication of 15-30 seconds can also be performed in conjunction with soaking. Loosen the black knurled cap holding the nozzle tip holder assembly. Carefully remove the assembly and wipe it and the connected polypropylene and tygon sample tubing with a sterile cotton or Dacron applicator swab soaked in ETOH.

   Nozzle sterilization

Note: Some tips, i.e. the Elite Sort Sense tip with a lens attached, can not be sonicated. PLEASE follow manufacturer's suggestions for particular instruments.

External sterilization

2. Thoroughly saturate a gauze pad with 70% ETOH. Carefully wipe the walls of the stream chamber, access door, waste collector tube, sort collection tube holder, changing the gauze pad frequently. Allow the ethanol to air dry on the wetted surfaces. Wipe the top inner surface where the nozzle tip holder assembly mounts. Be very careful not to bend the stainless steel stream charging wire. Dry the wire with a sterile applicator swab.

3. Remove high voltage plates and wipe or soak with ETOH. Remove the laser beam obscuration bars, and the fluorescence objective lens. Clean lenses by first flushing the lens surface with distilled H_2O and then EtOH to remove any dried salt crystals. Follow by blotting dry with lens tissue paper. Wipe the outer barrel of the objective with ETOH and dry with a sterile gauze. Use Kodak lens cleaning solution to clean the front optic.

4. Using a sterile cotton or Dacron applicator swab soaked in ETOH, wipe all corners and other hard to reach spots, stream lamps and lamp housings, obscuration bars and strobe light. Carefully replace all components when dry.

Reassemble

5. Replace the obscuration bars, high voltage plates, nozzle tip holder assembly and the nozzle tip. Check that the stream charging wire is making contact with the metal sample introduction tube.

Sterilization of the Fluidics

6. Pre-rinse a pressure vessel with H_2O_2 two times, refill with 500 ml of H_2O_2 and pressurize. Turn the fluidics switch to FILL, attach a large sterile syringe to the stop-cock valve and open the valve allowing the syringe to fill with H_2O_2. Shut off the stop-cock and with the fluidics switch still in FILL mode, press the NOZZLE FLUSH button for 30 seconds to draw fluid into the vacuum drain line of the nozzle holder assembly. Turn the fluidics switch to RUN and allow the H_2O_2 to back flush the sample line for 30 seconds.

Note: It is recommended that two additional Alloy Products or other specified pressure vessels of 4 liter capacity be purchased for holding the H_2O_2 and sterile water. If this is not practical, the sheath container may be repeatedly emptied and rinsed with H_2O_2 and sterile water during the respective sequence listed below.

Note: The incoming air pressure line from the house air supply has a 0.45 Gelman filter or equivalent. A 0.22 micron filter can also be used to protect the pneumatic components and pressure tanks from particulate matter. This filter should be inspected periodically and changed when necessary.

7. Wipe the O-ring and sample attachment port with an H_2O_2 saturated sterile applicator. Place a sterile tube filled with H_2O_2 onto the sample introduction port, pressurize the tube and increase the sample differential pressure to achieve a high flow volume. Allow to run for 10 minutes making sure the tube does not run dry.

8. After a minimum of 10 minutes, turn the fluidics switch to OFF, depressurize the tank and empty of the remaining H_2O_2 solution. Attach the tank containing sterile water or rinse the same tank thoroughly with sterile water twice, refill and pressurize the system. Repeat steps 1 and 2, using sterile water instead.

9. After sterile water has sufficiently flushed the system, add the sheath tank container containing sterile saline. The swinex plastic holder should have a millipore pre-filter element and a 0.45 micron filter. To the top of the swinex tubing port, a Millipore MILLI-FIL GS 0.22 micron Filter Unit can be added to filter out particles and reduce the chance of contaminants entering the fluidics via the pressure vessels. The placement of stop-cock valves at either end of the 0.22 micron unit allows for the sheath introduction line to be closed for maintaining sterility when a filter change is necessary.

10. Pressurize the sheath tank and turn the fluidics knob to FILL, attach a large sterile syringe to the stop-cock valve and open the valve allowing the syringe to fill with saline. Shut off the stop-cock and with the fluidics switch still in FILL mode, press the NOZZLE FLUSH button for 20-30 seconds to draw fluid into the vacuum drain line of the nozzle holder assembly. Run sterile saline for 10 minutes before beginning the optical alignment and QC check.

11. Prior to sorting, turn on the vacuum pump that evacuates aerosols from the stream chamber collection area. Should an aerosol be generated, the vacuum pump will limit its spread within the chamber to a more confined area.

Note: Equip vacuum exhaust port with a Whatman HEPA-CAP 0.2 micron membrane filter.

SORT

Post-sort   12. After conclusion of the sort, run a tube filled with 0.5% NaOCl at a high flow volume. If samples contained cells that produce mucin, use a tube filled with hyaluronidase enzyme instead, followed by a water flush and then 0.5% NaOCl. This breaks down the mucin. Follow with a sterile water rinse, allowing about 1 ml of water to remain in the tube when shutting off the pressure.

Note: Disinfection of stream chamber surfaces and components.

Note: The use of sturdy latex gloves is advised when cleaning the sample and stream chamber areas of the cell sorter, particularly after any sort involving human or other pathogenic specimens.

13. Spray the lower stream chamber, sink basin and plexiglass flip-down viewing door with either ETOH or H_2O_2 until the surfaces are sufficiently wet. Wipe with a wet gauze pad and allow the solution to remain in contact for several minutes.

14. Remove high voltage plates and clean. Wipe down stream chamber and inner surface of the door. Use applicator swabs for hard to reach areas, surfaces and obscuration bars.

15. Replace high voltage plates.

References

1. National Committee for Clinical laboratory Standards. Protection of laboratory workers from infectious disease transmitted by blood and tissue. Proposed Guidelines. NCCLS Document M29-P, Vol. 7 No. 9. Villanova, PA.: NCCLS; 1987. pp. 337-345, 358-359, 388.
2. National Committee for Clinical laboratory Standards. Protection of laboratory workers from instrument biohazards; Proposed Guideline. NCCLS publication I17-P (ISBN 1-56238-122-9). NCCLS, 771 E. Lancaster Ave, Villanova, PA 19085, USA., 1991. pp. 17-18, 36-39, 44-45.
3. Acquired Immunodeficiency Syndrome. Recommendations and Guidelines. Department of Health and Human Services, Public Health Service, Centers for Disease Control, Atlanta, Georgia 30333. 1988. pp 33-35.

High Speed Cell Sorting Using the Cytomation CICERO® and MoFlo® Systems

PETER LOPEZ

Introduction

High-speed flow cytometry can be described as any flow cytometric technique (analysis or sorting) performed at rates of over 10,000 events per second (eps). In the past, researchers only obtained this type of throughput using their own custom-built, one-of-a-kind instrumentation. Cytomation, Inc. has designed two product lines that bring the ability to perform high-speed flow cytometry to every lab.

CICERO® is a product introduced in 1988 that allows owners of standard cell sorters to upgrade their systems for high-speed sorting (15,000 pps maximum). This upgrade involves the replacement of the data acquisition computer and sort electronics with the CICERO interface and CyCLOPS® software, while retaining the original optical and fluidic components. Typically, the cost of this upgrade is less than 1/10th that of a new cell sorter.

In 1994, Cytomation launched the MoFlo® system, a complete Modular Flow cytometric platform designed from the ground up to allow routine high- purity sorting at 25,000 eps with performance capabilities well in excess of that rate. The MoFlo system includes a modular, multi-laser optimized optical bench, a high-pressure rated fluidics system, proprietary electronics, and SUMMIT™ software. High-performance sorting is realized through the use of patented parallel-processing sort electronics. This modular system was designed for easy upgrading and flexibility, and is capable of being operated at high speeds in any configuration from a single air-cooled laser system to a large multi-laser system. The MoFlo can be configured with the CyClone® X-Y table for high-speed deposition of single cells into microtiter plates or microscope slides, as well as with optional multi-stream (>2) sorting capability.

The other major cytometry instrument manufacturers currently offer options to their standard cell sorters which permit high-speed operation. Becton-Dickinson produces a TurboSort® option for its FACS® series of cell sorters, and the Coulter EPICS Elite ESP® cell sorter can be configured for high-speed operation. All systems capable of high-speed analysis and sorting are equally capable of operating at conventional speeds.

INSTRUMENT DESIGN CONSIDERATIONS

A number of practical considerations need to be addressed in the design of a system for high-speed flow cytometry. First, electronics for data acquisition and sorting must be able to accurately accept and process the high throughput of signals being generated. Fluidics design is the second important consideration. The system should be capable of generating a high droplet formation frequency, since the higher number of droplets lowers the probability of a droplet containing more than one cell, resulting in an aborted sorting event.. Fig. 1 illustrates this improvement in yield with increased droplet formation rate as predicted by Poisson statistics.

Systems utilizing optimally designed electronics and fluidics should be capable of this performance. High-speed sorting should not degrade the performance of the cytometer in this re-

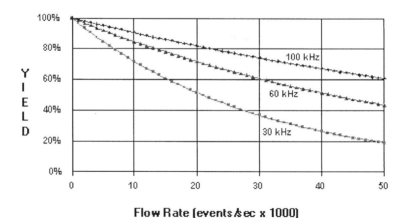

Fig. 1. Percentage yield at increased droplet formation rate

gard. The purity and recovery of high-speed sorted fractions should be similar to samples sorted at lower speeds. One final consideration is that the time required to set up the instrument for high-speed applications should not be different than for low speed; indeed, these set-ups should be identical.

Electronics

Signal processing and sort-decision electronics must be capable of fast, reliable, error-free operation for high-speed throughput to be realized. In addition, these operational parameters must be maintained in multi-laser, multi-parameter configurations. Mo-Flo electronics employ a patented parallel processing architecture to achieve these goals. This architecture permits the system to operate with a 5usec dead time in single as well as multiple laser configurations. Thus, high performance is maintained and is totally independent of the number of lasers, the number of detectors, or the use of non-rectilinear sort regions.

High-resolution droplet/event profiling is critical to high-speed cell sorting. Droplet/event profiling is a measure of the accuracy with which the location of a cell is known within the fluid stream. The more accurate this knowledge, the higher the purity of the sorted fraction. The MoFlo system is capable of resolving this profile to a $1/16^{th}$ droplet resolution.

Finally, sorted material should follow a well defined path to enter the sort collection receptical without fanning during high speed sorting. The proprietary adjustable step charge applied to sorted droplets on the MoFLo prevents fanning, and thus makes multiple stream sorting with more than 2 streams possible.

Fluidics

In order to achieve high sample throughput, a 25-meter-per-second flow rate, and a droplet formation rate of up to 200,000 droplets per second, the MoFlo fluidics system was designed entirely with high pressure HPLC tubing and fittings. So fitted, the system is capable of handling pressures in excess of 100psi. An in-line sheath fluid filter, also with this high pressure rating, is used to minimize particulates from causing nozzle clogs. In addition, a high-speed optimized flow cell/nozzle tip assembly allows a more gentle acceleration and hydrodynamic focusing of cells , resulting in excellent viability of cells sorted at high pressure.

APPLICATIONS

Typically, high-speed flow cytometry is considered when populations of interest exist at low frequency. Such applications include hematopoietic stem-cell analysis and purification, fetal cell detection and purification as found in maternal circulation, residual-disease monitoring, circulating dendritic cell isolation, cell-cycle analysis of rare populations, and multi-parameter data acquisition or sorting of transfected populations. These applications usually involve sorting, but analytical studies also benefit from the ability to process data at high speed. In addition, high-speed cell sorting is used simply for its high throughput, where adequate numbers of analysed or sorted events are required for subsequent assays. Examples include the separation of X- and Y-chromosome-bearing sperm for in-vitro fertilization, chromosome purification, and immunobiological applications such as the isolation of populations involved in T-cell development. Of course the time savings realized when cell-sorting experiments are completed in 1/10th the time normally needed is an obvious economic benefit. This time savings is also beneficial in that sorted material is available for subsequent bio-assays much faster.

Sample preparation for high-speed cytometry

Preparation of samples for high-speed flow cytometry is similar to that used for routine cytometry, with a few exceptions. All solutions involved in sample preparation which are made in-house must be well filtered prior to use to remove particulates. This preparation includes wash media, sheath fluid, buffers, and soluble dye solutions. Filtration using a 0.45 μm filter is adequate for most non-sterile applications, with 0.2 μm used for sterile work. If any particulates are visible in the solution after filtering, refiltering is needed.

Sample throughput may exceed 1×10^8 per hour, so sample preparation will be scaled up accordingly. For immunofluorescence staining, bulk processing of large quantities of cells in a large volume of buffer should be avoided, as this condition often results in less intense fluorescence staining. Aliquots containing 5×10^6 to 1×10^7 cells resuspended in a minimal volume of wash media (not more than 250 µl) may be processed using the appropriate quantity of antibody. Samples should be vortexed a few times during the incubation period. Final concentrations for high-speed cytometry should be on the order of $2\text{-}5 \times 10^7$ cells/ml. Samples must be filtered prior to flow to remove any cellular clumps and debris using a nylon mesh with a sieve size less than 1/2 the nozzle orifice diameter. A 30 µm mesh is adquate for use with a 70 µm nozzle. Viable samples should be kept on ice prior to analysis.

Set-up for sorting

System set-up for high-speed flow cytometry is similar to that used for routine cytometry. Filtered sheath fluid (< 0.5 µm) must be used in a well cleaned sheath tank along with a fresh in-line sterilizing sheath filter (Gelman Supor-200 0.2µm or equivalent). For sterile sorting, the in-line sheath filter assembly should be autoclaved or purged with 70% ethanol. Sheath fluid temperature should be held constant during the set-up and for the duration of the sort, as changes in sheath temperature can cause the droplet delay measurement to drift.

The selection of nozzle orifice diameter should be determined by cell size. For best viability and cell recovery, the nozzle orifice should be more than 5 times the diameter of the sorted cell population. For human lymphocytes, a 70 µm nozzle is appropriate. On the MoFlo, a 70 µm nozzle is typically used with a sheath pressure of 60psi.

Drop delay determination

A drop delay matrix is performed in two steps, either manually or using the automated CyClone X-Y table. Initially, 160 fluorescent microspheres are sorted into respective matrix positions on a glass microscope slide using incremented whole drop delay values within the range typically observed for that nozzle. When using a 70 μm nozzle on the MoFlo with 95-105,000 drops generated per second, the typical drop delay falls between 35 and 45 drops. By sorting 160 microspheres in each drop of this initial drop delay matrix, 1/16th drop delay accuracy is easily determined. The 160 sorted microspheres are typically distributed between 2 drop delay values, with each 10 microsphere multiple seen in an adjacent droplet representing a 1/16th drop difference. For example, if 140 microspheres are seen at the drop delay position of 39 drops, with 20 seen at the 40 drop position, then the drop delay is 2/16ths greater than 39, or 39 and 2/16ths. This delay value is confirmed by generating a second drop delay matrix slide using the 1/16th drop increment calculated above. Sorting of the experimental sample can now commence.

Sorting

When performing high-speed cytometry, cells must be at the concentrations mentioned earlier. Lower concentrations of $1 - 5 \times 10^6$ may be used for analysis and sorting at more conventional rates. High throughput should never be accomplished by sample over-pressure as this condition will degrade instrument performance due to widening of the sample stream core, causing both an increase in coefficient of variation (cv) and in cell coincidence (hard aborts). Samples for high-speed sorting should be initially run at minimal sample pressure and then brought up to speed by gradually increasing sample pressure, while observing light scatter and flourescence distributions. If significant degradation of light scatter distribution or fluorescence cv is observed at higher throughput, the cellular concentration should be increased.

Gentle agitation and cooling of sample should be maintained for the duration of viable cell sorts. Collection media should be supplemented with the appropriate additives including antibiotics when possible. Collected volumes of sorted material will be similar to that seen with conventional instrumentation, since droplets are of a similar size.

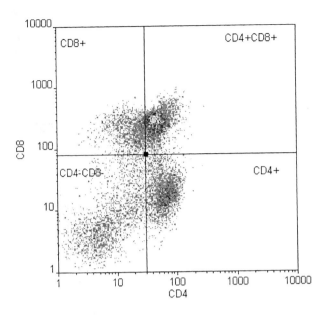

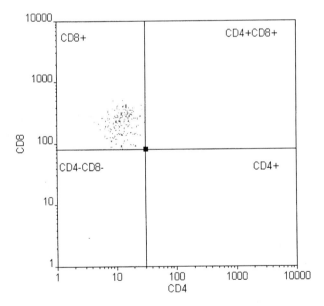

Fig. 2. CD8 single positive mouse thymocytes sorted at 30,000 cells per second using the MoFlo

Fig. 2 illustrates a sort of CD4 /CD8 immunofluorescence labeled mouse thymocytes for single CD8 positive cells performed using the MoFLo at a flow rate of 30,000 cells per second. Reanalysis of the sorted fraction, originally 4% of the total starting population, showed a purity of 99.3%.

Multiple stream sorting producing more than 2 sorted populations simultaneously (4Way ^{TM}) increases sorting throughput by a factor of two in addition to the increase made available by high-speed sorting.

Summary

High-speed flow cytometric equipment is now available from the major commercial instrument manufacturers. Cytomation systems offer the user the ability to complete experiments in 1/10 the time formerly needed. Instrument set-up is similar to that used for conventional instrumentation. In optimally designed instruments, operation at high speeds can be obtained without compromising yield or recovery.

Addendum

This chapter was prepared and submitted for publication in 1997. Since that time changes and technological advances occurred which are reported here.

1- The Cytomation MoFlo is now delivering high purity sorting at 50,000 eps.

2- Sample concentrations required to achieve this throughput are in the order of 1 x 10^8 cells/ml.

3- Coulter became Beckman Coulter, Inc. and released the EPICS® ALTRA, with a HyPerSort option for high speed work.

Please visit the "In Living Color" World Wide Web site for additional updates in this dynamic field at http://www.nlivingcolor.com.

Purification of Mouse Fetal Liver Hematopoietic Stem Cells

SEAN J. MORRISON

Introduction

This protocol is for the purification of mouse fetal liver hematopoietic stem cells (HSC). The fetal liver Thy-1.1^{lo}Sca-1$^+$Lineage$^-$Mac-1$^+$ population is a nearly pure population of multipotent hematopoietic progenitors, as described in PNAS USA 92:10302-6, 1995. Injection of fewer than 10 cells from this population into lethally irradiated mice results in long-term multilineage reconstitution by donor type cells. This population contains all long-term self-renewing multipotent progenitors in the fetal liver. Fetal liver HSC are particularly interesting in that they represent the best documented example of a stem cell population that appears to undergo daily self-renewing divisions. A drawback of this purification strategy is that the expression of Thy-1.1 and Sca-1 by HSC is specific to certain mouse strains, such as C57BL-Thy-1.1

Fetal liver Thy-1.1^{lo}Sca-1$^+$Lineage$^-$Mac-1$^+$ cells are isolated by FACS using four fluorescence and two scatter parameters. The protocol described below also provides for the pre-enrichment of Sca-1$^+$cells using MACS magnetic beads, but this step is optional.

Materials

Equipment

- Fluorescence activated cell sorter with four fluorescence, and two scatter parameters

- Mini-MACS columns, MACS avidin conjugated magnetic beads, and MACS magnet
- Centrifuge
- Dissecting microscope, fine forceps
- 1 ml syringes with 25 gauge needles
- Nylon screen for filtering cells
- Falcon 2058 (5 ml) tubes for sample preparation and FACS

Solutions

- Staining Medium: Hanks balanced salt solution, containing 2% calf serum
- Cocktail of antibodies against lineage markers including Ter119 (anti-erythroid antigens), KT31.1 (anti-CD3), GK1.5 (anti-CD4), 53-7.3 (anti-CD5), 53-6.7 (anti-CD8), 8C5 (anti-Gr-1), 6B2 (anti-B220), all suspended at appropriate concentrations in staining medium. Each of these antibodies should be titrated and used at the minimum concentration that clearly distinguishes positive from negative cells (typically 1/100 to 1/1000).
- Phycoerythrin conjugated anti-rat IgG (H+L) second stage antibody (Jackson Immunoresearch) suspended at an appropriate concentration in staining medium. This antibody should be titrated to determine the optimal concentration which distinguishes Lineage$^+$ cells from Lineage$^-$ cells, without staining non-specifically.
- Rat IgG (Sigma) suspended at 0.1mg/mL in staining medium
- Directly conjugated antibodies 19XE5-FITC (anti-Thy-1.1), E13-biotin (anti-Sca-1 or Ly6A/E), M1/70-APC (anti-Mac-1) suspended in staining medium. Antibody concentrations should be determined based on titrations.
- MACS avidin-magnetic beads (Miltenyi Biotec) suspended in de-gassed staining medium according to manufacturers instructions, depending on cell numbers to be selected. Typically 2×10^8 cells are resuspending in 0.4 ml of staining medium plus 0.1 ml of MACS magnetic bead solution.
- Avidin-texas red (Cappel), to be used at a concentration determined by titration.

– Propidium iodide (Sigma) dissolved in staining medium at 0.5mg/mL. This should be prepared fresh from a 1mg/mL stock stored in the dark at -20°C.

Preparation

Set up timed pregnancies of C57BL-Thy-1.1 mice. The morning on which the vaginal plug is noted, the day after mating, is counted as embryonic day 0.5. Fetal liver HSC have been purified between E12.5 and E15.5.

Procedure

1. Sacrifice the mother by cervical dislocation. Remove out the fetuses, and leave them in staining medium on ice. The liver can be cleanly dissected using fine forceps and a dissecting microscope. Tear away the epithelium, then lift out the liver. Pool the livers in staining medium on ice.

   Dissect the livers out of the fetuses from timed pregnant C57BL-Thy-1.1 mice between E12.5 and E15.5

2. Dissociate the livers into a single cell suspension by drawing them into a 1 ml syringe through a 25 gauge needle, then expel through a nylon screen to remove clumps. E12.5 livers yield 3×10^6 cells each. The liver doubles in size each day of development, reaching 30×10^6 cells on E15.5.

   Prepare a single cell suspension of fetal liver cells

Note: Prepare the cocktail of lineage marker antibodies by dissolving KT3.1 (anti-CD4), GK1.5 (anti-CD4), 53-7.3 (anti-CD5), 53-6.7 (anti-CD8), Ter119 (anti-erythroid antigen), 8C5 (anti-Gr-1), and 6B2 (anti-B220) at appropriate concentrations (determined by titrations) in staining medium.

Incubate the cells in a cocktail of antibodies against lineage markers

3. Spin down the cells (400g for 5 min), aspirate the supernatant (using a pipette hooked up to a trap under vacuum), and resuspend in the lineage marker cocktail at a concentration of 10^8 cells per mL. Incubate on ice for 20 minutes.

Suspend phycoerythrin conjugated anti-rat immunoglobulin second-stage antibody in staining medium. The concentration should be determined by titration.

Note: The Ter119 antibody is the most critical component of the lineage cocktail, since most cells in the fetal liver are erythroid progenitors or erythrocytes that stain with Ter119.

Incubate the cells in phycoerythrin conjugated anti-rat IgG antibody

4. Wash off unbound lineage markers by diluting the suspension of cells in lineage cocktail 5 fold with staining medium, then spin down the cells, and aspirate the supernatant. Resuspend in the anti-rat second stage antibody. Incubate on ice for 20 minutes.

Suspend rat immunoglobulin (Sigma) in staining medium at 0.1mg/ml.

Incubate the cells in rat immunoglobulin

5. Dilute, spin down the cells, and aspirate the supernatant. Resuspend in the rat immunoglobulin solution. Incubate on ice for 10 minutes.

Suspend directly conjugated antibodies (19XE5-FITC, E13-bio, M1/70-APC) in staining medium. Antibody concentrations should be determined by titrations.

Note: Rat immunoglobulin blocks unbound anti-rat IgG paratopes so that directly conjugated antibodies are not bound by the anti-rat second stage antibody on lineage+ cells.

Incubate the cells in directly conjugated antibodies (anti-Thy-1.1-FITC, anti-Sca-1-biotin, anti-Mac-1-APC)

6. Spin down the cells, and aspirate the supernatant. Resuspend the cells in directly conjugated antibodies. Incubate on ice for 20 minutes.

7. Wash the cells twice. Resuspend in 0.4 ml of staining medium plus 0.1 ml of MACS® avidin-magnetic beads per 10^8 cells. Incubate for 12 minutes at 4°C.
   De-gas 100 ml of staining medium by putting it under vacuum for 15 minutes. The de-gassed medium will be used to wet the mini-MACS columns and to resuspend cells prior to loading the column.

   Incubate the cells in MACS avidin magnetic beads

Note: Fetal liver HSCs can be pre-enriched prior to FACS by the selection of Sca-1$^+$ cells using MACS magnetic beads. Pre-enrichment reduces the amount of FACS time required to isolate HSCs. Enrichments of 4 to 10 fold are typical. However this step is not necessary. To skip MACS enrichment, incubate cells directly in avidin-texas red instead of magnetic beads, resuspend in propidium iodide and proceed to FACS. HSCs represent 0.04% of unenriched fetal liver cells from E12.5 to E14.5, and 0.015% at E15.5.

8. Add avidin-texas red to the suspension of cells in magnetic beads. The concentration should be determined by titration. Continue incubating at 4°C for an additional 12 minutes.
   Prepare Mini-MACS columns for the collection of the magnetic fraction of cells according to the manufacturer's instructions (wet the columns with degassed staining medium).

   Add avidin conjugated to texas red to the suspension of cells and beads. Incubate

9. Wash off unbound beads and resuspend cells in 0.5 ml of degassed staining medium per 2 x 10^8 cells or any portion thereof. Load 0.5 ml onto each mini-MACS column. After the cells drain into the column matrix, place the column into the mini-MACS magnet. This causes cells bearing the paramagnetic beads to bind the column matrix. Wash unbound cells out of the column by adding 0.5 mL of degassed staining medium to the top of the column and letting it drain through. Allow the flow-through to pass through the column twice more. Wash unbound cells out of the fluid phase inside the column by passing another 0.5 ml of medium through the column. Discard the unbound fraction. Elute the magnetic fraction by removing the column from the magnet and washing cells out by forcing 1 ml of degassed medium through the column (according to manufacturer's instructions).

   Enrich Sca-1$^+$ cells by collecting the magnetic fraction on MACS columns

Note: Column enrichment can be performed at room temperature, but caution should be exercised, as the columns can become

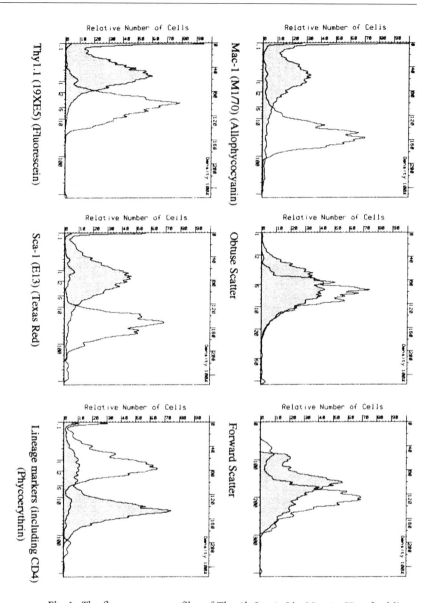

Fig. 1. The fluorescence profiles of Thy-1loSca-1+Lin-Mac-1+CD4- fetal liver hematopoietic stem cells are shown in the unshaded histograms based on a re-analysis of sorted cells. The shaded histograms show the fluorescence profiles of whole fetal liver cells

blocked when particularly warm. To prevent such problems columns can be run at 4°C.

10. Spin down the cells eluted from the column, and resuspend in 0.5 μg/mL propidium iodide in staining medium.

    Resuspend Sca-1⁺ enriched cells in propidium iodide solution

11. HSC are purified by two consecutive rounds of FACS. That is, the desired cells are selected by an initial round of FACS. Then contaminating cells are eliminated by running the selected cells through the FACS machine a second time. Cells may be sorted into staining medium.

    Purify Thy-1.1^{lo} Sca-1⁺ Lin⁻ Mac-1⁺ cells by FACS

Note: The fluorescence profiles of fetal liver HSCs relative to unfractionated fetal liver cells are depicted in PNAS USA 92:10302, 1995. A channel need not be dedicated exclusively to PI since PI⁺ cells will be gated out along with Lineage⁺ cells in the phycoerythrin channel.

Methods for Preparing Sorted Cells as Monolayer Specimens

ROBERT C. LEIF

Introduction

The major reason for sorting or separating Cells is to make a subsequent measurement on a purified population(s). Since, the quantity of cells separated by sorting and other techniques is often quite small, the conventional technique of making a smear on a glass slide produces a very sparse, essentially unusable, dispersion. In fact, the Centrifugal Cytology technique for preparing cells on a microscope slide, described below, was invented to permit conventional morphological analysis of separated cells. Initially, this was for density gradient separation of cells. Subsequently, the fluorescence activated cell sorter was invented, and Centrifugal Cytology was employed to determine the cellular composition of sorted fractions.

One of the best quality control techniques for both flow cytometry and cell sorting is to observe the cells with a microscope. Even prior to sorting, it very often pays to look at the cells. Centrifugal Cytology permits samples prepared and stained for flow cytometry to be deposited on a microscope slide for viewing. Since, many of the laboratories which perform cell sorting analyze and sort other groups' samples, it is very useful to be able to show the individuals supplying a sample what they really have. Although sorters and other cell separation techniques are often capable of separating a few good cells from many that are dead or in very poor condition, the old aphorism, garbage in equals garbage out, often still holds.

Conventional microscopic imaging of cells requires that they be supported on microscope slides. Current histochemical technique involves the application of stains to the cells after they have been placed on the slide. As stated above, this need not be the case for cells prepared for flow or sorting. Two requirements for cell preparation on slides are 1) the cells remain on the slides during staining or other processing and 2) the number of cells per unit area be in a range which is high enough to permit the visualization of a significant number of cells, yet low enough to minimize the proportion of cells layered over each other. Overlaid cells interfere with human visual or machine analysis.

ADVANTAGES OF MONOLAYERS VERSUS CONVENTIONAL SMEARS

The conventional smear has been the traditional way to place cells on a microscope slide. Since the distribution of cells in fresh or mixed stored blood is approximately random and the concentration of cells is approximately between 25 and 50%, monolayers have been made by smearing a drop on a slide. The drop is pulled across the slide, usually by a second slide. Air drying in the plasma ensures that the cells will remain glued to the slide. The surface tension and other drying effects reduce the thickness of the cells and spread them out. This significantly increases the visibility of the leukocyte granules and the thin strands of chromatin which connect the segmented parts of neutrophil nuclei. It often destroys the structure of both lymphocyte and monocyte chromatin. Most other cytology samples have a lower cell concentration than blood, and the cells are not homogeneously distributed. For instance, it has already been demonstrated that standard Pap smears do **not** provide a representative sample of exfoliated cells.[1,2]

Sorted cells and other samples where the cells are suspended in solution have significant advantages over traditional clinical cytological samples: 1) their concentration can be adjusted to produce dispersions approaching the optimum number of cells per unit area. 2) After the solution is mixed, a representative sample of the cells will be available for transfer to the slide. 3) The number of cells required to make an individual dispersion is minimized. 4) The cells can be processed and stained in suspension in a similar manner to present flow cytometry techniques. And 5) The capacity to have multiple preparations, each containing a representative sample of cells, permits the scientific method to be applied to optimize the staining and other parts of the specimen preparation procedure.

Subprotocol 1
Coating Methods for Glass Slides to Prevent Cell Loss

▓▓ Materials

Cleaning Solutions

- Liquinox detergent
- 1% Ammonium bifluoride

Adhesive Solutions

- 50-65% Mayers' albumin fixative in DH_2O
- 1 mg/ml Poly-Lysine
- 2% Aminopropyltriethoxysilane

▓▓ Procedure

Slide Cleaning

1. First clean them in conventional laboratory detergent such as Liquinox (Alconox inc.) in a small ultrasonic bath for about ten minutes. Ethanol also works.

2. Rinse thoroughly with distilled water.

3. Immerse in one percent ammonium bifluoride for two minutes (periods up to 12 min. have been used)

Note: Excess ammonium bifluoride can visibly etch the surface of a glass slide; however, if you do not see it, you have no need to worry about it.

4. Rinse and store in distilled water.

5. At time of use dry with lens tissue.

Note: A clean slide like any piece of clean glassware wets evenly with water without drop formation.

Adhesives   **Note:** For maximum adherence of centrifuged cells to glass slides, the slides must be coated.

6a. Mayers' albumin fixative. The slides must be coated by momentarily dipping them in a Coplin Jar which contains Mayers' albumin fixative solution (Harleco) diluted in distilled water to achieve a concentration between 50 and 65%.[3,4] Bake the slide, vertically, in a 70 °C oven for 10 minutes to insure a thin, even, tacky surface.

Note: The use of an albumin or other protein based adhesive has the advantage that the cells are literally glued to the slide. This is particularly important for cells with reduced adhesive properties, which often result from prior stabilization or fixation for shipping or the use of dissociation protocols. The use of Mayers' albumin has the very significant disadvantage of providing a very large number of sites for nonspecific stain binding.

6b. High molecular weight Poly-Lysine. Dip slides in 1 mg/ml Poly-Lysine and air dry.

Note: A high molecular weight Poly-Lysine solution (M.W. >70,000)[5] may be substituted for albumin.

Note: The use of this technique may reduce the number of cells which adhere to the slide, but it will diminish the amount of background debris. The homogenous D or L polymers should be used. Precoated slides are available from Sigma (Cat. # P 0425).

6c. Aminopropyltriethoxysilane.

Note: Rentrop et al,[6] have reported on the effect of various coating materials on the adhesion of frozen sections to glass slides. They found that 2% aminopropyltriethoxysilane in dry acetone was superior to: egg white, gelatin, collagen, and Elmer's White Glue. These authors speculate that the aminopropyltriethoxysilane forms covalent bonds with the aldehydes and ketones present in the tissues. Precoated slides are available from Sigma (Cat. # S 4651)

Subprotocol 2
Methods for Preparing Sorted Cells as Monolayer Specimens: Centrifugal Cytology

Preparation of Specimens for Image Cytometry or Human Observation

The three major factors controlling recovery of cells from a suspension onto a slide are 1) Cell losses in the apparatus; 2) The adhesiveness of the cell binding surface; and 3) the force pushing the cells onto the surface. The two basic methods for preparing monolayers are 1) pressure transfer where the cells are initially placed on a nonadhesive substrate and transferred by pressure to a slide which binds them.[7] And, 2) centrifuging the cells onto a slide. One variation of centrifugation is to have the cells settle at unit gravity.[8]

Monolayer Apparatus

Centrifugal Cytology

Centrifugal Cytology is the process where cells in suspension are centrifuged onto a substrate and then fixed concurrently with the application of centrifugal force. Thus, while the cells are being hardened by the fixative, they are simultaneously being pushed against the support. This fixation during centrifugation results in a very high recovery of the cells.[9] In order to ensure maximum recovery, the apparatus is configured to prevent fluid loss.[10]

Although presently, Leif Centrifugal Cytology Buckets or similar apparatus[11] are used for Centrifugal Cytology, a distinction must be made between the apparatus and the process. The latter is of scientific significance and is a direct outgrowth of modern cell biology. The Centrifugal Cytology process was originally based on electron microscopic technique and subsequently was evolved to follow Papanicolaou's wet fixation methodology.[12] Centrifugal Cytology can be used clinically to prepare cells for human screening[13] and shows great potential for automated clinical cytology.[14] It also is an extremely useful tool for monitoring the cell suspensions employed for flow cytometry or sorting.

One of the major uses of the Centrifugal Cytology Buckets is a generic protocol for viewing suspension preparations. The flexibility of the centrifugal cytology procedure and the Centrifugal Cytology Buckets, as well as the heterogeneity of biological cells precludes the absolute specification of many procedures. However, it is possible to provide rational starting points for development of optimized procedures.

Cells that have been prepared for flow analysis or subsequent to sorting can be monitored by centrifuging them onto a standard microscope slide, adding a coverslip, and viewing the wet mount with a microscope. Thus, virtually all flow preparation procedures are also centrifugal cytology procedures. Often, the flow staining procedures result in the chromophores being bound with sufficient strength to the cells, that they will remain localized during fixation, dehydration, and permanent mounting. In the text below, the rationale behind the selection of fixatives and staining procedures will be given with a few specific examples.

The Leif Centrifugal Cytology Bucket (Leif Bucket) is based on a swinging bucket rotor. The original aluminum bucket is a replacement for the standard swing-out cup of a swinging bucket rotor. A fluid tight chamber is formed by pressing and sealing an elastomeric sample block against a standard 3 by 1 inch microscope slide.[10] The slide serves as the base of the pyramidal sample chambers present in the block. The incline of the slanted chamber walls follows the radius emanating from the center of the centrifuge. Since cells and other particles follow a radial trajectory, this prevents their deposition on the chamber walls. The cell containing suspension is first placed in a chamber and then the cells are centrifuged onto the slide. Most of the supernatant is removed and a fixative is added in a manner that does not dislodge the cells. During fixation, the cells are pressed onto the slide by centrifugal force. After fixation, the slide is separated from the sample block and can then be processed by conventional staining techniques. The Centrifugal Cytology Bucket was designed to facilitate the cytological examination of cells from dilute biological fluids.

The first commercial embodiment of the Leif Buckets was used in conjunction with any universal laboratory centrifuge with a standard 1.875 inch rotor spacing. As described below, subsequent models have been designed for other centrifuges.

The Leif Bucket has been used to prepare the following tissues and body fluids for cytological examination: blood[9,15,16] bone marrow,[17] cervical scrapes,[18] body fluids including cerebral spinal fluid, nipple aspirate,[19] sputum,[20] urine,[21] eye fluids including tears and vitreous humor.[22] Centrifugal Cytology has been employed to quantitate biologically active lymphocytes: Jerne Plaque and rosette forming cells,[23] as well as natural killer cells.[24] Detailed information about these Centrifugal Cytology applications are given in the references including their use to collect sorted cells and prepare them for human observation.[25]

Description

The Centrifugal Cytology Bucket consists of the following components:

A. Carriers. Aluminum carriers (Fig. 1) for use in a centrifuge with a 1.875 inch rotor spacing. These carriers replace the rotor's sample cups and thus, are required to maintain dimensional stability at 500 times gravity. Many centrifuges now have swing out rotors which have cups that hold 500 ml. Since microscope slides are 3 by 1 inch, special plastic carriers (See also Fig. 5) have been designed for cups which have inner dimensions greater than 3 inches. The design of these plastic carriers was simplified because during centrifugation they are supported by the standard centrifuge rotor cup. Both the aluminum and plastic carriers support a foam pad, microscope slide, and the sample block during centrifugation. The hold downs and screws are used to press the sample block against the slide with sufficient pressure to form a fluid tight seal.

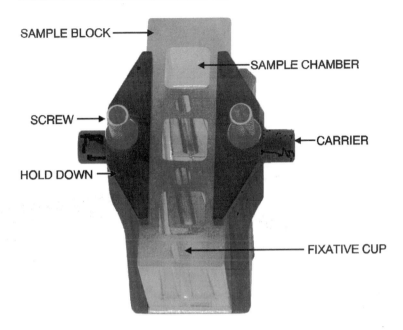

SAMPLE BLOCK

SAMPLE CHAMBER

SCREW

CARRIER

HOLD DOWN

FIXATIVE CUP

Fig. 1. Aluminum carrier version of the Centrifugal Cytology bucket. The carrier is supported by and pivots on the two cylindrical pins. Loosening of the two screws permits both of the hold downs to be rotated 180°. This permits insertion or removal of the sample block, microscope slide, and foam pad

B. Sample and Sorter Blocks. Presently, both three and four chamber sample blocks (Fig. 2) are available. Eight chamber and special sorter blocks (Fig. 3) have also been fabricated. Each sample chamber has an interconnected fixative cup. A plastic frit at the bottom of the fixative cup at unit gravity retards the flow of the fixative solution, while permitting the flow of fixative into the sample chamber during centrifugation. Rapid flow of any solution onto unfixed cells could dislodge them from the microscope slide. The sample blocks are molded from an elastomeric plastic, that is sufficiently hard to withstand the centrifugal force, yet soft enough to deform and thus seal to the surface of the microscope slide. For some work where the supernatant fluid must be removed with absolutely minimal disruption of the cell monolayer, a capillary (not shown in Fig. 2) which connects the top of the sample block to 1 mm above the slide has been added.

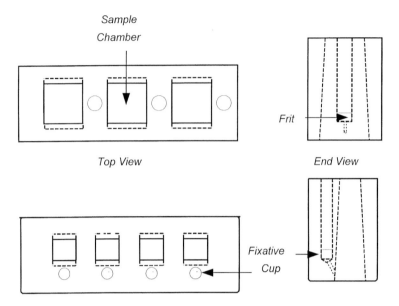

Fig. 2. Sample blocks. The top and end views of the 3 (top) and 4 (bottom) chamber sample blocks are shown. The hatched cylinder at the bottom of each fixative cup is fabricated out of a plastic frit, which retards the flow of solution at unit gravity. The capillaries, dashed lines, beneath the frit convey the fixative to about 1 mm above the slide. The chambers are sector shaped. In the top views, the dashed lines in the three chamber insert (top) and the four chamber insert (bottom) show the outline of the chamber at the base of the sample block, which is in contact with the microscope slide and delineates the area of the cellular dispersion

Top View

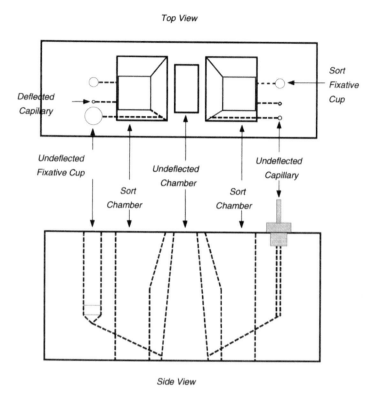

Side View

Fig. 3. Special Sample Block for Sorting. The top and side views of a special 3 chamber sorter sample block are shown. A chamber to collected the undeflected drops is located between the two sort chambers. The internal walls of the chambers and the capillaries in the sample blocks are shown by the dashed lines. The fluid retarding plastic frits at the bottom of each fixative cup are shown as hatched cylinders. The side view shows only the fixative cup, frit and capillaries for the undeflected chamber. The top view shows all three fixative cups, frits, and capillaries. The sort deflected fixative cup on the left and the deflected capillary on the right of the top view are located at the same height as their labelled counterparts. The undeflected fluid waste connector, which is shown at the top right in the side view, connects with the undeflected capillary and can also be connected to the sorter vacuum line

Each of the two sort chambers consists of two flattened pyramids. In order to provide the largest area to collect the sorted droplets, the top of the chamber consists of an inverted pyramid. This pyramid connects just above the level of the frits with a short conventional sector shape chamber. Any fluid that collects on the walls of the inverted pyramid will be driven by the centrifugal force into the standard sector shaped chamber. These standard sector shaped chambers are offset towards the back of the sample block in order to permit the undeflected fixative and fluid removal capillaries to connect with the undeflected chamber. The undeflected chamber is a sector shaped flattened pyramid.

C. Foam Pads. Three inch long by one inch wide, disposable foam pads are placed on the flat surface of carriers. These deformable pads support and cushion the microscope slides.

Materials

Equipment

Leif Centrifugal Cytology Buckets

Procedure

Operating Instructions

Bucket Assembly   The Centrifugal Cytology Bucket is assembled (Fig. 4-5) as follows:

1. Place a 3 X 1 in. foam pad, backing-side down, on the flat surface of the aluminum or plastic carrier. Do **not** remove the backing from the pad; since, the adhesive on the pad will glue the gasket to the carrier.

2. Place a plain, clean, unfrosted glass slide on top of the pad in the carrier (Fig. 4a).

3. Place a transparent sample block on top of the slide (Fig. 4b). The deformability of the sample block creates the seal between slide and sample block.

4a. For the single slide aluminum carriers, rotate the hold-downs 180° and tighten the thumbscrews by hand (Fig. 4c). Do not overtighten the screws.

Seal the sample block to the slide

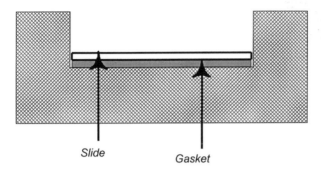

Fig. 4a. The gasket is first placed in the carrier followed by a cleaned slide

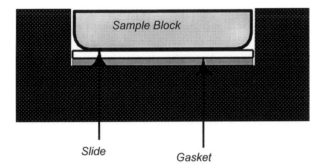

Fig. 4b. Compression of the sample block seals it to the slide, which is in turn supported by the gasket and carrier. The tops of the carrier and sample block have been removed

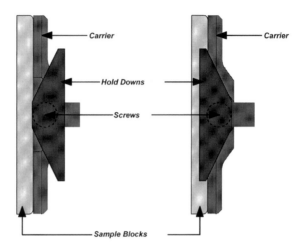

Fig. 4c. At the left, the hold down is shown rotated, which permits the sample block to be inserted or removed. At the right, the hold down is shown rotated 180°, which locks the sample block into place and seals the block to the slide

4b. For the two slide plastic carriers (Fig. 5), place the transparent top over the sample block. Check that the drill holes in the top are aligned with the sample chambers and the fixative cups. Tighten the single central screw by rotating the knob. This results in the transparent cover compressing the two sample blocks, each against its underlying slide.

Note: If necessary, cover the sample chambers with tape to avoid airborne contamination. The use of volatile buffers, such as bicarbonate, can result in unacceptably large pH changes.

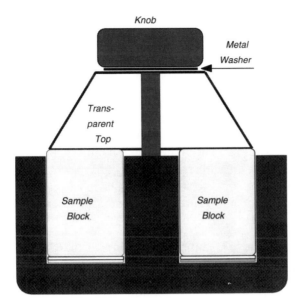

Plastic Dual Carrier

Fig. 5. The knob is attached to a threaded rod. The threaded rod is screwed into a metal insert (not shown), which is located in the center of the plastic carrier. Thus, rotation of the knob results in pressure on the flat metal washer, which distributes the force across the transparent top. The top, in turn, compresses the sample blocks, which seals them to the microscope slides. As before with the aluminium carrier, the slides are supported on gaskets. Removal of the top permits the sample blocks to be removed. Access to the sample chambers is through holes (not shown) in the transparent top

5. Take up the sample with a syringe, or pipette, and deliver it into one of the sample chambers.

6. Balance a pair of buckets according to the centrifuge manufactures' specifications.

Note: If it is necessary to increase the weight of one bucket, suspending fluid should be added to the center chamber of a three-port block, or to the two center chambers of a four-port block. Suspending fluid added to the outer chambers should be added equally to keep the slide truly vertical during centrifugation. Always balance a three-chamber sample block against another three-chamber sample block, and a four-chamber block against another four-chamber block.

7. Centrifuge the samples at 250 to 500 x gravity for 10 min.

8. After the centrifuge has stopped, gently tilt the bucket assembly and aspirate the supernatant from each sample chamber with a Pasteur pipette.

Note: Retighten the screws after the initial centrifugation; the pads can become irreversibly compressed.

Note: From step 9 on, fluid is removed from or overlaid onto a cell monolayer. A 21 gauge syringe needle or Pasteur pipette can be used or preferably a bidirectional auto-pipettor with a speed control should be employed to slowly add and remove fluid, so as to minimize disruption of the cell monolayer and maximize ist contact with fixatives.

9. Add fixative to the fixative cup with a syringe or pipette.

10. Recentrifuge the samples at 250 to 500 x gravity for a time appropriate to the fixation protocol you are using.

11. After centrifugation with fixative, draw off the supernatant as described in step 8.

12. Disassemble the bucket in the reverse order of its assembly.

13 a. For the single slide aluminum carriers, loosen the screws, rotate the hold-downs 180°, and lift the sample block straight up and out.

13 b. For the two slide plastic carriers, unscrew the single central screw from the two slide plastic carrier and remove the

transparent cover, and lift the two sample blocks straight up and out.

Note: Should the slide remain attached to the sample block, break the seal by inserting the edge of a thin metal spatula between the slide and the sample block.

14. To avoid carryover between samples, clean the sample block in a 5% solution of household bleach, in an ultrasonic bath (Branson B220 or equivalent) for 12 min. Remove the bleach from the buckets by rinsing them for 5 min. in tap water, followed by a 5-min. rinse in distilled water.

Cleaning Procedures

Note: Avoid the use of organic solvents other than ethanol with the sample block. Use ethanol only as a fixative.

15. Should sterilization be necessary, sterilize the sample block in either 70% ethanol or 10% formalin for no longer than 2 hours.

Sterilization

Note: *Do not autoclave the sample block!*

16. The sample blocks should be stored with their bottom side facing up. Handle the bottom surface with care; it serves as a seal to prevent the escape of the cell suspension during centrifugation.

Storage and Handling

17. Any time malfunctioning of the bucket is suspected, perform the following tests.

Bucket Malfunction

17 a. Assemble the bucket and fill all chambers with water. Centrifuge the bucket for 10 min. at 250 x gravity. Inspect for water loss, and empty the chambers.

17 b. Fill every other sample chamber with water and centrifuge at 250 x gravity for 10 min. Check for water loss from the filled chambers and water gain in the empty ones. This is a test of the isolation of the fluid chambers from each other.

Note: Make sure that the bucket is evenly balanced.

17 c. To check the proper functioning of the fixative cups, place water in the cups and allow the sample block to stand at unit gravity on a level surface. It should take approximately 2 minutes for the cups to drain into their respective sample chambers. Fill the cups again and centrifuge at 250 x gravity for 2 minutes to determine if the cups empty when centrifugal force is applied to them.

Subprotocol 3
Methods for Preparing sorted Cells as Monolayer Specimens: Fixation and Staining

Fixation

The selection of fixatives is based on the requirements of the procedure. For either transmission or scanning electron microscopy, the fixative of choice is glutaraldehyde. For flow cytometry fluorescence measurements, it is glyoxal. For fluorescence image cytometry where morphological integrity or comparison with flow cytometry are important, it is glyoxal. For absorbance staining where granular structures need to be emphasized or previous tradition must be adhered to, 95% ethanol with 5% polyethylene glycol average molecular weight 1,450 is a good choice.

Proper fixation for centrifugal cytology is significantly facilitated by employing fixatives that are denser than the fluid originally present above the cell monolayer. A very simple test to determine relative density is to add a small amount of an absorbance dye (usually Phenol Red or food coloring) to the fixative and then test whether it is denser than the original fluid. This test is also a very good training procedure for performing centrifugal cytology. One or more mLs of the supernatant fluid should be added to a small test tube. Then, the colored fixative should be delivered slowly to the bottom of the tube. The results of too fast delivery will be readily apparent. The colored fixative ricochets off of the tube bottom and flows up. If after successfully layering the fixative under the supernatant, the delivery device is removed too quickly, there will be a colored tail extending upwards.

If the aqueous fixative floats or does not effectively sink, then the density of the fixative can be increased by the addition of 5 to 10.0% (v/v) DMSO, the replacement of the water by 20% deuterium oxide, or both.[26]

Ethanol

Since ethanol is the fixative commonly employed for exfoliative cytology, it has been extensively used. The addition of polyethylene glycol with a Mol. Wt. of Ca. 1,450 helps stabilize the morphology of the cells. In fact, if one is very careful and has a strong hair drier, it is possible to air dry 95% ethanol 5% polyethylene glycol fixed blood leukocyte centrifugal cytology preparations (J. Hudson, unpublished results). For routine fixation with ethanol, 15 minute centrifugation at 250 x gravity appears to be adequate.

Scanning electron microscope studies of centrifugal cytology preparations which were air dried from xylene were essentially identical with conventional critical point dried preparations. The explanation is the surface tension of organic solvents is much lower than that of aqueous solutions.

Aldehydes

Three aldehyde fixatives are shown in Fig. 6. Aldehydes primarily fix by forming inter and intra molecular crosslinks. They also form adducts with amino groups, which eliminate the latter's capability of being positively charged at neutral pH. This decrease in charge decreases both the solubility of proteins and the binding of negatively charged dyes.

Formaldehyde

Formaldehyde, since it is a monoaldehyde, is a poor crosslinking agent. It can only crosslink by forming an aminal by condensing with two separate, but very close amino groups. A major advantage of formaldehyde is that it is a sufficiently poor fixative, that its reaction with the tissue can be reversed.[27,28] Optimal results of centrifugal cytology studies of erythrocytes were achieved with formaldehyde as the fixative. Note, a trace of bovine serum albumin must be present for most manipulations of erythrocytes.

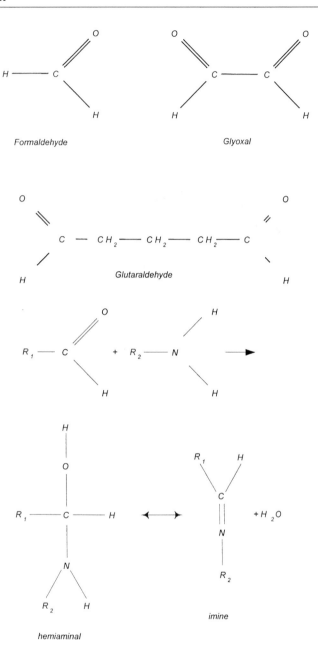

Fig. 6. Structure of three aldehyde fixatives. When the aldehyde with the R_1 functional group reacts with the amine with the R_2 functional group, the hemi-aminal is formed. The hemiaminal loses water to form the imine, with which it is in equilibrium

Glutaraldehyde

Dialdehydes, such as glutaraldehyde, can form two hemiaminals by reacting with two amino groups that can easily be located on separate macromolecules. Glutaraldehyde has one very significant disadvantage; it forms fluorescent adducts, which provide significant background fluorescence.

Glyoxal

The crosslinking activity of glyoxal, which also is a dialdehyde, is between formaldehyde and glutaraldehyde. Glyoxal has the major advantage of not forming fluorescent adducts which produce detectable background. Golomb and Monnier[29] demonstrated that in the case of glyoxal, the hemiaminals convert to imines (Schiff bases). They demonstrated that for twenty mM glyoxal, the time for maximum reaction with BSA was 30 minutes. Dimer formation with reduced RNase was strongly evident at the minimum time, 5 hours, as shown by these authors. Ueno et al.[30] demonstrated single strand breaks in rat liver DNA 2 hours after a single oral dose of 200 mg/kg body weight. After 9 hours, the DNA extracted from the other organs, except possibly for the spleen appeared to be unaffected. No crosslinking of the DNA was observed. Thus, Ueno et al. concluded that, glyoxal must be genotoxic in man.

Most of the work[16] on glyoxal was for flow analysis of fluorescent antibody stained leukocytes obtained after lysis of whole blood at neutral pH. Selective lysis of the erythrocytes at neutral pH does not have the problem of acid induced desorbtion of the antibody from the remaining leukocytes.[31] Examples 5 and 6 of the Leif et al. Patent[16] describe a procedure for multiparameter flow analysis. Example 7 of the same patent describes a preparation employing a Centrifugal Cytology Bucket. The prior exposure to saponin required very rapid stabilization of the leukocytes. This was achieved with 14 g/L (0.24 M/L) glyoxal and 146.7 g/L DMSO (1.88 M/L) DMSO, and Polyethylene glycol MW 1450 was also present at 16.7 g/L. For more conventional work, the DMSO concentration should be reduced to 50 to 100 g/L.

DMSO

Dimethylsulfoxide, DMSO, is an excellent fixation and stain enhancing agent, which facilitates the entry of the fixative and often of dyes, such as DAPI, into cells. It also stabilizes the hydrophobic bonds in cell membranes. Another major advantage of DMSO is it has a density of 1.10. The addition of 5 to 10% DMSO (v/v) to a fixative containing solution results in a density significantly greater than that of the original cell suspending solution. Thus the fixative, which underlays the cell suspending solution, comes into intimate contact with the cells, which are located on the surface of the microscope slide. Concentrations of DMSO greater than 10% (v/v) have been employed to freeze cells.

Evaluation of Fixatives

There are three excellent tests for fixation protocols. The first is to perform electron microscopy on the fixed cells. The second is to measure the electronic cell volume distribution prior to and after fixation. And, the third is to perform image analysis on the cells.[32] The goals of the automated cytology procedure will significantly influence the criteria for fixation quality. Procedures which are useful for fluorescence in situ hybridization nuclear spot counting may be very poor for automated analysis of exfoliated cytology specimens. For morphological studies and/or textural analysis of nuclear features, a reasonable starting place is a supravital DNA dye, such as Hoechst 33342.[33]

Thornthwaite et al.[15] employed the Automated Multiparameter Analyzer for Cells II focused flow electronic cell volume transducer to determine optimum fixation conditions. They established that even blood leukocytes from the same individual do not have the same osmotic properties. "The optimum concentration of glutaraldehyde (v/v) at a fixed concentration of 0.05 M. cacodylate buffer (pH 7.4), based on the position of the peak channel of the electronic cell volume distributions, was 3.8% for mastocytoma cells, 4.9% for human lymphocytes and 4.0% for human granulocytes and monocytes. The existence of a significant osmotic effect of glutaraldehyde indicates that a significant amount of this fixative is excluded during the period of fixation when the initial cross-linking takes place. Unfortunately, focused flow impedance cell volume instruments for the research market presently are not commercially available. Several of the Coulter clinical hematology instruments do employ focused flow for their special purpose electro-optical leukocyte transducers.

Staining

Slides may be stained with your choice of stain or other labeling agent. For surface immuno-labeling of cells, start with a procedure for flow analysis where the antibodies are applied in suspension prior to fixation. Standard, routine PAP stain protocols may be used. Hematoxylin and eosin stains are also appropriate. Air drying must be avoided at all stages where information concerning chromatin fine structure is to be acquired.

In order to avoid the presence of water droplets, which are immiscible in mounting media, aqueous staining procedures require that the cells be dehydrated through graded alcohols (all v/v): 50%, 70%, 95% and 100% ethanol. Usually, an immersion in 50% ethanol and 50% xylene or xylene substitute is included prior to immersion in xylene or xylene substitute, which is followed by the final step permanent mounting. The use of nonaqueous mounting media has the obvious advantage of eliminating nonradiative energy loses to water.

Procedure

Two to 4% (v/v) glutaraldehyde phosphate buffer (pH 7.2) or 0.05M/L Na-cacodylate brought to pH 7.4 with 2M HCl and containing up to 5 to 12% DMSO (v/v).

Glutaraldehyde

Note: The use of electron microscope grade of glutaraldehyde is highly recommended. In fact, the use of glutaraldehyde fixing solutions which have recently been used successfully by your local electron microscopist is highly recommended.

Note: For routine fixation with glutaraldehyde, 15 minute centrifugation at 250 x gravity appears to be adequate. For ultrastructural studies with scanning or transmission electron microscopy, 45 minute exposure to glutaraldehyde is required for proper fixation.

Glyoxal Concentrate Note: This formula is preliminary.1. Dissolve 10 grams Glyoxal trimer dihydrate (Cat. # G5754 Sigma) in 20 ml of distilled water.

Note: Heating in a 37°C bath probably will be needed to solubilize the glyoxal. Glyoxal and other aldehydes slowly transform into an organic acid(s), which can adversely effect the morphology of the cells.

2. Remove the acid components by adding 0.5 grams of a large bead such as, Mixed Bed Resin TMD-8 (Cat. # M 8157 Sigma) to remove the acid components, let sit for 30 minutes or until resin changes color from blue green to yellow. Since the resin beads are large the solution can be either filtered, centrifuged or centrifuged and decanted through filter paper. If the beads change color, repeat the addition of the resin.

Note: This resin treatment should also work with the commercially available 40% aqueous solution. Unfortunately, present commercial glyoxal often has a significant acid contamination.

0.1M Tricine Buffer
1. Dissolve 1.79 g (0.01 mole) of tricine (Cat.# 3077T Research Organics) and 0.10 g of NaN3 in 100 ml volumetric flask about a third full of deionized/filtered water.

2. Heat polyethylene glycol average Mol. Wt. 1450 to liquefy most of the contents of the container. A microwave oven works.

3. Add 5 g of the polyethylene glycol to the tricine solution.

4. Add 7.5 g DMSO to the tricine solution.

Note: With a viscous liquid, it is often easier to take the liquid up in a wide bore disposable transfer pipette and drip it into a volumetric flask which is placed on a top-loading balance.

5. Add deionized/filtered water to the 100 ml line on the flask.

Note: The 100 ml of solution can now be transferred to a polyethylene bottle.

6. Check the pH of the solution; it is usually slightly acid.

7. Add 10 M. NaOH to bring the pH to about 7.4-7.6. It is advisable to add small drops of the NaOH in successive small portions, checking the pH after each portion has dissolved.

8. Add 1 ml of the Glyoxal Concentrate to 9 ml of the 0.1 M Tricine buffer.

Glyoxal
Fixative
Solution

Subprotocol 4
Plaque Cytogram Assays

Centrifugal Cytology has been employed to enumerate biologically active cells. Thornthwaite and Leif[23] were able to enumerate and morphologically characterize both plaque-forming and rosette-forming cells. The technique described below can be modified to permit the visualization of any type of cell that can be effected by its neighbors in a manner that produces a result that can be made visible either by cytochemical or special optical techniques.

▪▪ Materials

Equipment

– Centrifugal Cytology Buckets

Note: For these studies, special 12 chamber Centrifugal Cytology Buckets were employed. Since at present, these are not commercially available, the volumes have been scaled up by a factor of 2, which is approximately the ratio of the dispersion area of the presently commercially availabe 4 chamber bucket (56 mm^2) to that of the previous 12 chamber bucket (25 mm^2).

– Multi-pipetting device

Note: When this procedure was developed, a special crank driven multi-pipetting machine was employed for fluid transfer beginning with step 6. This should be replaced by a suitably configured commercially available auto-pipettor.

– Centrifuge

Note: Almost any laboratory centrifuge that has a rotor for the Centrifugal Cytology buckets shown in Fig. 1 or a centrifuge with a swing out rotors with cups suitable for the buckets shown in Fig. 5.

Solutions

– Sheep red blood cells
– 4% glutaraldehyde in 0.05 cacodylate buffer (pH 7.4)

Note: The glutaraldehyde fixation provided excellent morphology. The addition of 5 to 7.5% DMSO (v/v) to the glutaraldehyde solution will both increase the rate of fixation and facilitate the over-layering. However, if fluorescence measurements are to be made, the glutaraldehyde should be replaced by glyoxal.

▨▨ Procedure

Prepare the spleen cells

1. Dilute the spleen cells to 2×10^5/ml.

2. Transfer 0.2 ml into a bucket chamber. Traditionally a 1 or 0.5 ml disposable syringe has been used.

3. Sediment the cells by centrifuging at 400 x gravity at 4°C for 5 min.

4. Slowly overlay the spleen cell monolayer with 0.2 ml of 4×10^7/ml sheep red blood cells SRBC.

5. Sediment the RBCs by centrifuging at 400 x gravity at 4°C for 5 min.

6. Remove the supernatant down to 1 mm above the cell surface

Add guinea pig complement

7. Slowly, (over a period of 5 min) layer 0.2 ml of 10% guinea pig complement into the bucket champer.

Incubate

8. Incubate for 20 min. in a level, 37°C incubator.

9. Remove supernatant down to 1 mm above the cell monolayer.

10. Very slowly (15 min), add 0.2 ml of 4% glutaraldehyde in 0.05   **Fix**
cacodylate buffer (pH 7.4). The end of the delivery system
should be at about 1 mm from the surface of the slide.

11. Let fix for 1 hr. at room temperature. The cells overlaid with
fixative can be stored overnight at 4°C.

12. The intact erythrocytes can stained with benzidine[17] and the   **Staining**
slides subsequently processed with Papanicolaou or any
other absorbance stain.

References

1. Hutchinson ML, Isenstein LM, Goodman A, et al. Homogeneous sampling accounts for the increased diagnostic accuracy using the Thin-Prep® Processor. Am J. Clin. Path 1994; 101: 215-219.
2. Tezuka F, Shuki H, Oikawa H, et al. Numerical counts of epithelial cells collected, smeared and lost in conventional Papanicolaou smear preparation. Acta Cytol. 1995; 39: 838-838.
3. Leif RC, Ingram DJ, Clay C, Bobbit D, Gaddis R, Leif SB and Nordqvist S. Optimization of the Binding of Dissociated Exfoliated Cervico-Vaginal Cells to Glass Microscope Slides. J. Histochem. Cytochem. 1977; 25: 538-543..
4. Wolley, RC, Dembitzer, HM, Hertz, F, et al. The use of a slide spinner in the analysis of cell dispersion. J Histochem. Cytochem. 1976; 24:11-15.
5. Jacobson, BS, Branton, D. Plasma membrane: rapid isolation and exposure of the cytoplasmic surface by use of positively charged beads. Science 1977 Jan 21; 195(4275): 302-4.
6. Rentrop M, Knapp B, Winter H, et al. Aminoalkylsilane-treated glass slides as support for in situ hybridization of keratin cDNAs to frozen tissue sections under varying fixation and pre treatment conditions. Histochemical Journal 1986;18:271-276.
7. Zahniser DJ, Hurley AA. Automated Slide Preparation System for the Clinical Laboratory. Cytometry (Communications in Clinical Cytometry) 1996; 26:60-64.
8. Knesel, Jr., EA. Roche Image Analysis Systems, Inc. Acta Cytologica,1996; 40:60-66.
9. Leif, RC, Easter, Jr. HN, Warters, RL, et al. Centrifugal cytology I. A quantitative technique for the preparation of glutaraldehyde-fixed cells for the light and scanning electron microscope. J. Histochem. Cytochem. 1971; 19:203-215.
10. Leif, RC. 1981. Swinging Buckets (Centrifugal Cytology). U.S. Patent 4,250,830. For information concerning Leif Centrifugal Cytology Buckets, contact Newport Instruments, 5648 Toyon Road, San Diego CA 92115, e-mail rleif@rleif.com. Similar Products are available from: International Equipment Corp. Needham Heights, MA, USA; Hettich, Tuttlingen, Germany; and StatSpin Technologies, Norwood, MA, USA.

11. Driel-Kulker AMJ, van, Ploem-Zaaijer JJ, Zwan-van der Zwan M, et al. A preparation technique for exfoliated and aspirated cells allowing different staining procedures. Anal Quant. Cytol. 1980; 2:243-246.

12. Papanicolaou GN, Chapter II, Technique, Atlas of Exfoliative Cytology. Published for the Commonwealth Fund by Harvard University Press, Cambridge, Mass, 1954; pp3-12.

13. Leif RC, Ingram DJ, Clay C, Bobbitt D, Gaddis R, Leif SB and Nordqvist S. Optimization of the Binding of Dissociated Exfoliated Cervico-Vaginal Cells to Glass Microscope Slides. J. Histochem. Cytochem. 1977: 25: 538-543.

14. Leif, RC, Chew, KL, King, EB, et al. The Potential Of Centrifugal Cytology Dispersions For Automated Cytology. in The Compendium on the Computerized Cytology and Histology (G. L. Wied, P. H. Bartels, D. L. Rosenthal, and U. Schenck Ed.). Tutorials of Cytology, Chicago lL. 1994.

15. Thornthwaite, JT, Thomas, RA, Leif, SB, et al. The use of electronic cell volume analysis with the AMAC II to determine the optimum glutaraldehyde fixative concentration for nucleated mammalian cells. In Scanning Electron Microscopy, Vol. II (R. P. Becker and O. Johari, eds.) 1978; 1123-1130, AMF O'Hare, Illinois, 60666.

16. Leif, RC, Ledis, S, and Fienberg, R. 1993. A reagent system and method for identification, enumeration and examination of classes and subclasses of blood leukocytes. U.S. Patent 5,188,935, assigned to Coulter Corp.

17. Leif, RC, Smith, S, Warters, RL, et al. Buoyant density separation I, the buoyant density distribution of guinea pig bone marrow cells. J. Histochem. Cytochem. 1975; 23:378-389.

18. Leif, RC, Gall, S, Dunlap, LA, et al. Centrifugal cytology IV: The preparation of fixed stained dispersions of gynecological cells. Acta Cytologica 1975; 19:159-168.

19. Leif, RC, Bobbitt, D, Railey, C, et al. Centrifugal cytology of breast aspirate cells. Acta. Cytologica 1980; 24:255-261.

20. Leif, RC, Silverman, M, Bobbitt, et al. Centrifugal cytology: a new technique for cytodiagnosis. Laboratory Management 1979; 17, September:38-41.

21. Bobbitt, D, Silverman, M, Ng, et al. Centrifugal Cytology of Urine. Urology. 1986; 28:432-433.

22. Stulting, RD, Leif, RC, Clarkson, J, et al. Centrifugal cytology of ocular fluids. Arch. Ophthalmol. 1982; 100:822-825.

23. Thornthwaite, JT, and Leif; RC. Plaque cytogram assay I. light and scanning electron microscopy of immunocompetent cells. J. Immunology. 1974; 113:1897-1908.

24. Leif, RC, Hudson, J, Irvin II, G, et al. The identification by plaque cytogram assays and BSA density distribution of immunocompetent cells. in Critical Factors in Cancer Immunology (J. Schultz and R. C. Leif Ed.) 1975; pp. 103-158. Academic Press, New York.

25. Barrett, DL, Jensen, RH, King, EB, et al. Flow Cytometry of Human Gynecologic Specimens using log Chromomycin A3 Fluorescence and log 90°Light Scatter. Journal of Histochemistry and Cytochemistry 1979; 27:573-578.

26. Dunlap, LA, Warters, RL, and Leif, RC. Centrifugal Cytology III: the utilization of centrifugal cytology for the preparation of fixed stained dispersions of cells separated by bovine serum albumin buoyant density centrifugation. J. Histochem. Cytochem. 1975; 23:369-377.

27. Overton, WR, and McCoy, JP Jr. Reversing the effect of formalin on the binding of propidium iodide to DNA. Cytometry 1994;16:351-356.

28. Overton, WR, Catalano, E, and McCoy, JP Jr. Method to make paraffin-embedded breast and lymph tissue mimic fresh tissue in DNA analysis. Cytometry (Communications in Clinical Cytometry) 1996; 26:166-171.

29. Glomb, MA and Monnier VM. Mechanism of Protein Modification by Glyoxal and Glycoaldehyde, Reactive Intermediates of the Maillard Reaction. J. Biol. Chem. 1995; 270:10017-10026.

30. Ueno H, Nakamuro K, Sayato Y, and Okada S. DNA lesion in rat hepatocytes induced by in vitro and in vivo exposure to glyoxal. Mutation Research 1991; 260:115-119.

31. Macey, MG, McCarthy, DA, Davies, C, and Newland, AC. The Q-Prep System: Effects on the Apparent Expression of Leukocyte Cell Surface Antigens. Cytometry (Communications in Clinical Cytometry) 1997; 30:67-71.

32. Poulin N, Harrison, A and Palcic B. Quantitative Precision of an automated image cytometric system for the measurement of DNA content and distribution in cells labeled with fluorescent nucleic acid stains. Cytometry 1994; 16: 227-235.

33. Krishan, A. Effect of drug efflux blockers on vital staining of cellular DNA with Hoechst 33342. Cytometry 1987; 8:642-645.

Fluorescent in Situ Hybridization Analysis on Cells Sorted onto Slides

KAREN CHEW AND ROBERT C. LEIF

Introduction

Fluorescence in situ hybridization (FISH) uses labeled DNA probes to determine the copy number of the specific DNA sequence targeted. After denaturation of the probe and target DNA, the probe is allowed to hybridize to an intact nucleus. The hybridization can then be visualized by direct conjugation with fluorescent dyes, or indirect conjugation, such as biotin-avidin systems.

FISH analysis is performed by counting the number of signals within an intact nucleus. If fresh tissue is available, material for analysis is easily prepared by touching the fresh tissue onto a slide. When only paraffin-embedded material is available, monolayer preparations of dissociated nuclei are ideal. Monolayer preparations offer the ability to prepare duplicate slides for multiple hybridizations. The number of nuclei placed on a slide can be controlled to avoid overlapping, and the flattened nuclei can be easily evaluated under the microscope. Since the purpose of this protocol is to detect small fluorescent spots rather than nuclear morphology, a CytoCentrifuge can be employed to produce an air dried specimen.

Recently, an alternative procedure has been described by Van de Rijke et al.[1] A version of the centrifugal cytology buckets was employed to produce monolayers of mononuclear cells (lymphocytes) for automated in situ hybridization absorbance spot counting. These authors state, "Also, cytocentrifugation using the Shandon Cytospin 2 proved suboptimal. Although providing a fixed location, the cells were not evenly distributed over the surface, leading to threefold longer screening times as compared with bucket cytocentrifugation". They reported that pH 9.3 borate 0.05 - 0.10 M borate solution swelled the cells and gave a good combination of homogeneous cell size and good hybridization results. Fifty percent acetic acid was employed as the centrifugation medium; and after disassembling the bucket, the resulting nuclei were air-dried for 5 min. Vrolijk et al.[2] reported on the successful automated brightfield spot counting of samples produced by this technique. The same group[3] previously reported success with a fluorescence system.

The following protocols describe a dissociation technique for obtaining whole nuclei from formalin-fixed paraffin embedded tissue, preparation of the nuclei onto a slide using a cytospin technique, and a FISH protocol. These protocols were developed[5,6,7] in the laboratory of Dr. Frederic Waldman at the University of California, San Francisco.

Subprotocol 1
Preparation of Nuclei

Materials

Equipment

- water bath with shaker
- vortex mixer
- glass conical 15 ml centrifuge tubes
- Parafilm
- pipette

Solutions

- xylene
- ethanol solutions: 100%, 95%, 80%, 50%
- distilled H_2O
- Pepsin (Cat. # P7012 Sigma)
- NaCl (pH 1.5)
- fetal bovine serum
- Phosphate buffered saline, PBS

▨▨ Procedure

Dewaxing Formalin-Fixed

1. Place two-to-four 50 μm tissue sections from each tissue block into a centrifuge tube.

Day 1

2. Add 2 ml xylene to each tube.

3. Incubate for 1 hour in a 55 °C water bath.

4. Remove the xylene and replace with 3 ml fresh xylene.

5. Cover the tubes with Parafilm.

6. Incubate overnight in a 55 °C water bath.

Day 2

1. Remove xylene and add 100% ethanol.

2. Incubate 1 hour at room temperature.

3. Remove 100% ethanol and add 95% ethanol.

4. Incubate 1 hour at room temperature.

5. Remove 95% ethanol and add 80% ethanol.

6. Incubate 1 hour at room temperature.

7. Remove 80% ethanol and add 50% ethanol.

8. Incubate 60 hours (2.5 days) at 4 °C.

Day 5: Rehydrate

1. Remove 50% ethanol and add distilled H_2O.

2. Incubate 1 hour at room temperature.

Dissociate

3. Prepare 0.5% pepsin: 25 ml 0.9% NaCl (pH 1.5) + 0.125 gm pepsin, dissolved in a 37 °C shaking water bath.

4. Remove distilled H_2O and add 0.5% pepsin.

5. Incubate 1 hour in a 37 °C shaking water bath.

6. Vortex the tissue and use a pipette to break up the tissue by aspirating and expelling the tissue vigorously.

7. Stop the pepsin activity by adding 2 ml fetal bovine serum.

8. Centrifuge for 5 minutes at 220 x gravity.

9. Remove the supernatant and add PBS.

10. Use a pipette to break up the tissue.

Note: Steps 10-12 can be repeated if necessary. Large pieces of remaining tissue may be removed.

11. Centrifuge for 5 minutes at 220 x gravity.

12. Remove supernatant and add 2 mls PBS.

13. Use a hemacytometer to determine the concentration of nuclei.

14. Adjust the concentration to approximate 1.0×10^6 nuclei/ml.

Subprotocol 2
Cytospin Preparation of Nuclei

Materials

Equipment

- Cytospin 3 Cytocentrifuge (Shandon Lipshaw, Pittsburgh, PA)
- Cytospin slide clips
- Cytospin disposable chamber with filter cards
- microscope slides

Procedure

1. Assemble the Cytospin chamber with a microscope slide in the slide clip. **Assemble the Cytospin**

2. Add 300 µl (300,000 nuclei) of the dissociated nuclei in PBS to the chamber. **Add dissociated nuclei**

Cytospin 3. Centrifuge at 500 rpms for 3 minutes.

4. Remove the slide and allow to air dry at room temperature overnight.

5. Store slides in nitrogen-filled bags at -20 °C for five days before FISH.

Subprotocol 3
In Situ Hybridization

▩▩ Materials

Equipment

- moist chamber
- water bath at 37 °C and 75 °C
- ice
- hot plate
- coverslip
- rubber cement

Solutions

- PBS
- denaturing solution (70% formamide (IBI Spectral grade, VWR) in 2X SSC, pH 7.0)
- Master Mix 2.1 (see recipe)
- Carrier DNA (Herring testes DNA (Cat. # D6898 Sigma) dissolved in water at 10mg/ml. Shear to small fragments by sonication)
- centromeric probe
- doubly distilled H_2O
- Carnoy's solution (3 parts methanol and 1 part acetic acid)
- ethanol solutions: 70%, 85%, and 100%
- proteinase K (1mg/ml solution in water)

- hybridization wash buffer (50% formamide in 2 x SSC, pH 7.0)
- 2 x SSC
- 0.1 μg/ml DAPI in antifade (see recipe)

Preparations

- Antifade:
    - 100 mg p-phenylenediamine dihydrochloride (Cat # P1519 Sigma) in 10 ml PBS.
    - Adjust to pH 8 with 0.5 M carbonate-bicarbonate buffer (0.42 g NaHCO$_3$ in 10 ml H$_2$O, pH 9).
    - Filter through 0.22 μm mesh.
    - Add to 90 ml glycerol.
    - Mix with DAPI stain and store in dark at -20 °C.
- Master Mix 2.1:
    - 5.0 ml formamide (final = 50%)
    - 1.0 ml 20 x SSC (final = 2 x)
    - 1.0 g dextran sulfate, average Mol. Wt. 500,000 (Cat. # D-8906 Sigma) (final = 10%) heat at 70 °C 1-2 hours to dissolve the dextran sulfate.
    - bring to pH 7.0 with HCl; bring volume to 7.0 ml with doubly distilled H$_2$O

Probe Mix

Item	Volume
Master Mix 2.1	7.0 ml
Centromere probe 1	1.0 ml
Centromere probe 2	1.0 ml (optional)
Carrier	0.5 ml
Double distilled H$_2$O	to total 10.0 ml

Table of Materials

Preparation of Nuclei	Cytospin Preparation	In Situ Hybridization
glass conical 15 ml centrifuge tubes	Cytospin 3 Cytocentrifuge (Shandon Lipshaw, Pittsburgh, PA)	moist chamber
water bath with shaker	Cytospin slide clips	water bath at 37°C and 75°C
Parafilm	Cytospin disposable chamber with filter cards	ice
vortex mixer	microscope slides	hot plate
pipette		coverslip
xylene		rubber cement
ethanol solutions: 100%, 95%, 80%, 50%		PBS
distilled H$_2$O		denaturing solution (70% formamide (IBI Spectral grade, VWR) in 2 x SSC, pH 7.0)
Pepsin (Cat. # P7012 Sigma)		Master Mix 2.1 (see recipe)
NaCl (pH 1.5)		Carrier DNA (Herring testes DNA (Cat. # D6898 Sigma) dissolved in water at 10mg/ml. Shear to small fragments by sonication)

Table of Materials (continous)

Preparation of Nuclei	Cytospin Preparation	In Situ Hybridization
fetal bovine serum		centromeric probe
		doubly distilled H_2O
		Carnoy's solution (3 parts methanol and 1 part acetic acid)
		ethanol solutions: 70%, 85%, and 100%
		proteinase K (1mg/ml solution in water)
		hybridization wash buffer (50% formamide in 2X SSC, pH 7.0)
		2 x SSC
		0.1 mg/ml DAPI in antifade (see recipe)

Procedure

1. Preheat moist chambers for incubation of slides in 37 °C incubator

   Preheat

2. Preheat 1 x PBS in 37 °C water bath.

3. Preheat denaturing solution in 75 °C water bath.

4. Prepare probe mix (10µl for each hybridization)

   Prepare probe mix

   Note: See Table in Materials for mixture.

5. Fix the cytospin slides in Carnoy's solution, 3 washes of five minutes each, room temperature.

   Fix the cytospin slides

6. Denature target DNA by immersing slides for 2.5 minutes in denaturing solution at 75 °C.

7. Transfer slides to 70% ethanol on ice for 2 minutes.

8. Dehydrate in 85% ethanol, followed by 100% ethanol, each for 2 minutes, room temperature.

9. Place slides in a solution of 0.05-10 μg/ml proteinase K in PBS at 37 °C for 7 minutes.

10. Dehydrate slides in 70% ethanol, followed by 85% ethanol and 100% ethanol, each for 2 minutes at room temperature.

11. Denature the probe mix in 75 °C water bath for 5 minutes.

12. Place the slides on a 37 °C hot plate, and place 10 μl of probe mix on each spot.

Coverslip and seal

13. Coverslip the slides and seal with rubber cement.

Incubate

14. Incubate in a moist chamber overnight at 37 °C.

15. The next day, remove the rubber cement.

Hybridization

16. Wash the slides in hybridization wash buffer, 3 changes at 45°C for 10 minutes each. (The coverslips should float off during the first washing.)

17. Wash the slides in 2 x SSC at 45 °C for 10 minutes.

18. Repeat the wash in 2 x SSC for 10 minutes at room temperature.

19. Rinse the slides in two changes of doubly distilled H$_2$O at room temperature for 5 minutes each.

20. Allow the slides to air dry.

Counterstain with DAPI

21. Counterstain with 8 μl DAPI and coverslip.

References

1. Van de Rijke FM, Vrolijk H, Sloos W, et al. Sample Preparation and In Situ Hybridization Techniques for Automated Molecular Cytogenetic Analysis of white blood cells. Cytometry 1996; 24:151-157.
2. Vrolijk H, Sloos WCR, Van de Rijke FM, et al. Automation of spot counting in interphase cytogenetics using brightfield microscopy. Cytometry 1996; 24:158-166.
3. Netten H, Young IT, Prins M, et al. Automation of fluorescent dot counting in cell nuclei. In: Proceedings of the 12th IAPR International Conference on Pattern Recognition. Vol. I, Conference A: Computer Vision and Image Processing (ICPR12, Jerusalem, Israel, October 9-13, 1994). Los Alamitos, CA: IEEE Computer Society Press, 1994: 84-87.

4. Thornthwaite JT, Leif RC. Plaque cytogram assay I. light and scanning electron microscopy of immunocompetent cells. J. Immunology. 1974; 113: 1897-1908.

5. Sauter, G, Carroll, P, Moch, H, et al. c-myc copy number gains in bladder cancer detected by fluorescence in situ hybridization. Am. J. Path. 1995; 146:1131-1139.

6. Sauter, G, Haley, J, Chew, K, et al. Epidermal-growth-factor-receptor expression is associated with rapid tumor proliferation in bladder cancer. Int. J. Cancer 1994; 57: 508-514.

7. Sauter, G, Moch, H, Moore, D, et al. F. Heterogeneity of erbB-2 gene amplification in bladder cancer. Cancer Research 1993; 53:2199-2203.

Section 8

Core Facilities

ROCHELLE A. DIAMOND

Core Flow Cytometry Facilities are becoming common place in many institutions and companies to provide a shared cost resource for investigators. How these facilities are set up physically and financially are discussed by Steve Merlin and Larry Sklar. Ingrid Schmid discusses the all-important aspect of biosafety in flow cytometry facilities.

Be sure to check out the World Wide Web sites listed in our appendix to discover the core facilities in various regions of the world and what they offer.

Core Facility Management

STEVEN MERLIN

Introduction

With the number of institutions and laboratories employing flow cytometric techniques, the formation of Core facilities has become a practical alternative to the large capital outlays required to acquire, set up and maintain flow cytometers on an individual basis. Key aspects of organizing a core facility involve laboratory management, quality control and instrument maintenance. By integrating the scheduling, quality control, maintenance and training within a centralized core facility, economies of scale can be archived for reducing associated expenditures. With instrumentation centralized, Quality Control procedures and instrument standardization can be better controlled.

Setting up a Core Facility

Many laboratories have need for using a flow cytometer but do not have the financial resources or work volume to justify the costs involved with acquiring and maintaining their own instrumentation. In other situations, a laboratory having their own flow cytometer may use it on a limited basis, resulting in excess capacity. For both these situations, it makes both economic and practical sense to set up a core facility in which financial resources can be combined to purchase and maintain instrumentation and share expenses.

A number of points need to be considered in determining whether a core facility will adequately serve the needs of the intended groups. First, prospective users and the type of techniques that they will use must be identified. Based on this information, system hardware, software and accessories can be identified. A group meeting should be held for presenting the information and technical specifications, as well as solicitation of comments from the participants. A common mistake when considering the purchase of a system is not taking into consideration future needs. System flexibility for future upgrading may be an important aspect before making a final decision on the purchase of the hardware and software that constitutes the system.

Assuming your institution has the number of users to justify setting up a centralized facility, its physical location and size will be two of the most important considerations. Ideally, a space that is located in a geographically central area with easy accessibility is highly desirable. Space is a major concern in most institutions. Often instruments are crowded into small rooms making working conditions sub optimal. As the number of users increases, an additional instrument may be added to the facility, therefore a room that will allow for future expansion is preferred.

Setting up a Cell Sorter Lab

Fluorescent Activated Cell Sorters have special requirements regarding infrastructure and utilities. Upon request, each manufacturer can provide documentation listing specifications. It is important that a representative of the in-house engineering staff be involved with initial planning. In most cases, renovations to the infrastructure and utility work will be contracted out. Therefore it is important that all parties involved in the setup of the core facility have the necessary specifications before beginning the work.

General considerations

Room size

The room should allow for easy access and comfort. Storage space and room for expansion should be factored into the calculations. Cell sorters usually require pressurized air, vacuum and if equipped with larger lasers, water-cooling systems. If the existing room does not have all the required utilities, installation of the required utilities in an adjacent "utility" room is preferable.

Walls should be roughly textured and covered with flat latex paint. Avoid smooth surfaces finished with enamel paint and non-painted metal air ducts and conduit.

Electrical outlets should be adequate in number and located in close proximity to where the instruments will be located. A telephone close to each instrument is highly desirable, especially during troubleshooting sessions. For computer systems, a modem or LAN connection is recommended. (see Data Management and Storage Section 2, Chapter 6)

Sink/Counter space is very important for sample preparation, sheath fluid processing and filtering, as needed, and a miriad of small appliances, such as a sonicator, centrifuge, and vortex. A deep sink basin for emptying the waste container will limit splashes.

Room Lighting

Fluorescent lighting is fine for instrument preparation and maintenance procedures, but should be avoided during sample acquisition. Incandescent lighting in which the intensity can be adjusted is recommended. A rheostat control for the incandescent lighting should be located on or near the instrument console for operator convenience. Light switches for the fluorescent lighting should be located close by as well. Remote control devices also exist for switching lights on and off.

Environmental requirements

Flow Cytometry systems consist of electronic circuitry found in the computer, as well as the sensor head of bench-top analyzers and the optical bench/console of cell sorters. The electronic hardware is designed to perform in a defined temperature and humidity range. A filtered, air-conditioned environment between $18\,°C$ ($65\,°F$) and $24\,°C$ ($75\,°F$) with non-condensing humidity is the ideal range for instrument operation.

Pressure/Vacuum

Air pressure must be clean, dry and oil free. The pressure must be regulated. A standard vacuum line is also required. Both lines should be fitted with hose barb fittings or hose coupling devices, preferably "quick connects". Check with the respective manufacturer for pressure and vacuum requirements, which may vary depending on the type of instrument installed. To prevent aerosols in the stream chamber of sorters, a portable electric vacuum can be purchased separately or obtained with the instrument. These pumps generally are noisy, so locating them in an adjacent utility room and bringing a vacuum hose through the wall will eliminate unpleasant noise.

Laser Cooling requirements

Ideally, an existing in-house closed loop water cooling system is a good choice. Generally, if the lasers will be connected to an existing system, boost pumps and temperature control monitoring will be necessary to comply with manufacturers' requirements. Where a closed-loop cooling system is not available, portable heat exchanger units can be purchased and installed, however it is highly desirable to locate them in an adjacent utility room, both for the heat and noise generated. Specific cooling, pressure and water hardness requirements for the types of lasers purchased are available from both the laser manufacturer and the service department of the flow cytometry manufacturer.

Electrical requirements

Check with the manufacturer for the minimum number of outlets, current rating and plug styles required. Standard 3-wire wall outlets with isolated ground are required. One line should be a dedicated circuit for the instrument console and computer system. For added electrical power integrity, you may want to investigate the need for a voltage regulator and/or uninterruptible power supply (UPS). This device is highly desirable in areas where the power may fluctuate or is affected by heavy equipment, i.e., centrifuge motors, compressors, etc. It is advisable to check voltage fluctuations, particularly in older facilities with a line monitor. These devices are plugged into an existing outlet and record voltage spikes and drops over a variable time period. Make sure ground circuits are adequate.

Lasers generally require higher voltage, three-phase wiring rated for 50 amps per phase. A main disconnect must be located in the same room as the system, preferably with the operator having easy access in case of an emergency.

Administration

Once the decision has been made to establish a core facility, its organization and administration will need to be defined. What department will have ultimate responsibility for its oversight? Where does it fit ideally within the organizational layout of the institution or company? Who will propose, formulate and approve of the policies dealing with use of the facility? These are just a few of the items that will need to be addressed in establishing the facility.

The type of institution in which the facility is located will likely influence the choice of a department that has oversight of the core facility. In a large academic institution, the core lab might be used by a very diverse group of users. This may include immunology, microbiology, Veterinary science, clinical research as well as outside groups. In an industrial setting, lab groups more closely related to a single discipline may share the facility. Obviously, use by more diverse groups will require more careful evaluation on where the core lab fits within the organizational scheme.

Since the facility will be used presumably by many individuals representing a diverse group of disciplines, a set of guidelines governing the daily operation and use of the facility is needed to prevent misunderstandings and avoid conflicts. A proposed list of topics should be drawn up and circulated for input and comments. After revisions have been made, review and approval of the document may be required at a specific administrative level having jurisdiction over all groups that will utilize the facility.

Guideline Content

- A brief statement indicating the purpose of the guidelines and who approved them.

- Facility and Equipment: Description of instrument hardware, software, accessories and general specifications of each.
  - Period of Operation: Days and times, special hours, holidays; Advance notice requirement for sorting.
  - Location
  - Procedure for requesting instrument or computer workstation time.

- Analysis/Sorting fees: broken down based on instrument type, with or without assistance for analyzer systems, single or multiple laser operation for cell sorter.

- Minimum charge, setup and material costs.

- How payment is to be made.

- Miscellaneous items such as biohazardous materials notification, handling, transport, etc.

Financial Considerations

As part of the initial discussion in establishing a core facility, a means of covering operational costs has to be defined. The type of method employed will most likely be governed by whether funding is available on a continuing basis. In situations where yearly funding is not an issue, most core facilities are run as "Service labs". In this capacity, users have access to equipment time without charge. In situations where little or no money is available to cover operational expenses, a fee schedule is established where users pay for actual time used. A variation of this is to charge a yearly fee up front with a guaranteed number of hours covered. With this variation, labs can better budget their expenses and the Core Facility receives a fixed amount of income up front.

Whichever financial arrangement is selected, costs associated with facility operation need to be determined so that either a user fee schedule or budget can be accurately formulated. In calculating costs, the following formula can be used or modified for determining the overall operating budget:

Operating Budget = [Salary] + [Benefits] + [Service Contracts] + [Supplies] + [Other]

Benefit costs vary with salary level, geographic region and type of institution. A general range is between 20-30 % of the base salary. The Human Resources department or Business Administration office can provide the actual percentage used.

What will it cost to run a particular instrument during the course of a year? This parameter uses the cost of a yearly service contract for a specific instrument configuration, type of laser(s) and additional accessories installed. The manufacturer can provide service contract fees. For facilities that may use the services of an in-house Biomedical Engineer or the facility manager to perform some repairs, costs for factory training, hardware replacement costs and hourly field service fees can be provided as well.

Funding for Core Facilities

A Manager's Manifesto: to Fund or to Die

LARRY A. SKLAR

Introduction

Flow cytometers are expensive instruments to buy, to operate, and to maintain. Because federal and institutional funds are tight in this era of health care reform and deficit spending, the shared flow cytometry facilities need to form partnerships with their institutions and their investigators to perform optimally. The facility managers and faculty principal investigators have a responsibility to insure that their administrators understand that it is typical for facilities to raise 30-70% of their costs in revenues. This article will consider some of the challenges confronting the managers of flow cytometry facilities and review some of the strategies we have used at University of New Mexico (UNM) to meet the challenges. Our focus is on research facilities that may have some clinical projects, but does not address the special needs of facilities whose prime responsibility is patient diagnostics and other clinical samples. Some of the advice will be obvious given the common experiences we share in our facilities, but it may be quite helpful for getting facilities started.

Funding and Revenues

I have been involved with three facilities: The Scripps Research Institute (TSRI) from 1981 to 1990 (Facility director from 1986-1990); UNM (Faculty PI) from 1990-present; and the National Flow Cytometry Resource at Los Alamos National Laboratory from 1990 to 1992 as Director and from 1992 as Co-Director. The TSRI facility grew under managers David Finney and Don McQuitty from one instrument to six (three sorters, two analyzers, one microscope) and three staff. The UNM facility with manager Larry Seamer grew to four service instruments (one each for sorting, analysis, microscopy, spectrofluorometry) and an equivalent number of research instruments for kinetics, spectroscopy, and video image analysis. Working with Seamer and me were one part time operator and several faculty, staff scientists, and students who provide expertise for collaborative work on one or more of the instruments.

These three facilities functioned on completely different funding systems. The TSRI facility was responsible for generating all of its funding from users and its function was almost exclusively dedicated to service for its investigators. At that time, TSRI operators were technicians. The LANL facility is charged with raising all of its funding from grants and its goals are research, development, collaboration, service, and training. The instruments are operated both by scientific and technical staff. The UNM facility is 70% supported by institutional resources and is involved in service, training, collaborative research and instrument development. The perspective of the UNM facility as a typical core lab will dominate this article.

I will present my ideas about funding sources for core facilities, new instruments and personnel. I will also discuss some approaches to building a base of institutional support for your facility. The pressing issue that we face every year is raising the funds to keep the facility open. Table I lists some of the types of funding which may be available for your facility to meet your recurring expenses.

Table I. Funding Sources

1. Institutional Support

2. Internal User Fees

3. Clinical Fees

4. Grant Projects

5. External Fees

6. Partnerships with external users

7. Special Opportunities

Once instruments have been purchased, the main costs of the facility are associated with salaries for personnel, instrument maintenance and a modest budget for material and supplies. Many cytometry facilities have several kinds of instruments: bench top or "clinical" analyzers which are user-operated flow cytometers; research sorters which are operator-dependent flow cytometers; and confocal microscopes. In our facility at UNM, we also make available fluorescence spectrometers, a lifetime fluorometer, a stopped-flow spectrometer, and a scanning spectrophotometer which are located in my research lab. Fees are negotiated for the latter instruments: sometimes they are available for free, and at other times they are used on a fee for service or a collaborative basis. These decisions depend on the nature of the project, the type of personnel support that is required, and whether the personnel are paid from research grants or institutional service sources.

Flow cytometer usage is weighted heavily toward afternoons so that the investigators have had the chance to do their experiments and prepare their samples. If the sign up schedule ends at 5PM, 3-4 hours of service represents a respectable day of activity on a machine. In an academic setting, with holiday periods and characteristic slow periods before them, 800-1000 hours a year on a bench top machine is reasonable. User cancellations are more common on the sorters which decreases the fraction of time the instruments are in use. Table II shows costs for a hypothetical two instrument facility with one senior manager. Additional revenues would need to be provided by the institution.

Table II. Hypothetical two instrument/one manager facility

Revenues		Costs	
Analyzer 800 hrs @ $40/hr	$32,000	Analyzer Service Contract	$12,000
Sorter 500 hrs @ $70/hr	$35,000	Sorter Service Contract	$30,000
		Manager Salary/Fringe	$60,000
		Misc. Supplies/Service	$10,000
Total	$67,000		$112,000

The facility described in the example would test the patience of a single manager. If he or she were also doing the billing and scheduling, it would seem very busy. The manager would interact with 4-5 separate investigators and experiments every day and assist with data analysis, experimental design, and instrument set up. It would leave little time for the manager to have professional growth. More likely, facilities are typically staffed with one person per instrument or two people per three instruments. By the time the budget is expanded to include travel, some opportunity for occasionally upgrading computers, an individual facility may generate revenues from users fees which represent only 30-40% of annual costs.

Institutional Support

It is often the case that flow facilities primarily serve a specific research group, a Department or an individual School. If your campus has two or more flow cytometry facilities, you are likely to have a harder time making the case for a broad base of support. Our facility is in a School of Medicine but financial support has been based almost entirely in our Cancer Center. In the last few years we have begun to educate our administrators in the Health Science Center that we serve a broader community. This includes faculty in the School of Medicine not affiliated with the Cancer Center, faculty in the Health Science Center in Schools of Medicine, Nursing, and Pharmacy, and faculty on the Arts and Science Campus not affiliated with the Health Science Center. The process of education may take several years, and not every administrative unit will eventually buy in. However, the process itself is important and should be actively pursued. The documents you acquire which describe the effectiveness of your facility will come in handy as you participate in grant applications with PI's.

The manager's role in the education process is to: 1) utilize user logs to make the case that users come from many disciplines; 2) explain that the facility is required on-site and that samples cannot simply be shipped around the country to commercial flow ventures; 3) show that continued funding of active investigators as well as new funding opportunities arise from the work done at the facilities; 4) demonstrate that the good science which is being done benefits the institution. Managers can encourage their PI's and faculty associated with their facility to lobby in their own Schools and Departments with documents showing grants dependent upon the facility and publications that would not have possible without your facility. In many cases, your institution cannot recruit the best faculty without modern instrumentation.

NIH funded institutions have programs where grant overhead is returned for new capital equipment or instrument upgrades. State schools often have legislatively mandated instructional budgets. We have found that our administrators are beginning to be very responsive to requests from the flow facility because our users are typically NIH funded and we can document serving many of them. Because institutional support is likely to be the biggest portion of your funding, try to spread the burden of the costs of your facility as well as the rewards as broadly as possible.

Often individual investigators do not know much about the capabilities of your facility beyond the specific assays they are doing. Many immunologists and cell biologists may not know you have recently acquired the ability to do intracellular cytokine assays or apoptosis. Advertising capabilities can be balanced between your web site, a local e-mail based user list, hard copy report (newsletter) of what's happening in your facility, and word of mouth. The e-mail user list seems a particularly good way of letting your users know when you are on vacation or travel, about new methods that are in the literature or what a local colleague has developed, about instrument breakdowns, or new policies affecting fees, hours, or safety. Think about having a yearly flow cytometry lecture and open house. Flow cytometry facilities can play an important role in recruiting new faculty to your institution and it is good to have someone represent the facility when new recruits visit.

User fees

Virtually all facilities charge hourly fees. When I was at Scripps, we had a user's group that was charged an annual fee at the beginning of each fiscal year in lieu of an institutional contribution to keep the hourly fees as low as possible. We all face the difficulty of setting fees commensurate with the cost of the activity, the value of the service, and the funding levels of the investigators. There is a fine line between fees that chase marginally funded investigators away and fees that give investigators the opportunity to compete for new funding. Have a faculty committee review user fees annually; let them help decide the balance.

Philosophically, we have taken the position that our facility is only as well funded as the investigators we serve. We have decided to judge ourselves not only by the number of dollars we raise in service fees, but by the number of hours we operate, the number of publications, and the increasing number of funded users. Taking a long-term perspective, we have decided to support two types of pilot projects. For a funded or fundable investigator, we will prototype developmental experiments for a reasonable number of hours. This allows the investigator time to decide if the approach is feasible and whether publishable or fundable data can be acquired.

New faculty, without funding, are often given the opportunity to develop projects during their first year, at minimal charge, if instrument time is available and it is within our budget. Our goal is to build good will with the new faculty and their Departments and to contribute to the ability of our faculty to achieve competitive grant funding.

We tend to have three types of cost structures: instruments that require an operator, instruments that require only the user, and data processing (whether we have a stand alone computer for each instrument or not). The requirement of the operator typically doubles the basal cost (~$30/hr); the computer use alone is typically about half the basal cost. We feel we have to charge for computer use to cover the cost of computer purchases, supplies, upgrades, maintenance, and site licenses. Clearly, there has to be a charge if data analysis precludes data acquisition on the same machine. However, we encourage users who are interested to purchase their own site licenses. After hours data analysis is a reality that many of us have come to appreciate.

Historically, our primary clinical role has been ploidy analysis. This was a particularly active area when Lynn Dressler was at UNM and spurred Larry Seamer to develop the tru-ploid analysis as a research tool. Our institution also has separate bench top flow cytometry facilities for clinical samples. An immunophenotyping lab for the Southwestern Oncology Group is directed by Cheryl Willman, MD and a flow lab for the UNM Hospital is directed by Catherine Leith, MD. My research lab has its own Analyzer which was purchased with my own grant funds. Basic researchers do not typically have access to these other labs unless the core facility instrument is out of service. We are therefore not involved in processing clinical samples specifically, but encourage investigators with clinical research projects to work with us. Fees for these projects are required to be the same. Flow facilities that enjoy support as part of a federally recognized cancer center have to give priority to investigators who are members of the Cancer Center, and may even have lower fees for them.

The biomedical community in New Mexico outside of our research institutions represents a relatively small number of business activities. We occasionally undertake short term fee for service projects, where the fees include the NIH indirect cost rate. Thus a $100 fee would have a $48 indirect cost charge added to make up for the fact that the external user was not contributing to the overhead costs which would normally come to our institution from a grant. We will consider below some more important revenue sources which involve partnerships between internal and external users.

Partnerships

The management teams of flow cytometry facilities should actively pursue specific alliances with investigators. It is my understanding that most academic facilities have a faculty PI and a facility manager. Between the two of them, there are a number of active roles that they can take to generate research funds for the facility

1. The most basic support for projects of individual investigators takes several forms: providing preliminary data, documenting standard methods, developing novel methods, co-authoring manuscripts, critiquing the technical as well as the scientific content of grant proposals, and writing letters of support. The letters provide a description of the services available and the fees charged by your facility. Encourage your PI's to write facility charges as line items in grant proposals and to make commitments of support for the period of the grants. Make sure that your investigators understand

when your facility is performing at a level beyond service so that your contributions can be recognized as full-fledged collaborations. Do not let the scientists take the contribution of the facility for granted.

2. Direct financial support for core facilities should be included as projects within Program applications. For example, when groups of investigators band together to prepare a Program for funding, it may include several projects and core facilities which support all of the projects. Core facilities may assume the role not only of performing specific assays or developing new methods. They may also take a larger role in sample preparation, staining, calibration, quantitation, or developing reagents. These funding situations provide the opportunity for a long term source of revenues for your facility. They may also help you to achieve institutional recognition and help you to develop a particular expertise in your facility. In our facility, we have been interested in time-based methods. We are partnering with the National Flow Cytometry Resource in kinetic analysis, automated sample handling, and subsecond sample delivery. This provides our on-site investigators specific support for issues involving ligand-receptor interactions, cell activation, and analysis of molecular assemblies.

3. Partnerships with external users are possible. Several novel funding mechanisms have become available that allow core facilities to match their specific skills with the needs of small business. The small business partnerships will be considered below under special opportunities. At one time or another, many of our core facilities have had the opportunity to serve as beta test sites. Under new federal grant rules, we can participate in development activities and also generate revenues.

Special Funding Opportunities

This is a time of profound change in our granting institutions and funding opportunities. While most of us have been accustomed to our investigators being the driving force funding basic and clinical investigations, there are some new opportunities. In the last few years, we have witnessed an expanding commitment to applications development and commercialization. Federal granting agencies are targetted to spend 2.5% of the grant budgets to support small businesses. With the downsizing and "conversion" of the defense expenditures in the U.S., there have also been grants for dual use (military/civilian) applications from Defense Advanced Projects Research Agency (DARPA) and National Institute of Science and Technology (NIST).

In recent years, there have been two types of partnerships with small business: SBIR (small business innovative research) and STTR (small business technology transfer). In the SBIR, university facilities collaborate with companies developing novel products. In the STTR, the small business can be involved in commercialization which actually requires a specialized expertise in your facility. For example, we have been interested in flow cytometric ligand-receptor or cell adhesion assays which could screen the activity of potential new products.

Most of the funding, approaching $1 billion/yr has been for SBIR. The applications are peer-reviewed and judged on scientific merit and the likelihood of producing a product. Phase I, a one-year $100,000 award is to obtain proof of principle. Phase II, with funding of $750,000 over three years, is for the product. A new "fast track" initiative combines the two phases for projects judged to be of exceptional scientific merit and commercial viability. Your facility can request up to 30% of the funds awarded. SBIRs are sponsored by most institutes at NIH, with application deadlines of April 15, August 15, and December 15. CDC and FDA have a December 15 deadline.

Defense Conversion funding sources have been in and out of favor in the 1990's. Gary Salzman at Los Alamos has lead a collaborative project with Biorad for the development of battlefield flow cytometers. The technology reinvestment program (TRP) from the Advanced Research Projects Agency (Dept. of Defense) concentrates on high technology dual use projects. The projects are mostly developmental and require industy to match funds. They fund things like remote sensing technology. The Advanced Technology Program (ATP) from the National Institute of Science and Technology (Dept. of Commerce) focuses on high risk, commercial projects of broad benefit. They have funded stem cell and virus projects along with electronics. For the latest opportunities, check www.darpa.mil and www.atp.nist.gov.

The Center for Alternatives to Animal Testing (CAAT) funds pilot projects and processes which provide alternatives to animal testing. Flow cytometry of cell viability, cell cycle, and apoptosis could fit into this category. Check out their web site http://info-net.jhsp.edu/~caat/.

Finally, point your investigators to a booklet prepared by Larry Prograis at NIAID. His booklet "How to Write an NIH Grant: How to Get Research Funding in Trying Times" is a gem for NIH funding and it provides application information for about 50 foundations that grant research awards. His e-mail address is lpi3r@nih.gov.

Costs

Because the major recurring costs in a flow cytometry facility are the salaries of the operators and the instrument service contracts, it is worth discussing whether there are any options to the way we normally operate.

Service contracts typically cost 10-12% of the price of a new instrument each year. The service contract guarantees rapid repair which may be crucial in a highly used facility where there is only single machine. Being down for a week would clearly be a disaster. If you have a number of complementary instruments or if the lab manager is skilled in trouble shooting and negotiating with technical support staff from the manufacturer, you may find it worthwhile to limit your commitment to service contracts for the machines and computers. We all know that there is no real option to laser repair, but these needn't be done on contract either. If you don't have service contracts, you need to budget $5,000-10,000 per instrument for maintenance and repairs.

Can you afford not to have a service contract?

In the 15 years I have been doing flow cytometry, in only a handful of years have our repair costs approached those we would have paid if we had a service contract on any single instrument. A large part of this record was the skill of the facility managers I have worked with. On the other hand, many institutions are willing to budget a service contract as a recurring expense. I think this is because the managers view instrument service as beyond their expertise or not cost effective in terms of instrument down time, lost revenues, or service time lost by the manager.

Raising funds for a new instrument

As our $300,000 multilaser instrument reached five years of age, we became very concerned about paying $30,000 a year for service. We approached our administrators with the following option. After six years, we could either have the same flow cytometer which would by then totally obsolete and not worth the service contract price. Or, we could contribute the annual service amount toward a lease-purchase of a fairly recent demonstration model flow cytometer which would then be ours. This new model would keep the facility current at no added expense. While our administrators weren't thrilled with the idea of buying a new instrument, they did reluctantly agree that it was a sensible use of funds that would otherwise be lost. The catch for us was that as a facility we had to keep the instruments running through our own initiatives, through new revenues or through the skill of our manager. If you intend to try this scheme, make sure you find out about the tax implications of capital equipment lease-purchase arrangements in your state.

When you need a new instrument, there are a number of possible strategies. Generally speaking, individual investigators cannot get $100,000 or more from an individual grant. However, granting agencies are sympathetic to providing partial support for an instrument if there is a matching institutional commitment to the rest. NSF, in fact, in the past has offered two types of instrumentation grants: one for infrastructure improvement and one for instrumentation development. When you are applying for infrastructure improvement, it is essential that your institution contributes a 50% match to the NSF amount.

Shared instrument grants at NIH from the National Center for Research Resources (NCRR) have also been a good source of instrument money. These grants are intended for purchases up to $400,000 and are appropriate for major flow cytometry purchases. The applications are judged largely on the quality of the investigations which are to be performed. It is a good idea for the instrument to support five or more investigators who are already funded by NIH and for the applications to clearly require the new instrumentation. A plan for managing the instrument is required so it helps if you already have a core facility organizational plan in place.

Cost effective personnel management

The 8AM-5PM work day is not particularly compatible with a flow cytometry facility where live cells are being handled and experiments tend to be done between 1 and 5PM. When machine time is at a premium, it is worth figuring out whether: 1) you could generate more revenues in your facility by being open longer hours; 2) you are limited by the amount of machine time available; 3) you are limited by the personnel available. My experience has been that case 2) is true the least often, while 1) and 3) are more often true. Staggering the shift time of the people in the lab and tending toward later starting times is one approach that can be used to increase the productivity of your facility. If later starting times are not compatible with the schedule of the people in your facility, try to see whether you could remain open one or two nights a week for an hour or two. If you have three people in your facility, flexible time schedules can effectively improve the output of your facility by about one person. Your administrators might be impressed enough by the decreased costs and increased revenues or even by saving the cost of a new instrument to raise everyone's salary.

At some institutions, capital expenses are easier to come by than recurring personnel expenses. Try to get that information before proceeding. In any event, the best reason for hiring additional people, is that the hours that you bill and the increased service that you provide will pay for themselves. I tend to believe that facility expansion should finance itself to the extent possible.

Windfalls and Shortfalls

There is nothing more discouraging than a big expense at the end of a fiscal year in a budget that is already spent. If your creditors or administrators are helpful, you may be able to push the costs forward. In some years, we have found no solution but to borrow from our users. Occasionally, a user will be in a windfall situation where he or she has money in a grant or a state account which can be used to solve our problem. Of course, you need to make sure that these arrangements are consistent with the regulations at your institution and the agencies which have granted the funds.

Windfalls, surprisingly, can be just as troublesome: an allocation for a service that wasn't needed; an unexpected period of high revenues from users, or a grant that came too late in a fiscal year to spend out. Rolling money forward into the next fiscal year can be just as treacherous as being short. Most government agencies and academic institutions have a "use it or lose it" philosophy. The best idea we have come up when faced with a surplus is to try to anticipate your needs for the next fiscal year: a computer upgrade, supplies and spare parts, paying bills early. We bill about every two months and our fiscal year often has an end date different from our investigator grants. Sometimes it is possible to move your year end collection into the next fiscal year of the facility. It helps in cases of both windfalls and shortfalls for the facility PI and the manager to have open and honest communication about what is being done, what is needed, and why.

Our facility at UNM has service, training, development, and collaborative functions. We have sensed that many of the faculty at our institution were not familiar enough with flow cytometry to profit from the facility. Our institution has recently received an institutional development award (IDeA) from NIH to make our faculty more competitive. In the flow faciliity, we now have the opportunity to actively seek out collaborations. We are looking forward to expanding the role of our manager to include scientific co-ordination of projects and more active collaborations.

Acknowledgements

Supported by the Cigarette Tax of NM to the UNM CRTC and RR01315.

Biosafety in the Flow Cytometry Laboratory

INGRID SCHMID

Introduction

Biohazards in the flow cytometry laboratory arise either from specimen or sample handling, or from the use of the flow cytometer. Specimens can contain known, e.g., Human Immunodeficiency Viruses (HIV-1, -2), Hepatitis viruses, or unknown pathogens. Most of the infectious agents that are encountered in a typical clinical or research flow cytometry facility are transmitted by percutaneous or mucous membrane exposure, or by ingestion; some, however, can be transmitted by aerosol inhalation (see table 1). For further information see ref. 1. In addition, many dyes used in flow cytometric assays are toxins, mutagens, or carcinogens. Protection of laboratory personnel, and in particular flow cytometry operators, from exposure to such occupational hazards is of ultimate concern.

I. Standard precautions

Guidelines for specimen handling are available from the Centers for Disease Control and Prevention (CDC) [2,3]and also from the National Committee for Clinical Laboratory Standards (NCCLS).[4] Laboratory personnel who handle human cells are required to follow procedures as outlined in the document: US Department of Health and Human Services Occupational Exposure to Bloodborne Pathogens.[5] Policies regarding the control of infectious disease hazards are generally selected by the laboratory director considering the specimens that are processed in the laboratory.

Table 1. Select human bloodborne agents associated with laboratory-acquired infections due to manipulation of biological samples[*]

Agent	Disease	Source of infection	Route of infection	Biosafety Level practices, safety equipment, & facilities
Hepatitis B, C, D virus	Hepatitis	blood, cerebrospinal fluid, urine, tissues	inoculation, exposure of mucosal membranes to aerosols, broken skin	BSL2; BSL3 in case of aerosol production, large quantities or high concentrations
Cytomegalovirus	Cytomegalic inclusion disease	blood, tissues,	inoculation, exposure of mucosal membranes to aerosols, broken skin	BSL2
Epstein-Barr virus (EBV)	Infectious mononucleosis, Burkitts lymphoma	blood, tissues, EBV-transformed lines		
Human immunodeficiency virus (HIV-1, 2)	Acquired Immunodeficiency Syndrome	blood, body fluids, tissues	inoculation, exposure of mucosal membranes to aerosols (containing concentrated virus), broken skin	BSL2; BSL3 in case of aerosol production, large quantities or high concentrations
HTLV-1, 2 virus	Leukemia	blood, body fluids, tissues	inoculation, exposure of mucosal membranes to aerosols (containing concentrated virus), broken skin	BSL2; BSL3 in case of aerosol production, large quantities or high concentrations
Neisseria meningitidis	Meningitis	pharyngeal exudates, bronchoalveolar lavage, cerebrospinal fluid, blood	inoculation, direct skin contact, aerosol inhalation	BSL2; BSL3 in case of aerosol production or high concentrations
Salmonella Salmonella typhi	Salmonellosis, Thyphoid fever	blood	inoculation, direct skin contact	BSL2; BSL3 for large quantities

Table 1. Continued

Agent	Disease	Source of infection	Route of infection	Biosafety Level practices, safety equipment, & facilities
Brucella	Brucellosis	blood, cerebrospinal fluid, tissues	inoculation, direct skin contact	BSL2; BSL3 for tissue cultures of infected cells
Toxoplasma	Toxoplasmosis	blood	inoculation, aerosol inhalation	
Trypanosoma	Sleeping sickness, Chagas disease			
Leishmania	Leishmaniosis	blood	inoculation, aerosol inhalation	
Plasmodium	Malaria			

* This table was adapted from US HHS Publication: Biosafety in Microbiological and Biomedical Laboratories, 3rd Edition, 1993.

Generate or adopt a biosafety operations manual covering the following areas:

1. Determine the risk of exposure for all employees: review the appropriate safety procedures for the pathogens studied in the laboratory (see table 1 and table 2, for further information see ref. 1), taking into account the presence of individual risk factors of the personnel, e.g., pre-existing diseases, compromised immunity;[6] if the flow cytometry operator is pregnant, consider the potential risk to the fetus.

2. Containment controls:
   - Universal precautions: treat all human blood and potentially-infectious materials as infectious; whenever possible fix all samples by choosing an appropriate inactivating agent for flow cytometric samples, e.g., 1% formaldehyde solution; perform all procedures to inactivate infectious agents very carefully, i.e., use effective concentrations and the necessary time of incubation, taking into account the infectious agent studied; samples that are thought to be properly fixed, but in fact are not, could pose a serious health risk.[7,8]
   - Engineering controls: use technology and devices to isolate or remove hazards from laboratory personnel, e.g., biosafety cabinets, capped tubes, mechanical pipetting devices,

closed flow cells and enclosed sample collection drawers on flow cytometers.

- Work practice controls: develop a manner of performing a task to reduce the likelihood of exposure, e.g., avoid the use of "sharps" by finding suitable replacements, perform all laboratory manipulations that generate aerosols, e.g., pipetting, vortexing, etc., in biosafety cabinets. For protection against exposure to droplets and aerosols produced by jet-in-air flow cytometers see sections II C and II D below.
- Personal protective equipment: use clothing and equipment to protect laboratory personnel from exposures, e.g., laboratory coats, gloves, masks.

3. Housekeeping practices: routinely decontaminate the work area after work is completed using appropriate disinfectants (see table 2 and table 3); establish rapid clean-up of spills as standard laboratory practice.

4. Waste disposal: dispose of all contaminated materials, e.g., tubes, pipets, pipet tips, gloves, etc., into appropriate puncture-resistant biohazard containers; decontaminate waste prior to disposal as appropriate, either through autoclaving or addition of concentrated household bleach. For disposal of chemical waste such as formaldehyde solutions and solutions containing toxins, e.g., propidium iodide, follow the chemical waste disposal regulations of your institution.

5. Labels and tags: when human cells are processed in the laboratory, put a label with the international biohazard symbol on all laboratory access doors. Store toxic substances such as DNA dyes separate from other materials in an area that is clearly labeled: Caution, Toxic Material. Label all solutions that contain toxic substances with the same warning label.

6. Training: educate employees in safe laboratory practices before work is started; strict adherence to the work practices appropriate for the agents studied in the laboratory is essential.

7. Vaccinations: vaccination against Hepatitis B is generally available and highly recommended.

8. Post-exposure evaluation and follow-up: follow the most recent guidelines for post-exposure prophylaxis against infection with HIV from CDC.[9]

9. Record keeping: maintain accurate training records and records of post-exposure management.

II. Biosafety considerations for the operation of flow cytometers

- Follow carefully all instructions and maintenance recommendations from the manufacturer of the instrument. Collect the waste fluid in a suitable container containing fresh concentrated household bleach in sufficient quantity to achieve a final concentration of 10% when the flask is full. Disinfect the fluid lines regularly by running disinfectants through the instrument lines, e.g., 10% bleach solution (see table 3); always follow with distilled water to rinse out the disinfectant.

- Whenever a sample tube is put on the instrument, take care that the tube sits securely in the sample introduction port, otherwise, it could be blown off once it is pressurized, and splash sample onto the operator.

- Mostly, samples that are analyzed on flow cytometers are fixed. However, certain flow cytometric applications, e.g., measurement of calcium flux, viable cell sorting, require that cells be run through the instruments unfixed. Flow cytometers are generally too large to operate within a biosafety cabinet, therefore, unfixed samples have to be handled in the open during these experiments. To compensate for the fact that aerosols are not contained within a biosafety cabinet, it is advisable for these experiments, particularly, when handling unfixed human cells, that the use of biosafety level 2 facilities and safety equipment be combined with biosafety level 3 practices (see table 2).[14,15]

- Generally, newer analytic flow cytometers have biosafety features to reduce the risk of operator exposure to instrument-generated sample droplets and aerosols, e.g., an enclosed flow cell, a droplet containment module to prevent sample splashing through back-dripping. In contrast, jet-in-air flow cytometers and deflected-droplet cell sorters pose a risk that operators be exposed to droplets and aerosols which could contain potentially infectious organisms from unfixed samples. In particular, cell sorters generate droplets during their normal operation which can be aerosolized.[10] Aerosol production can increase substantially during instrument failure modes, e.g., a partially clogged nozzle tip, air in the fluidic system. Usually, newer jet-in-air instruments contain features to reduce aerosol production and escape, e.g., a fluid stream catch tube connected to a vacuum line, an enclosed sample collection drawer. Because aerosols that escape into the room create an inhalation hazard to laboratory personnel, particularly, when experiments are done with known biohazardous samples, it is important to test the efficiency of aerosol control measures on jet-in-air flow cytometers and cell sorters. Procedures for testing aerosol containment are available using aerosolized bacteriophage and a detection system of settle plates with E. coli lawns which measure escape of aerosol droplets that rapidly settle from air.[10,11,12,13,14] In addition, procedures for testing room air for escape of aerosol droplets that are smaller than 5 μm using an air sampler are published.[13,14] These small droplets are of particular concern, because in contrast to the larger aerosol droplets which are deposited in the nasal area and pharynx during inhalation, they can reach the lung of the exposed individual. If aerosol containment is incomplete, modify the instrument to achieve such containment. Contact the manufacturer of the flow cytometer for instructions.

- Only experienced flow cytometry operators should perform viable and potentially biohazardous sorting experiments. Always wear personal protective equipment, i.e., a wraparound, disposable, long-sleeved laboratory coat, examination gloves, protective eye glasses, and a High Efficiency Particulate Air (HEPA) N-95 respirator mask whenever unfixed or known biohazardous samples are sorted or analyzed on a jet-in-air flow cytometer. A partially clogged nozzle tip is one of the major reasons for increased aerosol production on flow cytometers. Proper sample preparation, i.e., dispersion of clumps, pre-filtration of the sample, and installation of an in-line filter on the sample introduction line, can aid in

Table 2. Biosafety level 2 containment and level 3 practices[*]

Biosafety levels	BSL2	BSL2 using BSL3 practices
A. Laboratory facilities		
1. Ventilation	Negative pressure	Negative pressure
2. Posted hazard sign	Required	Required
3. Laboratory separated from the general public	Yes, while experiments are in progress	Yes, while experiments are in progress
B. Containment equipment		
1. Biosafety cabinets or other physical containment system	Required for all aerosol generating processes	Required for all work with infectious agents
2. Biosafety cabinet certification	Annually	Annually
3. Other physical containment	Appropriate physical containment devices, e.g. centrifuge safety cups, are used when procedures with a high potential for creating infectious aerosols are being conducted[a]	Appropriate physical containment devices, e.g. centrifuge safety cups, are used when procedures with a high potential for creating infectious aerosols are being conducted[b]
4. Freezers/refrigerators	Biohazard sign must be posted	Biohazard sign must be posted; all agents are stored separate in closed, labeled, containers
5. HEPA-filtered vacuum lines	Recommended	Recommended
6. Personal protective equipment (i.e. laboratory coats, gloves, etc.)	Required – gloves should be worn when skin contact with infectious material is unavoidable	Required – combinations of special protective clothing, gloves, respirator masks, protective eyeglasses, etc., are used for all activities with infectious materials that pose a threat of aerosol exposure[c]
C. Practices		
1. Public access during experiments	Controlled	Not permitted
Decontamination	Daily and upon spills; waste before disposal	Daily, upon finished work with infectious material, and spills; waste before disposal

Table 2. Continued

Biosafety levels	BSL2	BSL2 using BSL3 practices
3. Pipetting	Mechanical devices	Mechanical devices
4. Eating, drinking, smoking and application of cosmetics	Not permitted at any time	Not permitted at any time
5. Handwashing facilities	Required	Required
6. Minimization of aerosol production	Recommended	Recommended
7. Autoclave on-site facility	Must be available within the building	Must be available
8. Insect/rodent control program	Required	Required
9. Bench top work	Permitted	Permitted, not recommended
10. Transport of infectious material	Durable leakproof container	Durable leakproof container
D. Training		
1. Technical training	Required	Required
2. Medical surveillance (i.e. baseline serology)	Required when appropriate	Required when appropriate

[*] This table was adapted from "Working with Biohazardous Materials", Lawrence Livermore National Laboratory (1992)

[a] These procedures include centrifuging, grinding, blending, vigorous shaking or mixing, sonic disruption, opening containers of infectious materials whose internal pressures may be different from ambient pressures.

[b] These procedures include manipulation of cultures and of clinical or environmental material that may be a source of infectious aerosols

[c] Required with aerosol generating equipment, manipulation of high concentrations or large volumes of infectious materials; activity involving all clinical specimens, body fluids and tissues from humans or from infected animals.

clog prevention. Despite this measures, cells can still accumulate within the nozzle tip and cause a fluid stream obstruction. When this occurs, put the fluidic control into the off position. It is critical to keep the instrument's door closed until aerosol has cleared from the chamber before clearing of the clog is attempted. On most flow cytometers aerosol clearance will take approximately three minutes. Before the experiment is continued, make sure the stream exiting the nozzle is straight and stable. For further details on performing viable cell sorting experiments, refer to the recently generated guidelines for sorting of unfixed cells[14] which include recommendations for sample handling, operator training and protection, laboratory design, instrument setup and maintenance, and testing for instrument aerosol containment.

Table 3. Summary information on chemical disinfectants for decontamination[*]

| Disinfectant | Chlorine compounds | Alcohols | | Formaldehyde (Formalin) | Quaternary ammonium compounds |
		ethyl	isopropyl		
Practical requirements					
Use dilution	1/100 dilution of 0.71 M sodium hypochlorite, ~500 ppm (a)	70-85%	70-85%	0.2-8%	0.1-2%
Contact time to lipovirus	10 min	10 min	10 min	10 min	10 min
Broad spectrum	30 min	not effective	not effective	30 min	not effective
Inactivates					
Vegetative bacteria	✓	✓	✓	✓	✓
Lipovirus	✓	✓	✓	✓	✓
Non-lipid viruses	✓	(b)	(b)	✓	
Bacterial spores	✓			✓	

Table 3. Continued

Disinfectant	Chlorine compounds	Alcohols		Formaldehyde (Formalin)	Quaternary ammonium compounds
		ethyl	isopropyl		
Physical characteristics					
Type	Liquid	Liquid	Liquid	Liquid	Liquid
Stability (c)		✓	✓	✓	✓
Corrosive	✓				
Flammable		✓	✓		
Residue	✓			✓	
Organic material inactivated (d)	✓				✓
Potential application					
Surfaces (e)	✓	✓	✓	✓	✓(f)
Instrument surfaces and parts	✓	✓	✓		
Flow cytometer fluid lines	✓	✓			

(a) Available halogen
(b) Variable results depending on the virus
(c) Shelf life of greater than 1 week when protected from light and air
(d) Prior to decontamination cleaning with lipophilic detergent/disinfectant necessary
(e) Work surfaces, dirty glassware, decontamination of fixed or portable equipment surfaces
(f) Usually compatible with optics, but consider and effects on associated materials such as mounting adhesives
* This table was adapted from "Biohazardous Operations", Lawrence Livermore National Laboratory, 1995.

References

1. United States Department of Health and Human Services, National Institute of Health. Biosafety in Microbiological and Biomedical Laboratories 1999; Fourth Edition. Publication No. (CDC) 93-8395.
2. Center for Disease Control, Morbid Mortal Wkly Rep 1987; 36 (Suppl):2S, 3S-18S.
3. Center for Disease Control, Morbid Mortal Wkly Rep 1988;37:377-388.

4. National Committee for Clinical Laboratory Standards 1997: Protection of laboratory workers from infectious disease transmitted by blood, body fluids, and tissue, 2nd ed., document M29-A.

5. Occupational Exposure to Bloodborne Pathogens 1991. United States Federal Code Regulation No. CFR PART 1910.1030.

6. United States Department of Health and Human Services, National Institute of Health 1996; Guidelines for research involving recombinant DNA molecules.

7. Sattar SA and Springthorpe VS. Survival and disinfectant inactivation of the human immunodeficiency virus: a critical review. Rev Inf Dis 1991; 13:430-447.

8. Druce JD, Jardine D, Locarnini SA et al.. Susceptibility of HIV to inactivation by disinfectants and ultraviolet light. J Hosp Inf 1995; 30:167-180.

9. Center for Disease Control, Morbid Mortal Wkly Rep 45:468-472, 1996.

10. Merrill JT. Evaluation of selected aerosol-control measures on flow sorters. Cytometry 1981;1:342-345.

11. Giorgi JV: Cell sorting of biohazardous specimens for assay of immune function. In: Darzynkiewicz Z, Robinson JP, Crissman HA (eds.), Methods in Cell Biology 42, Flow Cytometry, Academic Press, New York, 1994; 359-369.

12. Ferbas J, Chadwick KR, Logar A et al.. Assessment of aerosol containment on the ELITE flow cytometer. Cytometry (Communications in Clinical Cytometry) 1995;22:45-47.

13. Schmid I, Hultin LE, Ferbas J.. Testing the efficiency of aerosol containment during cell sorting. In: Robinson JP, Darzynkiewicz Z, Dean P, Dressler L, Tanke H, Rabinovitch P, Stewart C, Wheeless L (eds.), Current Protocols in Cytometry, Suppl. 1, Unit 3.3, Wiley & Sons, New York, 1997.

14. Schmid I, Nicholson JKA, Giorgi JV et al.. Guidelines for sorting of unfixed cells. Cytometry 28:99-117, 1997.

15. Schmid I, Kunkl A, Nicholson JKA. Biosafety considerations for flow cytometric analysis of Human Immunodeficiency-infected samples. Cytometry (Communications in Clinical Cytometry) 38: 195-200, 1999.

Section 9

Unlocking the Spectrum of Possibilities

ROCHELLE A. DIAMOND

Many years ago, I heard Leonard Herzenberg of Stanford University give a talk to a user's group in San Diego on the history and future of flow cytometry. He ended by stating his goal of someday performing 12-color flow cytometry. "Wow!", I thought to myself, "what power to generate that kind of information from a single sample!" Now a days, most of us in core facilities are routinely performing 3 and 4 color, and some labs are up to 5 and 6 color with a little help from our friends and service providers. Herzenberg's Stanford group however, is making good on their visions and goals – they are now up to 8 colors.[1,2] Utilizing three spatially separated laser beams at 407 nm, 488 nm, and 595 nm, they have detected 8 fluorochromes simultaneously using 10 parameter technology to measure properties of cell populations in both the mouse and human systems.

Why are so many parameters important? Today's biological questions are complex and demanding. Not only do we want to know "who" these cells are, which in some cases may take up to 6 colors, but "what" are they all doing? The ability to perform comparative functional assays for intracellular cytokines, enzymes, and nuclear proteins, in combination with cell cycle analysis and stages of apoptosis, would lead to exponential leaps in our understanding of many biological systems.

What sets these Stanford laboratories apart from the rest of us? The intellectual drive and foresight of these flow pioneers have provided the vision to expand the envelope. They have seen several keys. One key is the direct conjugation of fluorochromes to monoclonal antibodies. Herzenberg's group has made extensive use of conjugation chemistry with not only the traditional fluorochromes – FITC, the phycobiliproteins, and Texas Red – but also with the newer tandem Allo-7 and

Cy5 dyes. Most commercial purveyors of reagents are only marketing direct conjugates of FITC, PE, and biotin in their mouse, rat, and human reagent lines. Several vendors are just beginning to sell the PE-Cy5 in their lines (PharMingen, Caltag, Sigma). In addition, there is limited access to both PerCP reagents (available only as human and a few mouse Mab's) and PE-Texas Red conjugates (only Life Technologies). Some commercial enterprises provide contract services to do these kinds of conjugations, if you provide the antibody or hybridoma cells. For users to push the envelope on a routine basis, we need to push the vision of our vendors to expand their product lines. We can also collaborate with user groups and contract to share costs for commercial conjugations to reagents of mutual interest. Or, we could get our own hands back to the bench and conjugate for ourselves. There are published protocols for conjugating all of these reagents. The priorities of time, cost, and availability will determine how fast we reach the multi-colored vision.

Another key to obtaining the multi-color goal is the use of multiple laser beams. Lasers are becoming cheaper and more versatile every year. Setting up multi-laser benches on commercial research-grade machines is not too hard, as some have room in the bed for 3 laser beams. In the 8 color case, Stanford only needed two lasers – one argon running in all lines and split between 488 nm and a dye head at 595 nm. The other laser was an UV-violet enhanced krypton laser set at 407 nm. The trick here was to have all 3 beams parallel but not collinear. This allows for the signal from each beam to be separated in time from each other and collected via delay timing circuits (gated amplifiers) on two of the beams. The data were then merged into one event. There were of course, other modifications on their hybrid machine as one might expect, to make this all work. Yet it all seems quite possible, especially when commercial enterprises get involved. In Stanford's case, they used Cytomation's Mo-Flo electronics system, which is capable of collecting 10 parameters with 3 separate timing circuits, in conjugation with BD's FACStar Plus Bench.

A third key is software capable of handling and analyzing the kind of data generated by 8 color staining, as well as compensating the complex 8 color matrices. The ability to think in 10 dimensions is mind-boggling. Adam Treister and Mario Roederer authored software generated to analyze in 10-D for the Stanford project. There have been other types of software generated through the years to help with the complexity, like cluster ana-lysis[3],[4] and Prism[tm] (Coulter Cytometry), Paint-a-gate[tm] and At-tractors[tm] (Becton Dickinson), classification systems and regres-sion trees,[5] but real visualization of the data is imperative for interpretation.

So, what does the future hold? With leaps in memory capacity, advances in networking and servers, on-line archives for all this data collection should just be a terminal away as we head into the 21st century.[6]

Tandem conjugations have been helpful in expanding our col-or capabilities, utilizing the cyanine dyes in combination with the phycobiliproteins, but there are other ways of getting around the problems of spectral separations for fluorochromes. I happened to hear Harry Crissman speak at the Southern California Flow user's group meeting in Irvine not long ago, and was inspired by the Los Alamos National Laboratory Flow Cytometry Re-source researchers' use of phase-sensitive detection.[7],[8] The Los Alamos group has built a flow instrument that resolves sig-nals from heterogeneous fluorescence and quantifies the signal component decay times directly on fluorochrome-labeled cells. Their instrument combines flow cytometry and fluorescence lifetime spectroscopy principles by adding an optical modulator, a frequency generator, a high speed detector/preamplifier, and phase detection electronics. In short, the laser intensity is modu-lated at a particular frequency and the modulated fluorescence signals are processed by the phase-sensitive detection electronics to obtain the phase-shift and fluorescence lifetime at the fre-quency used for each fluorochrome on the cell. The Los Alamos group has mainly been focused on DNA-binding fluorochromes such as the Hoechst dyes, DAPI, ethidium homodimer II, 7-AAD, mithramycin, TOTO, and YOYO, as well as FITC.[9],[10] They have found distinguishable signatures for all of these dyes. In addi-tion, they can change the signature with the presence of deuter-ium oxide in the staining buffer to increase fluorescence inten-sity and phase shift. They have also found interestingly enough, that the ability of the probes to interact with the DNA substrate changes these properties. Because one is able to extract this kind of information from a single detector, the combination of the 8-color system with this laundry list of fluorescence lifetime detect-able dyes makes the possibilities tantalizing.[11] Is this going to be the wave of the future?

The ability for high-speed sorting has come to town, not only in the Becton Dickinson TurboSort (25000 cells/sec), but also in Cytomation's dedicated Mo-Flo (100,000 cells/sec). Our ability to sort faster has already opened up new facets in research – detection of fetal cells in maternal blood, purifying pluripotent hematopoietic stem cells, and cloning out rare transfectants from haystacks of negative cells. Cytomation has developed and is commercializing a 4-way sorting system, effectively doubling the number of populations that can be sorted at once, at no increased cost or time spent. Can we increase simultaneous sorting in even more directions?

Other modes of sort collection will become increasingly important, especially for molecular biologists. Researchers have already requested the ability to prescreen gene libraries by sorting *E. coli* into 384-well plates that go directly into robotics for arraying clones on filters for massive screening. The cell size range for analysis and sorting has been expanding over the years. Orifices up to 400 microns are now available for sorting plant protoplasts and the like. Larger ones might be useful for sorting whole organisms like sea urchin plutei or blastula stages of embryos. Micromixing devices and injection devices for fast kinetic studies on flow cytometers are already available will become even more precise.[12] Will detectors eventually become more sensitive with CCD detectors[13] or will diode array technology ever eliminate the need for spectrum separation? Maybe only in our dreams. But dreaming is vision to some people. To unlock the spectrum of possibilities we need only to imagine the possibilities, because as we all know, "if we can see 'em- we can sort 'em".

References

1. Roederer M, DeRosa S, Bigos M et al. 8-color, 10-parameter FACS: elucidation of the complex heterogeneities in leukocyte subsets. Tissue Antigens 1996; 48:485.
2. Roederer M, De Rosa S, Gerstein R et al. 8-color, 10 parameter flow cytometery: elucidation of comple leukocyte heterogeneity. Cytometry 1997; 29: 328-339.
3. Cram LS, Martin JC, Steinkamp JA et al. New flow cytometric capabilities at the national flow cytometry resource. Proc. of IEEE 1992; 80:912-917.

4. Murphy, R. automated identification of subpopulations in flow cytometric listmode data using cluster analysis. Cytometry 1985; 6:302-9.

5. Beckman RJ, Salzman GL, Stewart CC. Classification and regression trees for bone marrow immunophenotyping. Cytometry 1995; 20:210-217.

6. Cell sorting pioneers integrate statistical techniques. Scientific Computing & Automation. February 1996; 14-16.

7. Steinkamp JA, Crissman HA. Resolution of fluorescence signals from cells labeled with fluorochromes having different lifetimes by phase sensitive flow cytometry. Cytometry 1993; 14:210-216.

8. Steinkamp JA, Yoshida TM, Martin JC. Flow cytometer for resolving signal from heterogeneous fluorescence emissions and quantifying lifetime in fluorochrome-labeled cells/particles by phase-sensitive detection. Rev. Sci. Instrum. 1993; 64(12):3440-3450.

9. Sailer BL, Nastasi JG, Baldez JA et al. Interactions of intercalating fluorochromes with DNA analyzed by conventional and fluorescence lifetime flow cytometry utilizing deuterium oxide. Cytometry 1996; 25:164-172.

10. Sailer BL, Nastasi JG, Valdez JG et al. Differential effects of deuterium oxide on the fluorescence lifetimes and intensities of dyes with different modes of binding to DNA. J Histochem Cytochem 1997; 45(2):165-175.

11. Steinkamp JA, Lehnert BE, Keij JF. Phase-sensitive detection as a means to recover fluorescence signals from interfering backgrounds in analytical cytology measurements. Optical Diagnostics of Biological Fluids and Advanced Techniques in Analytical Cytology. 1997; 2982:447-457.

12. Zenin VV, Aksenov ND, Shatrova AN et al. Micromixing-stirring device for flow cytometry studies. Optical Diagnostics of Biological Fluids and Advanced Techniques in Analytical Cytology. 1997; 2982:447-457.

13. Beisker W. Use of CCD sensors in flow cytometry for nonimaging applications. Optical Diagnostics of Biological Fluids and Advanced Techniques in Analytical Cytology. 1997; 2982:415-425.

Appendices

GFP

STEVEN KAIN

Products available from CLONTECH Laboratories, Inc.

Table 1.

Product	Cat. #
GFP Variant Vectors	
pEGFP Vector	6077-1
pEGFP-1 Promoter Reporter Vector	6086-1
pEGFP N-Terminal Protein Fusion Vectors	many
pEGFP C-Terminal Protein Fusion Vectors	many
phGFP-S65T Humanized GFP Vector	6088-1
pEBFP Vector	6068-1
pEBFP N-Terminal Protein Fusion Vectors	many
pEBFP C-Terminal Protein Fusion Vectors	many
pGFPuv Vector	6079-1
Wild-Type GFP Vectors	
pGFP Vector	6097-1
pGFP-1 Promoter Reporter Vector	6090-1
pGFP N-Terminal Protein Fusion Vectors	many
pGFP C-Terminal Protein Fusion Vectors	many
p35S GFP Plant Expression Vector	6098-1
Other GFP Products	
GFP-N Sequencing Primers	many
GFP-C Sequencing Primers	many
GFP Monoclonal Antibody	8362-1
GFP Polyclonal Antibody (IgG Fraction)	8363-1, -2

Fluorescent Proteins Newsgroup

STEVEN KAIN

A newsgroup for the discussion of fluorescent proteins has been created within the bionet hierarchy of Newsgroups. This newsgroup is intended to provide a forum for discussion of bioluminescence, to promote further development of reporter proteins obtained from bioluminescent organisms (e.g., GFP, luciferases, and *Aequorin*), and to facilitate their application to interesting biological questions. (A full copy of the newsgroup charter can be found in the BIOSCI archives.)

We hope you will find this newsgroup useful and encourage you to participate in the discussions.

Discussion Leaders: Steve Kain & Paul Kitts, CLONTECH Laboratories, Inc. *Administration:* BIOSCI International Newsgroups for Biology *Newsgroup Name:* bionet.molbio.proteins.-fluorescent *To subscribe:* If you use USENET news, you can participate in this newsgroup using your newsreader.

You can also access the newsgroup on the World Wide Web using the URL http://www.bio.net/hypermail/FLUORESCENT-PROTEINS/

To receive "The BIOSCI electronic newsgroup information sheet" which describes the BIOSCI newsgroups and gives instructions on how to subscribe via e-mail:

If you are located in the Americas or the Pacific Rim

- Send a mail message to the Internet address:
  biosci-server@net.bio.net

- Leave the subject line of the message blank and enter the following line in the mail message: info usinfo. This message will be automatically read by the computer and you will be sent the latest copy of the information sheet.

If you are located in Europe, Africa, or central Asia

- Send a mail message to the Internet address:
  biosci-server@net.bio.net

- Leave the subject line of the message blank and enter the following line in the mail message: info ukinfo. This message will be automatically read by the computer and you will be sent the latest copy of the information sheet.

Anderson's Timesaving Comparative Guides

Using the World Wide Web at www.atcg.com, researchers can find, compare, and order life science products from hundreds of suppliers and distributors.

According to Anderson Unicom Group, Inc. (AUG), the average researcher can spend between five and ten hours a week sourcing products. AUG helps researchers by cutting that time to just a few minutes to source and compare similar products and prices.

Designed by scientists for scientists, Anderson's Timesaving Comparative Guides allows researchers and purchasing professionals the opportunity to search multi-supplier listings for products by application, catalog number and by text. A link to the supplier's website is standard protocol just in case the researcher wants more information.

Researchers can easily locate products within categories. Examples include: Antibodies, Apparel, Broths and Media, Chemicals, Column Accessories, Columns, Equipment, Filters and Membranes, Gels and Gel Materials, Kits, Libraries, Modifying Enzymes, Nucleic Acids, Photographic Materials, Plasticware, Restriction Enzymes and Vectors.

AUG provides a system built around a database that stores all pertinent product information in a common location. This is the key to the whole system as it brings up product comparisons, pricing, and enough product information for the researcher to make a buying decision.

A typical search for an item takes place like this:

- A researcher enters the site at: www.atcg.com.

- Then clicks on Search for Products.

- Product searches are done by using a product name, choosing a category, or supplying text (example: Taq, EcoR I, etc.).

- A listing of comparable products comes up with pricing and technical data. An added bonus! If the institution allows it, an order can be immediately placed for the product that was just located.

The system can be customized as an Intranet or Internet ordering solution. User profiles are established and ordering processes are defined. Suppliers are added so an institution has access to all their contracted pricing. Products can be searched for by using all suppliers or choosing a specific supplier.

After reviewing and approving the product choice, the researcher can then place an order over the Internet. Flexibility is the key to this system as it allows a company or institution the ability to set up all purchasing protocols that are needed for order placement. After an order is placed, that order and product are stored in a prior order log for easy reordering. It is just that easy.

Save time, money, effort and order products the easy way. Use Anderson's Timesaving Comparative Guides. It does the work for you.

ARC Lamp Flow Cytometers

JEFF HARVEY

Arc lamps have been used as illumination sources in flow cytometers for a good thirty years or so. They are often referred to as
"white light" sources, since their output contains essentially all
of the color components of light, from the UV through the infra-
red. As we know from basic physics, white light is the result obtained by combining all colors of light. Having said that, it must
also be understood that these cytometers never use white light to
illuminate the sample! Instead, a filter is used to select one or
more of the color components of light from the lamp which
best match the absorption requirements of the fluorescent
probes used to label the sample.

One of the greatest strengths of an arc lamp system, in fact, is
that different illumination wavelengths may be selected, simply
by changing the characteristics of the excitation filter. Cells can
be labeled with the fluorescent probes that are optimal for a specific experiment and the illumination wavelength(s) matched to
those probes. Thus, a single cytometer, using a single illumination source can be used to analyze samples labeled with UV-excited fluorochromes or with fluorochromes requiring excitation
with any of the visible wavelengths of light. With the recent availability of double- and triple-bandpass filters, it is even possible to
simultaneously select more than one color of light. Dual excitation wavelength experiments can thereby be done with a single
light source.

The arc lamps that are most suitable for use in flow cytometers
are collectively described as short arc lamps, a term referring to
the small distance from the cathode to the anode in these lamps.
Figure 1 is a drawing of such an arc lamp. The short arc length
results in the very high intensity outputs that these lamps are
capable of producing (confining the arc to a small volume yields
extremely high light density). The arc itself is caused to form by a

high voltage pulse which ionizes the gas, typically xenon gas, mercury gas or a mixture of the two, inside the lamp, allowing current to flow between the cathode and anode. Do not be misled by the fact that the highest light densities are created by the 100W Hg lamp and the 75W Xe lamp respectively. The existence of 200W or 500W lamps reflects only that they consume more power, not that they create higher light densities.

The most commonly used lamps are also the brightest, i.e., the 100W Hg and the 75W Xe lamps are the most commonly used. The 75W Xe lamp has its peak output in the blue region of the visible spectrum, making it an excellent choice for use with fluorescent dyes such as fluorescein. Mercury lamps exhibit very high output at the emission lines that are characteristic of mercury vapor. These occur in the UV, visible violet, green and orange-red regions of the spectrum. These lamps have historically been used most frequently for applications requiring UV excitation. Xenon-Mercury lamps have as their output a combination of the broad emission spectrum of the Xenon vapor, with the strong emission lines of the Mercury vapor superimposed on it.

A characteristic of arc lamps that before worked against their use as an illumination source in flow cytometers was output in-

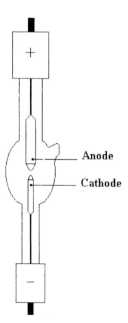

Fig. 1. Diagram of a Typical Arc Lamp

stability. The arc would flicker and/or "wander" (i.e., change position). This could and did lead to occasional poor data resolution and a requirement for frequent realignment. The primary cause of these characteristics was non-uniform degradation of the cathode material. As the position of the shortest path between the anode and the cathode would change, the arc would "jump" to this new position. This would sometimes manifest as flickering and would require aligning the system to the new arc position. These lamps were also short-lived, again due to the rapidity of the cathode degradation, typically averaging less than 200 hours of usable lifetime.

Within the past few years, advances in cathode materials technology have resolved these problems. The new "Super Quiet"^{TM} lamps from Hamamatsu, for example, have an output performance specification rated at +/- 1% in daily operation. These lamps are available in 75W Xe and 75W Xe-Hg forms and also have average lifetimes of 1000 and 2000 hours, respectively. The 75W Xe lamp is intended primarily for applications requiring excitation with visible blue light. The 75W Xe-Hg lamp, alternatively, is intended for applications requiring any of the wavelengths corresponding to the strong emission lines of Mercury vapor (most commonly UV, visible violet or green light). With the improvements in performance exhibited by these lamps, they represent an excellent alternative as a light source for a flow cytometer.

The output from an arc lamp, as previously noted, contains all of the color components of light. The light from the lamp is first collected and focused by a condenser lens. It may be focused through a slit, so that only the highest intensity output of the lamp, that from the arc itself, is allowed to pass toward the sample. A filter is then used to select the color component(s) of the light that will be used to illuminate the sample. This light then passes toward the objective that will focus it onto the sample. The latter is illustrated in Figure 2.

Arc lamp optics typically utilize an epi-illumination format. This means, simply put, that the same objective is used to focus light onto the sample and to collect fluorescent light from the sample. Because it is desirable to focus as much light as possible onto the sample and to collect as much fluorescent light as possible from the sample, the epi-illumination objective must have as high a numeric aperture (NA) as possible. This typically re-

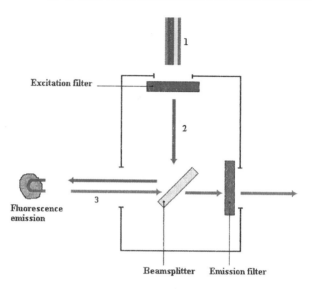

Excitation filter

Fluorescence
emission

Beamsplitter   Emission filter

Fig. 2. Excitation Wavelength Selection in an Arc Lamp Flow Cytometer [the white light emitted by the lamp is composed of different chromatic components (1). The excitation filter transmits only specified wavelengths (2) (in the example, light between 520 and 560 nm, i.e. green light). The beamsplitter is a dichroic filter which reflects the excitation light (actually, all light shorter than its specified wavelength – 570 nm) toward the sample. The fluorescent compounds used to label the sample are excited and fluorescent light emitted. By definition, the fluorescent light (3) has a longer wavelength; in the figure it is represented by the red light. This light is therefore able to pass through both the dichroic beamsplitter and the emission filter in the filter block (the latter transmits light of wavelengths › 590 nm), and is directed toward the fluorescence PMT's]

quires the use of oil immersion optics and, in the case of at least one commercial flow cytometer, a lens with an NA of 1.3. Such an optical configuration is illustrated in Figure 3.

The color component of light that will illuminate the sample is reflected toward the sample by a dichroic (two-color) mirror. The epi-illumination objective then focuses the light onto the sample. The fluorescent light from the sample is collected by this same epi-illumination objective and passed back toward the array of detectors for these signals. Since the fluorescent light is longer in wavelength than the excitation light, the dichroic mirror is designed to allow it to pass through and be directed to the appropriate fluorescence detectors. By using image forming optics, the fluorescent light can be focused through a slit that serves to reduce the amount of background light that is allowed

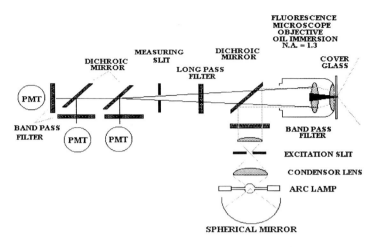

Fig. 3. Epi-illumination Optics in a Commercially Available Arc Lamp Flow Cytometer System (Remove parenthetical note, if possible)

to travel toward the fluorescence detectors. This improves signal-to-noise and facilitates the achievement of quite adequate fluorescence sensitivity with an arc lamp system.

In an arc lamp cytometer, light from the illumination source will be scattered by the sample particles, just as in any flow cytometer. However, because of the use of a high NA lens and immersion optics, measuring the desired angles of scattered light is somewhat more demanding in arc lamp systems than in laser systems. Nevertheless, successful methods for measuring narrow and wide angle light scatter have been developed and are available.

One of these designs was developed by Harald Steen, at the University of Trondheim, in Norway[1]. As Figure 4 illustrates, this design employs a unique darkfield illumination configuration. An appropriately sized dark spot is mounted in the secondary focal plane of the epi-illumination objective. The dark spot casts a conical dark field, one which completely covers the input aperture on the second objective, the one responsible for collecting the scattered light and focusing it toward the light scatter detector array.

This means that, ideally, the only light able to enter the light scatter objective will be the light which is scattered into the dark

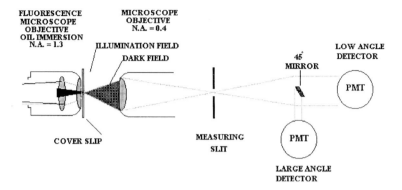

Fig. 4. Light Scatter Optics in a Commercially Available Arc Lamp Flow Cytometer

field by the sample particles. Although it may not be intuitively obvious, the light which is scattered in the large angles (correlating with particle shape or complexity, e.g., granularity) is scattered into the central region of the dark field. the light scattered in the narrow angles (correlated with particle size) falls in the periphery of the dark field.

This enables a quite elegant means for separating the narrow and wide angle scattered light and directing each to a separate detector. A small 45^0 mirror is positioned in the center of the light path, between the two light scatter detectors. This mirror reflects the light scattered in the large angles (and therefore in the center of the field) to one of the detectors and allows the narrow angle scattered light, in the periphery of the field, to pass around it to the other detector. This approach to narrow and wide angle light scatter measurements provides results which are at least equivalent to the common alternative of measuring wide angle light scatter with a separate objective, at 90^0 to the sample interrogation point[1].

As with fluorescence, background reduction is important in light scatter measurements. The dark field greatly reduces the amount of stray light that can possibly enter the path leading to the light scatter detectors. The use of a pinhole or, in this case, a slit, through which the scattered light is imaged, further reduces the amount of background light that is encountered. By combining these techniques with the use of a photomultiplier tube as the narrow angle light scatter detector (this is usually a simple photodiode), it is possible to improve light scatter sensitivity (i.e., signal to noise) to a degree that allows reliable detection of particles as small as 0.2-0.3 microns in size.

A block diagram example of an arc lamp cytometer using this configuration is illustrated in Figure 5. A final point on the optical design of arc lamp cytometers has to do with how the light is focused onto the sample. Most fluorescent microscopes, from which design arc lamp cytometers borrow heavily, use Koehler illumination, to ensure uniformity of illumination over the entire viewing field. With the greatly enhanced stability of the new generation of arc lamps, it has become possible to instead use critical illumination. This technique images the arc of the light source in the sample plane, maximizing illumination intensity. By this approach, one can achieve the best possible fluorescence sensitivity from an arc lamp system, sensitivity that rivals that from laser cytometers.

Arc lamp cytometers may be used for any application in flow cytometry. Historically, they have been most frequently used for DNA content analysis. This has been a consequence primarily of the availability of high intensity UV output from several types of these lamps, combined with the appeal of using one of the UV-excited DNA fluorochromes (DAPI and Hoechst being the most

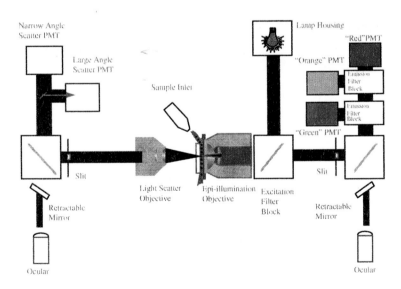

Fig. 5. Optical Layout of an Arc Lamp Flow Cytometer

common). With the recent availability of the Xenon lamps from Hamamatsu, arc lamp cytometers are more frequently being used for those applications which require visible blue light (i.e., 470-490nm) as the excitation wavelength. The appeal of arc lamp systems will continue to focus on the capability to select different excitation wavelengths, particularly as new fluorescent dyes continue to be developed and introduced.

References

1. Steen H B, "Light Scattering Measurement in an Arc Lamp- Based Flow Cytometer," Cytometry. 1990; 11: 223-230.

CompuCyte Laser Scanning Cytometer

SUSAN DEMAGGIO

The CompuCyte Laser Scanning Cytometer (LSC) is a micro-scope slide-based cytometer that provides data equivalent to flow cytometry. The LSC measures 4-color fluorescence and light scatter and records the position and time of measurement of each cell. Cells of interest can be relocated, visually confirmed, restained, remeasured, and photographed. This allows you the advantage of both multicolor fluorescence and morphological information on cells.

The LSC optical configuration is designed to produce a large depth of field with nearly collimated excitation to achieve accu-rate constituent measurements. Optical alignment is fixed and stable. The system is modular, so you can configure it to meet your needs.

Beams from an argon ion laser and/or a HeNe laser are steered to the computer-controlled scanning mirror, through a scan lens and a standard microscope, to create a scan line at the specimen on a microscope slide. Up to 4 photomultiplier fluorescence sen-sors utilizing filters for a variety of wavelengths of light are avail-able. A digital camera is used to visualize cells with bright field illumination or epi-fluorescence. Images can be captured by the PC and stored as BMP files. Light scatter is collected by a solid state sensor, which automatically moves out of the light path for viewing. The use of a computer controlled stage with absolute position sensors allows the stage to step in 0.5 μ increments.

Applications for which the LSC is suited include:

- FISH (Fluorescence in-Situ Hybridization). The LSC automa-tically and rapidly counts the number of FISH probe spots per cell. The instrument isolates and characterizes probe spots within the cell contours based on a second, third, or fourth sensor signal.

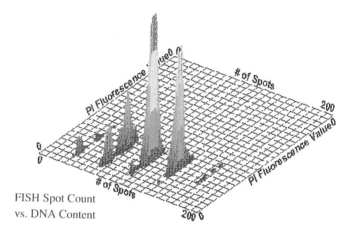

FISH Spot Count
vs. DNA Content

Fig. 1. A bladder cancer specimen was hybridized to a chromosome 17 probe conjugated to FITC and counterstained with PI. The abnormal spot counts correspond to the various spot numbers on the FITC sensor scan display

- Immunophenotyping. The LSC is ideal for performing Immunophenotypic analysis of hematological specimens. The instrument provides simultaneous 4-color immunofluorescent data directly analogous to that obtained via conventional flow

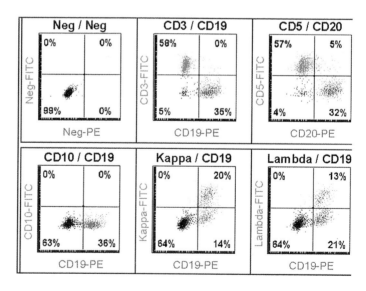

Fig. 2. Histogram displays are similar to conventional Flow cytometry

cytometry. Extremely small and / or hypocellular specimens (body fluids and fine needle aspirations biopsies) can be successfully analyzed. Cells can be microscopically examined at any time - before, during, or after automated immunofluorescent analysis. Specimen preparation techniques are less restricted and more cost-efficient.

- DNA Content. The LSC performs DNA content analysis directly on smear and other slide preparations of solid tissue tumors. Substantial quantities of tissue are not necessary, sample preparation is minimal, cells are available for visual examination both during and after analysis, and thousands of cells can be automatically analyzed so that results – including S phase fraction determinations – are highly significant. The LSC provides positive gating of cells if interest simply by immunofluorescent staining for a relevant antigen. Data is generated in less time and with less expense than other technologies. With extremely small samples, the LSC is able to generate excellent results which could not be obtained with other technologies.

- Apoptosis & Cell Cycle Analysis. The sensitivity, flexibility and ability to study the morphology of cell populations makes the LSC the instrument of choice for research in apoptosis.

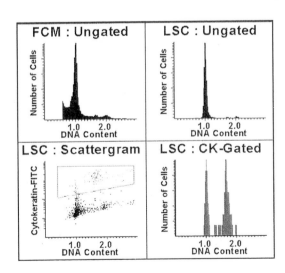

Fig. 3. DNA histograms compare very favorably with Flow Cytometry (FCM) data

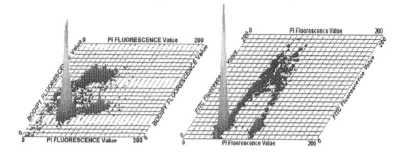

Fig. 4. The LSC has been used to study DNA replication by BrDU incorporation. L\HL-60 cells were pulse labeled with BrDU during cell culture, treated with an antibody to BrDU conjugated to FITC and stained with PI and RNase. Shown here is the resulting BrDU-FITC fluorescence VS. DNA isometric display. HL-60 cells were induced into apoptosis. DNA strand breaks were directly labeled by incubation in deoxynucleotidyl transferace (TdT) and BODIPY- conjugated dUTP. They were counterstained with PI and Rnase. An LSC determination of Bodipy and PI fluorescence is shown here as an isometric display

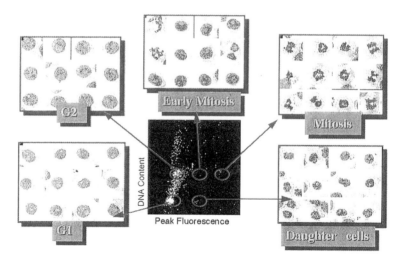

Fig. 5. Shown here is a scattergram of per cell DNA versus peak fluorescence of HL-60 cells stained with PI and RNase. The sample was stained with H&E and data points in each of the five regions were automatically relocated. The cell cycle progresses from recently divided "daughter cells" to resting cells at "G1". DNA is synthesized between "G1" and "G2" and mitotic cells appear in the region between "Early Mitosis" and "Mitosis"

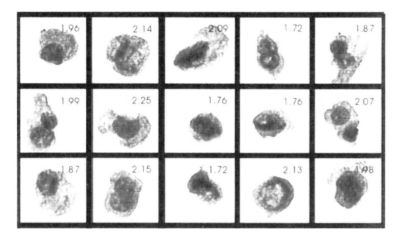

Fig. 6. Specimens were stained with PI for DNA as well as AE₁/AE₃ conjugated to FITC to measure cytokeratins. Images were generated by relocating cytokeratin positive cells with high DNA indices shown on each image. The specimen is from a FNA of a kidney from a patient with renal cell carcinoma. Images were obtained by staining the slide with hematoxylin and eosin, replacing the slide on the stage and relocating the cells of interest

- Cell Cycle & Mitotic Cells. The ability to characterize and confirm cell populations through sensitive immunofluorescence and image capture makes the LSC ideal for cell cycle studies.

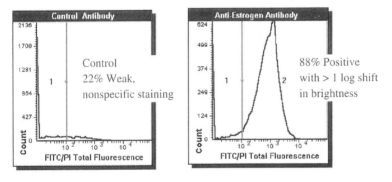

Fig. 7. Breast cancer tissue reacted with FITC-conjugated estrogen receptor (ER) antibody and counterstained with PI can be quantitated by taking the ratio of the amount of antibody staining to DNA content. With the control antibody, only 22% of the cells measured in the following example had nonspecific staining. Eighty eight percent of the patient cells measured displayed fluorescence intensity greater than 1 log shift brighter than the control

- Cytology. The LSC can be utilized in characterizing tumors with images of cells with abnormal ploidy.

- Immunofluorescence can also be used to view the presence of intracellular antigens in paraffin-embedded sections.

- Invasion and migration studies can be performed and quantitated directly on Boyden chamber fittus.

- The LSC can measure fluorescently-labeled secreted proteins and cytokineson Elispots.

- Microbial specimens can be analyzed by the LSC for studies in water quality, blood product contamination, pharmaceutical product sterility, and bacterial identification using FISH.

- Localization and visualization of intracellular constituents such as cytokines, viral particles, organelle specific probes, and nuclear antigens.

CompuCyte Corporation, 12 Emily Street, Cambridge, MA 02139-4507, USA, phone: 617-492-1303

Coulter Flow Cytometry Systems

SUSAN DEMAGGIO

Beckman Coulter EPICS XL Flow Cytometry System

The XL system features the capability to analyze up to 4 color of immunofluorescence from a single air cooled laser. Other multicolor applications include: multiparametric DNA analysis, platelet studies, reticulocyte enumeration, cell biology / functional studies and a broad range of research applications. The instrument is self contained and biohazard safe. The user retains the flexibility to change filter elements for versatility in research settings. For high volume users, the XL-MCL system offers walk-away sample handling for superior efficiency with the Multi Carousel Loader (MCL). Capable of throughputs of up to 100 samples per hour, the fully featured XL-MCL incorporates positive bar-code identification and true vortex mixing prior to aspiration. The XL remains the only flow cytometer to offer state of the art Digital Signal Processing for reliable linearity and drift-free amplification and compensation. The single laser design eliminates concerns regarding multi-beam stability, signal delay and alignment. Compensation is achieved by a unique digital compensation matrix. The new XL SYSTEM II Software fully automates instrument set-up and compensation for 2, 3, and even 4 color applications.

EPICS ALTRA Flow Cytometer Cell Sorter

The EPICS ALTRA flow cytometer is the first of a new generation of cell sorters from Beckman Coulter. It replaces the Coulter Epics Elite ESP cell sorter. This high-performance instrument is a powerful tool for innovative research in immunobiology, cell physiology, molecular biology, genetics, microbiology, water quality and plant cell analysis.

The ALTRA System offers unmatched fluorescence sensitivity with the unique quartz SortSense™ flowcell design, for sense-in-quartz, jet-in-air operation. The system features the capability to analyze and sort on up to 6 colors simultaneously while performing complex multi-parameter applications. These include DNA cell-cycle analysis, quantification, functional studies, chromosome enumeration and physiological measurements.

The standard ALTRA System allows rapid separation of large numbers of specific cell populations with high purity, recovery and yield. Sorting at data rates up to 10,000 cells per second (with air-cooled lasers) and up to 15,000 cells per second (with water-cooled lasers) is possible. The system features nine sort modes. Four modes optimize between purity and yield for two sorted populations, including one special enrichment mode. Three feature the new ALTRASort™ mixed-mode that isolates a high purity population to the left and captures the remainder of the same population to the right. This is ideal when working with rare or precious populations. The AccuSort™ modes are for accurate counting when sorting individual cells for cloning with the Auto-clone® sorting option or for sort matrix verification. The optional HyPerSort™ System incorporates new sorting technology that provides true high speed sorting, while maintaining sensitivity with the SortSense flowcell. HyPerSort technology uses high pressure and high frequency to maximize purity, recovery and yield at data rates up to 30,000 cells per second. The HyPerSort System can be purchased with the ALTRA System or added later to a standard ALTRA System.

Extremely versatile, the ALTRA Cell Sorting System is based on an industry-standard optical platform with an extensive range of laser options, allowing almost any combination of excitation wavelengths. This user-configurable optical design provides unmatched flexibility to satisfy cutting-edge research applications. The ALTRA System includes the Coulter FlowCentre™ workstation, a high performance multi-media platform incorporating Intel® Pentium® technology. The system features EXPO™ Software for instrument control, data acquisition and analysis in a Windows95 environment. The ALTRA System is the only cell sorter to offer all of following in one package:

- High-resolution SortSense analysis and sorting

- Windows98 acquisition and analysis

- HyPerSort System option

- Low-power air-cooled or high-power water-cooled laser flexibility

- Video sort set-up and monitoring

- Autoclone sort option

- Time-of-flight and PRISM parameters

Beckman Coulter EXPO Cytometer Software System

The Beckman Coulter EXPO Cytometer Software System mentioned above provides Windows™ based data acquisition for COULTER® EPICS® Flow Cytometers: all current. Standard Windows tools allow for easy copy and paste into your word processing documents or spread sheets. This research cytometry software works the way you expect, allowing integration of graphics, spreadsheets and text documents. The software has multi-tasking capabilities which let you acquire data and print experiment reports as background tasks while simultaneously analyzing other data or working in another application.

Acquisition and Analysis EXPO Cytometer Software provides full instrument control and data acquisition. Analysis capabilities include flexible protocol design, powerful Boolean gating, color eventing, batch analysis and customized reports. The software can be networked for listmode processing at remote workstations

Coulter EPICS Elite ESP High Performance Cell Sorter

The Elite ESP (Enhanced System Performance) features the capability to analyze and to sort up to 6 colors simultaneously while performing complex multi-parameter applications such as DNA quantification, functional studies, chromosome enumeration or physiological measurements. The cell sorting capabilities of the

Elite ESP allow for the rapid separation of large numbers of specific cell populations using one of the four pre-defined sort modes. A fifth sort mode offers custom sort setup. Greater then 99% purities can be achieved, even at data rates of up to 15,000 cells per second. Individual cell cloning is possible with the Autoclone sorting Option. Extraordinarily adaptable and versatile, the Elite ESP offers a wide-open, industry standard optical platform that is user-configurable. The flexible nature of the optical design, which was developed 7 years ago for the original Elite System, is still state-of-the-art, able to satisfy most every cutting-edge research application. Laser preferences, anywhere from the inexpensive, air cooled lasers to the top-of-the-line high powered water-cooled lasers, can easily fit on the optical bench, allowing almost any combination of excitation wavelengths. The system's ergonomic design occupies a mere 21 square feet of laboratory space. The Elite ESP is the only high speed sorter to incorporate all the following: low-power air-cooled lasers; high-resolution sense-in-quartz analysis and sorting; video sort set-up and monitoring; Autoclone sorting option; time-of-flight; PRISM and Panelizer, all with high-performance Penitum workstation for data acquisition and analysis.

Cytomation Products for Upgrading Flow Cytometers

SUSAN DEMAGGIO

Cytomation was founded in 1988 to provide upgrade systems for existing flow cytometers whose performance became limited by obsolete signal electronics and computers. These retrofit products challenge the performance of their respective manufacturers' latest instruments. Cytomation introduced the MoFlo high performance flow cytometer in 1994. Cytomation's philosophy has been to provide systems with the least risk of obsolescence though modular upgradability hence MoFlo (Modular Flow cytometer).

Summit™

Summit is Cytomations's "Universal" acquisition and analysis software package. Running on the industry standard Windows NT/95/98 operating system, Summit provides multiple layouts, color gating, overlays and data management with "drag & drop" capabilities. In addition to incorporating today's expected "cut & paste" feature for word-processing and spreadsheet reports, it also allows for transparently reading and writing all major PC or Mac based FCS listmode files. It provides complete compensation capabilities with a 8 x 8 matrix. Summit can be used offline for analysis of data acquired from any of the leading manufacturers' flow cytometers.

CICERO

CICERO is a retrofit package that includes the Summit software running on an Windows NT platform. The package includes a customized instrument interface designed to accommodate most of the flow cytometers in common use with minimal differences in the Standard Operating Procedures for each cytometer.

CICERO electronics utilize 16 bit ADCs and run with a dead time of less than 15 U seconds. CICERO's high speed data acquisition, in excess of 15,000 events per second, is most advantageous in the analysis mode by dramatically data loss when studying rare event phenomena.

MoFlo Electronics Upgrade

The FACStar/FACStar Plus fluidics and optical bench can be coupled with the same high performance electronics as used in the MoFlo. This hardware acquisition system, coupled with a Windows NT based user interface unit, replaces any previous computer and provides superior data acquisition, data analysis and control of the FACStar optical bench.

It is possible to add many additional features only found on the MoFlo such as 4Way^{TM} sorting, high density cloning and ratiometric sorting. Through it's use of patented parallel processing technology it provides the FACStar system with the capability for up 3 laser, multi-parameter (10+) analysis.

Glossary

Absorption
Receiving and retaining light of a certain wavelength as in energy or light.

7-aminoactinomycin D (7-AAD)
fluorescent analog of the antibiotic actinomycin D which complexes specifically with G-C rich regions of DNA. Ab_{max} 550, E_{max} 660.

ACF - antibody conjugated fluorospheres
dye encapsulated beads which are conjugated to specific antibodies and are used to reveal cell surface antigens.

Acridine orange - AO
metachromatic dye which binds nucleic acids differentially. Single stranded binding emits a red fluorescence, double stranded binding emits a green fluorescence.

ALLO-7 - allophycocyanin/cyanine7
tandem conjugate dye used for multicolor analysis. $Ab_{max}650$ $E_{max}750$.

Allophycocyanin
fluorescent molecule from algae excited by HeNe laser. $Ab_{max}650$ $E_{max}660$.

AMCA - 7 amino-4methylcoumarin-3acetic acid
a uv excited blue fluorescent dye $Ab_{max}350$, $E_{max}455$.

Annexin V -
detects phosphatidylserine residues on the plasma membrane which have become extenalized due to apoptosis or other disruptive mechanism.

Antibodies
A class of proteins secreted by sensitized B lymphocytes following contact with antigen. Antibodies bind specifically to the antigen that induced their formation.

Argon laser
Most common laser used in commercial flow cytometers. Air cooled argon lasers are of fixed wavelengths (488 nm). Water cooled high output argon lasers are tunable and capable of uv wavelengths as well as visible spectral bands. May be used for pumping dye lasers.

Autofluorescence
fluorescence exhibited by unstained cells due to internal flavin molecules and other nonspecific background fluorescence.

Azide
sodium azide, an inhibitor of metabolism that prevents the cell from capping and shedding cell surface receptors bound with antibody.

Band pass filter
Optical filter which transmits a fixed small spectral band of light.

Barrier filter
Optical filter which blocks a spectral band of light.

Bernoulli drive
disc storage device for computers.

BODIPY
derivative dyes of boron dipyrromethane. Green fluorochrome similar to fluorescein but more pH and photo stable.

Biotin/avidin
the vitamin biotin binds the molecule avidin with high affinity. Biotin is usually conjugated to antibodies or ligands and then revealed with avidin conjugated to fluorchromes.

BrdU- bromodeoxyuridine
nucleic acid base which can be incorporated in place of thymidine into DNA by cells in S phase of the cell cycle. Used as a reagent to determine activated cells and to measure DNA synthesis by quenching other DNA dyes or as an antigen for anti-BrdU antibodies.

BSA - bovine serum albumin
a protein used in staining buffers as a carrier.

BSL - bio-safety level
classification for the type of biology work in relation to hazardous procedures.

Cascade Blue-
reactive derivative of pyrene $Ab_{max}390$, $E_{max}415$ perferred to AMCA which causes cross-talk with fluorescein.

Calcium Flux
the binding of a ligand to a cell surface receptor provokes a rapid influx of
Ca^{2+} ions accompanied by the release of intracellular membrane bound
Ca^{2+} resulting in a temporary rise in free cytoplasmic Ca^{2+}.

Channel
The number value of a parameter which usually signifies the intensity of a
signal. The bin number which the digitized signal is assigned according to
intensity.

Chromomycin A3
fluorescent antibiotic which binds to G-C rich regions of DNA.

Coefficient of Variation (CV)
The standard deviation of the data divided by the mean of the data, typically
expressed as a percentage. When applied to channel data measured on a
population of cells, the CV is a measure of variation independent of the
population mean.

Compensation
The amount of signal to be subtracted from the spectral overlap of signals
between two photodetectors. See fluorescence compensation.

Confluent
cells grown to stationary phase by crowding in tissue culture.

Conjugated
joined or connected by chemical derivatization.

Conjugated Antibody
An antibody bound to biotin or a fluorochrome.

Contour Plot
A graphical presentation of two parameter data in which contour lines show
the number of events. Similar to a topographical map, the contour lines
show event frequencies as peaks and valleys.

CRBC
circulating red blood cells.

Cy3 - cyanine3
derivative of indocarbocyanine, fluorescent dye $AB_{max}545$, E_{max} 565.

Cy5 - cyanine 5
derivative of indocarbocyanine, fluorescent dye $Ab_{max}640$, $E_{max}670$. Used in
tandem conjugate with R-PE to make an excitable dye at 488 which emits at
670.

Cy7 - cyanine7
derivative of indotricarbocyanine, fluorescent dye used in tandem conjugates with either R-PE or APC. Ab_{max} 750, E_{max}800.

Cytosolic rise
rise of ions in the cytoplasm of a cell.

DAPI - 4',6-diamidino-2-phenylindole
A-T specific DNA stain excited by UV light Ab_{max}345 E_{max}460.

Data Acquisition
The electronic and software function of collecting data from a cytometer's sensors, filtering the data (live gating), and storing it in the cytometer's computer.

Data Analysis
The software function of numerically and graphically manipulating data to generate statistics and graphics.

Data File
A collection of measured values from a single sample combined with text describing the sample, that have been stored to disk. See also List Mode File.

Dendritic Cell
immune cell known to present antigens.

Dichroic Mirror
A mirror made of rare earths which will transmit and reflect light of particular wavelengths.

$DiIC_1$ (3), $DiOC_6$ (3)
orange and green fluorescent dyes used to measure membrane potential.

DMSO – dimethyl sulfoxide
organic solvent which makes membranes more soluable.

Dot Plot
A graphical means of presenting two parameter data. Each axis of the plot represents values of one parameter. A dot represents each event (particle).

Doublet
Two beads or cells in close proximity or adhering to each other. Some cytometers use electronics and software to distinguish doublets from large single cells.

Drop delay
The distance as judged by time for a cell to travel from the point of laser interrogation to the droplet breakoff point in a cell sorter.

Eosin Y
Red acid dye used to measure viability by microscopy. The intact cell membrane of live cells excludes this dye.

Ethidium bromide - EB-
DNA dye which intercalates between bases of double stranded nucleic acids.

Epitope-
antigenic determinant, a molecular region recognized by an antibody.

Event
A unit of data representing one particle detected by the cytometer. Each event may consist of several parameters measured for that particle.

Event Rate
The number of particles detected per second. If the threshold is set to zero, the event rate is equal to the number of noise pulses plus number of particles flowing past the laser beam each second.

FACS- FCM
fluorescent activated cell sorter, flow cytometer.

FACS-Gal
assay for the reporter gene lac-Z which produces β-galactosidase.

FALS - forward angle light scatter
Light that is measured as a particle passes in front of a laser beam. Measures relative cell size and is related to the cell cross section and refractive index.

FBS -
fetal bovine serum.

Fc-receptor
cell surface receptor which binds the Fc portion of the antibody molecule.

FCS
Flow Cytometry Standard. A standard format for flow cytometer data files.

FDG - fluorescein di-galactoside
substrate used to measure the amount of β-galactosidase enzyme which cleaves FDG into fluorescein and galactose in reporter transfected cells.

Ficoll-Hypaque
polymer gradient medium for separating live from dead cells by centrifugation.

Filter
An optical element that allows only light of selected wavelengths to pass through. Typically a glass disk. A filter may work by absorbing (colored glass filters) or reflecting (dichroic or interference filters) unwanted wavelengths. Short-pass filters pass light below a specified wavelength. Band-pass filters pass light within a specified narrow range of wavelengths. Long-pass filters pass light above a specified wavelength.

FISH
fluorescent in situ hybridization.

Fixed
to make cells rigid and preserve them from autolysis.

Flow Cell
The structure that contains the sample stream. The flow cell is designed to hydrodynamically focus the stream, transmit the incoming laser beam and, exit scattered fluorescent light.

Fluorescein
fluorochrome excited at 488 nm which emits at 530 nm.

FITC - fluorescein isothiocyanate
reagent used to conjugate fluorescein to proteins.

Flow Cell
The structure that contains the sample stream. The flow cell is designed to hydrodynamically focus the stream, transmit the incoming laser beam and, exit scattered fluorescent light.

Fluorescence
The phenomenon of light emission that occurs when a fluorochrome absorbs a wavelength of light which excites its electrons to drop to a lower energy level.

Fluorescence Compensation
An electronic circuit used to reduce the unwanted fluorescence of one fluorochrome from another fluorochrome's signal output.

Fluorochrome
A fluorescent dye. A molecule capable of absorbing light energy and then emitting light at a longer wavelength (fluorescence) as it releases this energy.

Gain
Amplification of a signal. Increasing the gain of an electronic device results in a larger output signal for a given input signal.

Gate
Numerical or graphical boundaries that define a subset of data from a cytometer. Gates may be single or multi-dimensional. You may set gates after data acquisition to analyze the data by subpopulation. You may set gates before or during data acquisition to select data (see Live Gate). You may set gates to sort cells which fall within it.

G_0
resting stage of the cell cycle.

G_1
activation stage of the cell cycle prior to DNA synthesis.

GFP - green fluorescent protein
green jelly fish protein used as a reporter gene product.

HBSS/BSA
Hank's balanced salt solution with bovine serum albumen.

Histogram
A graphical means of presenting single parameter data. The horizontal axis of the graph represents the increasing signal intensity of the parameter, and the vertical axis represents the number of events (particles).

Hoechst 33258
UV excited blue fluorescent dye which binds A-T rich regions of DNA.

Hoechst 33342
same as 33342 except that the dye penetrates living cells and can be used for cell cycle analysis of viable cells.

Immunofluorescence
Fluorescence from fluorescent antibodies attached specifically to antigen sites on a cell.

INDO-1
UV excited dye used to measure calcium flux.

Ionomycin
calcium ionophore used to trigger calcium flux in cells.

Isotype Control
A sample stained with an irrelevant antibody matched to the fluorochrome of the specific antibody as well as its species, class, and subclass of immunoglobulin. It is run in conjunction with a panel to assess fluorescence due to nonspecific staining and autofluorescence.

Laser
Light Amplification by Stimulated Emission of Radiation. A light source that is highly directional, monochromatic, coherent, and bright. The emitted light is in one or more narrow spectral bands, and in most lasers is concentrated in an intense, narrow beam. The argon-ion laser commonly used in cytometers emits blue light at 488 nm.

LDS-751
Styrl 8 - DNA stain excited by 488nm emitting ›640nm. This dye will enter intact cells to a lesser extent than dead cells for live/dead discrimination.

List mode file
A file of unprocessed data containing all measured parameters for each event (cell) in the sample according to the order that the data was collected.

Linear Amp
A linear amplifier produces a signal output proportional to the input signal amplitude. For example, a linear amp could have output varying from 1 to 5 V as the input signal varies from 0.01 to 0.05 V.

Live Gate
A type of numerical or graphical boundary defining a subset (region) of data. A live gate is applied to a data plot before acquisition. You may choose to acquire and store data inside or outside a live gate.

Log Amp
A logarithmic amplifier producing a signal output proportional to the logarithm of the input signal amplitude. A four-decade log amp, for example, will have an output varying from 0 to 10 V as the input signal varies by a factor of 10,000. Log amps are useful when analyzing samples containing some cells hundreds of times brighter than others.

Logical Gates
Combinations of one or more previously defined regions which define a subset of the total sample population. You use the logical operations AND, OR, and NOT, along with parentheses, to create logical gates.

Long Pass
Optical filter which blocks light shorter than a particular wavelength and transmits light greater than that wavelength.

MACS -
super para-magnetic bead/column system for enriching or separating cell populations of interest.

Mean Channel
The average channel of a sample from a population of cells.

Monoclonal Antibody
Laboratory-produced antibodies all having identical molecular structure and specificity.

Parameter
A measurement of a cell property as it passes through the cytometer's laser beam. Each parameter is the output of a single photomultiplier or photo-diode, measuring fluorescent or scattered light.

PBS
phosphate buffered saline.

PE - phycoerythrin
fluorochrome which is excited at 488nm and emits at 580nm. Used as an absorber in tandem conjugates.

PE-Cy5
tandem conjugate of the two dyes R-PE and Cyanine5 which excites at 488 nm and emits at 670 nm. Also known as Quantum Red, Cychrome, and Tricolor.

PE-Texas Red
tandem conjugate of two dyes R-PE and Texas Red which is excited at 488nm and emits at 613nm.

Peak
The maximum number of events in a single channel, measured within an analysis gate or series of gates. "Peak channel" refers to the channel number at which the peak is located. Peak also refers to the maximum value of a signal pulse.

PerCP- peridinin chlorophyll protein
Fluorochrome with Ab_{max} 490, E_{max}680. Photobleaches rapidly but little crosstalk with R-PE.

Photodiode
A solid-state device for measuring light intensity. Like PMT's, photodiodes generate an output current proportional to the incident light intensity. Though smaller, cheaper, and simpler to use than PMTs, they are not as sensitive.

Photomultiplier Tube (PMT)
A device for measuring light intensity. PMTs produce an output current proportional to the intensity of incident light. In some applications, PMTs are sensitive enough to count individual photons of light.

Phycobiliprotein
family of pigments found in algae.

PI - propidium iodide
nucleic acid intercalating dye used for DNA measurement in conjuction with RNase.

PKH
tracking dye - lipophilic dyes which bind to cellular membranes to track cells and their offspring.

Pulse
A signal that momentarily changes value. As a particle moves through the cytometer's laser beam, the resulting fluorescent and scattered light will appear as pulse signal. This photodetector signal will have low values as the cell enters and leaves the beam and a peak value when the particle passes the beam center.

Pulse Area
The mathematical integral of a single event pulse; the area under the pulse curve. The pulse area represents the total light emission, and will be proportional to the amount of fluorochrome on a particle's surface.

Pulse Height
The peak height of a single event pulse. If the illuminating beam width is greater than the particle size, the pulse height will be proportional to the amount of fluorochrome on the cell surface.

Pulse Width
The width of a single event pulse. The pulse width is proportional to the time the particle traverses the beam and is a measure of particle size and stream velocity.

Region
A boundary drawn around a subpopulation of events on a plot. Combine regions to create logical gates.

Resolution
A measure of a cytometer's ability to distinguish between two populations of particles with differing fluorescent or scatter intensity.

Rhodamine 123
cationic dye known to stain mitochondria and measure membrane potential.

RPMI
medium used for bone marrow derived cells.

Sample
The cells or particles being measured by the flow cytometer.

Sensitivity
A measure of a cytometer's ability to distinguish particles from background noise. Sensitivity is often expressed in terms of a minimum number of fluorochromes per particle required to clearly distinguish a stained particle from an unstained particle. Sensitivity will depend on the instrument, the dye, and the preparation method.

Sheath fluid
Cell-free fluid used in flow cytometer, which surrounds the central sample stream. Sheath fluid is usually a buffered saline solution.

Short Pass
Optical filter which blocks light at wavelengths longer than a particular wavelength and transmits light shorter than that wavelength.

Side Scatter (SSC)
Also called 90-degree, orthogonal, right-angle, or wide-angle scatter. Light scattered by a particle at approximately 90° from the incident beam. Intensity of side scattered light is related to the internal structure (granularity) of the particle.

Snarf-1
Intracellular pH indicator fluorochrome.

Stem Cell
primal cell from embryonic or self-renewing sources which give rise to cells that differentiate to their end stage of development.

Thiazole Orange
dye known to bind RNA, used especially to measure reticulocytes.

Threshold
A level discrimination to electronically eliminate unwanted data. Some of the particles passing through the cytometer laser beam may be debris or cell fragments. You can set a threshold to avoid collecting data for these events. Only events with parameter values above the threshold will be collected.

Thymocytes
lymphocytes (white blood cells) residing in the thymus.

TO-PRO
monomeric cyanine dyes that stain nucleic acids.

TOTO-1
DNA binding dye - analog of thiazole orange.

Trigger Signal
The parameter used to indicate the presence of a cell or particle at the interrogation point. Forward scatter is typically used as a trigger signal. Signals from other detectors are acquired only when the trigger signal is activated. A threshold signal must be attained to trigger the signal.

Trypan Blue
Blue acid dye used to measure viability by microscopy. The intact cell membrane of live cells excludes dye.

Zip drive
disc storage device which holds 100 Mbytes of data.

Lasers in Flow Cytometry

PAUL GINOUVES

Introduction

Flow cytometry and lasers are practically synonymous. Although flow cytometers can be very useful and commercially successful without a laser – the simple and ubiquitous Coulter Counter is undoubtedly the best example – virtually all of today's research-grade flow cytometers incorporate at least one laser. In these cutting-edge systems, the laser provides the fundamental pathway for information extraction. Unfortunately, the laser is usually the single most expensive and perhaps most misunderstood system component.

This chapter deals with the laser in the modern research flow cytometer/cell sorter. Among the topics covered are fundamental laser physics, laser types, output wavelengths, operating characteristics, and simple maintenance. It is not the intent of this section to present a comprehensive overview of either laser technology or the dynamics of laser-fluorochrome interactions. There are a number of fine texts available that cover those subjects (see Appendix). It is, however, a brief overview designed with the biological user in mind, i.e., what fundamental laser knowledge is useful to a beginning flow cytometrist? Further, even though solid-state lasers are discussed, the focus is on ion lasers, since this type of laser represents over 99% of the installed base of flow cytometers. Lastly, no purchase recommendations for a specific type of laser or cell sorter are made – these are complex matters which are better decided after detailed discussions with your cytometer manufacturer of choice.

Background

The acronym LASER stands for Light Amplification by Stimulated Emission of Radiation. It was coined in 1957 by Gordon Gould, a co-inventor of the technology, and key patent holder for a variety of fundamental laser applications. As the date would indicate, laser history is not deep. The ion laser has been commercially available for only about 30 years. Viable low-power argon-ion lasers suitable for flow cytometry have a much shallower history, as they emerged only in the past 15 years or so. As such, in the 1980's it was quite common to have high-power water-cooled ion lasers in clinical settings. Now, the air-cooled platform completely dominates clinical instrumentation.

Almost all lasers in use in flow cytometry today are ion lasers, where ionized noble gas is the gain medium which produces coherent radiation. Other laser types include diode lasers, diode-pumped solid-state lasers, and dye lasers. Solid-state lasers are just beginning to gain a foothold in cytometry, most notably as a replacement for the 632.8 nm line of the Helium-Neon laser.

Fundamental Laser Operation

The primary components of an ion laser are the power supply, the laser head, and the plasma tube/resonator structure. The power supply transfers energy, in the form of electricity, to the plasma tube, which contains inert gas. The laser head provides a stable platform for the plasma tube and resonator. The resonator structure allows for the amplification of photons into a coherent beam.

As you know from basic physics, electrons orbiting the nucleus of atoms occupy discrete energy levels. Electrons normally occupy the lowest available energy levels – this is the atom in its ground state configuration. Electrons can occupy higher energy levels, but this leaves the lower energy states either sparsely populated or completely vacant. Electrons can make a jump in level with the absorption of energy. This absorbed energy can arise from a number of sources – electrical or light, for example.

When an electron absorbs energy from a photon, the electron jumps to a higher, excited energy level. The amount of energy required to move an electron from one energy level to another is quanticized – only a specific amount of energy is accepted by the receiving electron. Thus only photons which match this specific energy are absorbed.

Atoms in their excited state do not want to stay that way. They find a way to return to the ground state. The laser takes advantage of one mechanism of relaxation – spontaneous photon emission. Excited atoms release photons which have energy exactly equal to the difference between the excited and non-excited states.

The laser capitalizes further on spontaneous emission by allowing a spontaneously-emitted photon to interact with another excited atom, liberating another photon. The original photon remains, so where there was one photon, now there are two. These two photons then can trigger more photons. This process is known as stimulated emission.

In the laser, stimulated emission causes amplification of photons along a single optical axis. This axis is established by creating an optical pathway, or cavity, which is flanked by reflective mirrors (Fig. 1). If the number of photons traveling along the optical axis is sufficient, lasing occurs. In the optical cavity, the

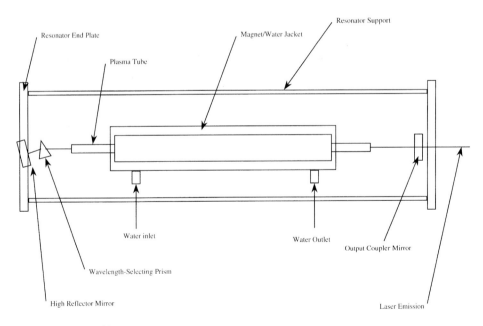

Fig. 1. Drawing of laser resonator

high reflector (rear mirror) is designed to fully reflect wave-lengths of interest. The output coupler (front mirror) is purposely designed to "leak" a small percentage of the amplified light. This leakage is the laser beam which exits the laser head.

Electrical energy is provided to the plasma tube to both initiate and sustain the emissive process. Ion lasers are notoriously inefficient – they typically convert only a small fraction of input power into light energy. The rest is dissipated as heat, which must be removed either with cooling water, or in the case of the air-cooled laser, with air.

Laser Variety

As there are just three major commercial suppliers of high-end research-grade cytometers, namely Becton Dickinson, Beckman Coulter, and Cytomation, it's fairly easy to quantitate the breadth of product offerings for laser excitation. Water-cooled lasers have traditionally dominated the research market because of their tunability, high output power, spectral bandwidth, and general flexibility. In addition, traditional cell sorter designs utilize a jet-in-air sensing scheme to allow for droplet formation and subsequent deflection in an electric field. Sheath flow rates of tens of meters per second in these designs mean that the transit time in the beam is quite short (on the order of a few microseconds), and high laser power is required to generate the best fluorescence signal. With the development of a quartz-enclosed interrogation point which also allows for sorting, air-cooled primary excitation became practical in a cell sorter.

It's rare to see a flow cytometer that is not equipped with an argon-ion laser. This is no real surprise, since the most frequently used fluorochromes are excitable with the 488 nm emission line of argon. In commercial cytometers, 488 nm takes the form of either the low power Uniphase or Spectra-Physics air-cooled system (nominally around 15-20 mW output power), the Coherent Enterprise II 621 simultaneous UV/488 (150 mW 488 nm), the Coherent Innova 305 (1.5 W 488 nm), or the Coherent Innova 70 (1.3 W 488 nm).

Laser Characteristics

What is a laser beam? A laser produces a narrow beam of mono-chromatic coherent light. Coherence (waves traveling in phase) is the best known property of the laser beam. (Stimulated emission is in phase with the photon that accomplishes the stimulating, and as stimulation builds, it remains in phase. Coherence is a laser beam is not perfect, but it is much better than other light sources.)

A laser beam disperses only over large distances. Thus, even a low-power beam can be useful (and potentially hazardous to the eyes and skin) over significant distances. Some lasers are designed to emit only a single wavelength – the common red Helium Neon laser at 632.8 nm is a good example (although HeNes are available at other wavelengths, one laser cannot deliver more than a single wavelength). Other lasers, like the water-cooled argon laser, deliver many wavelengths, with some packages delivering UV and visible wavelengths simultaneously.

Although the argon laser can produce many wavelengths simultaneously, it is mainly useful for flow cytometry only when a single wavelength is lasing. This is accomplished by the insertion of a low-loss prism assembly into the optical cavity. The prism refracts the light in the cavity such that only a single wavelength satisfies the condition of being perpendicular to the face of the high reflector mirror. Only this wavelength then lases.

Fixed single-wavelength lasers have mirrors which are optimized for one output wavelength. A tunable laser, like the water-cooled argon, has mirrors which allow lasing at most of the transition wavelengths. With very few exceptions, UV wavelengths and visible wavelengths are achieved with two different sets of mirrors which are interchangeable by the user.

Multi-wavelength mode (sometimes called multi-line mode) can be useful in flow cytometry when a dye laser is used. Optical pumping of a continuously-tunable dye laser is usually accomplished by an argon laser which produces most of its power at several closely-spaced wavelengths. The output power of the dye laser depends more on the input power of the pumping laser than its wavelength when the majority of the pumping power is close to the absorption maximum of the lasing dye. Thus the user gains efficiency with the pump laser in multi-line mode.

Multi-line UV (350.1-363.8 nm) is also commonly used in the excitation of a variety of DNA-specific probes and other UV-excited fluors such as the calcium probe INDO-1. For probes of this type, there is usually little resolution penalty for excitation at more than one wavelength simultaneously.

Another multi-line setup which is in common use is simultaneous lasing at UV and 488 nm. Maximum usefulness from this type of laser is achieved by separating the two wavelengths externally to the laser head. All three major manufacturers offer such an arrangement for the Coherent Enterprise II 621 laser, which is specified at 50 and 150 mW at multi-line UV (350.1-363.8 nm) and 488 nm respectively.

The laser beam from most lasers used in flow cytometry has a nearly gaussian transverse emission mode – this is otherwise known as a TEM_{00} profile. (Laser resonators have two distinct types of modes: transverse and longitudinal. Transverse modes affect the cross-sectional beam profile – its intensity pattern. Longitudinal modes correspond to discrete resonances along the length of the laser cavity. These modes occur at different frequencies or wavelengths within the gain bandwidth of the laser.) Any TEM_{00} beam is diffraction-limited, that is, it can be focused to nearly the smallest theoretical spot size for that wavelength. This theoretical smallest spot size is always smaller than the focused beam profiles required for flow cytometry. Beam parameters affect the performance of your cytometer, as beam focusing optics in flow cytometry are designed to match the laser wavelength, beam diameter, and intended use. Since not all beam-shaping optics are useful for all laser wavelengths and beam diameters, be sure to consult your cytometer manual for details about the beam-shaping optics for your particular cytometer.

Table 1. Output Wavelengths and Representative Powers for Different Laser Types

Argon		Krypton		Mixed-Gas		HeNe	
λ	Power (W) (1)	λ	Power (W) (2)	λ	Power (W) (3)	λ	Power (W)
1090.0	0.050	793.1-799.3	0.030	752.5	0.030	632.8	0.010
528.7	0.350	752.5	0.100	647.1	0.250	612.0	0.003
514.5	2.000	676.4	0.150	568.2	0.150	594.1	0.003
501.7	0.400	647.1	0.800	530.9	0.130	543.5	0.003
496.5	0.600	568.2	0.150	520.8	0.130		
488.0	1.500	530.9	0.200	514.5	0.250		
476.5	0.600	520.8	0.070	488.0	0.250		
472.7	0.200	482.5	0.030	476.5	0.100		
465.8	0.150	476.2	0.050	457.9	0.030		
457.9	0.350	413.1	0.300	350.7-356.4	0.050		
454.5	0.120	406.7	0.200				
363.8	0.140	356.4	0.120				
351.1	0.140	350.7	0.250				

HeCd		Dye		Diode-Pumped Solid State		Diode	
λ	Power (W) (4)	λ	Power (W) (5)	λ	Power (W) (6)	λ	Power (W)
442	0.030	570-650	1.250	1064	0.080-6.0	980	2.000
354	0.030			532	0.005-10.0	808	15.000
325	0.030					670	0.500
						650	0.015
						635	0.010

NOTES:

All wavelengths are in nanometers.

1. Figures representative of 5.0 W multi-line output power.

2. Figures representative of 1.0 W multi-line output power.

3. Figures representative of 2.5 W multi-line output power.

4. Representative commercially-available power.

5. Based on 25% conversion of 5.0 W pump energy.

6. Figures represent available continuous-wave power range. Not necessarily from the same device.

The Air-Cooled vs. Water-Cooled Debate

Over the past 5 to 7 years, one of the biggest questions for re-searchers considering the purchase of a cell sorter has been: air-cooled or water-cooled lasers? There are a number of tangible advantages for each either, so how does one decide? The answer is, unfortunately, not so simple – the laser is but one component in a complicated system. The user needs to consider many factors in the purchase of a cytometer, not the least of which include data display and processing, sorting capability, number of parameters, upgradability, etc.

As a rule, you should use the lasers which have been qualified, tested and incorporated by the instrument manufacturer of choice. The purchase of a laser outside this framework is almost always problematic, as the instrument manufacturer will not provide technical support or field service for that laser. Additionally, since all the major manufacturers provide special mounting hardware and optics which optimize the laser to the cytometer, you will need to concoct these on your own if you buy directly from an unsupported manufacturer. This can be a time-consuming and costly venture.

With a very few exceptions (and none of them are in use in flow cytometry), air-cooled lasers are not tunable. They have fixed mirrors which provide for one (ex 448 nm), or perhaps a narrow spectrum of output (ex. 488-514.5 nm). A single-wavelength design has some advantages. It allows the laser manufacturer to design a laser with fixed mirror mounts and a sealed cavity which makes for completely hands-off operation and absolutely no maintenance. Further, air-cooled lasers are in general inexpensive, reasonably reliable, and because they plug into a standard wall outlet, ideal when there's no access to three-phase 208VAC and cooling water.

Low power air-cooled lasers must be used in conjunction with an optical scheme which provides very high sensitivity, as air-cooled lasers produce much less excitation power than their water-cooled brethren. Many researchers feel that air-cooled lasers limit experimental flexibility. This is probably true if the lab likes to experiment with a variety of fluorochromes, develop new protocols from scratch, and/or sort cells at high speed. High-power water-cooled lasers have the advantage of tunability, and of course, power to spare. End-users need to consider their situation carefully to determine whether or not air-cooled or water-cooled lasers represent the best solution for their particular usage pattern.

Air-Cooled Ion Lasers

Low-power air-cooled argon-ion lasers in use in flow cytometry are manufactured by Uniphase or Spectra-Physics. The Cyonics and Spectra Physics lasers have a well-deserved reputation for longevity and reliability. These air-cooled devices are nominally 25 or 30 mW lasers at full operating current, but are specified and run by the cytometer manufacturer at 15 mW, both to increase stability and lifetime, and because the fluorescence collection optics are sensitive enough to allow for the lower laser power. (As mentioned previously, low-power air-cooled lasers are generally useful only when a sense-in-quartz optical configuration is employed.)

Helium-Neon (HeNe) lasers are available at 632.8, 612, 594, and 543 nm, although only the 632.8 nm wavelength has found wide acceptance into flow. This is mainly due to the fact that the 612, 594, and 543 nm lines are extremely low-gain, and thus provide only a few mW of output. The major manufacturers of HeNes are Spectra Physics, Uniphase and Melles Griot. Becton-Dickinson and Cytomation both offer the Spectra Physics Model 127 HeNe which provides 35 mW at 632.8, since a lower power HeNe would be of little use in a jet-in-air configuration.

Air-cooled Helium-Cadmium (HeCd) lasers are available from Melles Griot and LiCONiX, and have long been used as a substitute for argon UV. The HeCd emits in the UV at 325 or 354 nm, nominally at 35 mW output.

Water-Cooled Ion Lasers

Virtually all water-cooled ion lasers sold into the flow cytometry market are manufactured by Coherent. Coherent established its reputation in the flow market with the venerable Innova 90, as more than 1000 were installed into the flow market during the 1980s. Water-cooled lasers used to have a reputation for poor plasma tube reliability, and the reputation was perhaps deserved early in the 1980s, when UV plasma tube lifetimes could often be measured only in hundreds of hours. With advances to plasma tube design and construction over the intervening years, water-cooled plasma tubes now have average lifetimes of greater than 6,000 hours for visible operation, and greater than 3,000 hours when run continuously in the UV.

A variety of water-cooled lasers are available to suit the needs of the flow user. The most popular water-cooled laser sold today is the Coherent Enterprise II 621. This laser emits a single beam of UV and 488 radiation, useful for routine immunophenotyping and UV-excited DNA analysis, or both simultaneously. The Enterprise is a hybrid of sorts, since it incorporates many of the convenience features of an air-cooled laser: no required maintenance and hands-off operation. Although water is required to cool the plasma tube, the amount of heat produced is about 5 kW, which is small enough to dissipate with a compact water-to-air heat exchanger. This eliminates the need for an external water source. Further, the Enterprise, unlike almost every other water-cooled laser, requires only single-phase 208 VAC, which ensures its worldwide compatibility.

The Innova 300 is popular for high-power tunable UV and visible excitation, and it is available in both argon and krypton versions. Krypton models allow accessibility to the violet lines 406.7 and 413.1 nm, as well as 568.2 and 647.1 nm. The Innova 300 is available in packages with up to 10 W of output power, although flow users seldom require more than 6 W multi-line.

The Innova 70 is a utility-grade laser which is useful for basic excitation or for dye laser pumping. It is available in a mixed-gas (argon+krypton) configuration, the Innova Spectrum, which is tunable across the argon and krypton lines.

Dye Lasers

Dye lasers were for years the excitation source for Texas Red, as they could be easily tuned to the 596 excitation maxima of that fluorochrome. The popularity of the dye laser has in recent years waned as more researchers chose red-excitable dyes over Texas Red. Dye lasers utilize the strong fluorescent emission of a dye within a resonant cavity to produce a laser beam. They require optical pumping with a high-power ion or solid-state laser, and are continuously tunable (utilizing a birefringent element) over a fairly broad range. The most common laser dye used in flow cytometry, Rhodamine 6G, is tunable from approximately 575 to 625 nm when pumped by the visible lines of an argon ion laser. Conversion efficiency is about 25%, in other words, 5 W of argon in delivers 1.25 W of 596 nm from the dye head. The only dye head being sold commercially into the flow cytometry market is the Coherent CR599.

Solid-State Lasers

Diode lasers are just beginning to find a foothold in cytometry. The advent of higher powered (~10 mW) 635 and 650 nm packages with circular or custom beam profiles makes diode lasers practical as an alternative to the red HeNe.

Commonly-available diode-pumped solid-state lasers at 532 nm have not yet penetrated the flow cytometry market in any great numbers due to their relatively high cost per mW, and an emission line which is not useful for the excitation of fluorescein.

Diode-pumped solid-state lasers utilize a crystalline gain medium. Most frequently, this is Neodymium-doped Yttrium Aluminum Garnet (Nd:YAG). This gain medium is optically pumped by one or more infrared-emitting diode lasers. The primary lasing wavelength of Nd:YAG is 1064 nm, which is obviously not particularly useful for flow cytometry. The more useful 532 nm is obtained by a process called frequency doubling. In a frequency doubled laser, the fundamental wavelength interacts with a non-linear crystalline medium, producing a beam of twice the frequency, and half the wavelength.

Frequency-doubled solid-state lasers at 532 nm are commonly plagued by a phenomenon called "chaotic green noise". This random amplitude fluctuation is caused by competition among several longitudinal modes inside the laser cavity. There are two solutions to this noise problem. One is to make a very large cavity which lases in many modes. Operation in this manner tends to cancel the chaotic green noise. Another solution, which is indeed a cure rather than a band-aid, is to force the laser to run in only is single longitudinal mode (single frequency). This is achieved by employing a single-frequency tight-angle resonator (Coherent Compass 532) or by making the cavity extremely short (microchip laser). As solutions to "green chaos" are expensive and technically difficult, most low cost ($<$\$10,000) 532 nm solid-state lasers will exhibit this chaotic green noise, making them unsuitable for high-resolution flow work.

Solid-state lasers have their obvious appeal. They are many times more efficient at converting electrical energy into light. Thus, they require only standard wall power, and they dissipate modest amounts of heat which is either handled conductively, or with a small muffin fan.

Although numerous attempts have been made at commercializing a near-488 nm solid-state laser, the technical hurdles are daunting, and to-date, no commercial product exists. There's been much discussion of late about blue diode lasers, but they are still many years away from commercial availability, and will be offered at very low powers and wavelengths closer to 400 than 488 nm.

In the first decade of the second millenium, solid-state lasers will be available at wavelengths which will work with fluorochromes in use today. Until then, the ion laser will remain the preferred source of excitation, unless the current fluorochrome paradigms change.

Laser Safety

The greatest concern when using a laser is eye safety. In an instrument application such as flow cytometry, there are often many smaller reflected beams present at various angles near the laser. These beams are formed by specular reflections from polished surfaces such as lenses, filters, beam-shaping optics, flowcell faces, etc. While these beams will be weaker than the primary beam, they may still be sufficiently intense to cause eye damage.

Laser beams will also burn skin, clothing, or paint. They can ignite volatile compounds such as alcohol, ether, or other solvents, and can damage light-sensitive elements in photomultipliers and photodiodes on direct exposure.

Several pointers are indispensable when working with lasers:

- Never disable safety interlocks which are designed to protect you.

- Never sight down the beam axis. Keep your head above or below the level of the beam. The best way to prevent accidental exposure is to wear protective eyewear for the wavelength in question.

- Wavelengths shorter than 400 nm (UV) are invisible. Take extra precautions to avoid exposure when working with UV.

- Do not remove protective covers from the laser head unless required for normal maintenance, and then only after reading the operator's guide carefully. When laser covers are removed, hazardous reflections and voltages are present. Circuits in the laser head operate at lethal currents.

- Do not stare at reflected laser light.

- Do not use the laser in the presence of flammables or volatile solvents.

- Read the operating manual for your laser(s). It contains valuable information on the correct operation of the laser.

Laser Maintenance

Many lasers in use in flow cytometry require no routine maintenance, other than to disperse accumulated dust on the outside of the output coupler. This is accomplished with a short burst of clean, optics-grade compressed air. However, high-power water-cooled lasers like the Innova 70, 90, or 300 series may need an occasional optical cleaning to maintain optimum performance. The frequency of such maintenance depends on the cleanliness of the ambient environment.

Laser maintenance is not difficult, and cytometer manufacturer's training programs usually cover most aspects which would be relevant to the flow user. The important thing to remember about laser maintenance is that it should be done only when necessary. Cleaning optical surfaces too often increases the possibility of scratching, which will shorten the lifetime of the optic and decrease performance.

A laser maintenance kit consists of hemostats, two small medicine droppers, bottles of reagent-grade methanol and acetone, and high-grade lens tissue (Kodak). Maintain a sufficient supply of reagent-grade methanol, acetone and Kodak lens tissue – there are no substitutes which are adequate for laser maintenance. The use of cotton swabs or Kim-Wipes for mirror or window cleaning will do more to destroy the optics than clean them.

Laser optics, like many of the optics in the cytometer, should be handled with utmost care. The slightest scratch, trace of dirt, or film will severely diminish the laser's efficiency. Before you handle any laser optics, thoroughly wash your hands, and provide a clean, soft working surface.

Remember: optics and optical coatings are fragile. Always grasp the optic by the outer edge and never touch the optical surface.

There are several items in a high-power laser which may require cleaning; mirrors, Brewster windows, light-regulation-circuit beam pick-off, and intracavity prism. Two methods are established for cleaning laser optics: the "drop and drag" method, and the "hemostat and lens tissue" method. The drop and drag method is used for optics that are easily removed from the laser head, such as the laser mirrors. The drop and drag method exerts a minimal abrasive force because no pressure is exerted on the optical surface. The hemostat and lens tissue method is excellent for cleaning the hard-to-reach surfaces such as Brewster windows and the intracavity prism. Refer to your laser operating manual for specific details on how each optic in your system should be cleaned, and the appropriate procedure for each. Laser optics are expensive, and therefore, cleaning mistakes can be costly.

Laser Consumables

All lasers fail. A failure in the laser industry is frequently defined by the laser's inability to achieve its customer output specification. In flow cytometry, the laser has failed when it no longer provides enough light for useful analysis, or as a result of old age, is optically noisy enough to disrupt useful analysis.

It is important for the lab manager to discuss with your cytometer provider the expected lifetime of a laser in that particular cytometer, and the costs associated with its repair. Repair can take the form of plasma tube replacement in the larger models, head-swap in the smaller models, or outright new purchase for items like HeNes. Ask your provider to discuss whether or not repairs are made in the field or whether the laser needs to return to the factory. Most importantly, you should budget for laser repair, either separately, or under the umbrella of your cytometer's service contract. A cytometer without a functioning laser is only an expensive boat anchor.

A number of 3rd-party repair houses offer ion laser refurbishing. Although these outfits offer cut-rate prices and attractive promises, the resulting product is no longer covered under either the cytometer manufacturer's service policy or the laser manufacturer's. Plasma tube refurbishers are only placing band-aids on plasma tubes which have been obtained on the used market.

The modern water-cooled metal-ceramic plasma tube design does not allow for the replacement of the bore-defining elements, which sputter and erode under normal operating conditions. Any refurbished plasma tube you purchase comes with the bore erosion from the previous user, prior to its acquisition by the refurbisher. Also, a refurbished tube may be of a vintage many times removed from the current state-of-the-art. A refurbisher cannot deliver the advances which ultimately deliver the longest possible lifetime. Numerous advances in Brewster window material and coatings, cathode technology, gas pressure dynamics, and manufacturing technology insure that a plasma tube manufactured in 2000 is going to last longer than one purchased perhaps just a few years earlier. Over the long haul, manufacturer's original replacement tubes are the better bet for cost-effectiveness, power, low noise, and mode quality, and this translates into better cost management, and higher quality flow data.

Resources for Reagents

Linscott's Directory of Immunological and Biological Reagents is in its 10th edition. This directory is worth every penny it costs to have as a resource in a core flow cytometry facility or a laboratory that regularly purchases immunological and biological reagents. The 10th edition listed sources for over 22,600 monoclonal antibodies, 17,700 conventional antigens and conjugates, and more than 24,000 additional products such as molecular biology reagents, transduction reagents, lectins, cytokines, immunoassay kits, cell biology reagents, enzymes, peptides, infectious agents and their derivatives, contract/custom services, and more. The directory also includes World Health Organization international reference standards, and many reagents from the National Institutes of Health and the American Type Culture Collection. It saves not only time and energy, but serves as a marvelous search engine for new products as it is available not only in hard copy but on diskette.

Linscott's Directory can be purchased on diskette or hardcopy.

Linscott's Directory
4877 Grange Road
Santa Rosa CA 95404
phone: 707-544-9555
fax: 415-389-6025

Flow Cytometry Internet Access and World Wide Web Sites of Interest

This list is in no way complete and will get out of date quickly but it will give you a place to start looking for help. The sites that will remain constant and dependable and a source for all other web addresses are the Purdue University Cytometry Site and the International Society for Analytical Cytometry (ISAC) Site. It is there that you will find information on the Cytometry Users List-server that is maintained and updated daily. Check out both these areas for a body of supporters who will gladly answer all kinds of questions and help with stickly problems and mysteries.

J. Paul Robinson, Purdue University Cytometry Labs
Professor of Immunopharmacology
robinson@flowcyt.cyto.purdue.edu
PH:317-494 6449
FAX:317-494 0517
web *http://www.cyto.purdue.edu*
cytometry@flowcyt.cyto.purdue.edu

The mailing list archives contain all the messages sent over the listserve and provide not only answers to your questions but a clue about who is doing the type of research you are - and who to contact for help in certain areas. Point your browser to *http://www.cyto.purdue.edu/hmarchive/index.htm*

Purdue maintains a list of websites for academic, research, and clinical flow cytometric laboratories as well as commercial cytometry suppliers at
http://www.cyto.purdue.edu/flowcyt/websites.htm

This group of people have also produced some wonderful CD-roms which are available and contain great teaching aids, example data files and images, for both cytometry and image analysis.

The ISAC WWW Home Page has a number of additions and updates. Among them are:

- updated information of ISAC Congresses and other meetings

- additional links to other Internet resources in cytometry

- an updated section on the flow cytometry data file standard and related issues

- a new jobs vacancies/wanted section

- the new Electronic Congress Hall

Dave Coder
Editor, ISAC WWW Home Page
http://www.isac-net.org/
dcoder@u.washington.edu
    This Site has a Compendium of Cytometry Sites and also has a Reference Data and Standards site for information on quality control and standardization. In addition, the first of committee-specific sites has been added. The pages for the Biosafety Committee are at:
http://www.isac-net.org/committees/biosafety/biosafety.html
    You can access relevant documents concerning biosafety, links to related sites, as well as contact information for the committee chair and all members.
    The current ISAC newsletter is on-line as well at
http://www.isac-net.org/newsltr.html
    The Clinical Cytometry Society now has a Home Page. Visit its Web Site at:
http://www.cytometry.org
    Comments and suggestions welcome to
John W. Parker, M.D., President Clinical Cytometry Society
parker@pathfinder.hsc.usc.edu
or to our Web Master, Roy Overton at *overton@umdnj.edu*
    The National Laboratories of the Department of Energy and the University of California at Los Alamos National Laboratory are resources available to Flow Users in all areas of flow cytometry
http://www-lsdiv.lanl.gov/NFCR/
    A distribution list has been started with the goal to serve as a platform where you can get information on aquatic flow cytometry by posting questions on the technology, applications, literature, meetings, instrumentation, methodology etc. The purpose of this list is to improve communication between users

and developers of flow cytometry in aquatic sciences, both for marine and fresh water applications. The list is maintained by Richard Jonker (mailto:rjonker@aquasense.com) AquaSense, Amsterdam, The Netherlands. The name of the list is aquatic@-flowcytometry.org. It has been set up at the http://www.flowcytometry.org website, which is the site of the EC-AIMS project on automated identification of microbial subpopulations. To post a message: send it to *mailto:aquatic@flowcytometry.org*

To subscribe: send an email to mailto:request-cytometry @flowcytometry.org with the message subscribe. The home page of this project can be found at
http://www.flowcytometry.org/aquatic.

Mario Roederer at Stanford has put together a series of web pages which provide protocols for the conjugations of FITC, PE, Cy5PE, Cy7PE, Texas Red, APC, Cy7APC, Cascade Blue, and Biotin to immunoglobulins, as well as a few of these to Annexin V (as a model for conjugation of non-Ig proteins). Some of the pages are still "under construction", but it is reasonably complete. He has a wealth of knowledge in flow.
Roederer@Darwin.Stanford.Edu.
http://cmgm.stanford.edu/~roederer/abcon

A Catalog of free software is available at:
http://www.bio.umass.edu/mcbfacs/flowcat.html

A free softeware to explore data files is available at
http://wwwmc.bio.uva.nl/~hoebe/fexplore.exe

A list of publications in cytometry is available through the Wiley site - both Journal articles in Cytometry and books - Check them out at:
http://journals.wiley.com/cytometry/

Don't forget to look at the websites from other users. Commonly, they will have the answers to technical questions you might have. Here are just a few:

- Salk Institute, LaJolla CA
  http://pingu.salk.edu/fcm.html

- University of Washington
  http://nucleus.immunol.washington.edu/Research_facilities/ cell_analysis.html

- University of Georgia
  http://www.rserv.uga.edu/cellan/home.htm

- Roswell Park Medical Center
  http://members.tripod.com/~buff_flow/index1.html

- Stanford University
  http://facs.stanford.edu/

- Purdue University Cytometry Labs
  http://www.cyto.purdue.edu/index.htm

- University of California, Irvine Optical biology Core Facility
  http://calweb.bio.uci.edu/OBC/index.htm

- TSRI (The Scripps Research Institute) Flow Cytometry Core Facility
  http://facs.scripps.edu

- National Flow Cytometry Resource at Los Alamos National Laboratory (LANL)
  http://www-lsdiv.lanl.gov/NFCR
  NFCR (LANL) Flow Cytometry Data File Repository consists of flow cytometric data in list and histogram files and images of plot data.

Lastly, for updates and additional information associated with this manual, point your web browser to
http://www.nlivingcolor.com

Users Groups Around the Country

- Irvine, CA; **Southern California Research Flow Users Group**
  Contact: Susan DeMaggio, phone: 714-824-4110,
  email: suedemag@uci.edu

- Los Angeles, CA, Cedars-Sinai; **Los Angeles
  Flow Cytometry Users Group**
  Contact: Sharron Kelly, phone: 310-855-5571

- Sacramento, CA, **Sacramento Area Flow Cytometry
  Users Group**
  Contact: Linda Frey, phone: 916-734-3041,
  Ramona Anderson, phone: 916-733-1725

- San Diego, CA, **San Diego Area Users Group**
  Contact: Eileen Bessent, phone: 619-552-8585 x 6185

- San Francisco @ Univ. of CA, **Northern California
  Cytometry Group**
  Contact: Bill Hyun, phone: 415-476-2632,
  email: hyun@ cc.ucsf.edu

- San Francisco @ Univ. of CA **Bay Area Flow Cytometry Group**
  Contact: Eric Wieder 415-695-3807,
  email: eric_wieder.givi@quickmail.ucsf.edu, Bill Hyun,
  phone: 415-476-6703, email: hyun@dmc.ucsf.edu

- Denver, CO, **Colorado Flow Users**
  Contact: Barb Goffin, phone: 303-321-6027, fax: 303-320-4769,
  email: jgerdes@biotechnet.com

- Miami, FL, **South Florida Cytometry Group**
  Contact: Dr. Philip Ruiz, phone: 305-585-7344,
  email: pruiz@mednet.med.miami.edu

- **Tampa/St. Petersburg, FL Users Group**
  Contact: Sue LaRose, phone: 813-892-8614

- Atlanta, GA, **Greater Atlanta Area Flow Cytometry Users Group**
  Contact: Dr. Robert Bray, phone: 404-712-7308,
  Carol Hopkins, phone: 404-501-5253

- Des Moines, IA, **Iowa Flow Cytometry Users Group**
  Contact: Molly Bradshaw, phone 319-335-4500 x 2405,
  email: mollyhatlerbradshaw@uhl@uiowa.edu

- Chicago, IL, **VA Lakeside Medical Center**
  Contact: Dr. Charles Goolsby, phone: 312-943-6600 x 3651

- Boston, MA, **Boston Users Group for Cytometry**
  P.O. Box 850636, Braintree MA 02184
  email: bugbytes@shore.net, phone: 617-632-3179

- Baltimore, MD, **Chesapeake Cytometry Consortium**
  Contact: Patricia Echeagaray, phone: 301-228-2170,
  email: echeagaray@sri.org

- Detriot, MI, **Wayne State University**
  Contact: Dr. Alex Nakeff, phone: 313-577-1844

- Detriot, MI,, **Great Lakes International Imaging Flow Cytometry Association**
  Contact: Dr. Carleton Stewart, phone: 716-845-8741,
  email: stewart@sc3101.med.buffalo.edu

- Indianapolis, IN, **Indianapolis Flow Users**
  Contact: Phil Marder, phone: 317-276-5071

- Minneapolis, MN, **Upper Midwest Users Group**
  Contact: Linda Setterlund, phone: 612-813-7521

- Research Triangle / Durham / Winston- Salem, NC,
  Research Triangle Cytometry Association
  Contact: Marie Iannone, phone: 919-483-9486,
  email: mai49583@glaxo.com, Dr. Michael Cook,
  phone: 919-613-7819

- New York Metro, New Jersey, Connecticut,
  NY **New York/New Jersey Flow Cytometry Users Group**
  Contact: Joanne Thomas, phone: 516-444-2373,
  email: jthomas@path.som.sunysb.edu

- Allentown, PA, **Mid-Atlantic Flow Cytometry Users Group**

- San Antonio, TX, **San Antonio Flow Cytometry Users Group**
  Contact: Dave King, phone: 210-567-4096

- **South Central Flow Cytometry Association**
  Contact: Bill Hinson, Sec.Jefferson AR, phone: (501)543-7598,
  email: whinson@nctr.dfa.gov

Commercial Suppliers for Flow Related Products

Accurate Chemical and Scientific Corp.
300 Shames Dr.
Westbury, NY 11590
(800) 645-6264
(516) 997-4948 FAX
www.accurate-assi-leeches.com

Aldrich Chemical Company, Inc.
1001 West Saint Paul Ave.
Milwaukee, WI 53233
(800) 336-9719
(800) 368-4661 FAX
http://www.sigma.sial.com/safinechem

American Type Culture Collection (ATCC)
10801 University Blvd.
Manassas,VA 20110-2209
(800) 638-6597
(703) 365-2745 FAX
http://www.atcc.org

Amersham Pharmacia Life Science, Inc
2636 S. Clearbrook Dr.
Arlington Heights, IL 60005
(800) 323-9750
(800) 228-8735 Fax
http://www.apbiotech.com

Applied Cytometry Systems, Ltd.
Dinnington Business Centre
Outgang Lane, Dinnington
Sheffield. S25 3QX. UK.
+44 (0)1909 566982
+44 (0)1909 561463 FAX
Email: info@appliedcytometry.com

Bangs Laboratories, Inc.
9025 Technology Dr.
Fishers, IN 46038-2886
(317) 570-7020
(317) 570-7034 FAX
http://www.bangslab.com

Becton Dickinson Immunocytometry Systems (BDIS)
2350 Qume Drive
San Jose, CA 95131-1807
(800) 223-8226 (Ordering information)
(800) 448-BDIS (Customer support)
(800) 954-BDIS FAX
http://www.bdfacs.com/

BD Biosciences
2350 Qume Dr.
San Jose, CA 95131-1807
http://www.bdbiosciences.com
(408) 954-2347 FAX

BioErgonomics, Inc.
4280 Centerville Road
St. Paul, MN 55127-3676
(800) 350-6466
(612) 426-5740 FAX
http://www.bioe.com

Biological Detection Systems, Inc. (BDS)
see Amersham Pharmacia Bioproducts

Biomeda Corp.
1155-E Triton Dr.
Foster City, CA 94404
(800) 341-8787
(415) 341-2299 FAX

Bio-Rad Laboratories, Inc. (USA)
2000 Alfred Nobel Drive
Hercules, CA 94547 (800) 4BIORAD
(800) 879-2289 FAX
http://www.bio-rad.com

Boehringer Mannheim Corporation
(Roche Molecular Biochemicals)
9115 Hague Rd.
P.O. Box 50414
Indianapolis, IN 46250-0414
(800) 262-1640
(800) 428-2883 FAX
http://biochem.boehringer.com

Bruker Instruments Inc. (Spectrospin)
19 Fortune Drive
Billerica, MA 01821
Phone: (978) 667-9580
Fax:(978) 667-3954
http://www.bruker.com

Calbiochem-Novabiochem Corp.
10394 Pacific Center Court
San Diego, CA 92121
(800) 854-3417
(800) 776-0999 FAX
http://www.calbiochem.com

Caltag Laboratories
1849 Bayshore Hwy. #200
Burlingame CA 94010
800-874-4007
(415) 652-9030 FAX
http://www.caltag.com

Cedarlane Laboratories Ltd.
5516 8th Line RR #2
Hornsby, ONT L0P 1E0
Canada
(905) 878-8891
(905) 878-7800 FAX
http://www.cedarlanelabs.com

CLONTECH Laboratories
1020 E. Meadow Circle
Palo Alto, CA 94303-4230
(800) 622-2566
(650) 424-1604 FAX
http://www.clontech.com

Cold Springs Harbor Manual Source Book
Cold Springs Harbor Laboratory Press
10 Skyline Drive
Plainview, NY 11803-2500
(800) 843-4388
(516) 349-1946 FAX
http://www.cshl.org/
http://www.biosupplynet.com

Coherent, Inc.
Laser Group
5100 Patrick Henry Drive
Santa Clara, CA 95054
(800) 527-3786
(800) 362-1170 FAX
http://www.cohr.com

CompuCyte Corp.
12 Emily Street
Cambridge, MA 02139
(800) 840-1303
(617) 492-1301
http://www.compucyte.com

Corning Glass Works
Corning's color glasses are now made by
Kopp Glass, Inc. and sold through distributors.
Mooney Precision Glass Co.
P.O. Box 174
Enfield, CT 06083

Beckman Coulter Corporation
11800 S.W. 147th Avenue
P.O. Box 169015
Miami, FL 33196-2500
(800) 635-3497
http://www.coulter.com/Coulter/cytometry

BioCytex
140, Chemin de L'Armée d'Afrique
13010 Marseille - FRANCE
Phone: +33 (0)4 91 94 29 39
Fax: +33 (0)4 91 47 24 71
http://www.biocytex.com

Chromaprobe, Inc.
Independence Avenue, Building 4C
Mountain View, Califonia 94043
Telephone / Fax 650-964-8410
Toll Free Orders: 888-964-1400
http://www.chromaprobe.com

Cytomation, Inc.
400 E. Horsetooth Rd
Fort Collins, CO 80525
(303) 226-2200
(303) 226-0107 FAX
http://www.cytomation.com

DAKO Corporation
6392 Via Reel
Carpenteria, CA 93013
(800) 235-5743
(805) 566-6688 FAX
http.//www.dakousa.com

Electron Micoroscopy Sciences
521 Morris Rd.
P.O. Box 251
Fort Washington, PA 19034
(800) 523-5874
(215) 646-8931 FAX

Enzyme Systems Products
486 Lindbergh Ave.
Livermore, CA 94550
(888)449 – 2664
(510) 449-1866 FAX

Evergreen Laser
9G Commerce Circle,
Durham, CT 06422
860 349-1797
860 349-3873 FAX
http://www.EverygreenLaser.com

Exalpha Corporation
P.O. Box 1004
Boston, MA 02205-1004
(617) 558-3265
(617) 969-3872 FAX
http://www.exalpha.com

Exciton Chemical Company, Inc.
P.O. Box 31126
Overlook Station
Dayton, OH 45437
(937) 252-2989
(937) 258-DYES FAX
http://www.exciton.com

Flow Cytometry Standards Corporation
P.O. Box 194344
San Juan PR 00919-4344 USA
(800) 227-8143
787-758-3267 FAX
http://www.fcstd.com

Flow Cytometry Support
PO Box 3450
Saratoga, CA 95070-1450, USA
Voice: 408/735-5061
http://www.hooked.net/users/racox/fcs.

Gen Trak, Inc
5100 Campus Drive
Plymouth Meeting, PA 19462
(215) 825-5115 or (800) 221-7407
(215) 941-9498 FAX

Genzyme (See R&D Systems)
One Kendall Square
Cambridge, MA 02139
(800) 332-1042 or (617) 252-7760
(617) 252-7759 FAX
http://www.genzyme.com

Hamamatsu, Corp.
360 Foothill Rd.
P.O. Box 6910
Bridgewater, NJ 08807-0910
(800) 524-0504 or (908) 231-0960
(908) 231-1218 FAX
http://www.hamamatsu.com

Immune Source
101 First Street Suite #495
Los Altos, California 94022
Phone: (888) 466-8637 Toll-free:
888-IMMUNE-S
(650) 947-0377 FAX
http://www.immune-source.com

Ion Laser Technology
3828 S. Main Street
Salt Lake City, UT 84115
(801) 262-5555

Jackson ImmunoResearch Laboratories, Inc.
P.O. Box 9
872 W. Baltimore Pike
West Grove, PA 19390-0014
(800) 367-5296
(610) 869-0171 FAX
http:www.jacksonimmuno.com

Laser Innovations
668 Flinn Ave. #22
Moorpark CA 93021
(805) 529-5864
(805) 529-6621 FAX
http://www.laserinnovations.com

LeincoTechnologies
359 Consort Drive
St. Louis, Missouri 63011
Phone: 314-230-9477
Fax: 314-527-5545
http:/www.leinco.com

Lexel Laser
48503 Milmont Drive
Fremont, CA 94538
(800) 527-3795
(510) 651-6598 FAX
http://www.lexellaser.com

Liconix
3281 Scott Blvd.
Santa Clara, CA 95054
(800) 825-2554
(408) 492-1303 FAX
http://www.liconix.com

Life Technologies/Gibco/BRL
One Kendall Square
Grand Island, NY 14072-0068
(800) 828-6686
(800) 331-2286 FAX
http://www.lifetech.com

Linscott's Directory
4877 Grange Road
Santa Rosa, CA 95404
(707) 544-9555
(415) 389-6025 FAX
http://iaffairs.unl.edu/foreign_students/LINSCOTTSDIREC-
TORY.html

Luminex
12212 Technology Blvd
Austin, TX 78739
(888) 219-8020 toll-free
(512) 219-8020 direct dial
(512) 258-4173 fax
http://www.luminexcorp.com

Martek Biosciences Corporation
6480 Dobbin Road
Columbia, MD 21045
(410) 740-0081
(410) 740-2985 FAX
http:/www.martekbio.com

Melles Griot, Laser Group
2011 Palomar Airport Road, Suite
Carlsbad, California 92008
1-800-645-2737 / (760) 438-2131
FAX: (760) 438-5208
http://www.mellesgriot.com

Molecular BioSciences, Inc
4699 Nautilus Court, Boulder, Colorado 80301
(303) 581-7722
(303) 581-0575 FAX
http://www.molbio.cim

Molecular Probes, Inc.
4849 Pitchford Avenue
Eugene, OR 97402
(800)438-2209
(541) 344-6504 FAX
http://www.probes.com

Novacastra Laboratories
U.S. Distributor Vector Laboratories
30 Ingold Rd.
Burlingame, CA 94010
(800) 227-6666
(650) 697-0339 FAX
http://www.vectorlabs.com

Omnichrome
13620 Fifth Street
Chino, CA 91710
(909) 627-1594
(909) 591-8340 FAX

Oriel Corp.
250 Long Beach Blvd.
P.O. Box 872
Stratford, CT 06497-0872
(203) 377-8282
(203) 378-2457 FAX
http://www.oriel.com

Omega.Optical, Inc.
3 Grove St.
P.O. Box 573
Brattleboro, VT 05301
(802) 254-2690
(802) 254-3937 FAX
http://www.omegafilters.com

One Cell Systems Inc.
100 Inman Street
Cambridge MA 02139
(617) 868-2399 x 309
(617) 492-7921 FAX
http://www.onecell.com

Ortho Diagnostic Systems, Inc.
Route 202
Raritan, NJ 08869
(800) 322-6374

Partec USA
c/o Walter C. Farley
P.O. Box 329
Harpswell, ME 04079
207 833-6793
612 833-7831 FAX
http://www.partec.de

PEL Freeze Inc.
PO Box 68
Rogers, AR 72757
(800) 643-3426
(501)636-3562 FAX
http://www. pelfreez-bio.com

PharMingen (Becton Dickinson)
10975 Torreyana Road
San Diego, CA 92121
(800) TALK TEC (825-5832) (Technical support)
(801) 848-MABS (6227) (Customer service)
(619) 677-7749 FAX
http://www.pharmingen.com

Phoenix Flow Systems
11575 Sorrento Valley Road, Suite 208
San Diego, CA 92121
(619) 453-5095
(619) 259-5268 FAX
http://www.phnxflow.com

Polysciences, Inc.
400 Valley Rd.
Warrington, PA 18976-9990
(800) 523-2575
(800) 343-3291 FAX
http://www.polysciences.com

Prozyme
1933 Davis Street, Ste 207
San Leandro, CA 94577-1258
(800) 457-9444
(510) 538-6919 FAX
http://www.prozyme.com

R&D Systems
614 McKinley Place N.E.
Minneapolis, MN 55413
(612) 379-2956 or (800) 343-7475
(612) 379-6580 FAX
http://www.rndsystems.com

Research Organics
4353 East 49th Street
Cleveland, OH 44125
(800) 321-0570
(216) 883-1576 FAX
(217) http://www.resorg.com

Riese Enterprise
BioSure Division
12301 Loma Rica Drive, Suite G
Grass Valley, CA 95945-9355
(800) 345-2267
(530) 273-5097
http://www.riese.com

Schott Glass Technologies Inc.
Member of the Schott Group
400 York Ave.
Duryea, PA 18642-2026
(570) 457-7485
(570) 457-6960 FAX
http://www.schottglasstech.com

Seradyn
Particle Technology Division
1200 Madison Avenue
Indianapolis, IN 46225
(317) 266-2956 or (800) 428-4072 x 2956
(317) 266-2918 FAX
http://www.seradyn.com

Sierra Cytometry
3150 Susileen Drive
Reno, NV 89509
(702) 329-9772
72607.155@compuserve.com

Sigma Chemical Co.
P.O. Box 14508
St. Louis, MO 63178
(800) 325-3010
(800) 325-5052 FAX
http://www.sigma.sial.com

Soft Flow Hungary, Ltd.
Debrecen, Wessele'nyi 17.
Hungary, H-4024
(36) 52 46599 Phone/FAX
Soft Flow, Inc. (North American office)
11513 Galtier Drive
Burnsville, MN 55337, USA
(800) 956-0100
(612) 895-0900 FAX
http://www.wavefront.com/~soft-flow/

Spectra-Physics
1335 Terra Bella Avenue
Mountain View, CA 94039
(650) 961-2550
(650) 968-5215 FAX
hppt://www.splasers.com

Spherotech, Inc.
1840 Industrial Drive, Suite 270
Libertyville, IL 60048-9817
(708) 680-8922 or (800) 368-0822
(847) 680-8927 FAX
http://www.spherotech.com

Sterogene Bioseparations Inc.
5922 Farnsworth Ct.
Carlsbad, CA 92008
(800) 535-2284, (760) 929-0455
(619) 929-8720 FAX
http://www.sterogene.com

Synthegen, LLC
10590 Westoffice Drive, Suite 200
Houston, TX 77042
(800) 949-8903
(713) 952-9370 FAX
http://www.synthegen.com

Tetko Inc. is now Sefar America, Inc.
525 Monterey Pass Road
Monterey Park, CA 91754
(626) 289-9153
Fax (626) 282-8111

Tree Star, Inc.
20 Winding Way
San Carlos, CA 94070
800-366-6045
http://www.treestar.com

Uniphase Corp. (Cyonics is now part of Uniphase)
163 Baypoint Parkway
San Jose, CA 95134
(408) 434-1800
(408)954 0760FAX
http://www.uniphase.com

Vector Laboratories
30 Ingold Rd
Burlingame, CA 94010
(800) 227-6666
(650) 697-0339
http://www.vectorlabs.com

Verity Software House
P.O. Box 247
45A Augusta Road
Topsham, ME 04086
(207) 729-6767
(207) 729-5443 FAX
http://www.vsh.com

Worthington Biochemical Corporation
730 Vassar Ave
Lakewood, NJ, 08701
(800) 445-9603
(800) 368-3108
http://www.worthington-biochem.com

Zymed Laboratories
480 Carlton Court
South San Francisco, CA 94080
(800) 874-4494
(650) 871-4499 FAX
http://www.zymed.com

Data Analysis for Cell Census Plus™ System

REBECCA POON

Subprotocol 1
ModFit LT v2.0 Alternative

ModFit LT v2.0 and the Proliferation Wizard

This is an upgraded alternative to ModFit v5.2 and the Cell Proliferation Model (see Subprotocol 2). The ModFit LT v2.0 from Verity Software House, Inc. was designed for the deconvolution analyses of complex histogram data in cell cycle, DNA ploidy and cell proliferation studies. In contrast to the earlier ModFit v5.2, ModFit LT v2.0 has the Proliferation Wizard, an equivalent of the Cell Proliferation Model, as an integral part of the program. ModFit LT v2.0 requires a 32 bit operating system such as Windows 95 or Windows NT. It is available for both PC and Macintosh platform. It will be assumed in this manual that the user has already installed the ModFit LT v2.0 program according to the instructions provided by the manufacturer, Verity Software House, Inc. The user is referred to the manufacturer's manual for additional information.

▓▓ Procedure

Analyzing the first listmode data file

1. Double click on the icon to execute the program.

2. Click on the file folder icon on the tool bar.

3. The "Open Data" dialog box will appear (Fig. 1). Highlight the list mode file to be analyzed. Click on the **Open** button.

Note: The first file to be analyzed should be the time zero, non-proliferated control to establish the position of G1.

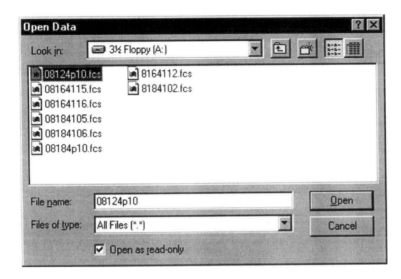

Fig. 1. Open Data

4. A dialog box will appear with the heading "Choose Parameter for Analysis" (Fig. 2). Select the appropriate histogram parameter for the deconvolution analysis, e.g. FL-2 for PKH-26. Click on **OK**.

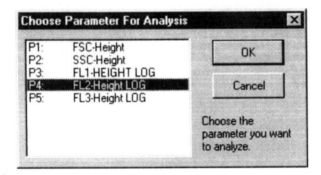

Fig. 2. Choose Parameter for Analysis

5. A dialog box will appear with the heading "Define Gates" (Fig. 3). Activate the **Gate On** button. Click on the **Define Gate 1** button.

Note: This screen allows the definition of two gates, Gates 1 and 2. They may be activated one at a time or together with an additive effect i.e. events must satisfy both gates.

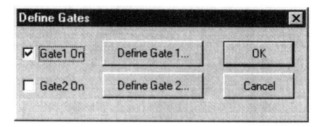

Fig. 3. Define Gates

6. A dialog box will appear with the heading "Define Gating Parameters" (Fig. 4). The correct gate name (1 or 2) should be displayed. Select by highlighting the appropriate parameters. Click on **OK**.

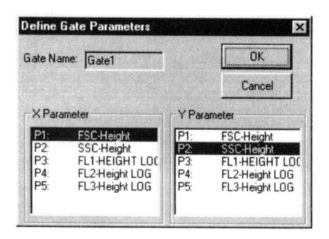

Fig. 4. Define Gate Parameter

7. The "Define Gating Region" dialog box will appear with a histogram of the gating parameters (Fig. 5). Click on **R1** to activate the region. The name R1 may be moved by clicking on it, holding and dropping at the desired location. The shape of the polygonal region may be compressed vertically or horizontally by moving the handle bars. Alternatively, fine adjustments may be made by dragging each vertex one at a time. The entire region may be moved by clicking on the cross in the center. Click on **OK** to close the dialog box.

Fig. 5. Define Gating Region

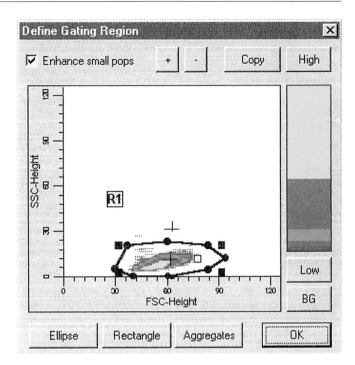

8. ModFit has returned to the "Define Gates" window (see Fig. 3). The above operations (Step 5 - 7) can be repeated now to activate and define Gate 2. Otherwise, Click on **OK** to complete the operation and return to the main screen.

9. The screen now displays the PKH fluorescence histogram (see Fig. 6, page 758). The gating histogram created in step 7 may be seen by scrolling downwards. Pull down the "Analysis" menu. Go to the "Proliferation Wizard" and click on the **Create or Edit Model** button.

10. Click on the button for "Start from scratch (Fig. 7, page 758). Create a new cell proliferation model". This button is highlighted by default unless the model has been running.

Note: The "Proliferation Wizard" dialog is displayed offering tabs for different aspects of the model: **Start, Parent, Generations, Other.** When starting a Proliferation Wizard session, select each tab to adjust any settings to your data on the parent population (steps 11 to 14). Once the appropriate model is created, the settings should be kept constant for analysis of proliferated samples of the same parent population.

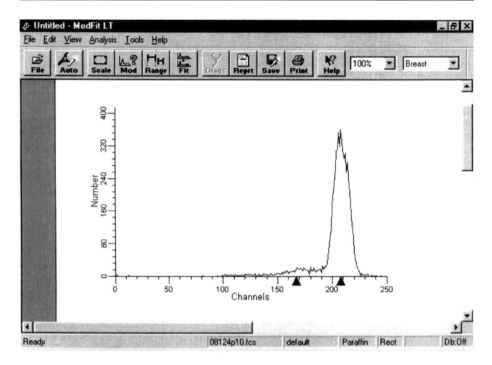

Fig. 6. Main screen with day zero control file opened

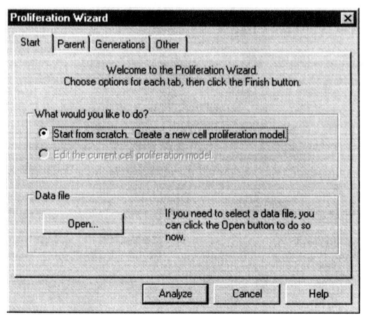

Fig. 7. Proliferation Wizard: Start Tab

11. Use the **Parent** tab to check the location of the parent population (Fig. 8). The program automatically identifies the rightmost peak in the histogram. This position can be adjusted by moving the cursor or by editing the number in the "Parent channel" box.

12. The **Generations** tab displays edit boxes for the number of generations, log decades and channel spacing between generations (Fig. 9, page 760). When a new value is entered for any one of these parameters, the dotted lines indicating the position of the daughter generations and the spacing between generations will be automatically re-computed as appropriate.

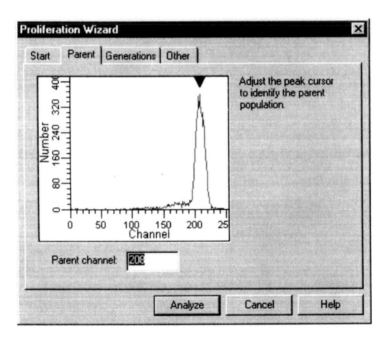

Fig. 8. Proliferation Wizard: Parent Tab

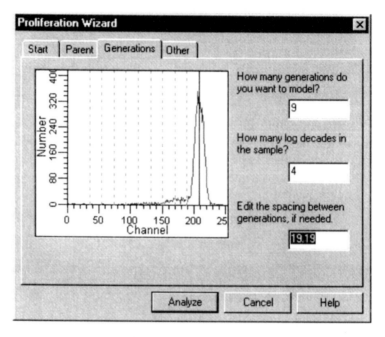

Fig. 9. Proliferation Wizard: Generations Tab

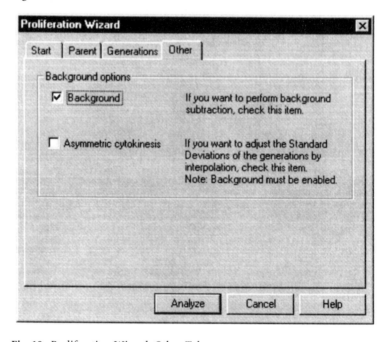

Fig. 10. Proliferation Wizard: Other Tab

Note: The default number of generations to be modeled is 9, the present number of log decades in the data is 4 and the default channel spacing between generations is 19.19. The graphics display a PKH-26 histogram with a red line identifying the position of the parent generation and dotted lines for each daughter generation.

13. The **Other** tab presents background options (Fig. 10, page 760). Under the **Other** tab, enable **Background** checkbox only if rectangular background substraction is desired.

14. When the **Asymmetric cytokinesis** checkbox is enabled, the standard deviations of the generations are broadened by interpolation along the background rectangle.

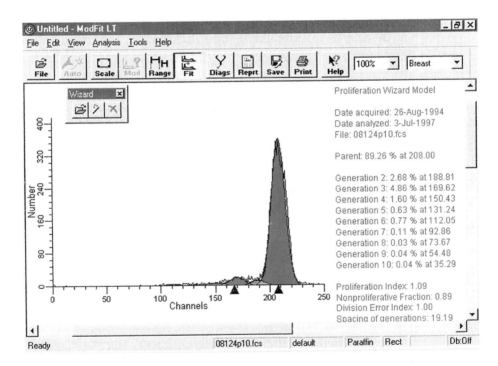

Fig. 11. Deconvoluted histogram of PKH-26 labeled human PBL on Day Zero, unproliferated control parent population

Note: This additional option becomes available only when Background is enabled.

15. Click on the **analyze** button to start analysis of the data.

Note: The deconvoluted histogram will appear with the "Info List" containing the numerical data (Fig. 1, page 761). A "Wizard" toolbar will be presented allowing you to open files, return to the wizard or exit the Wizard. The screen is fully interactive. The gating region may be edited directly on the graphics displayed here. All of the graphics (deconvoluted PKH histogram, the gating histograms) and the "Info List" may be sized and moved to different parts of the screen to create a printout using the printer icon on the toolbar.

Note: An active Proliferation Wizard session can be saved in a report with the specific settings for the proliferation model created. The report must be saved with the Wizard toolbar visible to be able to re-enter the Wizard dialog when the report is re-opened.

Subsequent data analysis of proliferated samples

The particular model that fits your analysis requirements have been constructed in the previous section. The following steps for the analysis of the next data file is a repeat of steps 3 to 8 of the previous section.

1. Click on the file folder icon on either toolbar.

2. The "Open Data" dialog box will appear (see Fig. 1 above). Highlight the list mode file to be analyzed. Click on the **Open** button.

3. A dialog box will appear with the heading "Choose Parameter for Analysis" (see Fig. 2 above). Select the appropriate histogram parameter for the deconvolution analysis, e.g. FL-2 for PKH-26. Click on **OK**. The parameter should be the same as for the control file above.

4. The "Define Gates" dialog box will appear (see Fig. 3 above). Clicking on **OK** in the "Define Gates" dialog will execute the proliferation model with the parameters created in the previous section for the control data file. The screen will then display the deconvoluted PKH histogram of the proliferated sample file with the gating histograms and the **"Info List"** (Fig. 12).

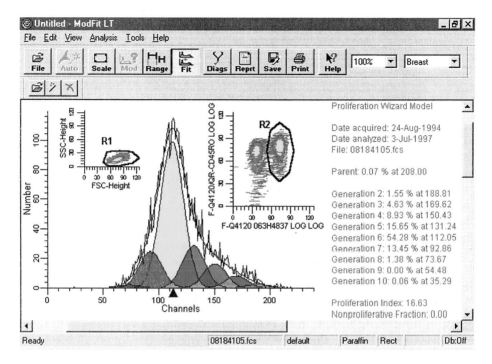

Fig. 12. Deconvoluted histogram of CD4+/CD45RO+ subset in human PBL activated in culture with 2.5 µg/ml PHA for 6 days. Insets show gates set by light scatter and immunophenotype

Note: If adjustments of gating region or additional gating parameters are needed for the proliferated sample, click on the "Define Gate" button(s) and repeat Steps 5 to 8 of previous section.

5. To begin the analysis of unrelated data files or a different parent population, re-enter the Proliferation Wizard dialog by clicking the Wizard button on the Wizard toolbar, or selecting **Proliferation Wizard -> Create or edit model** from the "Analysis" menu. Review and edit the model by using the tabs.

6. The program will remain in the Proliferation Wizard mode until you exit by selecting **Proliferation Wizard -> Quit** from the "Analysis" menu or by clicking on X on the Wizard toolbar.

Subprotocol 2
ModFit v5.2 Alternative

ModFit v5.2 and Cell Proliferation Model

The ModFit v5.2 program from Verity Software House, Inc. was originally designed for the deconvolution analysis of complex histogram data in DNA cell cycle and ploidy studies. ModFit v5.2 is designed to run under Microsoft Windows 3.1 and can analyze list mode files from a variety of flow cytometers. Cell Census Plus™ System provides a Cell Proliferation Model on a 3.5 in. diskette. This model can be installed into the ModFit v5.2 program for the deconvolution analysis of PKH-26 fluorescence data in cell proliferation measurements. This manual will assume that the user already has the ModFit program operational. The user is referred to the ModFit manual for details regarding general features of the program.

▓▓ Procedure

Installation of the Cell Proliferation Model into ModFit v5.2

1. To install the model into the ModFit environment, run the Setup program from the Cell Proliferation Model diskette through the File Manager in Windows. Place the Cell Proliferation Model in the 3.5" diskette drive. Specify the drive path in the File Manager in Windows.

2. Double click on the Setup.exe file to begin the program. The message will appear that the program is initializing.

3. After initialization, the next screen will present the options of **Continue, Exit** or **Help**. Choosing **Continue** proceeds to the next screen that displays a prompt to specify path destination. The default is **C:\ModFit**.

4. After specifying the path destination choose Continue. Installation will automatically be completed to the path specified.

Note: The directory installed in ModFit for the Cell Proliferation Model is named **Celldiv**. This directory is organized into four operation files: **Simple.Mod, Complex.Mod, Celldiv. equ** and **Demogr.equ**. Besides the Celldiv directory, two sample data files are copied into a directory named **Samples**. The first time user can become familiarized with the following analysis procedure using the sample data files provided.

1. Execute the ModFit program.

Note: This step will bring up the default ModFit display area (Fig. 13). This screen contains six windows: **Range, Scaling, Outside In, Inside Out, Model** and **Info List**.

Analyzing the
first listmode
data file

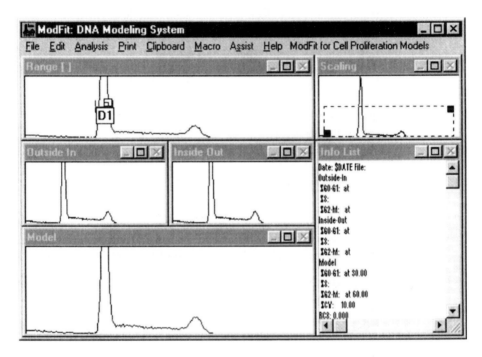

Fig. 13. Default Screen

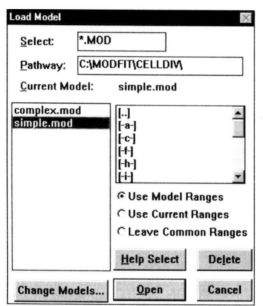

Fig. 14. Load Model Screen for Step 2

2. Using the File menu, select and load the version of the Cell Proliferation Model desired from the Celldiv directory: Simple.Mod or Complex.Mod (Fig. 14).

Note: The name of the selected model will appear in the title bar of the Model window (see Fig. 20 and Fig. 22).

3. Go to the File menu and choose the Select File option. Click on the list mode file to be analyzed (Fig. 15).

Note: The first file to be analyzed should be the time zero, non-proliferated control to establish the position of G1.

4. Click on the Set Format box (Fig. 15). Step 4 opens to Edit File Format screen (Fig. 16).

5. Activate the Reduce resolution to 256 channels box.

6. Select the correct type of flow cytometer/file format.

7. Click on OK.

Note: The program will return to the Select File screen, Figure 15.

8. Click on Open in the Select File screen (Figure 15) to load the list mode file.

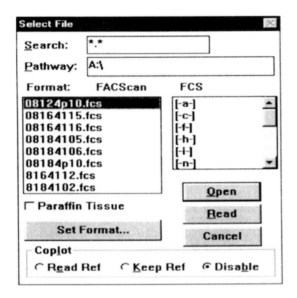

Fig. 15. Select File Screen for Steps 3 und 4

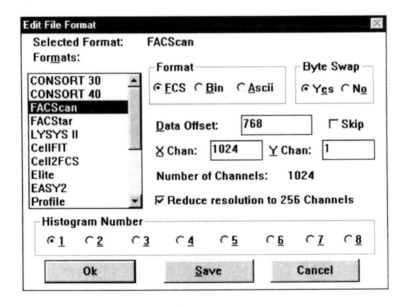

Fig. 16. Edit File Format Screen for Steps 5-7

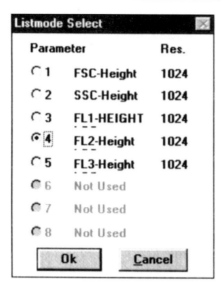

Fig. 17. Listmode Select for Step 9

9. The List mode Select screen with appear (Fig. 17). Specify the histogram parameter to be analyzed in the model (e.g., FL1 for PKH67 and FL2 for PKH26). Click on OK.

10. The Gate Setup screen will appear (Fig. 18). Select the desired gating parameters. Click on OK to proceed to the next screen (Fig. 19).

Fig. 18. Gate Setup for Step 10

Note: Gating can be set on any combination of parameters, e.g., light scatter and/or fluorescence. Fluorescence parameter(s) can be used when the cell sample is simultaneously labeled with a surface immunophenotyping reagent.

11. In the Set 2P Gate Screen, draw a bitmap around the population to be analyzed for proliferation. Click on OK to begin the reading of the data file.

12. ModFit will return to the main screen with the time zero control file opened (Fig. 20). Go to the Scaling Window, manually scale the data by dragging the corner black squares or handles, to cover channel zero to channel 255 (Fig. 21).

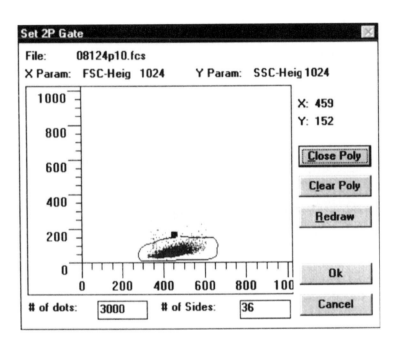

Fig. 19. Set 2P Gate Screen for Steps 11

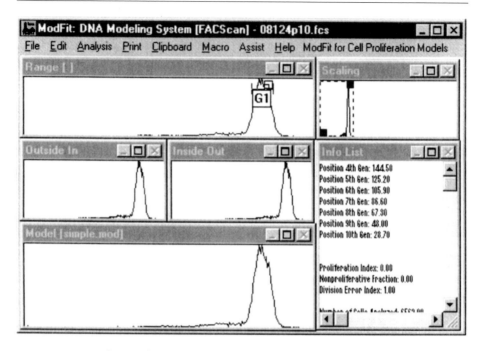

Fig. 20. Time Zero Control File on Main Screen

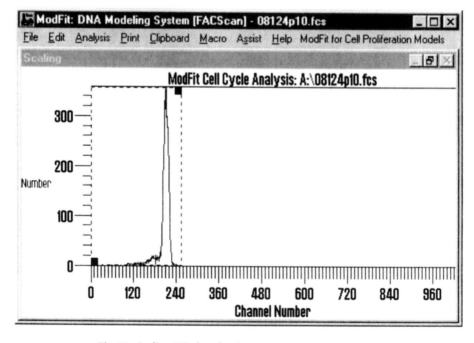

Fig. 21. Scaling Window for Step 12

Note: The height of the Y-scale should approximate the height of the histogram. It is important to keep the full 256-histogram scaling range, especially when using the Complex model.

13. Adjust G1 range in the Range window, if necessary, to bracket the first or parent generation (see Fig. 20).

Note: The G1 range should cover approximately 19 channels.

14. Identify the position of the first generation by pressing the Ctrl and Z keys simultaneously or use the pull-down menu and click on "Execute Outside-In and Inside-Out".

Note: When analyzing the time zero non-proliferated control the shaded histogram in the "Outside In window" should match the shape and position of the raw histogram; otherwise check the position of G1 range in step 13. This position of the G1 range should be maintained in subsequent analysis of proliferated samples from cultures of the same parent population (see "Edit Model Parameter" below).

15. Perform the deconvolution of the histogram of the parent generation By simultaneously pressing the CTRL and X keys or by using the pull-down menu.

Note: The analysis results will be presented in the model data box and the "Info List". The PI or Proliferation Index should be <1.1 and the distribution in G1 should be >90% (Fig. 22, page 772).

Edit Model Parameter

Once the G1 position of the parent population is determined above, the following steps are taken to keep this value constant for analysis of all related data files of proliferated samples.

1. Enter the Edit Model Parameter dialog box (Fig. 23, page 773). Examine the top of the window named "Model Components". Make sure that Model Component 1 (1st population) is highlighted.

2. In the window named **"Component Detail"** on the lower half of the screen, highlight **"Mean"** under **"Description"**.

3. Activate the button for Dependency (**Dep.**). The options for the **"Dep."** parameter will be displayed on the upper right corner of the screen.

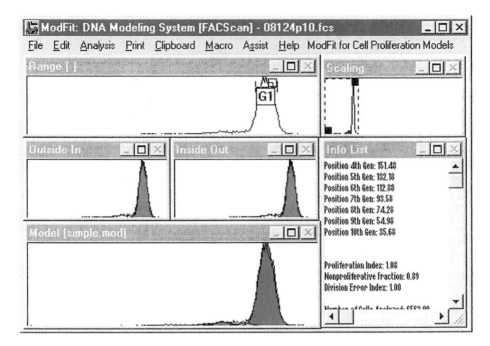

Fig. 22. Deconvoluted histogram of PKH-26 labeled human PBL on Day Zero, unproliferated control parent population

4. Choose **Static**.

5. Activate this choice by clicking on the **"Replace Detail"** button. This step replaces the default setting of "float" for **Dep.** in the highlighted bar for **"Mean"** under **"Component Details"**.

6. Activate the button for **Estimates**. Option menu will be displayed on the upper right corner.

7. Choose Neither from the option menu.

8. Activate this choice by clicking on the **"Replace Detail"** button. This step replaces the default setting of **"Outside-in"** for Estimates in the highlighted bar for **"Mean"** under **"Component Details"**.

9. In the **"Component Detail"** window, highlight **"SD"** under **"Description"**.

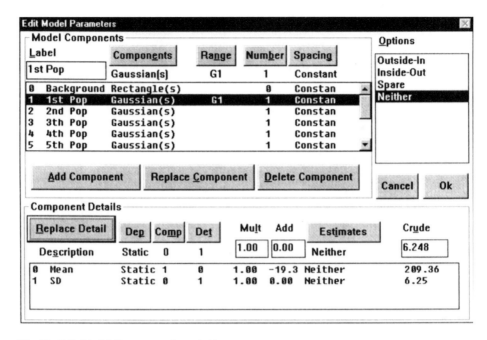

Fig. 23. Edit Model Parameter, Steps 1-11

10. The same operation is performed to fix the value of "SD" by repeating steps 3 to 8.

11. Click on OK to close the dialog box. The model parameters have now been edited. The mean and SD, standard deviation, of the parent population have been fixed for the analysis of related proliferated data files.

Note: Caution: These parameters must be returned to the default settings before analyzing unrelated data files. The edited settings are not saved when you exit the program.

The above procedure ensures that the positions of G1 through G10 are precisely fixed as predicted from the Time Zero control. Continue with the analysis of proliferated samples by loading the data file.

Subsequent data analysis of proliferated samples

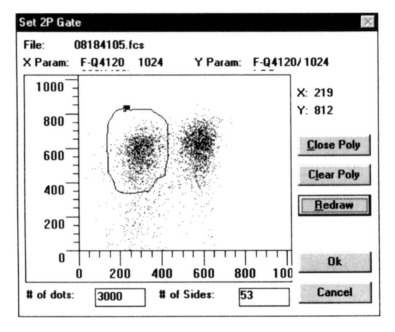

Fig. 24. FL-1 (CD4) and FL-2 (CD45RO) histogram of human PBL activated in culture with 2.5µg/ml PHA for 6 days. Bitmap is shown drawn around the CD4-/CD45RO+ subset

1. Go to the File menu and choose the Select File option. Click on the list mode file to be analyzed (see Figure 15).

2. Open the data file.

3. Specify the histogram parameter to be analyzed in the model (e.g. FL-2 for PKH26) (see Figure 17). This should be the same as for the Time Zero control file.

4. Set a gate by selecting the desired gating parameters (see Figure 18) and drawing a bitmap around the population to be analyzed for proliferation. An example for two color immunofluorescence gating is given in Figure 24. Click on OK to begin reading the data file. The program will return to the main screen.

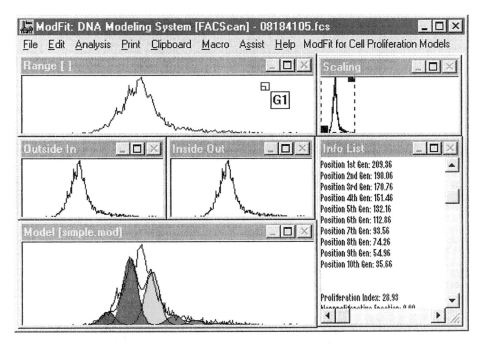

Fig. 25. Deconvoluted histogram of CD4 -/CD45RO+ subset in sample described in Figure 24

5. Use the pull down Analysis menu and choose **Execute Model.**

Note: The deconvoluted histogram will appear in the model data box and the "Info List" will be updated to show the numerical values for the sample (Figure 25). The positions of G1 through G10 should remain the same.

Monoclonal Antibody Production

GARY WOO

Introduction

The easiest way to produce monoclonal antibodies is to have someone else make them for you. Or, more accurately, have something else make them for you. The ubiquitous laboratory mouse is the most common host used for producing monoclonal antibodies. Other small animals such as rats and rabbits are also common, but mice are ideal for novel antibody production because of their availability, low maintenance and well characterized immune system. Using a host animal and its immune system obviates many obstacles of de novo synthesis of an antibody. Besides the correct sequence, conformation is integral for correct binding of antibody to antigen. The antigen presented to the mouse can be expressed on the surface of a cell or purified. Antigen conformation is vital to antibody binding and the immune system will respond to whatever antigen form is presented. Cell surface antigen presentation is desireable for antibody production against a native form. If the antibodies will ultimately be used to label cell surfaces, it makes most sense to immunize with cells that express the antigen. When the cell makes the antigen, there is no need to worry about whether or not the correct conformation has been synthesized. Large, isolated antigens such as proteins or carbohydrates can also be directly injected into the mouse. Antibodies against drugs or cytokines can be made this way. Smaller peptide antigens such this are too small for the immune system to recognize. An adjuvant can help present the antigen and facilitate an immune response. The peak immune response occurs between three and five days post-immunization. The majority of antibodies produced will be of the IgM isotype. IgG isotype formation is favored by waiting longer to harvest. However, after seven days fewer splenocytes will be actively making antibodies to the introduced antigen.

Hybridomas Are Fused Lymphoid Cells That Make Monoclonal Antibodies

The technique of cell fusion followed by selection is widely used in the production of monoclonal antibodies. Each normal B-lymphocyte from a spleen is genetically programmed to produce a single antibody. When a mouse is injected with an antigen, B-cells that recognizes that antigen are stimulated to grow and produce antibody. In the animal, each stimulated B-cell forms a colony of cells in the spleen or bone marrow. Unfortunately, normal lymphocytes will not grow indefinitely in culture and so cannot be used to establish an immortal, productive clone.

The mortality of normal B-cells is sidestepped by fusing them with myeloma cells. Myeloma cells are cancerous lymphocytes derived from a host animal. Fusion with myelomas will impart immortality to the B-cells. Many different lines of myelomas from mice and rats have been established. From these, mutant myeloma cell lines that have lost the salvage pathways for purines have been selected. These mutants are negatively selected when cultured in media containing hypoxanthine (a purine), aminopterin, and thymidine (HAT). When these mutant myeloma cells are fused with normal antibody producing cells from a mouse spleen, hybridoma cells result. Like myeloma cells, hybridoma cells can grow indefinitely in culture. Hybridomas also have the purine salvage pathway of normal lymphocytes that allows growth in HAT media. If a mixture of fused and unfused cells is placed in HAT medium, the unfused mutant myeloma cells and the unfused spleen cells die. This leaves a culture of immortal hybridoma cells, each of which produces a single antibody. Clones of hybridoma cells can be tested separately for the production of a desired antibody. Thereafter, the clones producing that antibody can be cultured in large amounts.

MUTANTS IN SALVAGE PATHWAYS OF PURINE AND PYRIMIDINE SYNTHESIS ARE GOOD SELECTIVE MARKERS

This metabolic pathway has been particularly useful in cell-fusion experiments. Most animal cells can synthesize the purine and pyrimidine nucleotides de novo from simpler carbon and nitrogen compounds, rather than from already formed purines and pyrimidines. The folic acid antagonists amethopterin and aminopterin interfere with the donation of methyl and formyl groups by tetrahydrofolic acid in the early stages of de novo synthesis of glycine, purines, nucleoside triphosphates, and thymidine triphosphate. These antagonistic drugs are called antifolates. In addition to the biosynthetic pathways, enzymes in most cells also can use purines and thymidine directly by salvage pathways that bypass the metabolic blocks imposed by antifolates.

A number of mutant cell lines have been discovered that are unable to carry out one of the salvage steps. For example, cell cultures lacking thymidine kinase (TK) have been selected, and cultures have been established from humans who lack adenine phosphoribosyl transferase (APRT) or hypoxanthine-guanine phophoribosyl transferase (HGPRT). These different types of salvage mutants become useful partners in cell fusions with one another or with cells that have salvage pathway enzymes but that are differentiated and cannot grow in culture by themselves. The selective medium most often used to culture such fused cells is called HAT media, because it contains hypoxanthine (a purine), aminopterin, and thymidine. Normal cells can grow in HAT media; salvage mutants cannot, but their hybrids with normal cells can.

CLONING

In the fusion trays, several different hybridoma colonies may be growing in a single well. Cloning is the process where individual hybridomas are plated and grown in separate wells. A colony that arises from a single cell can be considered monoclonal. A monoclonal colony of hybridomas can produce monoclonal antibodies that are pure and of diagnostic value. In theory, cloning requires that one cell be placed into each well of a culture plate and allowed to grow. In practice, it is often necessary to seed the wells with more cells to get a reasonable amount of growth. Wells that produce multiple colonies can be cloned again if it is determined that one of the colonies is producing antibodies. Cloning can be accomplished by sterile sorting with a flow cytometer or by limited dilution.

ANTIBODY TESTING AND HARVESTING

After cloning, the wells with hybridomas that appear to be monoclonal are tested for positive antibody production. A monoclonal colony will appear as a circular mass of cells with cell density gradually decreasing away from the center. The most common test methods for specific antibody detection are ELISA and cytotoxicity to assess the presence of antibodies. There are numerous ELISA kits and reagents available commercially. For whole cell immunizations, antibodies can be screened via cytotoxicity. Since cells express a wide range of membrane antigens, it is necessary to use a panel of cells. The panel will contain cells known to be either positive or negative for the antigen in question. This will allow discrimination of antibodies that react non-specifically from those that react to the desired antigen. If the hybridomas are growing robustly, diluting fewer cells per well may give more monoclonal cultures. Monoclonal colonies that are still producing antibodies are then chosen for expansion. After hybridomas have been grown in unchanged media, antibody can be harvested directly from the conditioned media. Harvest media before significant cell death has occurred to prevent any released proteolytic enzymes from degrading the antibodies. Cultures of hybridomas can be injected back into a mouse of the same strain to produce ascites fluid. With inbred mice strains, hybridomas introduced back into the mouse will be immunologically invisible. The hybridomas will grow and produce ascites fluid. Within one or two weeks the mice will have produced sufficient fluid in their abdominal cavity to harvest. This approach is requires more animals, but will produce antibodies at a much higher titer than from cultured supernatant.

▨ Materials

Equipment

- 3 ml syringe
- 27 1/2 gauge needle
- Autoclaved instruments:
    - one package containing 2 scissors and 3 forceps
    - one package containing 1 cell strainer, 1 glass pestle and 1 forceps

Solutions

RPMI MEDIA PREP
- R0: to a 500 ml bottle of RPMI, add
  - 5 ml glutamine pen-strep (200mM)
  - 5 ml sodium pyruvate (100 mM)
  - (optional) 0.5 ml b-mercaptoethatnol (BME) 5 X 10-2 molar
  - (optional) 0.5 ml penicillin-streptomycin
- R10: add 55 ml of fetal bovine serum (FCS)to a 500 ml bottle of R0.
- E10: add 55 ml of enriched calf serum (formula fed bovine serum) to 500 ml R0
- 1 x phosphate buffered solution (PBS)
- Cold (4°C refrigerator) STERILE HYPOTONIC BUFFER (to lyse red blood cells)
  - 10 x stock soln. is:
    - NH4Cl 1.55 molar
    - NaHCO$_3$ 0.1 molar
    - EDTA 1 millimolar
  - pH to about 7.15. To use, dilute to 1 x and aliquot in 4 ml tubes.
  - Store at 4°C.
- polyethylene glycol (PEG 4000)
- 100 x HAT Stock

Fusion Media
- 10% enriched calf serum (or formula fed bovine serum),
- 1% 100 x HAT (for a 1 x working dilution),
- 89% RPMI media

Cloning Media
- 10% Freshly thawed hybridoma cloning factor (HCF)
- 15% Fetal bovine serum (FCS)
- 1 % Hypoxanthine thymidine (HT) in 100 x conc.
- 74% RPMI

Preparation for Splenocyte Isolation

Under sterile hood:
- Filter 100 ml of RPMI media (all media for fusion should be filtered as needed).
- Fill one 50 ml conical tube with freshly filtered RPMI media (media tube).
- Prepare one 50 ml conical tube to receive splenocytes (fusion tube).
- Place 1 sterile pipet (plastic or glass pasteur) into each of the 50 ml tubes.
- Pour few about 5 ml of RPMI media in petri dish to receive whole spleen.

Preparation of Myeloma cell line

Cultured 6531 myeloma:
- two 250 ml ml flasks grown for 3-4 days or until confluent and heathly (check under microscope before beginning fusion)
- Prepare for Fusion

Materials
- Digital timer
- Warm serum free RPMI media (in 25°C water bath): approx. 300 ml
- Warm (37°C incubator) polyethylene glycol (PEG 4000): 1 ml
- Enriched calf serum (ECS) or formula fed bovine serum
- 100 x hypoxanthine, aminopterin, thymidine (HAT)Fill media tube with freshly filtered RPMI media and keep warm by placing in a plastic cup filled with warm (37°C) water from water bath.

Prepare Fusion Media (E10 1 x HAT)
- Fusion media is:
  - 10% enriched calf serum (or formula fed bovine serum),
  - 1% 100 x HAT (for a 1 x working dilution),
  - 89% RPMI media
- Filter fusion media and feed control wells with 100 µL.
- Multi-well cell culture plates

Procedure

Immunization

Inject mice with antigen

1. Suspend cells in about 0.5 ml of PBS. First, inject a small amount into each of the rear foot pads of the mouse. This will act as a timed release delivery. Inject the remaining suspension into two intra-peritoneal sites.

Note: Two intra-peritoneal injections introduce the bulk of the antigen. Injecting into two sites reduces the chance that the antigen will be delivered into a fat deposit, intestine or other region that will prevent recognition by the immune system.

Note: Peptides too small for immunologic response must be delivered in an adjuvant. Emulsify the antigen in an adjuvant such as complete Freund adjuvant or alum immediately prior to injection. Inject both foot pads and intra-peritoneally as with cell suspension.

2. Wait 3-5 days

Note: Booster immunizations can follow the primary immunization at intervals of days to weeks. A sustained delivery of antigen favors IgG production. This approach is popular since the IgG isotype is generally more stable and easier to chemically manipulate.

B-Cell Harvest

Preparation for Splenocyte Isolation

3. Under sterile hood:
- **Filter** 100 ml of RPMI media (all media for fusion should be filtered as needed).

- **Fill** one 50 ml conical tube with freshly filtered RPMI media (media tube).

- **Prepare** one 50 ml conical tube to receive splenocytes (fusion tube).

- **Place** 1 sterile pipet (plastic or glass pasteur) into each of the 50 ml tubes.

- **Pour** about 5 ml of RPMI media in petri dish to receive whole spleen.

4. The quickest way to sacrifice the mouse is by cervical dislocation. To maintain optimum health and viability of splenocytes, avoid harsh anesthetics and other chemicals. After sa-
crificing mouse, lay mouse with left side facing up. **Sterilize**
mouse by spraying with 70% ethanol. The spleen will be in the
left hypochondriac region posterior and lateral to the sto-
mach. Begin spelectomy by cutting through abdominal
wall to reveal the spleen directly under the muscle layer. It
will appear as a large (~15 mm) dark red, oblong organ. Since
the spleen is a fragile organ, use care not to rupture it during
removal. Once inside the abdominal cavity, blunt forceps
should be used in lieu of scalpels or scissors. The fine mesen-
teric tissue will easily pull away and release the spleen.

Remove the Spleen

Note: All subsequent procedures to be conducted under sterile
hood

Splenocyte Isolation

5. Place cell strainer in a fresh petri dish. Using forceps, lift
spleen and **wash** by dripping few mL of RPMI media with pipet from fusion tube. Then place spleen in cell strainer. Add
few mL of RPMI media and begin **grinding** spleen with glass
pestle. Use pipet from fusion tube to **remove** RPMI media
with ground spleen cells. Make sure spleen is completely
ground before transfering to fusion tube. Add few ml of fresh
RPMI media and **repeat grinding** and transfering until RPMI
media from petri dish is mostly clear (3-4 times). To ensure
maximum cell yield, wash bottom of the cell strainer, pestle
and petri dish. Should have 15-25 ml of spleen cell suspension.
Centrifuge at 500 x g for 5 minutes.

6.

- **Decant** supernatant from fusion tube.

- **Add** 4 ml of cold hypotonic buffer solution (lyses red blood cells).

- **Start** timer for 1 min.

- **Vortex** tube until pellet resuspends.

- At end of 1 min, **dilute** buffer by gradually adding 45 ml of RPMI media.

- **Centrifuge** spleen cells at 500 x g for 5 minutes.

- **Decant** supernatant from fusion tube.

- **Resuspend** pellet in 10 ml RPMI media and remove any tissue that will not disperse.

Myeloma Preparation

7.

- **Harvest** myeloma by shaking or tapping flasks and pouring into 50 ml conical tubes.

- **Centrifuge** myeloma and spleen cells at 500 x g for 5 minutes.

- **Aspirate** supernatant from myeloma tubes. Using 10 ml pipet, **resuspend** and **combine** myeloma cells in one tube with 10 mL RPMI media.

8. Count cells

- **Count** both myeloma and spleen cells using a hemacytometer. An average immunized mouse spleen will yield 120 to 180 million (M) lymphocytes.

- **Calculate** number of fusion trays to plate. Small volume, multiple well plates are the best choice for fusion trays. Each fusion tray should have about 20 M spleen cells.

- **Calculate** number of myeloma cells to use. Spleen cells and myeloma should be mixed in a 3:1 ratio, but can be fused in a 4:1 or 5:1 ratio if myeloma count is low.

9. Before combining spleen and myeloma cells be sure to **plate** the control wells. Add 7 μl of each cell suspension to each control well.

Plate the Control Wells

Note: With HAT selection, cells in these wells should die off in about 7 days.

10. Use a graduated pipet to **combine** proper ratio of spleen and myeloma cells and **centrifuge** at 500 x g for 5 minutes.

Combine Myeloma with Splenocytes

11. **Aspirate** supernatant from fusion tube and remove as much liquid as possible without disturbing cell pellet. Place fusion tube in the cup of warm water.

Fusion of Myeloma and Splenocytes

12. **Fuse** cells by adding 1 ml of warm PEG 4000 and stirring with pipet vigorously for 2 minutes. Gradually dilute the PEG by adding 1 ml of RPMI and shake tube for 30 seconds. Add another 2 mls of RPMI and shake for another 30 seconds. Add 4 mls of RPMI and shake for a final 30 seconds. Finally, add the remaining RPMI and centrifuge at 500 x g for 5 minutes.

Note: This step must be done with fusion tube immersed in warm water.

13. **Filter** fusion media and feed control wells with 100 μL.

Prepare Fusion Media (E10 1 x HAT)

14.
- **Aspirate** supernatant from fusion tube and remove as much liquid as possible without disturbing cell pellet.

Resuspend the Pellet of New Hybridomas

- Gently **pour** fusion media into tube without disturbing pellet. Use pasteur pipet to gently resuspend pellet. Suck pellet up from bottom and gently break apart on the side of the tube. Remove any tissue that will not disperse. Pour cell suspension into clean petri dish. Dispense 100 μl into each well by using a multi-channel pipetman. Be sure not to add any of the fusion mixture to the control wells on tray 1. When finished dispensing, place trays in 37°C, 5.0 % CO2 incubator.

Note: DO NOT VORTEX TUBE. Newly fused hybridomas are very unstable at this stage.

Testing for Antibodies

15. See individual commercial kit instructions

Note: The most common test method for specific antibody detection is ELISA. There are numerous kits and reagents available commercially. For this application, the trays are coated with antigen.

Note: For whole cell immunizations, antibodies can also be screened via cytotoxicity. Since cells express a wide range of membrane antigens, it is necessary to use a panel of cells. The panel will contain cells known to be positive and negative for the antigen in question. This will allow discrimination of antibodies that react non-specifically from those that react to the desired antigen.

Cloning

Preparations   **16.** From screening test data, **determine** which wells of the fusion trays look like candidates for cloning. Under the microscope check each of these wells to make sure that there are hybridomas growing.

Note: Hybridomas colonies will be comprised of round cells that may overlap toward the center of the colony.

Cloning Media   **17. Thaw** a fresh bottle of HCF. Each clone will require about 5.5 ml of media. From this, **calculate** the total volume of media needed (make a little extra in case of mistakes i.e. calculate 6 ml for each clone).

Count the Cells   **18.** Count the cells in the wells to be cloned using a hemacytometer.

Calculate the Cell Dilution   **19.** Divide the total number of cells needed by the hemacytometer cell count. This dividend will be the cell dilution.

Note: Cloning can be accomplished by sterile sorting with a flow cytometer or by limited dilution.

Note: Use 96 well culture plates or other multi well culture format. Ideally, monoclonal colonies require one cell per well or 96 hybridomas evenly distributed among the 96 wells of the tray.

When cloning by limited dilution however, a higher number of cells per well is usually needed to establish colonies. An estimate of 3-5 cells per well usually promotes good colony growth. So 3 cells per well in 96 wells gives a total of 288 cells needed.

20.

Dispense the Cells

- Use the multi-channel pipetman to dispense the cells.

- Use a pipetman to thoroughly mix the media and cells in the fusion well to be cloned.

- Use a pipetman to take the correct cell dilution from the fusion well and dispense into a tube of media. **Vortex** the tube and pour into a sterile media trough.

- Set the multi-channel pipetman to **dispense** 100 µl into each of the wells.

21. When finished place trays in a 37°C, 5.0 % CO2 incubator and grow for 7-10 days before testing.

Clone Test

22. Use ELISA or cytotoxicity assay to assess the presence of antibodies as in the initial screening.

Note: After cloning, the wells with hybridomas that appear to be monoclonal are tested for positive antibody production. A monoclonal colony will appear as a circular mass of cells with cell density gradually decreasing away from the center.

Note: If a supernatant tests positive for antibody, but has more than one colony, it may be recloned. Follow the same procedure as the initial cloning. If the hybridomas are growing robustly, diluting fewer cells per well may give more monoclonal cultures. Monoclonal colonies that are still producing antibodies are then chosen for expansion.

23. From the wells of the 96-well plate, transfer hybridomas to a small flask in about 10 mls of RPMI. As the cells grow, gradually add more media. Eventually the small flasks may be split or transfered to larger flasks.

Clone expansion

Note: New hybridomas will grow best when gradually expanded.

Antibody production   **24.** After hybridomas have been grown in unchanged media, antibody can beharvested directly from the conditioned media.

Note: Harvest media before significant cell death has occurred to prevent any released proteolytic enzymes from degrading the antibodies.

Ascites fluid antibody production   **25.** Cultures of hybridomas can be injected back into a mouse of the same strain to produce ascites fluid. With inbred mice strains, hybridomas introduced back into the mouse will be immunologically invisible. The hybridomas will grow and produce ascites fluid. Within one or two weeks the mice will have produced sufficient fluid in their abdominal cavity to harvest. This approach is requires more animals, but will produce antibodies at a much higher titer than from cultured supernatant.

References

1. Molecular Cell Biology 2nd Ed., Darnell, et al, 1990.

Subject Index

Printing: Druckhaus Beltz, Hemsbach
Binding: Buchbinderei Schäffer, Grünstadt